EDIBLE STARCHES AND STARCH-DERIVED SYRUPS

EDIBLE STARCHES
AND
STARCH-DERIVED SYRUPS

Nicholas B. Petersen

NOYES DATA CORPORATION

Park Ridge, New Jersey   London, England

1975

Copyright © 1975 by Noyes Data Corporation
No part of this book may be reproduced in any form
without permission in writing from the Publisher.
Library of Congress Catalog Card Number: 75-10358
ISBN: 0-8155-0584-1
Printed in the United States

Published in the United States of America by
Noyes Data Corporation
Noyes Building, Park Ridge, New Jersey 07656

FOREWORD

The detailed, descriptive information in this book is based on U.S. patents since 1968 relating to edible starches and starch-derived syrups. Where it was necessary to round out the complete technological picture, some earlier, but very relevant patents were included.

This book serves a double purpose in that it supplies detailed technical information and can be used as a guide to the U.S. patent literature in this field. By indicating all the information that is significant, and eliminating legal jargon and juristic phraseology, this book presents an advanced, commercially oriented review of edible starches and starch-derived syrups as depicted in U.S. patents.

The U.S. patent literature is the largest and most comprehensive collection of technical information in the world. There is more practical, commercial, timely process information assembled here than is available from any other source. The technical information obtained from a patent is extremely reliable and comprehensive; sufficient information must be included to avoid rejection for "insufficient disclosure." These patents include practically all of those issued on the subject in the United States during the period under review, there has been no bias in the selection of patents for inclusion.

The patent literature covers a substantial amount of information not available in the journal literature. The patent literature is a prime source of basic commercially useful information. This information is overlooked by those who rely primarily on the periodical journal literature. It is realized that there is a lag between a patent application on a new process development and the granting of a patent, but it is felt that this may roughly parallel or even anticipate the lag in putting that development into commercial practice.

Many of these patents are being utilized commercially. Whether used or not, they offer opportunities for technological transfer. Also, a major purpose of this book is to describe the number of technical possibilities available, which may open up profitable areas of research and development. The information contained in this book will allow you to establish a sound background before launching into research in this field.

Advanced composition and production methods developed by Noyes Data are employed to bring our new durably bound books to you in a minimum of time. Special techniques are used to close the gap between "manuscript" and "completed book." Industrial technology is progressing so rapidly that time-honored, conventional typesetting, binding and shipping methods are no longer suitable. We have bypassed the delays in the conventional book publishing cycle and provide the user with an effective and convenient means of reviewing up-to-date information in depth.

The Table of Contents is organized in such a way as to serve as a subject index. Other indexes by company, inventor and patent number help in providing easy access to the information contained in this book.

15 Reasons Why the U.S. Patent Office Literature Is Important to You —

1. The U.S. patent literature is the largest and most comprehensive collection of technical information in the world. There is more practical commercial process information assembled here than is available from any other source.

2. The technical information obtained from the patent literature is extremely comprehensive; sufficient information must be included to avoid rejection for "insufficient disclosure."

3. The patent literature is a prime source of basic commercially utilizable information. This information is overlooked by those who rely primarily on the periodical journal literature.

4. An important feature of the patent literature is that it can serve to avoid duplication of research and development.

5. Patents, unlike periodical literature, are bound by definition to contain new information, data and ideas.

6. It can serve as a source of new ideas in a different but related field, and may be outside the patent protection offered the original invention.

7. Since claims are narrowly defined, much valuable information is included that may be outside the legal protection afforded by the claims.

8. Patents discuss the difficulties associated with previous research, development or production techniques, and offer a specific method of overcoming problems. This gives clues to current process information that has not been published in periodicals or books.

9. Can aid in process design by providing a selection of alternate techniques. A powerful research and engineering tool.

10. Obtain licenses — many U.S. chemical patents have not been developed commercially.

11. Patents provide an excellent starting point for the next investigator.

12. Frequently, innovations derived from research are first disclosed in the patent literature, prior to coverage in the periodical literature.

13. Patents offer a most valuable method of keeping abreast of latest technologies, serving an individual's own "current awareness" program.

14. Copies of U.S. patents are easily obtained from the U.S. Patent Office at 50¢ a copy.

15. It is a creative source of ideas for those with imagination.

CONTENTS AND SUBJECT INDEX

INTRODUCTION

The general molecular formula for starch, a high-polymeric carbohydrate, is $(C_6H_{10}O_5)_n$ where n varies from a few hundred to over one million. Starch, insoluble in cold water and relatively resistant to hydrolysis by enzymes, occurs in the form of white granules or spherocrystals and is usually made up of both a linear polymer (amylose) and a branched polymer (amylopectin).

Starch was probably first separated from wheat flour in the ancient world and was well-known as a food product to the Romans and Greeks. This carbohydrate is widely found in nature and the starch obtained from grains such as corn and sorghum and roots and tubers such as tapioca, arrowroot or potato are of considerable industrial importance.

In 1973, the per capita consumption in the United States, as reported by the U.S. Department of Agriculture, was 1.9 lb of cornstarch, 18.2 lb of corn syrup and 4.8 lb of corn sugar. Most of these materials are used by the processed food industries (baking, confectionery, brewing, etc.) with limited amounts purchased directly by the final consumer. In the United States, corn is the major source of starch and the terms cornstarch, corn syrup and corn sugars have become common usage for the products.

This review covers 217 processes disclosed in 232 patents which have been issued since 1968 on the preparation, purification and use of edible starches and their derivatives and hydrolysates. The arrangement of this review is by product produced rather than by process used whenever possible.

The first two chapters cover the isolation or separation of starches from the raw materials, and the treatment of raw starch to remove impurities. Improvements in gelatinization of starch and use of pregelatinized starches are covered in the third chapter. Starch derivatives, their esters and ethers, are described in the fourth and fifth chapters.

Recent developments in enzyme isolation and production have provided improved methods for preparing the important starch degradation products. The processes for producing amylose from amylopectin now involves the use of α-1,6-glucosidase as covered in the sixth chapter. Enzymatic methods are also used in the processes described in the chapter on the preparation of dextrin and low DE hydrolysates.

Corn syrup has had increasing importance in recent years because of the high prices for sucrose. Corn syrups and sugars (including dextrose, maltose and fructose) are being used more frequently now as substitutes or partial replacements for the more expensive sucrose. Fructose production has increased tremendously in the last five years. The production of the corn syrups, maltose and fructose are covered by the final four chapters of this review.

DEFINITION OF TERMS AND ABBREVIATIONS

Characteristic — See Descriptive Ratio.

CIV Cooking Test — This test comprises suspending 60 g of crosslinked starch acylate in 1,100 g of a 35% (by weight) aqueous solution of sucrose and then lowering the pH of the suspension to 3.5 with citric acid. The suspension is then placed in a Corn Industries Research Foundation Viscometer equipped with a heating jacket which is maintained at 201°F. The viscosity is recorded at its peak and at 10, 15 and 40 min after the suspension has been placed in the viscometer.

D — Dextrose content which can differ from dextrose equivalent.

DE — Dextrose Equivalent refers to the total reducing sugar content of the dissolved solids in a starch hydrolysate expressed as percent dextrose. One method of determining this is by the Luff-Schorl method, NBS Circular C-40, page 195.

DP — Degree of Polymerization, or the number of saccharide units in the molecule. Dextrose = DP_1, Maltose = DP_2, Maltotriose = DP_3, and so on.

DS — Degree of Substitution is the average number of substituents per glucose unit. The maximum possible DS is 3, but most commercial starch derivatives have low DS numbers such as 0.1 or less.

DS — Dry Solids or Dissolved Solids.

DSB — Dry Solids Basis.

Derivatized Starches — Refers to chemical derivatives of starch such as the esters, phosphates, ethers, etc.

Descriptive Ratio — The sum of all the percentages on a dry basis of saccharides having a DP of 1 to 6 divided by the DE.

FE — Fermentable Extract refers to the total concentration of dextrose, maltose and maltotriose in the starch hydrolysate.

GSP — Granule swelling power is a measure of the extent of granule inhibition (or crosslinking) and may be defined as the amount of swollen or hydrated paste which is formed by the cooking in water under specific conditions of 1 g of dry starch as divided by the weight of anhydrous starch in the swollen paste.

Granular Starch — A cold water swelling starch with intact granules. See also page 62.

Inhibited Starch — Crosslinked starch, particularly the crosslinked esters and ethers. The starch granules are toughened so they are resistant to rupture on cooking.

MS — Molar Substitution, of derivatized starch.

Syneresis — Water loss or extrusion on standing, as in cooked puddings.

Texturized Starch — A cold water swellable, nonbirefringent substantially completely fragmented starch.

ENZYMES

Many enzymes or enzyme systems are used on starches to hydrolyze, isomerize or cause reversion. Recently developed or improved enzymes have been used to produce amylose, maltose and fructose. The following are simplistic definitions of the principal starch enzymes and their use in starch chemistry.

α-Amylase, also called the liquefying enzyme, cleaves at random the α-D-(1→4) bonds of the large molecules of starch and the long chains of dextrins formed by the primary and secondary cleavages. α-Amylase can also produce some maltose and glucose, at a slow rate, but does not hydrolyse the α-D-(1→6) bonds in amylopectin.

β-Amylase, the saccharifying enzyme, produces maltose from starch, glycogen and dextrins by selectively removing maltose units at the nonreducing ends of the molecule chains. β-Amylase activity also stops at the α-D-(1→6) branching linkages of amylopectins leaving polysaccharides that are called limit dextrins.

Glucoamylase, also termed amyloglucosidase, glucamylase or gamma-amylase, hydrolyzes starch directly to D-glucose. It is capable of slowly hydrolyzing the α-D-(1→6) bonds of amylopectin as well as the α-D-(1→4) bonds of amylose and amylopectin.

α-1,6-Glucosidase, also known as pullulanase, amylo-1,6-glucosidase or isoamylase, selectively hydrolyzes the α-D-(1→6) bonds of amylopectin. It is used for converting amylopectin to amylose and in combination with other enzymes to saccharify starch.

The abbreviations used to designate enzyme depositories and their catalog numbers are:

ATCC	American Type Culture Collection, Rockville, Maryland
FRI	Fermentation Research Institute, Inage, Chiba City, Japan
IMRU	Institute of Microbiology, Rutgers University, New Brunswick, N.J.
NRRL	Northern Regional Research Laboratory, U.S. Department of Agriculture, Peoria, Illinois

ISOLATION AND USE OF RAW STARCHES

STARCH FROM CORN AND SORGHUM

Unmodified cornstarch is generally produced by the wet milling process which separates the grain into the four main parts: germ, hulls, gluten and starch. After the shelled corn has been cleaned, i.e., by air drafts, screens or magnets, the corn is softened by steeping in warm water acidified with, for example, sulfur dioxide. After steeping, the softened kernels are passed through degerminating or attrition mills which free the germ and loosens the hulls.

The lighter, oil-containing germ can be separated by flotation or hydroclone separation. The resulting mixture of starch, gluten and hulls is then finely ground and the hulls removed by a screening operation. The gluten is next separated from the heavier starch particles by centrifuging and the starch washed and dried by various processes.

Heat Processing Steeped Sorghum Kernels

An improved process has been developed by *R.E. Bailey; U.S. Patent 3,498,796; March 3, 1970* for processing grain sorghums such as kafir, yellow maize, hegari and milo for use as foods.

The process includes the steps of moistening the grain to increase the internal moisture content as well as to add surface moisture and heating the grain to convert the starch to a softer form without popping or charring the grain, and to volatilize a substantial portion of the oils and acids along with excess moisture. The grain is then capable of being milled in conventional flour mill machinery.

The flour produced when grain sorghum is milled in accordance with this process is especially beneficial in that substantially all of the vitamins, minerals, and protein originally found in the grain is also present in the flour. A substantial portion of the oil of the germ is driven off such that the percentage of oil is comparable to that of wheat flour, reducing the tendency of the flour to become rancid. The process produces cracked grain products which are substantially coarser than flour but which are more palatable than the cracked grain presently used as a food product in many of the less-developed countries of the world since the bitterness is removed.

In the overall process the grain is cleaned, sized and tested for moisture and for the total content of horny endosperm and aleurone. The amount of the endosperm and aleurone present determines the temperature at which the grain can be heated without carmelization or popping.

The grain is then moistened, suitably by soaking in water or by tempering. The water content of the grain after moistening is suitably 16 to 22% by weight. Thereafter, the grain is subjected to heat at a temperature of at least 300°F for a time sufficient to convert the starches of the soft endosperm and the horny endosperm into a more digestible form of carbohydrate, without popping, charring or carmelizing the grain.

The term "subjected to heat" means a heated atmosphere in contact with the grain as the temperature of the heat source will normally be greater and the temperature of the grain will be less due to the thermal mass of the grain and cooling of the grain by evaporation of water. During the heating, a substantial portion of the tannic acid and oil is removed from the grain.

It has been found that the optimum temperature and time required for heating is a function of the percentage of the grain comprising the horny endosperm and the aleurone. Further, it has been found desirable to use the maximum temperature that can be used without popping or carmelization.

Thus, for milo having approximately a 20% horny endosperm and aleurone, a temperature of 360°F applied for 3 min has been found to convert the soft endosperm and the horny endosperm to a modified form which is extremely soft and also to volatilize virtually all the tannic acid and oils such that when the kernel is milled the resulting product will not possess the bitter taste normally characteristic of products milled from grain sorghums and the oil content and moisture content of the flour will be comparable to that of products milled from wheat.

It is important that the grain be subjected to high temperatures for short periods of time rather than use substantially low temperatures for longer periods of time if best results are to be obtained. The milling can be accomplished utilizing conventional equipment to obtain the desired product, whether a cracked grain or a fine flour.

Steeping Cracked Corn Kernels

Starch is obtained from cracked corn kernels in the process developed by *D.L. Gillenwater, G.B. Pfundstein and A.R. Harvey; U.S. Patent 3,597,274; August 3, 1971; assigned to Grain Processing Corporation.* This improved process provides very short steeping periods resulting in a greatly increased rate of starch production; recovery of a floury starch stream separate from the end-of-the-mill starch stream; reduced equipment and energy requirements; and more efficient recovery from the steep water of soluble corn components. In this process, whole kernel corn is passed through cracking rolls where the corn hulls are cracked. The clearance between the cracking rolls should be such as to fracture the hull of the kernels but should not damage the germ. Roll clearances of about 3 or 4 mm are generally satisfactory to crack the hulls in several places. If wet corn is used apparatus suitable for cracking such is employed. The cracked corn is then transferred to a steep tank (or series of steeping tanks can be used in countercurrent relationship, if desired) with an aqueous solution of 0.1 to 0.3% sulfur dioxide. Steeping is conducted at a temperature of 120° to 130°F, preferably 122° to 126°F. Short steeping periods are used and the steeping is completed in such short periods as 2 hr and generally in periods of 2 to 16 hr.

After steeping, the contents of the tank are removed and washed over a screen or screens. One particularly preferred type of screen used at this stage is a DSM screen. The DSM screens are characterized as high-capacity, high-efficiency wet screens of stationary design which use a concave screen of wedge bars and are capable of screening in the 50 to 75 micron range.

The undersized discharge from the screen comprising the steep water and soluble corn components (proteins, carbohydrates of low molecular weight and ash) as well as finely divided floury starch is passed to a cyclone separator to separate the floury starch product from the steep water.

The floury starch product can be further purified by separating minor amounts of gluten and fiber by conventional means. A comparison of the results of steeping cracked corn for relatively short periods with steeping whole kernel corn is set forth in the table below, the processing of the end-of-mill starch being the same in all cases.

	- - - - Recovery as Percent of Dry Corn Weight - - - -				
	- - - - - Cracked Corn - - - - - - -			- Whole Kernel -	
Steeptime in Hours	16	8	2	30 to 36	
Germ	7.9	8.0	9.3	4.0	5.1
Floury starch	6.0	5.6	3.4	—	—
End-of-mill starch	53.2	51.9	50.7	53.6	56.5
Protein in starch	2.03	1.72	2.45	2.03	2.08
Total starch	59.2	57.5	54.1	53.6	56.5
Hulls plus protein	28.5	36.5	29.7	27.0	25.9

As seen from the above, steeping of the cracked corn for as little as 2 hr yielded a floury starch fraction which can be separated as a separate and valuable product. The yield of the floury starch product by this process is generally within the range of 3 to 10% by weight of the total available starch. The floury starch is loosely deposited in the crown of the corn kernel and the granules thereof are round in shape.

In contrast, the starch deposited in the glutinous material of the corn kernel is the horny or end-of-the-mill starch. This horny starch becomes compressed as the corn kernels mature and the gluten dehydrates resulting in angular shaped starch granules. This horny starch ordinarily requires prolonged steeping and intensive milling to free it from the gluten matrix. However, such conditions are believed to cause material degeneration of the floury starch present in the crown region of the kernel.

By means of this process, floury starch is recovered immediately after steeping for relatively short periods and before being damaged by further processing (milling, etc.).

Multistage Steeping Process

To overcome the disadvantages of the stationary steeping process for corn, *H.J. Vegter; U.S. Patent 3,595,696; July 27, 1971; assigned to Honig NV, Netherlands* has developed a multistage process for improving the production of cornstarch.

This corn steeping process comprises the following steps: continuously passing corn through a series of successive steeping stages; continuously passing a steeping liquor through the same series of successive steeping stages in opposite sequence to the corn; contacting corn with liquor on entering each steeping stage and passing them concurrently through such steeping stage to extract soluble substances from the corn by the liquor; collecting a mixture of corn and liquor on leaving each steeping stage; screening such mixture outside the steeping stage for separation into corn and liquor; passing the separated corn to the next steeping stage of the series; and passing separated liquor to the preceeding stage of the series.

In the whole system there is only one corn feeding point and one corn discharging point. Moreover, steeping liquor is only fed to the system in one point and discharged in another point. This results in a considerable simplification of feeding and discharging means and controls. The whole system may be controlled automatically and adapted to the demand for steeped corn in subsequent stages of the plant.

Further, the whole space of the reaction vessels is used because of the continuous flow of corn therethrough. This means a gain in useful volume of 15 to 30%. The useful life of the reaction vessels has also been lengthened by the continuous flow of corn-liquor mixture. The fact that all mixtures of corn and liquor are screened outside the vessels permits a simpler construction of the vessels, and there will be no fluctuations in water demand and consequently large additional quantities of water for balancing are not required.

Finally, the contact between corn and liquor in each reduction vessel will be intimate and very effective, because they flow concurrently through such vessel. The apparatus and process is shown in detail in Figure 1.1. In the arrangement of Figure 1.1 there are reaction vessels **1, 2, 3** for the steeping operation together with a feed bunker **4** and a discharge bunker **5**. Curved screens **6, 7** and **8** for separating corn and liquor are positioned above reaction vessels **2** and **3** and discharge bunker **5**, respectively.

Fresh purified corn is fed to bunker **4** by a conveyor **9** at such a rate that the bunker always remains filled. In the feed bunker, the corn grains are contacted with a stream of heated steep liquor from line **10** to bring them to the desired steeping temperature. This stream of steep liquor has been withdrawn from the upper part of vessel **1** by a pump **11** and has passed a heat exchanger **12** provided with temperature control **13** before entering the bunker **4**. From this feed bunker the liquor flows downward around the grains of corn and returns to the vessel.

The corn grains leave the feed bunker **4** through a pipe **14** and enter vessel **1**
just on top of a mass **14** of previously fed corn grains. When they have arrived
in the vessel, the corn grains are contacted with a steeping liquor fed through
line **16** and travel gradually downward to the bottom part of the vessel in concurrent relation to the steeping liquor. During this downward flow of corn and
liquor, an intimate contact is ensured and part of the soluble substances in the
fresh corn are extracted by the liquor. The reaction vessel comprises plates **17**
which are constructed in such a way that a sufficiently constant contact time between corn and liquor is effectuated but do not substantially obstruct the downward flow thereof.

On arriving at the bottom part **18** of the reaction vessel, the corn-liquor mixture
enters a line **19** and is passed continuously to screen **6** above reaction vessel **2**
by pump **20**. The mixture may be diluted, if desired, with an additional amount
of steep liquor from line **21** to improve its pumpability. On its way to the
screen, the mixture of corn grains and liquor passes a heat exchanger **22** with
temperature control **23** for maintaining a correct temperature. On the screen it
is separated into steep liquor which passes through the screen and corn grains
which do not pass through it.

Similar operations are also carried out in reaction vessels **2** and **3**. On leaving
vessel **3**, a mixture of corn and liquor enters line **33** and is passed by a pump **33'**
to screen **8** for separation into liquor and grains. An additional amount of liquor
to improve pumpability may be added through line **34** and a heat exchanger **35**
with temperature control **36** is present between pump and screen in order to
maintain a correct temperature of the mixture.

FIGURE 1.1: MULTISTAGE STEEPING PROCESS

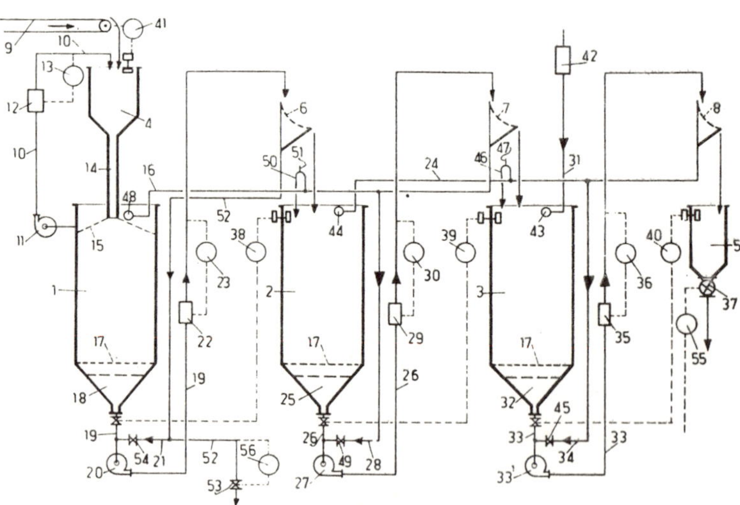

Source: H.J. Vegter; U.S. Patent 3,595,696; July 27, 1971

The corn separated on screen **8** flows into bunker **5** from where it is discharged by means of a sluice **37**. This corn has been freed then of a substantial part of its soluble substances and may be passed to a starch-recovering plant. In order that vessels **2** and **3** and bunkers **4** and **5** are always filled with corn to a substantially constant level during operation, these vessels and bunkers are provided with level controls **38, 39, 40,** and **41**.

Level controls **38, 39** and **40** are arranged to vary the passageway in discharge lines **19, 26** and **33** on variation of the corn level in vessels **2** and **3** and bunker **5**, while **41** varies the operation of feeding conveyor **9** on variation of the corn level in bunker **4**. It will be seen by following Figure 1.1 that the steeping liquor passes through the whole apparatus in opposite sequence to the corn while it flows concurrently with corn through each reaction vessel.

Screening Process for Starch Particles

V.P. Chwalek; U.S. Patent 3,813,298; May 28, 1974 has developed a continuous process for separating starch particles and fibers from an aqueous slurry of, e.g., wet milled corn. The process comprises the following steps:

(1) Passing a supply stream of the slurry over a screen bend having a slot width of such dimensions that a separation of starch from fiber is effected to obtain a first fraction comprising an aqueous slurry containing starch particles substantially free of fiber, and a second fraction comprising an aqueous slurry containing substantially all the fiber and a minor portion of the starch particles;

(2) Separating the second fraction into two components by centrifugally forcing the second fraction against a screen having openings of a dimension such that an aqueous component containing starch and fine fiber is separated from the fiber;

(3) Continuously recycling the aqueous component containing starch and fine fiber of step (2) to the supply stream of step (1); and

(4) Recovering the first fraction of step (1) and the fiber separated in step (2).

This process is of particular interest because formerly it would not have been obvious that the aqueous component of step (2) containing starch and fine fiber could be recycled to the screen bend. More particularly, since it was known that a screen bend used to separate starch and fiber would pass substantially no fiber, recycling a slurry of starch and fine fiber to the screen could cause the fine fiber to be caught in a closed circuit resulting in a runaround of the fine fiber, that is, the fine fiber in the aqueous component resulting from centrifugally screening the fiber, on recycling to the screen bend, would pass with the fiber overflow back to the centrifugal screen, and out again to the screen bend, and so on.

Apparently, therefore, it was thought that such recycling could result in a build-up of fine fiber in the system causing the system to become overloaded and inoperable, or the fine fiber could run around in the system until the fiber was physically broken down to a particle size that would pass the screen bend thereby contaminating the starch. This process may be better understood with reference to Figure 1.2 where embodiments are shown somewhat schematically.

FIGURE 1.2: SCREENING PROCESS FOR STARCH PARTICLES

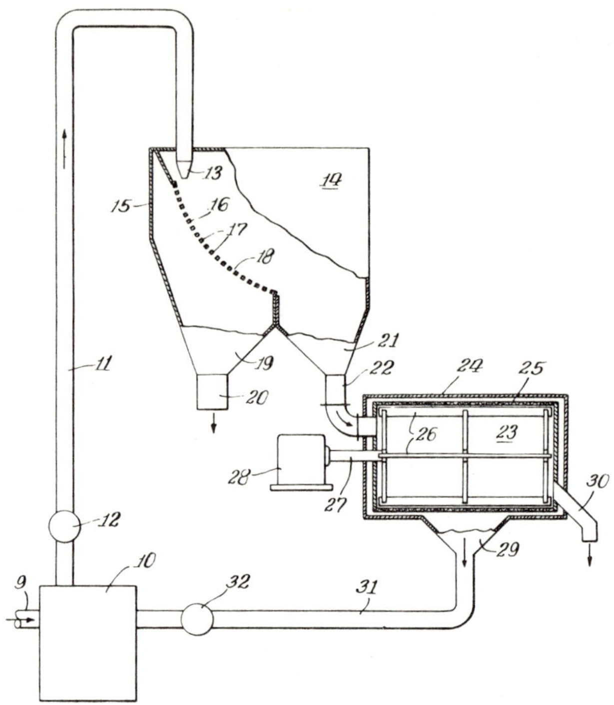

Source: V.P. Chwalek; U.S. Patent 3,813,298; May 28, 1974

In Figure 1.2 there is shown a feed stream **9** and a holding tank **10** which contains the liquid slurry of fine particulate matter and coarse matter. The slurry is passed along a conduit **11** with the aid of a pump **12** through a nozzle **13** into the screen bend **14**. This screen bend comprises a housing **15** and a screening deck **16**, over which the slurry is supplied. This screening deck is constructed of a plurality of bars **17** spaced to form slits **18**. The screening deck is curved so that the bars form the generatrices of cylinder surface.

Fine particles and liquid pass through the slits and collect in the receiving chamber **19** from which they are carried off through exit port **20**. The convex side of the screening deck is entirely shut in by the receiving chamber so that no losses can occur.

Coarse particles move along the screening deck and upon reaching the delivery end of the deck, are carried into a collection reservoir **21**. The coarse matter is then passed through the pipe **22** into a centrifugal paddle screen **23** comprising a housing **24**, a cylindrical screen **25**, and a plurality of paddles **26** rotatably mounted therein. The paddles are disposed on a drive shaft **27**, powered by a driving means **28**.

The overflow from the screen bend **14** is partially dewatered in the paddle screen **23** by forcing the water, as well as fine and some coarse particulate matter, through the screen **25**. The major portion of the coarse material delivered to the paddle screen will not pass through the screen **25**, but will be forced to exit through pipe **30**. The fraction of material which does pass through the cylindrical screen **25** is collected in the reservoir **29** and is recycled to the feed stream of the screen bend, by passing it through a conduit **31** with the aid of a pump **32** into the feed stream reservoir **10**.

In operation, as the liquid slurry flows along a tangential path to the screening deck **16**, a thin layer of the slurry is scraped off on each of the bars from the current of suspension. As can be deducted from the results, the thickness of the layer normally amounts to about one-fourth of the width of the slit between two bars. A solid particle which is at least half immersed in the layer will be entrained and passed through the slit. The biggest particle that can get into such a slit would, therefore, have a diameter of twice the thickness of the scraped-off layers, which implies that this diameter is normally equal to about one-half the width of the slit.

The efficiency of this process in separating starch from fiber, while simultaneously dewatering the fiber to a relatively low water content, was demonstrated in an experiment where the starch-fiber separation effected in both a screen bend and a paddle screen alone, was compared to a starch-fiber separation effected by this process. In this experiment, a separation of 50 microns was carried out.

The efficiency of the separation was determined by measuring the amount of material having a particle size greater than 43 microns in the starch stream and by measuring the water content of the fiber stream. When the separation was effected by using only a screen bend, a starch-fiber mixture having a water content of 74% was fed into a screen bend. Upon analysis, the starch stream was found to contain between 7 and 8 grains/gal of material having a particle size greater than 43 microns. The fiber stream was found to have a water content of about 75% by weight.

When a similar aqueous suspension was fed into a paddle screen, the starch stream was found to contain 150 grains/gal of material having a particle size in excess of 43 microns. The fiber stream from the separation had a water content of 65% by weight. When both a screen bend and a paddle screen were used in accordance with this process, the starch stream recovered from the screen bend contained 16 grains/gal of material having a particle size in excess of 43 microns and the fiber stream collected from the paddle screen had a water content of 62% by weight.

STARCH FROM WHEAT

The process for manufacturing starch and gluten from a dough of wheat flour and water differs entirely from all other starch manufacturing processes. The reason for this is the rather special physical properties of the wheat gluten. This gluten, about 8 to 17% of the flour, swells during the mixing process of wheat flour with water and forms a network of a cellular pattern where the gluten forms the cell walls and the starch fills the cells.

The hydrated or swollen gluten is very elastic and this property is used when leavening the dough for bread baking. When the yeast transfer some of the starch granules into alcohol, carbon dioxide gas is formed and expands the cells. The elastic gluten walls prevent the gas from escaping and a fluffy and more easily digestible bread can be baked.

When separating the gluten and starch from the wheat dough, water is used for washing the starch away. In the unwashed dough, the gluten is dispersed as a very thin network. Whereas the starch particles have a defined size and shape, the gluten has neither of these qualifications. It is, however, very sticky in relation to other gluten particles, which property makes it possible to collect the gluten in the form of gluten lumps in screening means, while the starch is washed through the screen by means of the washing liquid.

Knowing these facts it is understood that when starch and gluten are to be extracted from a dough of wheat flour and at the same time separated from each other, two contradictory conditions must be taken into consideration: first, that the gluten should accumulate as quickly as possible in order that the gluten losses may be kept small, and second that the gluten network mentioned above should be ruptured in order that starch granules embedded in this network can be treated with water.

These contradictory conditions have so far necessitated a labor-consuming and lengthy, and therefore costly, treatment, as well as a raw material of high quality or considerable sacrifices have had to be made in the yield of gluten and in the quality of the starch produced.

Removal of Gluten from Prepared Dough

A process for separating gluten from starch in a pretreated dough made from wheat flour and water is disclosed by *K.E.B. Plaven; U.S. Patents 3,575,710; April 20, 1971; and 3,489,605; January 13, 1970.* Known methods permit obtaining a high-grade gluten either by a slow, noncontinuous and thus costly treatment, or by a considerably quicker treatment which, however, may involve heavy gluten losses with the wash water, especially when wheat flours with weak gluten are used.

These drawbacks are eliminated by this process which comprises converting the gluten into a spongy shape in which the gluten still coheres in sufficiently large aggregates suitable for the following treatment but offers large surfaces in contact with the water, which facilitates and accelerates the extraction of the starch from the gluten. For this purpose the apparatus used in this process comprises perforated plates inserted for instance at the pump outlets, in the piping, or at the screen inlet, through which plates the gluten is pressed by the aid of the wash water passing through the treatment line.

Water is mixed with the flour at the weight ratio of 40 to 80 parts of water to 100 parts of flour to obtain a plastic dough, which possibly after a certain set swelling time, is treated with water for washing out the starch from the dough. The dough which eventually consists of mainly gluten maintains its plastic but undivided, i.e., cohering, nature throughout the washing treatment. During the washing the dough is treated in contact with water, with or without water spraying, by kneading or other known action, so that also the interior is exposed to the water. The treatment mentioned above is the pretreatment, after which the dough is subjected to the treatment as shown in Figure 1.3a. After being prewashed and divided into smaller lumps, the dough is transported through a pipe **1** ending in a pump **2** driven by motor **3** via shaft **4**. At or after the outlet of the pump there are two chambers **5** and **6**, divided by a perforated plate **7** through which the gluten lumps are pressed by the aid of the water as shown in Figure 1.3b. The lumps are discharged through pipe **8**. The only object of the pump is to press water and the gluten lumps through the perforated plate.

Water in this description may also mean water-containing starch obtained in the last preceding stage and/or the following stages of the washing or starch refining process. When being pressed through the perforations of the plate **7** the gluten lumps are converted into a spongy or waste-cotton-like shape, mainly without the formation of discrete very small gluten particles. The spongy lumps offer large surfaces exposed to the water, thus facilitating an efficient extraction of the starch granules in the water flow in pipe **8** and other pipes described below. Through pipe **8**, to which another pipe **9** may be connected for addition of extra water, the spongy lumps are transported to a screen **10** where gluten and water are separated.

FIGURE 1.3: REMOVAL OF GLUTEN FROM PREPARED DOUGH

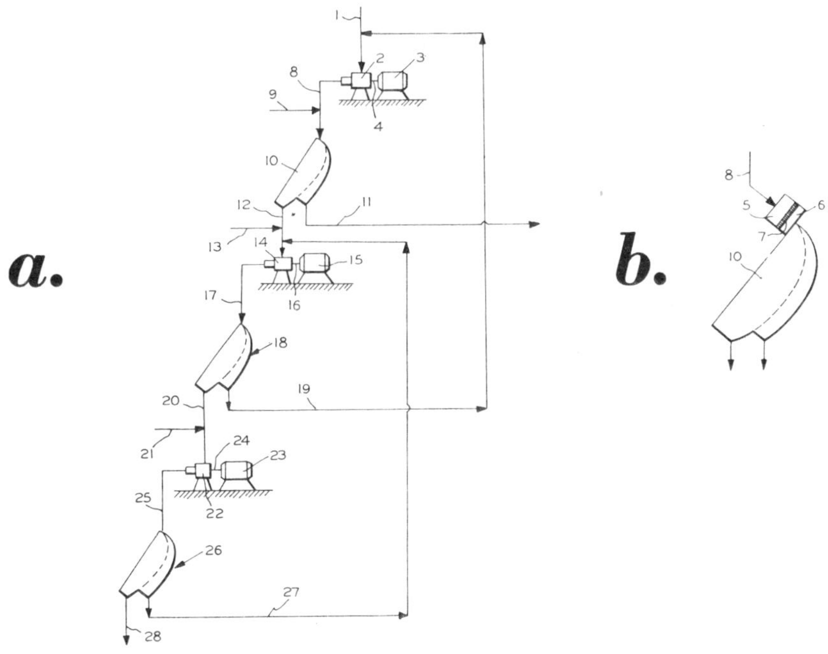

Source: K.E. B. Plaven; U.S. Patent 3,575,710; April 20, 1971

The water is discharged through pipe **11** and added to the prewashing section at a suitable point. Together with a minor amount of water not separated by the screen **10** the gluten is discharged through pipe **12** to which pure water or water from a following stage may be added through pipe **13**. A pump **14** with chambers and perforated plate, driven by a motor **15** via a shaft **16** presses the water and gluten through the perforations of the plate in the pump, further converting the gluten into a spongy shape and transporting the slurry through pipe **17** to another screen **18**, where the major part of the water is separated from the gluten.

The water is discharged through pipe **19** and is added to pipe **1** at a suitable point. The gluten together with remaining water is discharged through pipe **20** to which pipe **21** for water is connected. The pipe **20** is connected to pump **22**, driven by motor **23**, via shaft **24**. The pump **22**, which is of the same kind as the pump **2**, presses the gluten and the water through another perforated plate, thereby further converting the gluten into a spongy shape, and through a pipe **25** to a screen **26**.

From this screen the water is discharged through a pipe **27** to a suitable point in the pipe **12**. When leaving the screen **26**, through the pipe **28**, the gluten contains very little starch, but if desired it can be treated in further washing stages incorporating pumps, pipes, perforated plates, and screens as described above. Because the water separated off in one stage is added to a preceding stage, small gluten particles possibly carried by the water are returned to the process and agglomerates with the larger gluten lumps. Hereby practically no gluten is lost. Furthermore, it has been possible to carry out the washing in a rapid and fully automatic manner, labor being necessary for supervision only.

Dough Mixer for Separating Starch and Gluten

Mixing apparatus and methods have also been developed by *E. Plaven; U.S. Patents 3,669,739; June 13, 1972; and 3,506,485; April 14, 1970* for separating starch and gluten from wheat flour.

This process is characterized in that the lumps of gluten, to uncover starch granules embedded in each lump, are subjected to a repeated cutting and spreading operation in various directions within a perforated troughed treatment zone and of a nature such that the gluten lumps are shredded, cut and flattened and the gluten network ruptured, whereby small gluten particles or lumps in the perforations of the zone, and not otherwise caught, then stick to and are absorbed by or reformed with the gluten or gluten lumps being pushed through the treatment zone.

This enables the added or recycled wash liquid to discharge from the treatment zone without forming any noticeable liquor or wash liquid level. The gluten is allowed to largely resume its lump shape after the various cutting and spreading operations. The improved process embodies an apparatus characterized by the fact that the treatment means are in the form of improved stirring rods, scrapers and sabres arranged and used in an improved manner, and that the trough is perforated substantially throughout the extension of the sections, where the trough is concentric with the shaft of the scrapers and sabres and thus broadly parallel with the paths of rotation of the rods and the scrapers. The apparatus overall comprises an elongated treatment tank having inlet and outlet means at opposite ends. Figure 1.4 shows the improved stirring rods, scrapers and sabres in a perspective view from above one end of a trough comprising part of the process apparatus.

FIGURE 1.4: DOUGH MIXER FOR SEPARATING STARCH AND GLUTEN

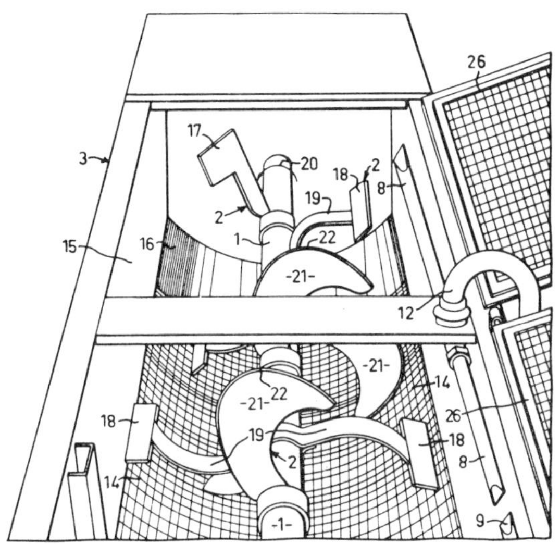

Source: E. Plaven; U.S. Patent 3,669,739; June 13, 1972

During the washing treatment in the trough **3**, the dough may be sprayed with fresh water, although it is more economically sprayed with so-called starch milk, i.e., water which has already been used one or several times for washing the dough in the trough or possibly in following washing devices.

At the feed end the treatment means **2** are in the shape of stirring rods **17** radially secured to the shaft **1**. In the perforated sections **14** of the trough, the treatment means are in the shape of sabres **21** and of blade-type scrapers **18**, or transfer and lifting means, the blades of which are mainly parallel and radial to the shaft. The direction of the scraper blades may, however, deviate somewhat from the parallelism with the shaft in order that the blades may have a slight pushing action on the dough or gluten towards the discharge end.

The combined length of the blades **18** and **21** in their extension parallel to the shaft corresponds mainly to the extension of section **14** in the same direction. For practical reasons, several scrapers and sabres are arranged along the shaft to cover the full length of the extension of the perforated section in parallel with the shaft. During their rotation in the direction shown by arrow **20** the scrapers **18** catch gluten lumps and push them along the curve of the perforated plate and lift them until they fall back onto the trough bottom.

During the treatment, the gluten lumps are caught several times by the scrapers **18**, lifted and returned to the bottom of the trough while being slowly conveyed towards the discharge end due to the slight angularity of the blades. When the gluten lumps are pushed along the curve of the perforated plates, due to their sticky characteristic, they absorb gluten particles which are too small to be caught by the scraper blades and such gluten particles as are sticking in the perforations.

Through this action, the perforation is kept open so that the wash liquid can be discharged without accumulating in the treatment area. It is very important that the wash liquid be drained off continuously from the perforated plate, for if the gluten lumps pushed along the curve of the screen plate are submerged in water, they will not have sufficient absorption power to catch and merge with the small particles sticking in the perforations or suspended in the wash liquid. Moreover, it is important not to break up the dough too much and to avoid a suspension of fine particles in a large amount of liquid.

The scrapers **18** are preferably arranged at an angle of about 90° with respect to the nearest stirring rods so as not to interfere with the stirring and tearing action of the stirring rods. The scrapers are very important for a trouble-free working of the apparatus. The scrapers **18** are connected to the shaft **1** by means of arms or connecting rods **19** bent to the form of an arc, the arc being situated in front of the scraper blade in relation to the direction of rotation as shown by arrow **20**. This shape of the connecting rods prevents the gluten lumps from sliding along the rods down to the shaft and winding themselves around the shaft and makes them fall off between the shaft and the trough sides.

Sabre-like treatment means **21** are also fastened to the shaft. The edge **22** of the sabres is turned towards the semicircular bottom of the trough and runs at an angle to the shaft, the main planes of the sabres forming an acute angle to the curve of the bottom of the trough. The sabres may be fitted both in the perforated sections **14** and the imperforate sections **16**.

The mode of operation of the sabres is entirely different from that of the scrapers, in that the sabres cut, shred and/or pinch the gluten lumps between the edge and trough bottom **14** and **16**, thus shredding the gluten lumps so that the lumps are opened and better exposed to the washing liquid, the result being that starch granules embedded in the lumps are uncovered and can be washed away from the gluten by means of the wash liquid fed to the trough through the spray pipes **8** and **9**. Other parts shown in Figure 1.4 include connecting pipe **12**, chamber side **15** and cover **26**.

Separation of Starch with Aqueous Ammonium Hydroxide

A nonsticky process for separation of starch and gluten from wheat flour has been developed by *P.H. Johnston and D.A. Fellers; U.S. Patent 3,574,180; April 6, 1971; assigned to the U.S. Secretary of Agriculture.* The coherence and stickiness are obviated by the use of ammonium hydroxide, which is volatile and therefore readily removable from the products. In contrast, various agents which previously have been advocated for the purpose, e.g., lime, sodium hydroxide, salt, malt extract, etc., are nonvolatile substances which necessarily remain in the final products. A particular disadvantage in using a fixed alkali such as potassium or sodium hydroxide is that this base causes a modification or denaturation of the protein (gluten) component whereby it cannot be used in bakery products. Ammonium hydroxide, on the other hand, does not cause this difficulty, the protein recovered by this process retains its native or vital qualities, so that it can be added to doughs for bakery products without causing any adverse effects such as diminished loaf volume. In practice, a mixture of wheat flour, water and ammonium hydroxide is formed and subjected to agitation or shear forces to form a smooth slurry. In preparing the slurry, generally 1 to 2 parts of water are used per part of flour.

For flours derived from hard wheats (high-protein) a preferred ratio is 1.5 parts of water per part of flour. For flours derived from soft wheats (low-protein), a preferred ratio is 1.25 parts of water per part of flour. Enough ammonium hydroxide (or ammonia gas) is added to the mixture to provide a pH of 6.5 to 9.5, preferably 7.5 to 9.0. It should be noted that without addition of NH_4OH, the slurry will generally have a pH of 5.8 to 6.0. The slurry is ordinarily prepared at room temperature (25°C) as being convenient and giving excellent results; however, temperatures somewhat lower or higher, e.g., 15° to 45°C, may be used if desired.

Having prepared the slurry of water, flour and NH_4OH, this slurry is subjected to centrifugation and there is formed a dense bottom phase and a supernatant liquid phase. The bottom phase contains essentially all the prime starch from the flour in almost a pure state. This prime starch phase being in a granular condition, completely free from stickiness, can be readily separated from the remainder of the system, and can be readily processed as by washing with water and drying to prepare a high-grade starch for any desired use. Any ammonium hydroxide remaining in the starch fraction will, of course, be vaporized when the starch is dried.

Essentially all the protein from the flour is contained in the supernatant liquid phase. Since this phase is free from stickiness and gumminess, it pours readily and thus can be easily removed from the starch phase. The supernatant liquid is dried to produce a protein product useful for many purposes, for example, as a protein supplement in bread or other foods, as an additive for mixed animal feeds, as a base for preparing milk-like beverages, etc. It is obvious that during the drying step ammonium hydroxide is vaporized so that the protein product has its native properties.

Example 1: The starting material was a hard red spring, straight-run, unbleached flour having a protein content of 14.25% (measured on the undried basis, moisture content was 13.3%). A mixture was made of 1,250 g of the flour, 2,031 g of water, and enough ammonium hydroxide to provide a pH of 8.0. The mixture was blended for 10 min in a 1-gal capacity Waring Blendor. The resulting smooth, easily pourable slurry was then centrifuged for 5 min at 1,500 rpm (about 500 g).

It was observed that there was formed a compact starch phase and a supernatant liquid phase. No problem with stickiness or coherence was encountered; the supernatant could be easily poured off the lower starch phase. The supernatant liquid was freeze-dried to a moisture content of 2.15%. A yield of 498 g of protein concentrate was obtained. Analysis indicated a protein content of 38.5% on a dry basis. The starch fraction was freeze-dried without washing. A yield of 720 g of starch was obtained. Analysis indicated it contained only 0.96% protein, on a dry basis.

Example 2: A series of runs were made similar to those of Example 1 except that the amount of ammonium hydroxide added to the slurry was varied, and in one instance (control run) no ammonium hydroxide was added at all. In each run the flour was derived from hard red spring wheat. The water-to-flour ratio was 1.625 to 1. The slurry was at 30°C and blended for 3 min, then centrifuged for 5 min at 1,500 rpm. The viscosity of the slurry was measured in each case. Also, the handling character (pourability) of the supernatant after centrifugation was observed. The results obtained are tabulated below.

Run	pH of slurry	Viscosity (Brookfield) of slurry, centipoises	Pourability of supernatant
Control	5.9	3,540	Thick, glutinous, difficult to pour.
1	6.52	3,415	Thick, but better than control.
2	7.0	3,150	Slightly thick, but pourable.
3	7.5	3,000	Do.
4	8.0	2,950	Smooth, excellent pouring quality.
5	8.5	2,925	Do.
6	9.0	3,050	Do.
7	9.5	3,370	Pours well but some clotting.
8	10.0	(1)	Gelled, could not get separation.

[1] Too thick to measure.

The starch fractions obtained in several of the runs were dried and analyzed for protein content. The results are as follows.

Run	pH of slurry	Yield of starch, percent, dry basis	Protein in starch, percent, dry basis
2	7.0	59	1.0
3	7.5	59	0.9
4	8.0	58	0.9
5	8.5	58	1.1
6	9.0	56.5	0.9

STARCH FROM POTATOES

Low Water Consumption in Preparing Potato Starch

An improved process for the production of starch, especially for plants on a large industrial scale, which requires a minimum of wash water has been disclosed by *H. Hemfort, H. Huster and F. Heimeier; U.S. Patent 3,756,854; September 4, 1973; assigned to Westfalia Separator AG, Germany.* The process relates especially to the production of starch from potatoes, manioc roots or the like, in which the pulp from the rasping machine is separated in a preliminary stage by means of a decanter into vegetable water and solids, the vegetable water is removed, the starch is washed out of the solids in a washing section consisting of a plurality of sieve stages, and is delivered to the refining section.

The process is characterized in that wash water is fed countercurrently through the washing section of the plant, and the starch-containing wash water running from each sieve is separated by centrifugal separators into a concentrated starch milk and starch-free water. The concentrated starch milk is fed from each or at least several of the centrifugal separators to the refining section of the plant, and the starch-free water is delivered to the preliminary stage and, after passing through all the stages, is fed back to the input of the decanter of the preliminary stage, or it is united with the vegetable water separated from it, or it is carried out of the process separately. In this manner, large quantities of wash water containing dissolved protein and finer solid particles are prevented from getting into the refining section of the plant. This makes the purification of the extracted starch considerably simpler, and a better quality is achieved in the end product. Furthermore, the water consumption is reduced to a minimum.

The process is outlined in Figure 1.5 where the heavy lines are for washing liquid, the medium lines for washing liquid containing extracted starch and the light lines for the solids. The starting material is shredded in the presence of water in the rasping machine R_1. The pulp is separated in decanter D into vegetable water and solids. The vegetable water is carried out through line L_1 and can be further processed in a protein-recovery system.

The solids removed from the decanter D, in a relatively dry consistency, consist of starch and vegetable cell tissue. They pass through line L_2 to a first washing stage comprising the first sieve S_1 where a first washing of the starch from the cell tissue is performed by the wash water brought in through line W_1. The solids left on sieve S_1 pass through line L_3 to a second washing stage comprising the second sieve S_2 on which a second washing takes place with the addition of water from the line W_2. In the example represented, the solids pass from sieve S_2 through line L_4 to a second rasping machine R_2.

FIGURE 1.5: LOW WATER CONSUMPTION IN PREPARING POTATO STARCH

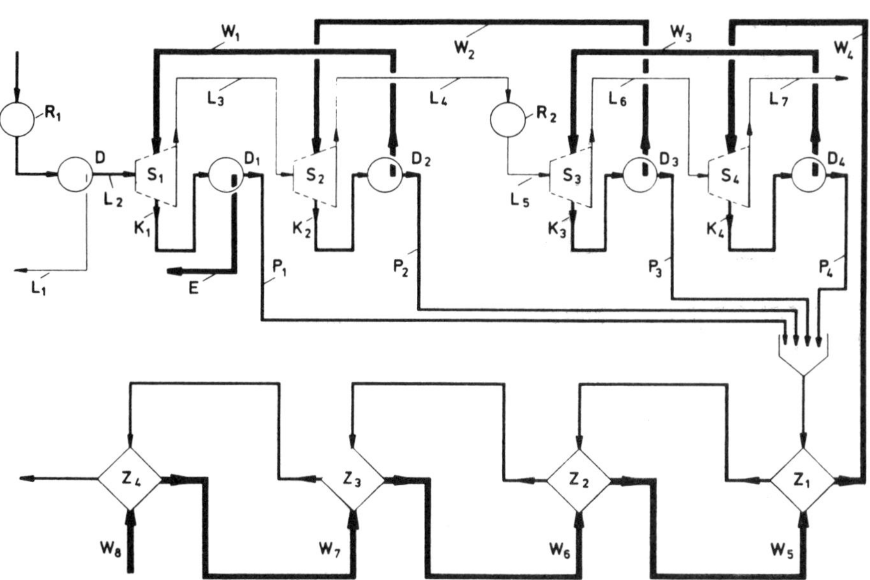

Source: H. Hemfort, H. Huster and F. Heimeier; U.S. Patent 3,756,854; September 4, 1973

The recomminuted solids are washed in two additional washing stages comprising respectively, sieves S_3 and S_4 with the addition of water from lines W_3 and W_4 and are finally carried out of the washing section through line L_7. The starch-containing wash water flowing from sieve S_1 passes through line K_1 to the centrifugal separator D_1, of the first washing stage, which divides it into a concentrated starch milk and a washing liquid lean in starch, e.g., starch-free water.

The starch-free water leaves the separator through the line E. It can be returned to the decanter D or to the rasping machine R_1, or it can be combined with the vegetable water running out through line L_1, or it can be removed separately from the process. The concentrated starch milk flows through line P_1 to the first separator Z_1 of the refinery section of the plant. Likewise, the starch-containing water passing through sieves S_2, S_3 and S_4 is separated in separators D_2, D_3 and D_4 into concentrated starch milk and starch-free water. The starch milk concentrated in these three stages flows through P_2, P_3 and P_4 also to separator Z_1. In the final stage of the washing section (sieve S_4 and centrifugal separator D_4) the separator can be omitted due to the low starch content in the water in line K_4.

The starch-free water emerging from the separators D_4, D_3 and D_2 serves as washing liquid in the washing section and flows through the wash stages countercurrently. The extraction section of the plant with sieves S_1 through S_4 can be operated countercurrently by itself. In this case fresh water is fed through line W_4.

The entire plant operates countercurrently. Fresh water is delivered to the final refining separator Z_4 through line W_8, flows countercurrently to the starch being refined through separators Z_4 through Z_1 to the final sieve S_4. After passing through sieves S_3 through S_1 countercurrently, the wash water is carried out through line E.

The countercurrent flow of the wash water keeps the proteins dissolved in it and the fine solid particles suspended in it from getting into the refining section of the plant. Therefore a high quality can be achieved in the end product at lower cost. The yield of protein, if protein recovery is performed, is increased by the combining of the wash water with the vegetable water.

Low Waste Potato Starch Process

R.L. Shaw, Jr.; U.S. Patent 3,813,297; May 28, 1974; assigned to U.S. Secretary of Agriculture has developed a method of producing a high-quality potato starch that eliminates the waste problem and substantially reduces the amount of water needed in the process. The process consists essentially of the following steps:

(1) Cutting and dehydrating potatoes in such a manner that heat damage of the starch is prevented. Since heat damage is directly related to moisture content, temperature and duration of exposure, a satisfactory product can be obtained by use of a number of cutting and drying conditions;

(2) Fine grinding the cut and dried potatoes in a pin or other impact-type mill;

(3) Air classifying the fine ground material from step (2) to obtain a starch-enriched fraction and protein-enriched fraction;

(4) Washing the starch-enriched fraction in a countercurrent fashion to remove protein, sugars and amino and other organic acids to obtain the desired starch product; and

(5) Combining the protein-enriched fraction from step (3) with the wash water from the countercurrent washing operation of step (4) and drying the combined material to obtain a relatively high-protein (20 to 30%) product.

Countercurrent washing minimizes the amount of water needed in this process. In fact, it requires only about 10% of the amount of water needed in a conventional starch plant. In addition, this wash water is combined with the protein-enriched fraction from the air-classification step and dried to yield a product containing about 25 to 30% protein. This product is an excellent animal feed supplement and eliminates wastes.

Example: Potatoes were dried at 85°C, coarse ground at 9,000 rpm in a pin mill and air-classified in a commercial air classifier at a feed gate setting of 15 and a fin setting of 10°. The starch and protein fractions represented 91.2 and 8.8%, respectively of the recovered material. The starch fraction was fine ground at 19,000 rpm in a pin mill and air-classified at the same settings as before to obtain a starch fraction and a protein fraction that represented 95.8 and 4.2%, respectively, of the recovered material.

The starch fraction had a protein content of 6.3% which was reduced to about 0.5% by the countercurrent washing procedure to obtain a high-quality potato starch. The wash water was combined with the protein fractions from the air-classification steps and dried to obtain a product containing about 25% protein.

STARCH FROM OTHER MATERIALS

Starch from Cellulose

J.R. Harvey; U.S. Patent 3,658,588; April 25, 1972 produces hydrolyzed products of celluloses by phosphorylyzing cotton, wood, leaves, grass, paper and other sources of cellulose with concentrated phosphoric acid and then hydrolyzing that phosphorylyzed cellulose. The phosphoric acid is removed by water or solvent washing the resulting starches and sugars. If phosphorylyzed cellulose is dehydrated by concentrated H_2SO_4 in the presence of glycol or glycerol, an epoxy results.

The yield of starches or sugars from pure cellulose exceeds 100% due to the water added by hydrolysis of the glucoside bonds and the controlled attack of the acids used. When hydrolyzed by diastase the starch prepared from cotton yields a maltose of about 15% which has a molecular weight comparable to starches and gives the iodine-starch color complex. The β-glucoside-linked starch is not reactive with iodine unless concentrated H_2SO_4 is added incrementally to the phosphoric acid and cellulose solution until a peak in viscosity has been passed.

The cotton-prepared starch or the beta-glucoside-linked starch is prepared by treating the appropriate cellulose with concentrated phosphoric acid for 4 hr at room temperature with agitation. If wood or paper containing lignin is used as the source of cellulose, the lignin will remain insoluble in the phosphoric acid and can be removed on a noncellulose suction filter before the hydrolyzing step. In the hydrolysis reaction water replaces the phosphoric acid radical on the cleaved glucoside bonds of the cellulose molecule making a starch. The starch is then freed of the phosphoric acid by washing on a fine sintered filter. The water will contain the phosphoric acid which can be concentrated by physical means.

Cellulose in concentrated phosphoric acid at approximately 70° to 71°C can be converted into glucose sugars by sequential phosphorylyzation and hydrolysis. Higher temperatures have been used for faster results. The mechanism of sequential phosphorylyzation and hydrolyzation alternates the quenching of the dehydration potential of concentrated phosphoric acid at 70°C between the cleaving of glucoside bonds and the hydration by water. The solution must be agitated and water must be added incrementally within defined limits.

Example 1: Two parts of cotton are mixed at room temperature (20° to 30°C) with 170 parts of orthophosphoric acid (85%). After mixing for 4 hr, 200 parts water are added and mixed for 30 min. The solution is cooled with mixing to approximately 4°C and filtered with mixing on a fine Buchner sintered suction funnel. The resulting starch is washed with water at approximately 4°C and agitated on the filter until free of phosphoric acid. The starch yields a maltose on the order of 15% when treated with diastase. The product also gives the iodine-starch color complex.

Example 2: Ten parts of cellulose are mixed with 200 parts of orthophosphoric acid (85%). The solution is agitated until a clear viscous liquid results. The temperature is then brought to 70° to 71°C with agitation. One part of water is added when a slight carmel color develops or when the carmel color intensifies. Any addition of water in excess will stop the mechanism of degradation at a viscosity plateau and may form gel lumps. Failure of a test sample to precipitate upon dilution with water after 24 hr, indicates formation of glucose sugars from the cellulose.

If there is a precipitate, the mechanism of degradation is continued. When no precipitate is present in a test sample after 24 hr, ethyl ether is added to the solution with mixing until miscibility with phosphoric acid is achieved. At this point ethyl ether is added until all the glucose sugars precipitate. The glucose sugars are removed by filtration, which must be rapid to prevent evaporation of ethyl ether, and mixed with 20 parts absolute methanol. Again ethyl ether is added to alcohol and sugar solution until all the glucose sugars are precipitated. The sugars are removed by filtration again. The washing procedure using ethyl ether and methanol may be repeated until the glucose sugars are freed of the phosphoric acid.

Refining Starch from *Saponaria vaccaria*

The plant *Saponaria vaccaria* is native to the western United States (e.g., Montana). It produces substantial amounts of seed and is generally considered to be a noxious weed. The plant is sometimes referred to as cow cockle, cow soapwort, or cow fat. If *Saponaria vaccaria* seed is to become a cash crop, as opposed to its present status as a noxious weed, a commercially useful refining process must be developed for the starch contained in the seed. *D.L. Brelsford and K.J. Goering; U.S. Patent 3,622,389; November 23, 1971; assigned to Montana*

Agricultural Research Corporation have developed an improved process for treating *Saponaria vaccaria* seed to obtain a refined starch meeting commercial standards (e.g., standards of whiteness). The process involves dehulling and degerming *Saponaria vaccaria* seed. This can be accomplished by conventional wet milling techniques or, less preferably, by dry milling techniques. In either event, an aqueous slurry of dehulled, degermed seed (i.e., a crude starch slurry) will be obtained or prepared as a starting material for the refining part of this process.

Next, the starch slurry is treated to dissolve the protein matrix of the dehulled, degermed seed. This is accomplished by raising the pH by alkali addition (e.g., to a pH of 11). As the protein is dissolved, alkaline insoluble pigment impurities are released from the seed and these impurities can be separated from the slurry by gravity. Under quiescent conditions, the insoluble pigment impurities float to the surface as a scum and settle as a sediment.

The protein solution containing suspended starch granules is then separated from the pigment impurities and the starch granules are removed from the protein solution (e.g., removed by centrifuging). The protein-rich solution is then treated by conventional processing to obtain a dry protein concentrate (e.g., 70% isolated protein). The separated starch granules (separated as a starch-rich slurry) are subsequently washed and dried. The resulting starch is a white, low-protein, commercially acceptable material.

The process can be further understood by reference to Figure 1.6. The feed material is, for example, a crude starch slurry obtained by the wet milling of *Saponaria vaccaria* seeds (i.e., it is a dehulled, degermed product). The slurry contains approximately 10% solids (i.e., crude or unrefined starch). The slurry is introduced into a protein dissolution tank **1** which is equipped with an agitator **2**. A reservoir **3** is provided for storing alkaline solution (e.g., aqueous sodium hydroxide) which is introduced into protein dissolution tank **1** in an amount sufficient to raise the pH of the slurry to, for example, 11.

Dissolution of the protein is accomplished with agitation in the tank in approximately 1 hr. The resulting alkaline slurry is then passed to settling tanks **4** and **5** which are operated alternately. In these settling tanks, the alkaline starch slurry is permitted to rest so that the alkaline insoluble pigment impurities can separate from the slurry by gravity (i.e., settling or floating). The time allowed for separation is 1 hr. The remaining aqueous phase (now free of pigment impurities) is then passed to centrifuge **6** where the alkaline slurry is separated into a concentrated starch slurry and a protein-rich solution.

The protein-rich solution (now substantially free of starch granules) is then passed to a treating chamber **7** where the protein in solution is precipitated by addition of dilute sulfuric acid to lower the pH to the isoelectric point. The protein precipitate is then processed by washing and separating wet protein which is thereafter dried to produce an isolated or concentrated protein.

The concentrated starch slurry obtained from centrifuge **6** is then processed by repeated washing and centrifuging in centrifuges **9** and **10**. The wet refined starch from centrifuge **10** is then dried in the starch dryer **11** to form a refined starch.

FIGURE 1.6: REFINING STARCH FROM *Saponaria vaccaria*

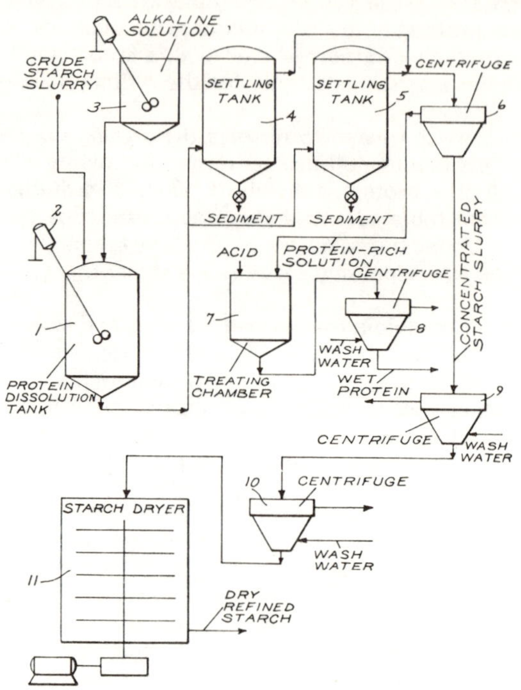

Source: D.L. Brelsford and K.J. Goering; U.S. Patent 3,622,389; November 23, 1971

UNMODIFIED STARCHES IN FOOD PROCESSING

Starch-Yeast Dietetic Tablets

H. Griffon and G. Tixier; U.S. Patent 3,378,377; April 16, 1968 have found that dietetic yeast-starch tablets can be readily prepared by thoroughly mixing fresh untreated yeast of 70% moisture content and powdered starch, dehydrating the mixture until the moisture is approximately 11% in order to place the mixture in equilibrium with the atmosphere (which corresponds to 9 to 13%) so that there is little or no tendency of the mixture to pick up or lose moisture to the atmosphere.

The dehydration can be carried out in any suitable manner such as by forming the mixture into a relatively thin layer on drying plates or equipment and/or subjecting the mixture to a mild current of warm air. The dehydrated mixture is then granulated in any suitable manner, and the granules are finally subjected to sufficient compression on a standard tableting machine to form tablets while

at the same time killing at least a predominant proportion of the yeast cell. By
using relatively high tableting pressures all or substantially all yeast cells can be
killed. The effect of tableting pressure on the number of living yeast cells is
well illustrated by the fact that when the compression of tableting amounts to
1,600 kg/cm^2 for a tablet of 6.3 mm the number of living yeast cells per gram
is 1.8 billion. When the compression is increased to 2,600 kg/cm^2 the number
of living yeast cells per gram is 260 million and pressures above 2,600 kg/cm^2
and, for example, up to 4,000 kg/cm^2 kill 100% of the yeast cells.

This process is predicated on the discovery that yeast cells can be killed by a
purely mechanical phenomenon, namely compression, so that while the yeast
cells lose their vitality they retain their other properties, notably as to color,
odor and taste, in contrast with other methods which involve relatively drastic
heating and dehydrating conditions thereby modifying adversely the organoleptic
properties and producing products which are brown in color and have the taste
of peptone.

This process is further predicated on the discovery that by subjecting the dehydrated yeast-starch granulation to sudden or violent instantaneous mechanical
shock which is correlated with thermoshock, then ideal tablets result for present
purposes. In carrying out the necessary compression with modern tableting
machines, the tableting occurs in a period of time of the order of $^2/_{10}$ of a second.

Example: A quantity of fresh untreated yeast having a moisture content of approximately 70% is thoroughly mixed with dry powdered starch in such proportions that the starch constitutes 75 to 25% of the mixture prior to dehydration.
The yeast-starch mixture is then dehydrated in any suitable manner until the
moisture content of the mixture is approximately 11%.

Prior to dehydration the relative proportions of fresh untreated yeast and starch
are preferably in the ratio of 100 g of yeast to 60 g of starch but variations from
this ratio are permissible within the percentage ranges set forth above. The dehydrated mixture having a moisture content of 11% is subjected to dry granulation in a known manner, and is then tableted on a standard tableting machine
adjusted to produce tablets of the requisite size and thickness. The effect of
compression on the number of living cells is shown by the following table.

Compression (kg/cm^2)	Thickness of Compressed Tablets (mm)	Total Number of Cells	Number of Living Cells	Number of Dead Cells	Percentage of Living Cells
		- - - - - Number of Cells per Gram (in millions)- - - - - -			
0	—	10,000	4,400	5,250	44.0
800	7.4	10,000	4,000	6,000	40.0
1,000	7.1	10,000	4,000	6,000	40.0
1,200	6.9	10,000	3,500	6,250	35.0
1,400	6.6	9,000	2,500	6,350	27.8
1,600	6.3	8,800	1,800	7,000	20.5
1,800	5.9	8,350	700	7,650	8.4
2,100	5.7	8,180	370	7,820	4.5
2,600	5.4	8,000	260	7,740	3.25
2,800	5.2	8,000	0	8,000	0
3,300	5.2	7,250	0	7,250	0
4,000	5.2	7,250	0	7,250	0

The above table illustrates how the number of living cells per gram of yeast can be reduced to zero or any other desired value. It is preferred to kill all the yeast cells so that in the compressed tablets there are no living yeast cells. The starch used may be of any suitable nature such as cornstarch, rice starch, potato starch, or cereal starch. The yeast which is used is preferably the ordinary fresh untreated commercial yeast having an initial moisture content of approximately 70%.

Starch-Based Multicoating for Fried Foods

A method has been developed by *S.L. Klug, G. Finkel and M.B. Sherain; U.S. Patent 3,669,674; June 13, 1972; and U.S. Patent 3,514,294; May 26, 1970; both assigned to DCA Food Industries, Inc.* whereby a foodstuff is prepared for frying by alternately applying dry powder coatings and batter coatings. Each coating contains 60 to 90 parts starch, 1 to 9 parts leavening, 1 to 15 parts sugar, and 1 to 15 parts salt. In addition, the batter contains 85 to 200% water based on the weight of the dry ingredients. The coatings may contain additives such as dry milk solids.

This process is based on the discovery that a superior fried comestible is produced by applying to the food a first powder coating containing starch in a major amount and a sugar, salt and leavening in a minor amount. To the powder-coated food is then applied a first batter coating comprising water and a major amount of starch and minor amounts of sugar, salt and leavening. A second powder coating of the nature of the first powder coating, is then applied, as well as a second batter of the nature of the first batter.

The first and second powder coating may be of identical composition as may be the first and second batters. Moreover, the dry ingredients of the batter may be the same as those of the powder coating. In addition to the above ingredients the powder and the batter may contain other materials as, for example, whey and nonfat dry milk.

The powder coating advantageously contains between 60 and 90 parts of starch, at least 1 and up to 15 parts of sugar (sucrose), at least 1 and up to 15 parts of salt, and between 1 and 9 parts of leavening. The batter contains the above powder ingredients in the ranges set forth and between 85 and 200% water based on the weight of the dry ingredients. The various proportions noted are given by weight.

It has also been found that a synergism is experienced in relationship to fat absorption, crispness and color, a low fat absorption, high crispness and optimum color resulting when, in a batter containing starch, sugar, salt, leavening and water, the ratio of sugar to salt is between 1 to 8 parts of sugar to 2 to 12 parts of salt, the ratio of salt to leavening being between 2 to 12 parts of salt to 1 to 9 parts of leavening, and the ratio of sugar to leavening being between 1 to 8 parts of sugar to 1 to 4 parts of leavening.

It is important to note that the effect of the individual components of a combination on a particular characteristic is radically less outside the ratio or in the absence of the other component than when in the respective combination within the ratio. The starch is advantageously present in the form of a flour, preferably a mixture of wheat flour and corn flour. A small percentage of the starch (up to 5%) is preferably gelatinized. Thus the following is the preferred range of ingredients of the coating powder.

Ingredients	Parts by Weight
Wheat flour	40 - 50
Corn flour	30 - 40
Gelatinized flour	0.25 - 2.5
Sugar (sucrose)	6 - 9
Leavening	1 - 4

In addition, other edible ingredients may be present, for example, up to 5 parts of whey or nonfat dry milk, up to 3 parts of dried egg, and up to 5 parts of monosodium glutamate, and other additives. The batter is a mixture of the dry ingredients within the ranges set forth above, and between 85 and 200% of water based on the weight of the dry ingredients. The following is a specific example of the coating powder.

Ingredients	Pounds	Ounces
Wheat flour	46	2
Corn flour	34	5
Gelatinized corn flour	0	7
Sugar	7	14
Nonfat dry milk	1	13
Salt	5	12
Leavening	3	11

The batter is produced by mixing the powdered ingredients listed above with 140% by weight of water. In preparing the comestible, a food substrate such as a fish stick, is coated with a first layer of the dry powder by rolling the food substrate in the powder or by dusting. The powder-coated food is then dipped into the batter, or the batter applied by means of a batter curtain, to form a first batter overlaying the first powder layer.

A second powder layer is then applied overlying the first batter layer, and a second batter layer is applied overlying the second powder layer. The food coated as described, may then be fried, either by full or partial immersion in hot frying oil and is ready for consumption, or the fried product may be frozen and stored for future use. In preparing the frozen product for consumption it may be reheated in any convenient manner.

The fried comestible produced by this method is neither soggy, soft nor gritty but has a crisp envelope and unusually good flavor, whether consumed immediately after frying or following the reheating of the fried, frozen and stored product. The coated comestible exhibits an unexpectedly low fat absorption on frying and no rendering of the fat occurs during the reheating of the previously fried product, nor does the product stick to the reheating vessel, a drawback which characterizes the conventionally coated and fried products. Furthermore, the flake-off is minimized and the end product is of relatively greater volume by reason of the high expansion of the coating during the frying operation.

Fatty Acid Polyphosphate Thickeners for Starch

Improved gelling and texturizing agents for starch have been disclosed by *R.B. Fearing and J.C. Sourby; U.S. Patent 3,597,232; August 3, 1971; assigned to Stauffer Chemical Company*. These agents comprise phosphates having the following formula:

$$R-O\!-\!\!\left[CH_2-\underset{\underset{R'}{\overset{|}{O}}}{\overset{|}{C}}H-CH_2-O-\right]_m\!\!\left[CH_2-\underset{\underset{R''}{\overset{|}{O}}}{\overset{|}{C}}H-CH_2-O-\right]_n\!\!\underset{\underset{O-(R'''O)_oH}{\overset{|}{|}}}{\overset{\overset{O}{\|}}{P}}-O-R'''-OH$$

where R can be hydrogen or a long chain fatty acyl group having 7 to 50 carbons and R' can be hydrogen or a long chain fatty acyl group having 7 to 50 carbons, R" can be hydrogen or a long chain fatty acyl group having 7 to 50 carbons, R''' is an alkylene group having at least 2 carbons but less than 10, but when R" is hydrogen at least R or R' must be a fatty acyl group, m can be an integer between 1 and 10, n can be an integer between 1 and 10, o can be an integer between 1 and 10.

The phosphates are formulated by reacting a long chain fatty acid di- or polyglyceryl ester such as diglyceryl monopalmitate, triglyceryl distearate, diglyceryl monooleate, and the like with polyphosphoric acid such as pyrophosphoric acid to produce a fatty acid ester of di- or polyglycerol monophosphate. Some diphosphate or even polyphosphate may be produced depending on the number of free OH groups per molecule or in other words OH groups left unesterified by the fatty acid.

This material is then treated with alkylene oxide such as ethylene, propylene, butylene oxides to produce the oxyalkylated product. More than 2 mols of alkylene oxide per mol of fatty acid ester of the di- or polyglyceryl mono- and diphosphate are used. The fatty acid esters that can be used to start the reaction can be selected from condensed glyceryl derivatives of palmitate, stearate, oleate, laurate, myristate and mixtures thereof. It has been found that from 0.1 to 2% by weight can be added to the foodstuff and phase separation of the water will be essentially eliminated or considerably reduced.

Example 1: A sample of transesterified diglyceryl monopalmitate in an amount of 570 g was melted at 80°C. To this molten product was added 213.7 g of molten pyrophosphoric acid by increments. After 2 hr at 85°C, the reaction mixture was diluted with 800 ml of benzene. This left an insoluble residue, 203 g, which was washed several times with benzene.

The total benzene layer remained homogeneous. A sample of this product indicated 85% phosphorylation had been achieved. Thereafter, ethylene oxide was bubbled into this product at 52°C for about 4 or 5 hr. No condensate appeared in a Dry Ice-cooled trap on the effluent tube indicating total absorption. The ethylene oxide introduction was continued until condensate appeared in the trap. The benzene was then stripped. Analysis of the end product indicated a yield of 764 g of polyoxyethylene palmityl polyglycerol phosphate.

Example 2: A starch base was prepared by admixing together 94% by weight water, 5% starch and 1% by weight of the material of Example 1. A second starch base was prepared by admixing together 95% by weight water with 5% starch. Each starch base was heated to 95°C and then cooled to 25°C. The viscosity of each base was measured with a Brookfield viscometer.

The second formation without the material of Example 1 had a viscosity of 7,360 cp, while the other base had a viscosity of 3,360 cp. The samples were then stored in a refrigerator at 5°C. After 3 days storage, a total amount of 3 to 5% by weight of the water in the formation without the material of Example 1 had separated therefrom. No water separated from the formulation containing 1% of the material of Example 1.

Starch plus Lipid Ester in Cooked Puddings

J.L. Hegadorn, R.R. Ferguson and B.J. Bahoshy; U.S. Patent 3,619,209; Nov. 9, 1971; assigned to General Foods Corporation have produced a nonsticking pudding mix that requires less than normal stirring during cooking. It has been found that by adding a minor but effective amount of lipid ester base surface active agent a good degree of scorch resistance can be imparted to a cooked pudding. The pudding mix composition may contain a conventional raw, amylaceous ingredient such as cornstarch, potato starch, tapioca starch, and the like and combinations thereof. Amylaceous flours, e.g., tapioca, wheat and rye, modified amylaceous ingredients, e.g., modified starches, and mixtures may also be included.

The amount of lipid ester surface active agent which may be used to bring about the need for less stirring during preparation and the desired scorch resistance may be varied. Thus, depending upon the type of surface active agent and the kind and amount of amylaceous ingredient or ingredients as well as of other ingredients such as sugars which are used, the amount, by weight, may be from 0.05 to 2.0% based on the total pudding mix composition.

When polyoxyethylene (20) sorbitan monostearate (also referred to as Polysorbate 60) is used, which is preferred, the amount may range from 0.1 to 0.5%, by weight, of the pudding mix. Sorbitan tristearate and/or sorbitan monostearate may be included therewith to provide approximately 1% total, based on the weight of the pudding mix, of surface active agents.

Example: A powdered vanilla pudding mix is prepared by blending in a ribbon blender 60 parts of sucrose, 10 parts of dextrose, 28 parts of raw cornstarch, 1.5 parts of salt, and color and flavor as desired. To the blend is then added 0.15 part of Polysorbate 60 and the mix is further intimately blended. 3 to 3½ ounces of the resulting composition are added to 2 cups of whole milk. The admixture is cooked to a boil over medium heat with only a minimum amount of stirring.

The cooked pudding is then allowed to cool to develop its set. It was noted that the pudding texture and flavor were not affected but that scorch resistance had been achieved. Moreover, it was observed that tendency toward sticking to the cooking vessel was not apparent.

Raw Starch-Gum Mixtures for Puddings

Conventional starch-containing puddings oftentimes have a pasty and heavy consistency and are highly susceptible to liquid separation or syneresis on aging. The problem of syneresis becomes especially apparent when starch puddings have been heat-treated to make them shelf-stable or when the puddings have been subjected to freeze-thaw conditions.

It has been disclosed in U.S. Patent 3,507,664 that a pudding composition which exhibits a relatively high resistance to syneresis on aging can be produced containing a natural gelling system and potassium and calcium metal salts but no starch. The presence of these metal salts, however, has a deleterious effect on the taste of the pudding. It has also been found that these compositions are not resistant to syneresis and break down upon being frozen and thawed.

According to this process developed by *M. Glicksman and E.H. Farkas; U.S. Patent 3,721,571; March 20, 1973,* a starch pudding composition, which has a high resistance to syneresis especially during heat treatment and during freeze-thaw conditions, is prepared from naturally occurring materials including non-chemically modified starch, milk or milk solids and gums, without requiring the presence of any metal salts. The gums which have proven successful in this process are xanthan gum and combinations of xanthan gum and locust bean gum and xanthan gum and tara gum. Nonchemically modified starches may be either raw starches or physically modified starches (e.g., pregelatinized).

It is contemplated that these pudding compositions may be fully prepared puddings such as canned or frozen puddings or the instant-type puddings such as those prepared from powdered mixes with the addition of either milk or water. Typical compositions for the prepared puddings of this process may be defined by the following percent weight ranges.

	Percent by Weight
Water	45 - 80
Carbohydrates	10 - 20
Fat emulsion	5 - 15
Milk solids	3 - 10
Nonchemically modified starch	2 - 10
Xanthan gum	0.08 - 1
Flavor/color	As desired

Typical compositions for the powdered pudding mixes which are adapted to be added to either milk or water are defined by the following percent weight ranges.

	Percent by Weight
Carbohydrates	35 - 60
Dried fat emulsion	25 - 50
Nonchemically modified starch	5 - 12
Xanthan gum	0.4 - 2
Dry milk solids	0 - 20
Color/flavor	As desired

Example 1:

	Grams
Water	660
Sucrose	118
Fat emulsion	90
Nonfat milk solids	50
Raw tapioca starch	48
Xanthan gum	0.6
Locust bean gum	0.6
Flavor/color	As desired

Example 2:

	Grams
Water	670
Sucrose	197
Fat emulsion	66
Nonfat milk solids	64
Raw cornstarch	40
Xanthan gum	1.0
Flavor/color	As desired

Each of the above compositions is prepared by forming a liquid fat emulsion by melting together a fat and an emulsifier, combining the dry ingredients and dispersing them in the water component which is then brought to a full boil, adding the liquid fat emulsion to the water phase and blending until uniform. The compositions are then homogenized and allowed to set for 15 min to produce puddings having a smooth creamy texture.

Samples of each of these puddings were separately packaged and stored in the following manners: the pudding was sealed in a metal can, autoclaved at 200°F for 30 min and stored at ambient conditions for 6 wk; the pudding was packaged in a plastic container and refrigerated at 40°F for 3 wk; and the pudding was packaged in a plastic container, placed in a freezer at 0°F for 6 wk and then thawed. In every instance the puddings retained their smooth creamy texture and were substantially free from the effects of syneresis.

Unmodified Tapioca Starch in Frozen Puddings

Frozen puddings which exhibit reasonably good freeze-thaw stability due to the use of chemically modified starches as described in U.S. Patent 3,669,687, page 152, are susceptible to textural changes thought to be due in part to rupturing of the starch granules, when the puddings are either held at or cycled through a temperature of about +15°F. Unfortunately, cycling through this +15°F temperature has been found to occur quite frequently during the distribution and storage of frozen puddings in commercial freezer units. For some reason, each time the puddings pass through the specific temperature of +15°F the occurrence of textural changes such as syneresis is more and more likely.

It has been found according to the disclosure by *M. Glicksman, B.N. Wankier and J.E. Silverman; U.S. Patent 3,754,935; August 28, 1973; assigned to General*

Foods Corporation that tapioca starch is unique in that it is possible to produce frozen puddings having exceptionally good freeze-thaw stability including cycling through the +15°F temperature range. These puddings contain as their sole thickener or gelling agent chemically unmodified tapioca starch. Typical compositions for the frozen puddings of this process may be defined by the following percent weight ranges.

	Weight Percent
Water	45 - 70
Carbohydrates	10 - 20
Vegetable fat	3 - 12
Milk solids	3 - 12
Chemically unmodified tapioca starch	3 - 10
Emulsifiers	0.1 - 1.0
Flavor/color	As desired

The following example will illustrate a specific formulation for pudding.

Example:

	Grams
Water	1,985
Sucrose	432
Hydrogenated vegetable oil (Wecotop A)	195
Nonfat milk solids	183
Tapioca starch, raw	180
Sodium stearyl lactylate (emulsifier)	12
Dextrose	9
Sodium caseinate	9
Salt	6
Polyoxyethylene sorbitan monostearate (emulsifier)	7.2
Flavor, vanilla extract	12.0
Color, vanilla	0.6

The pudding is prepared by melting all the fat and emulsifiers together in a water bath. The rest of the dry ingredients are then dispersed in water, the vanilla extract is added and the water mixture is brought to a full boil. The boiling water mixture is poured into a Waring Blender and blended at a low speed. The melted fat phase is then slowly added and blended until uniformly mixed. The mixture is then homogenized in an Eppenbach Colloid Mill for 5 min and poured into plastic containers.

The pudding composition was then thoroughly frozen and subjected to nine freeze-thaw cycles with each cycle ranging from below +15° to +75°F. No syneresis or other textural breakdowns were found to occur in any of the pudding samples prepared in accordance with this example. If other raw starches (e.g., potato, arrow root, corn, rice, wheat, waxy maize, sorghum and waxy sorghum) were used as substitutes for the raw tapioca starch in this pudding, noticeable syneresis and other textural breakdown were found to occur after only 2 or 3 of the freeze-thaw cycles.

Expanded Sugar-Based Granules

A process for preparing expanded sugar-based granules which are especially suited
for incorporation in various chocolate and confectionery articles has been developed by *W. Rostagno; U.S. Patent 3,835,226; September 10, 1974.* In this
process, drops of a syrup containing a sugar and at least one expanding agent
are deposited on a bed of powdered edible material and the drops are heated to
produce expansion and drying of the drops while they are in contact with the
edible material.

In one form of the process, drops of a syrup containing an expanding agent, ammonium bicarbonate, for example, are deposited on a flat horizontal support
which has been coated with a layer of edible powdered material. The edible
powdered material provides a deformable recipient closely conforming to the
shape of the drops at the moment of impact and at the same time facilitates
their drying by absorption. The drops of syrup resting on this bed of powder
are then heated to a temperature of between 140° and 160°C for 3 to 4 min.

The granules obtained after cooking are hollow dry capsules or particles of dry
expanded material, of essentially spherical shape, and their volume is about five
times that of the original drops. These granules are then separated from their
powdered support, for example, by screening. In a second form of the process,
a coating of edible powdered material is deposited on the drops before cooking.
The whole surface of each drop is thus in contact with the powder, a part of
which constitutes a surface film surrounding each of the granules after cooking.

The edible powdered material will generally be chosen having regard to the desired characteristics of the expanded granules and various powdered materials
may be selected to confer a particular color, flavor and/or texture to the finished
granules. Preferred edible materials include, for example, starch and its hydrolysis
products, cocoa, powdered coffee extract, milk powder, cereal flours, etc.

The influence of the powdered edible material on the chemical composition of
the expanded granules is illustrated in the example. Thus, although the syrups
used contain no starch, the granules obtained by cooking the drops of syrup in
contact with a bed of powdered starch have a starch content which varies between 13.5 and 48.4% by weight. These amounts of starch are related to the
chemical composition of the syrup used, its solid matter content, the quality of
the starch, as well as to the manner in which the drops are coated.

On the other hand, the starch in contact with the surface of the drop is capable
of forming, by gelatinization during cooking, an elastic film imparting a regular
substantially spherical shape to the expanding granule. In the same way, agree-
ably flavored granules may be obtained by using, for example, coffee extract or
cocoa powder as a support.

Example: A syrup of the following composition is prepared.

Sucrose	51.25%
Ammonium bicarbonate	0.50%
Citric acid	0.25%
Coloring	0.50%
Strawberry flavor	2.50%
Water	45.00%

This syrup is deposited in the form of drops on a belt conveyor previously coated with a layer of powdered starch. The deposited drops are then covered with a second layer of powdered starch and carried by the belt into a tunnel oven in which the temperature is between 144° and 152°C. The drops are maintained at this temperature for 3½ min. The resulting dry expanded granules have a specific gravity of 0.200 and a starch content of 48.4%.

TREATMENT OF STARCHES

REMOVAL OF LIPIDS

Starches derived from certain plant sources contain lipids (including fatty acids) which are intimately physically bound into the starch granule. The extraction of these lipids proves to be rather difficult as they are inaccessible to the mechanical or physical purification methods used during the refining of the starch as well as to subsequent solvent extraction procedures wherein, for example, typical hydrocarbon solvents may be used. The presence of these lipids has a considerable bearing on the rheological behavior of the starches which are of considerable importance in the preparation of foods, particularly gum-like confections.

Known methods for extracting lipids have involved solvent extraction of the starch by refluxing, at atmospheric pressure, with such solvents as methanol, dioxane, methyl Cellosolve and diethylamine. These procedures have not, however, provided effective removal since they require prolonged and repeated refluxing in order to remove a substantial portion of the lipid content.

In an attempt to overcome these disadvantages, these extraction procedures have also been conducted at elevated temperatures and pressures. Although the latter techniques have improved both the speed and the efficiency of the extraction, the necessary presence of water in such high temperature, high pressure reactions produces undesirable physicochemical or morphological changes in the granule so that the resulting starch displays severe rheological changes which limit the applications in which such starches can be used. U.S. Patent 3,695,933, page 232 also discloses defatting low DE syrups and U.S. Patent 3,583,874, page 81 uses a pregelatinized defatted starch in the preparation of puddings.

Solvent Extraction Using Controlled Moisture Conditions

F.J. Germino and J.R. Caracci, Jr.; U.S. Patents 3,717,475; February 20, 1973; and 3,586,536; June 22, 1971; assigned to CPC International Inc. have developed a process for reducing the bound fat content of granular starch to not more than 0.15% by weight by heating a mixture of the starch, a suitable solvent, and not more than 15% water, in a confined zone, at 225° to 300°F for not more than 30 minutes. The conditions are selected so as to avoid incipient gelatinization of the starch. Starch treated with this process is characterized by a fluidity of 30 to 80 and is an exceptionally fine congealing agent in the manufacture of gum confections.

Briefly, the process is as follows. Granular starch is treated, in a confined zone, with a suitable solvent such as methanol or ethylene glycol monomethyl ether, the total starch-solvent system containing not more than 15% moisture, and preferably not more than 12% moisture, by weight. The treatment takes place at a temperature from 225° to 300°F, preferably 250° to 275°F, for a time sufficient to reduce the fat content of the starch to 0.15% or less, the time not to exceed 30 minutes.

Care must be taken to avoid incipient gelatinization of the starch during the treatment; by incipient gelatinization is meant that more than 2% of the starch granules have been damaged, i.e., have swollen hilums or complete loss of birefringence as viewed under a polarizing microscope. In order to avoid such incipient gelatinization the conditions of moisture content of the system and time and temperature of the treatment, which are interdependent, must be properly adjusted.

If one is starting with, say, a wet starch filter cake containing about 40% water the starch may be dried prior to the treatment; alternatively, however, the drying step may be eliminated by using an anhydrous or nearly anhydrous solvent in sufficient quantity to result in a total mixture containing 15% water or less. This effect will be illustrated in the following example.

Example: This example illustrates the effect of varying temperatures, retention times, and water-contents of the system. A series of runs was made using an acid modified thin-boiling cornstarch (of 67 fluidity), which contained 0.56% bound fat, with methanol as the solvent. The starch was slurried with the solvent and placed in a confined zone, the temperatures raised to a specified point, held for a specified time in certain cases, and then brought back down to room temperature. The treatment conditions are set forth in the following table. In the case of Samples 1 through 15, 3 minutes were required to bring the temperature up to that specified and 3 minutes required to bring the slurries down to room temperature; 30 minutes were required to bring Samples 16, 17 and 18 to temperature and 20 minutes to cool them to room temperature; Samples 19 through 25 were heated and cooled almost instantaneously (within about 40 seconds).

After treatment, pastes were prepared from some of the samples and the strength of the resultant gels was determined in the following manner. 15% starch pastes were prepared on a boiling water bath for 30 minutes with hand agitation for the first four minutes. After cooking, the pastes were poured into wide-mouth four-ounce jars and 3.5 ml of paraffin oil placed over the surface to prevent shrinkage.

The samples were held at room temperature, and tested on a Bloom Gelometer (manufactured by Precision Scientific Co.) each hour over a period of four hours. The Bloom Gelometer values, in grams, are reported in the table under the heading Bloom gel strength. A number of the samples were examined under a microscope for granule damage, i.e., anything from a swelling of the hilums of the granules to a complete loss of birefringence. Incipient gelatinization was considered to be positive if more than 2% of the granules showed such damage.

As can be seen from the table, gum drops prepared from those starches which were defatted to 0.15% bound fat or lower and which further showed no incipient gelatinization could all be sanded (coated with sugar) within an hour or less (most could be sanded within 45 minutes or less) and were rated excellent or superior by the observers.

Sample Number	Treatment conditions				Characteristics of starches							Gum drop characteristics		
	Total H_2O in system (percent)	Sol-vent/ starch ratio	Tem-pera-ture (° F.)	Hold-ing time (min-utes)	Fat content (per-cent)	Scott vis-cosity (g./sec.)	Incip-ient gelatini-zation?	Bloom gel strength (gms.), hours				Sanding time		Evaluation
								1	2	3	4	Min-utes	Hours	
Control (67 fluidity corn starch).	12				0.56	50/63	No	109	239	292	300		48–72	Poor.
1	5.25	1.6/1	170	0	0.36	50/61	No	109	239	292	300		48	Do.
2	5.25	1.6/1	200	0	0.30	50/67	No	109	239	292	300		12	Fair.
3	5.25	1.6/1	250	0	0.09	50/71	No	109	239	292	300	45		Superior.
4	5.25	1.6/1	200	5	0.21	50/67	No	109	239	292	300		4	Good.
5	5.25	1.6/1	250	5	0.06	50/85	No	109	239	292	300	45		Superior.
6	5.25	1.6/1	170	10	0.25	50/65	No	109	239	292	300		6	Good.
7	5.25	1.6/1	210	10	0.17	50/71	No	109	239	292	300		1.5	Very good.
8	5.25	1.6/1	250	10	0.05	50/86	No	109	239	292	300	45		Superior.
9	2.5	4.5/1	250	10	0.06	50/81	No	291	432	465	519	45		Do.
10	15.0	5/1	250	10	0.06	50/145	Yes	264	321	389	359		12	Fair.
11	0	2/1	250	10	0.08	50/86	No	290	466	498	541	45		Superior.
12	0	5/1	250	10	0.07	50/87	No	281	441	495	544	45		Do.
13	8.7	4.5/1	250	10	0.04	50/125	No					45		Do.
14	2.9	4.5/1	250	10	0.04	50/107	No	309	457	498	530	45		Do.
15	7.4	4.5/1	295	30	0.03	50/320	Yes	230	290	361	368		12–15	Fair.
16	2.6	4.5/1	295	30	0.02	50/162	No	271	382	398	490		1	Excellent.
17	4.0	4.5/1	240	30	0.09	50/95	No	320	497	548	604	45		Superior.
18	2.0	4.5/1	300	2	0.06	50/120	No	299	424	506	564		1	Excellent.
19	7.9	4.5/1	300	2	0.03	50/198	Yes	204	347	376	411		4	Good.
20	13.6	4.5/1	300	2	0.06	30/67	Yes	137	184	208	216		16	Fair.
21	2.9	4.5/1	250	2	0.05	50/69	No	382	560	673	>690	30		Superior.
22	7.0	4.5/1	250	2	0.04	50/73	No	384	518	587	615	30		Do.
23	11.0	4.5/1	250	2	0.05	50/90	No	311	498	523	561	35		Do.
24	15.1	4.5/1	250	2	0.07	50/102	No	308	450	491	506		1	Excellent.

Solvent Extraction with Flash Heating

In a similar process, *T.J. Schoch; U.S. Patent 3,630,773; December 28, 1971; assigned to CPC International Inc.* has used flash heating to remove materials soluble in a hydrophilic fat solvent from starches.

It was found that if a suspension of corn or other cereal starch is heated very rapidly, to an elevated temperature, in intimate contact with an appropriate hydrophilic fat solvent, the fatty material dissociates almost immediately from the starch and dissolves almost quantitatively in the solvent. The prompt separation of the extracted starch material from the solvent yields a substantially fat-free product and prevents readsorption of the fat back into the starch.

The hydrophilic fat solvents ordinarily are certain alkanols and certain ethers. Examples of these materials are methanol, either in the anhydrous form or with added water, ethyl alcohol, 80% dioxane, and ethylene glycol monomethyl ether. Of these, methanol is by far the preferred solvent, because of its low cost, low explosive hazard, ease of recovery, and high efficiency for fat extraction.

Ordinarily, for practicing the process, the starch material is suspended in the hydrophilic fat solvent, by some technique that produces a smooth dispersion.

Ordinarily, the upper limit on the starch content of the slurry is about 40% by weight. Generally, the proportion of starch material in the slurry will be determined empirically.

While the presence of some moisture is desirable, the preferred slurry composition ordinarily is made up with a liquid vehicle, including any moisture in the starch, that comprises 85% by weight of methanol and about 15% by weight of water. Commercial starch frequently has from 10 to 12% by weight of moisture, and this amount should be taken into consideration in calculating the total amount of water present in a slurry.

The process is advantageously practiced on a continuous basis. A stream of slurry is heated as rapidly as possible to an elevated extraction temperature. Ordinarily, satisfactory results are obtained at temperatures up to 300°F. A preferred operating range is from 170° to 260°F. The heating is accomplished in a confined zone, where the pressure is maintained at a level such that the liquid vehicle of the slurry is maintained in the liquid phase. Prolonged digestion at the elevated temperature is avoided, by prompt separation of the extracted starch material from the solvent vehicle, following the rapid heating step.

Immediately following the flash heating of the starch slurry, the extracted starch material is separated from the liquid vehicle. For example, the hot slurry can be subjected to continuous filtration, under autogenous pressure. Alternatively, a cooling chamber may be used to reduce the temperature of the starch slurry below the boiling point of the liquid vehicle at atmospheric pressure, to facilitate handling. The cooled slurry can then be separated by the use of a filter or by the use of a centrifugal separator.

One preferred method involves flash cooling of the hot suspension, by its injection into a vessel that is equipped with some kind of condenser for capturing the vaporized solvent. Still another alternative involves the use of a pressure centrifuge, which eliminates the need for cooling in advance of centrifugal separation.

Example: Defatting Cornstarch — A lot of commercial dry cornstarch was found to contain 11.0% moisture, 0.56% total fat, and 0.30% protein. This starch was suspended in anhydrous methanol to make up a 32% starch slurry, that is, in the proportion of 1 kilogram of starch for each 4 liters of methanol.

A stream of this suspension was passed through a steam-jacketed stainless steel tube at the rate of 228 liters per hour. The average time of passage through the tube, from one of its ends to the other, was from 2.4 to 2.7 minutes. The tube was fitted with a discharge valve for maintaining the suspension under autogenous pressure developed by heating the methanol above its boiling point.

Three separate demonstrations were conducted, at three different extraction temperatures by adjusting the pressure of the steam in the jacket about the heat exchanger tube. Temperatures at the discharge end of the tube were in the three different runs, 196°, 243°, and 257°F, respectively.

The suspension was discharged from the heat exchanger tube into an enclosed vessel that was maintained substantially at atmospheric pressure, and that was equipped with a condenser. A portion of the methanol flashed off immediately upon injection into the flash evaporator, to reduce the remainder of the suspension to a temperature below the boiling point of the methanol. The partially cooled suspension was then immediately filtered to separate the extracted starch from the liquid vehicle.

The filter cake was then washed, first with hot condensed methanol from the flash evaporator, then with water. It was finally tray dried. Protein analyses were then run by the conventional Kjeldahl method, and the fat content was determined by the standard acid hydrolysis procedure, as described in the text, *Methods in Carbohydrate Chemistry,* vol. 4, pages 56-61, Academic Press, New York, 1964. The data relating to these three demonstrations of the process is summarized in the table below.

	Demonstration		
Text	1a	1b	1c
Extraction temperature, ° F	196	243	257
Percent fat content of extracted starch	0.20	0.10	0.085
Percent protein content of extracted starch	0.23	0.21	0.20

Defatted Starch in Gelled Products

An improved process for preparing gelled products (i.e., gumdrops, jelly beans, etc.) has been developed by *T.J. Schoch, D.F. Stella and H.J. Wolfmeyer; U.S. Patent 3,446,628; May 27, 1969; assigned to Corn Products Company.*

The improvements are obtained by the use of thin-boiling cereal starches which have been solvent-extracted to reduce the content of fatty acid below 0.3% on dry starch basis, and preferably below 0.15%. Some additional advantage may be realized by reducing the fat content to the vanishing point (i.e., below 0.05%), although such advantage may be offset by the cost of such extreme extraction procedures.

In general, it is found that the major advantages of this process are realized with cereal starches whose fat content is in the range of 0.05 to 0.15%. For purposes of comparison, the fat content of cornstarch (and of its thin-boiling modifications) is normally in the range of 0.60 to 0.65%, wheat starch contains 0.50 to 0.65%, and rice starch may run 0.6 to 0.8%. Defatting may be accomplished by any of the various known processes.

Obviously, the starch used in the process must be of a type normally used for gum confections and the like, that is, an amylose-containing cereal starch having a fluidity within the range of 30 to 80. The important features of this process are the use of starches having a fluidity of 30 to 80 and their fat content reduced below 0.3% and preferably below 0.15%.

For the manufacture of gum confections, the mixture of starch, sugar (and/or corn syrup) and water may be cooked by either the older batch process or by more modern techniques of continuous high-temperature cooking. In either case, the special benefits of a defatted starch are realized, to a degree dependent on the extent of defatting of the starch. This is illustrated by the following example.

Example: The following formulations were prepared, using an ordinary 67-fluidity cornstarch as control (fat content = 0.6%) and a solvent-defatted starch prepared therefrom (fat content = 0.06%):

	A	B	C	D
Regular 67-fluidity starch, lb	1.0	--	--	--
Defatted 67-fluidity starch, lb	--	1.0	0.9	0.8
Water, lb	1.25	1.25	1.25	1.25
Sucrose, lb	2.4	2.4	2.4	2.4
Corn syrup, 43°Bé, 66 DE, lb	5.6	5.6	5.6	5.6

The resulting slurries were preheated to boiling (106° to 108°C) and then pumped through a steam injection cooker accurately controlled to give a temperature of 129° to 135°C. Cooking time at this temperature was 26 to 28 seconds.

Identical amounts of citric acid, lemon flavor and yellow food color were then added to each batch. Refractometer readings on the hot syrups indicated actual solids contents in the range of 77 to 79%. The syrups were then deposited in conventional molding starch beds. Batch A (containing regular nondefatted 67-fluidity gumdrop starch) required 42 hours in the hot conditioning room before the confection pieces could be removed from the starch mold and sugar-sanded. Within ten minutes after depositing batches B, C and D (containing the same 67-fluidity confectionery starch, but defatted to 0.06% fat content), the gumdrops had attained sufficient gel structure to permit removal and sugaring without further holding time, despite the fact that the gumdrops were still too hot to touch. Indeed, when the temperature of the cooked syrup dropped below about 82° to 87°C, the confection tended to set up to a pectin-like gel; when the temperature was raised above 82° to 87°C, this gel reliquefied to its former consistency.

As a rigorous test of storage properties, portions of the gumdrops from each batch were held at room temperature for 24 hours, then sugar-sanded, and stored in sealed glass jars for 12 days at about 43°C. Only batch A (containing nondefatted starch) showed evidence of moisture-sweating and stickiness. Physical appraisals of the four batches of gumdrops after 12 days were as follows:

> Batch A: Normal body and gel structure, smooth but slightly stringy texture. Fairly good storage stability.

> Batch B: Body is much firmer and harder than batch A, with a structure similar to a pectin gel. Excellent storage stability.

> Batch C: Body is somewhat firmer than Batch A. Sugar coating is superior to any of the samples tested. Excellent storage stability.

> Batch D: Body and texture comparable with Batch A, but more pectin-like in structure. Excellent storage stability.

Thus, in addition to a very rapid congelation, it is apparent that the defatted starch is more efficient as a gel-producer, since equivalent body of the finished gumdrops can be realized with only 0.8 lb of the defatted starch, as compared with 1.0 lb of the normal nondefatted starch.

DESOLVENTIZING PROCESSED STARCHES

Steam or Humid Air for Solvent Removal

In a process described by *F.E. Kite and J.F. Stejskal; U.S. Patent 3,578,498; May 11, 1971; assigned to CPC International Inc.,* starch which has been treated with a hydrophilic organic solvent such as a low molecular weight alcohol, ketone, or the like, and which, as a result, retains relatively large quantities of the solvent, is desolventized by contacting it with a hot, humid gas, such as, for example, steam or moist air, while maintaining the starch at a temperature of about 160° to 320°F, and at a moisture level of about 1 to 15%. The process is applicable to any type or variety of starch, in either nongelatinized or pregelatinized form; the process does not alter or otherwise affect the properties of the starch. The following examples illustrate the process.

Example 1: This example illustrates methanol removal by humid air treatment of a heated bed of starch and the effect on solvent removal rate as the relative humidity and subsequent starch moisture level is increased at a constant starch bed temperature. A cold water swelling (i.e., pregelatinized) cornstarch containing 4,600 ppm of methanol was placed in a fluid bed column and heated to 217°F. A flow of hot, humid air was passed through the starch bed at sufficient rate to cause fluidization (0.2 cfm). The rates of methanol removal at various humidity levels are as follows.

Temperature (° F.) at which air was saturated	Calculated RH (percent) of the air in contact with the starch	Treatment time (minutes) to obtain residual methanol levels of (p.p.m.)—				Final starch moisture (percent)
		500	100	50	25	
175	44	18	37	50	71	6.0
185	55	12	27	35	46	7.4
194	65	5	13	19	27	10.0

These data illustrate that solvent removal rate is increased as humidity level of the fluidizing air is increased at a constant starch bed temperature. Turbulent flow in the fluid bed was readily maintained with cold water swelling starches up to relative humidity level of 75% with respect to bed temperature when starch moisture did not exceed 12 to 15%. At relative humidities above 75% and starch moistures above 12 to 15%, the rate of solvent removal was significantly slower due to a loss of turbulent flow, i.e., the cold water swelling particles began to ball up, cohere and agglomerate, and channeling plus plug flow occurred in the bed.

Example 2: This example illustrates the removal of various nonaqueous solvents from solvent-treated cornstarch products by hot, humid air treatment of the heated starch in a fluid bed. The treatment conditions were as follows: starch bed temperature — 217°F; humid air — 56% RH (saturated at 188°F).

Solvent	Solvent (p.p.m.) before treatment	Time of treatment, min.	Solvent (p.p.m.) after treatment
Acetone	800	30	300
Ethanol	11,400	60	280
Isopropanol	375	30	94

Heating in Presence of Gel Retarding Salt

In related work by *F.J. Germino, J.F. Stejskal and W.E. Schaffrath; U.S. Patent 3,619,291; November 9, 1971; assigned to CPC International Inc.*, starch which has been treated with a hydrophilic organic solvent, such as a low molecular weight alcohol, ketone, or the like, and which, as a result, retains relatively large quantities of these solvents, is desolventized by mixing it with a gel retarding salt, and heating the mixture to a temperature above about 150°F.

The gel retarding salt includes, for example, sodium sulfate, magnesium sulfate, disodium phosphate, sodium tartrate, sodium citrate, and the like and mixtures of such salts. In general, however, any inorganic salt having the characteristic of retarding gel formation under temperatures between 150° and 210°F while permitting removal of the solvent, is a suitable gel retarding salt for the purposes of the process.

Example 1: This example illustrates methanol removal from starch employing a sodium sulfate solution as the gel retarding salt. One thousand grams of a 50% starch cake containing in excess of 20,000 parts per million of methanol are dispersed in 1,000 grams of a 2-molar sodium sulfate solution that is heated to a temperature of 190°F in a constant temperature bath. The sample is held for a period of 1 hour at 190°F and aliquots are removed at specific time intervals. At the end of 15 minutes, the methanol content of the starch product is 180 parts per million. At the end of one hour, the starch product contains only 100 parts per million of residual methanol.

The desolventized starch product is formed into gumdrops and has an acceptable sand time equivalent to 45 minutes. This indicates that the desolventization process of this system effectively removes methanol from the starch without causing granular damage to the starch.

Example 2: This example illustrates the employment of magnesium sulfate as a gel retarding salt in the reduction of the methanol content of the nongelatinized starch.

A pre-water-washed starch cake comprising 6,100 grams of which 3,350 grams consist of starch is admixed with an aqueous salt solution comprising 2,010 grams of magnesium sulfate and 5,370 grams of water at a temperature of 190°F. As before, aliquots are taken at predetermined time intervals and the methanol content of the aliquot portions is measured. The results are set forth below in the following table.

Salt	Molar Conc.	Temp., °F	Residence Time (min)	Residual Methanol (ppm)
Magnesium sulfate	2.0	190	5	109
			15	90
			30	66
			45	63
			60	62

The results in the table illustrate that an even lower residual methanol content is provided by the employment of magnesium sulfate than is possible under the same conditions employing sodium sulfate. Additionally, less magnesium sulfate is required to achieve the desired methanol level.

NONDEGRADATIVE ENZYME MODIFICATION

The following processes cover enzyme treatment of starch to alter other materials in the starch or nonsaccharide groups attached to the starch molecules.

Enzyme Treatment of Atypical Units of Starch

Atypical fragments of a natural product molecule are altered by enzymic reaction in a process disclosed by *R.L. High and S. Rogols; U.S. Patent 3,488,256; January 6, 1970; assigned to The Keever Company.* For example, a carbohydrate molecule such as starch is reacted with a proteinase to alter atypical protein fragments. Similarly, a protein molecule such as an enzyme is reacted with lipase. The alteration effects a unique activation of the natural product molecule.

These methods selectively use as energy of activation some of the free energy made available by the selective acceleration of the degradation, or bond rupture, of such atypical molecular fragments. Referring to specific types of natural products, the process deals with methods of altering the nature of the reactivity of enzymes, starch and starch hydrolysates, dextrins, sugars, proteins, lipids, nucleic acid-containing substances, cellulose, chitin, dextran, hormones, gums, mucopolysaccharides, glycogen, and like substances of natural origin by a selective enzymic reaction involving atypical materials inherently associated with the molecules of these substances.

Such a use would also include the pretreating, or altering, a partial hydrolysate of starch before subjecting it to further hydrolysis for the production of dextrose. Although the process covers use of one, two or three enzymes simultaneously or sequentially, the example selected from the many in the complete patent shows the simultaneous use of three different enzymes on a starch.

Example: 200 grams potato starch were slurried in 1,700 ml distilled water. The pH was adjusted to 7.60 using 0.5 N NaOH. The temperature was stabilized at 40°C and the pH at 7.60 for one hour prior to any additions. Simultaneously, the following enzymes were added at the stated levels:

Amount (ml)	Enzyme Solution	Level (% of Starch)
32	Wheat germ lipase	0.009
150	α-Chymotrypsin	0.052
100	Phosphodiesterase	0.040

The enzymes were prepared as follows:

   (a)   Wheat germ lipase: 0.05 gram (2X crystalline, Worthington Biochem.) were dissolved in 100 ml 0.075 M $NaHCO_3$ buffered at 7.40 with NaH_2PO_4 and allowed to stand one hour in the cold prior to use.

   (b)   α-Chymotrypsin: 0.104 gram of the enzyme (3X crystalline, the salt free form from methanol at 11,000 μ/mg) were dispersed in 150 ml of a tris buffer at 8.00 pH.

   (c)   Phosphodiesterase: 0.08 mg of a lyophilized phosphodiesterase from *Crotalus atrox* venum were dispersed in 100 ml of 0.1 M magnesium hydroxide. The pH was adjusted to 8.80 using magnesium acetate.

The reaction proceeded 12 hours after which time 2 ml of 10% trichloroacetic acid were added. The product was filtered, washed with 2 liters of distilled water, then with 0.5 liter acetone, and was then air dried. A parallel reaction was carried out which contained all the reagents except the enzymes. The product of the parallel reaction has been referred to herein as washed potato starch. A similar run was made using cornstarch with the results of simultaneous treatment by lipase, proteinase and exonuclease on both the corn and potato starches being shown in the following table.

Starch	Percent Protein	Percent Ash	Percent Lipid
Potato starch, commercial	0.188	0.250	0.200
Potato starch, washed	0.161	0.240	0.190
Potato starch, treated	0.007	0.255	0.140
Cornstarch, commercial	0.367	0.370	0.260
Cornstarch, washed	0.313	0.365	0.190
Cornstarch, treated	0.091	0.084	0.130

Lipase Altered Starch

Carbohydrates having improved fermentation rates have also been produced by *S. Rogols and R.L. High; U.S. Patent 3,652,295; March 28, 1972; assigned to A.E. Staley Manufacturing Company*. This has been accomplished by adding a lipase-altered starch to a composition such as bread dough which is to undergo yeast fermentation. In the making of bread, for example, a typical carbohydrate composition comprises about 96% wheat flour and about 4% sugar. To this is added a small percentage; usually up to about 10%, but preferably about 3 to 6%; of a lipase-altered starch.

In addition the gassing power, or fermentation rate, of lipase-altered wheat starch can be increased by adding a small percentage; e.g., about 1 to 5% of the starch weight; of a calcium salt to the altered starch. It has been found also that ammonium persulfate, e.g., 10% of the starch weight, added to the altered starch contributes to the grain and texture of bread products. The latter also contributes to a change in the nature of the sugar materials present following fermentation.

Example: A standard steapsin solution was prepared by dispersing 0.05 gram of the enzyme in a calcium buffer at a pH of 7.4 and consisting of 0.0075 M calcium phosphate/calcium acetate.

A wheat starch slurry was adjusted to pH 8.0 and 0.05% hydrogen peroxide was added to destroy residual SO_2 in the starch. After 30 minutes the pH was adjusted to 6.80. Calcium chloride (0.5%) was added from a solution thereof. The resulting mixture was allowed to stand for one hour; at which time steapsin (from the above preparation) was added at the 0.004% level. The mixture was allowed to react for 5 hours. The solution was again adjusted to a pH of 6.80. Calcium chloride (2%) was added and the mixture was allowed to stand for one hour. The lipase-altered starch was then filtered out, washed with distilled water, and dried. The gassing power of this starch (Test Starch 1) when fermented is shown in the following table. A sample of altered cornstarch was prepared in the same manner as described above and its gassing power is shown also in the following table. Various controls and modifications are included in this table for comparison. Referring to this table; Control 1 is unaltered starch with no additions; Control 2 is unaltered starch plus the calcium salts; Test Starch 2 is steapsin-altered starch with no added calcium salts other than those in the steapsin preparation and Test Starch 3 is steapsin-altered starch with no calcium salts added at any point. It will be noted that the steapsin-altered starches, with or without added calcium salt, had a greater gassing power than the controls.

			CO_2 Produced*			
Hours	1	2	3	4	5	6
Wheat starch:						
Control No. 1	-	2	3	5	7	8
Control No. 2	5	7	13	17	20	23
Test starch No. 1	20	29	67	87	97	119
Test starch No. 2	7	26	43	53	62	67
Test starch No. 3	18	41	58	69	77	81
Cornstarch:						
Control No. 1	0	1	3	3	3	4
Control No. 2	1	3	4	7	9	10
Test starch No. 1	2	4	5	7	11	12
Test starch No. 2	8	10	17	23	25	30
Test starch No. 3	8	10	15	19	20	21

*Expressed as mm Hg.

OXIDATION OF STARCH

The alcohol groups in starch can be oxidized to form the aldehyde, keto or carboxyl groups. Use of alkaline hypochlorite produces mainly carboxyl groups along the starch molecule. Oxidants such as periodate react at the adjacent secondary hydroxyls, rupturing the bond with the formation of two aldehyde groups. This product is known as dialdehyde starch.

The hypochlorite oxidized starch has found use as a batter starch or breading starch which has good adhesion to the foodstuffs to which it is applied. The following three processes are directed toward the development of an improved batter starch. Reference can be made to U.S. Patent 3,786,159, page where oxidized starches can also be used to prepare dehydrated alcohol products.

Combined Hypohalogenous Acid-Halogen-Peroxide Oxidation

A method of preparing an improved starch batter composition has been discovered by *R.R. Gabel, R.M. Hamilton and W.E. Dudacek; U.S. Patent 3,607,393; September 21, 1971; assigned to CPC International Inc.* In the broad sense, this process comprises the steps of treating an aqueous raw or granular starch slurry with a mixture of hypohalogenous acid and halogen gas, followed by addition to the thus treated starch slurry of an oxidizing agent such as hydrogen peroxide. This starch batter displays excellent adhesion when applied to frozen or fresh foodstuffs and also exhibits superior cohesion in terms of film strength.

The treatment of starch with hypohalogenous acid and halogen gas may be carried out in a number of ways. For example, an alkali metal hypohalite such as sodium hypochlorite may be added to the aqueous starch slurry. As mentioned above, the slurry may be already acidified, or appropriate amounts of acid may be added to the starch slurry during hypohalite addition. Typical alkali metal hypohalites useful here include sodium, calcium, lithium, potassium, etc.; hypochlorite, hypobromite or hypofluorite. Preferred for use in the process, due to advantages of cost and availability is sodium hypochlorite. The amount of alkali hypochlorite added to the starch starting material usually should be that amount sufficient to supply or release 0.05 to 5.50% halogen gas based on starch solids and calculated as chlorine. Details of the process and product use are shown in the following examples.

Example 1: A granular starch was prepared as an aqueous 22° Baumé slurry containing 100 grams starch, dry basis. To the agitated slurry of granular starch was added 0.25 gram of chlorine in form of sodium hypochlorite. During the hypochlorite addition dilute hydrochloric acid was added to maintain the pH between 3.0 and 4.0. After addition of sodium hypochlorite the slurry was held at 48°C with agitation for one-half hour. Then a solution of hydrogen peroxide was added in an amount approximately equal to remove the residual chlorine. Here, approximately one milliliter of a 3% aqueous solution of hydrogen peroxide was added. The pH was then adjusted to between 5 and 7 by the addition of powdered sodium carbonate.

Example 2: The food batter starch described in Example 1 was tested for its adhesion, particularly in connection as a food breading batter for fish according to the procedure set out below.

Ninety grams of the above starch was mixed with 120 ml of water and six grams of sodium chloride to constitute a smooth slurry. Three blocks of frozen fish were uniformly and completely coated with the slurry by dipping. The batter-coated fish was then rolled in a cracker crumb breading mix. The breaded fish pieces were placed in a wire basket and then submerged in a deep fryer at 185°C. The breaded fish was fried until it came to the surface. Ten seconds additional frying was allowed before the basket was removed from the fryer, excess oil drained, and the fish placed on dry absorbent toweling.

Each piece of the cooked fish was stood on its long edge and cut through from top to bottom. The cut piece was opened like a book, and the meat fibers scraped from the coating. Adhesion was determined by noting the relative surface area that retained the meat fibers on all three test pieces of fish. The adhesion rating was reported as the percent of the area of the coating to which the fish fibers adhered. In the majority of the situations the starch batter of Example 1 had a fish-breading adhesion rating of greater than 90%, well over the accepted minimum standard.

Example 3: A large scale run was made in accordance with this process. 20,100 pounds of starch (dry substance) was prepared as an aqueous slurry having a 22.6°Bé and a pH of 4.8. To the starch, held at a temperature of 45°C, was added 4.0 gallons of hydrochloric acid to adjust the pH of the starch to 2.05. The time of acid addition was approximately 8.0 minutes.

5.2 gallons of sodium hypochlorite was diluted with water to a total of 250 gallons and then added to the acid starch slurry over a total time of 48 minutes. Sufficient sodium hypochlorite was added to produce 0.33% available chlorine. In order to maintain the pH between a range of 3.0 and 3.5 during the sodium hypochlorite addition 5.75 gallons of additional muriatic acid was added. The pH at the end of the sodium hypochlorite addition was 3.2. Thereafter, 2.0 gallons of a 17.5% hydrogen peroxide solution was added. The hydrogen peroxide was added after holding the sodium hypochlorite treated starch for approximately 14 minutes after all the sodium hypochlorite had been added. The pH after addition of hydrogen peroxide was 3.0 and this was then adjusted to 6.0 with soda ash. The above food batter starch exhibited excellent adhesion to fresh and frozen foodstuffs such as fish, and also had good cohesion property.

Oxidation of High Protein Starch

C.S. Campbell; U.S. Patent 3,655,443; April 11, 1972; assigned to American Maize Products Company surprisingly discovered that if the starch which is oxidized has a high protein content, it will consistently produce a batter starch having superior adhesion characteristics and it is not necessary to follow the stringent conditions previously required for oxidation of the starch. The normal level of protein in commercial starch is 0.2 to 0.4% by weight. If the protein level is adjusted to about 0.7% or more by weight of starch, the starch can be oxidized in a pH range anywhere from 2 to 10 and with an active chemical oxidant range of 0.2 through 20% by weight of starch or higher and the resulting batter will have consistently good adhesion characteristics. The protein used to adjust the protein level of the starch can be of either animal or vegetable origin. As a general rule, however, it is preferred to use a protein derived from the same material as the starch, e.g., add corn protein to cornstarch.

Oxidizing materials which can be used in this process include the edible non-toxic oxidizing agents such as hypochlorite, peroxide, peracetic acid, persulfate, chlorite, and permanganate. The preferred oxidizing agents are chlorine compounds, e.g., hypochlorous acid and sodium hypochlorite. The amount of chlorine to be added can be readily determined by one skilled in the operation. As a general rule, however, there is a relationship between the protein present and the amount of chlorine used. Optimum adhesion and best results are achieved when the ratio of active chlorine to protein is at least 0.2:1 and preferably not over 2.5:1.

The pH during oxidative treatment of the starch may be anywhere in the range from 2 to 10. Existing methods of starch oxidation require acid conditions, generally in the range of 5.8 through 6.3. Not only does the high protein starch broaden the useable pH range on the acid side, but it also permits the reaction to be carried out under basic conditions. Although excellent adhesion is obtained throughout this wide pH range, a pH range of 5 to 9 is preferred especially when the oxidizing material chosen is a hypochlorite. Hypochlorite oxidations conducted below a pH of 5 liberate considerable amounts of chlorine gas and hypochlorite oxidations conducted above a pH of 9 yield products with somewhat darker color.

Example 1: 200 grams of commercial cornstarch (protein content 0.3%) was slurried in 225 cc of tap water at a temperature of about 70°F. The density of the slurry was adjusted to 23° to 24°Bé by the addition of water. Filleted codfish portions of approximately 3 to 4 ounces in size were thoroughly dipped in the starch batter slurry, drained and then quickly breaded with breading mix. The breaded fillet was immersed into a deep fat fry bath with an oil temperature of 375° to 380°F and was fried until it floated. The fried fillet was drained, and the adhesion of the batter to the codfish was tested by scraping a fork lightly along the four sides of the fillet. The percent of the battermix which adhered to the fish was only 45%.

Example 2: Corn protein was added to commercial cornstarch to raise the protein content of the starch to 0.8% by weight. The starch was then tested as a battermix base by the procedure of Example 1. The percent batter adhesion was 50%.

Example 3: Example 2 was repeated but the high protein starch was oxidized in conventional manner with 0.3% hypochlorous acid by weight of starch. The modified starch was tested as a battermix base by the procedure of Example 1. The percent batter adhesion was 90%.

Examples 4 and 5: Example 3 was repeated with the starch oxidized at levels of 0.6 and 0.9% active chlorine. The percent batter adhesion was 100 and 80% respectively. The results of Examples 1 through 5 are summarized in the following table.

Example	%Protein	Chlorine Treatment Level (% Active Chlorine)	%Batter Adhesion
1	0.38	0.0	45
2	0.8	0.0	50
3	0.8	0.3	90
4	0.8	0.6	100
5	0.8	0.9	80

Examples 1 through 5 show results which are comparable to what is obtained in commercial production. Contrary to the nonconsistent results obtained with unoxidized normal commercial starch, the oxidized high protein starch consistently yields a batter starch which gives a high degree of batter adhesion.

Aged and Oxidized Batter Starch

A batter starch and process of making same is disclosed by *J.C. Fruin; U.S. Patent 3,767,826; October 23, 1973; assigned to Anheuser-Busch.* The starch is mixed dry with an oxidizing agent, preferably calcium hypochlorite and preferably aged before being mixed with water to form a batter for coating frozen food products, particularly fish. The crumb coated fish are fried and refrozen to form a consumer product, or may be frozen raw for final preparation immediately prior to use.

This process involves a slight or mild oxidation treatment of the starch. This preferably is done in the starch plant and involves the dry mixing of the starch with a hypochlorite in the range of 0.1 to 5.0 parts hypochlorite to 1,000 parts starch by weight. If substantially less hypochlorite is used, the oxidation effect is achieved much more slowly; and if substantially more hypochlorite is used, the oxidation is so strong that a good batter starch is not produced. While hypochlorites are preferred as the oxidizing agent, other oxidizing agents can be used.

After treatment with the hypochlorite in the dry form, the treated starch may be allowed to age. Accelerated aging can be used and longer time and higher temperatures also are effective.

The dry blending is important because it gives a faster and more certain reaction. It also is desirable that the starch be stored at room temperature or higher after treatment.

Example: Five pounds of granular dry cornstarch having 10% moisture was mixed with 0.005 pounds of calcium hypochlorite. The mixing was done in a Twinshell mixer for 20 minutes. The hypochlorite treated starch is then stored one week at room temperature. To make the batter, 150 grams of the treated aged starch is mixed with 150 grams white corn flour obtained by conventional milling techniques and 400 grams water is combined with this mixture of starch and flour to make the batter. The fish are a lot of frozen cod portions weighing 3⅓ ounces a piece. The fish are coated with the batter starch and then coated with bread crumbs. The crumb coated fish are deep-fat fried at 375°F for 3½ minutes.

The table below shows examples of results obtained in deep fat frying of fish, dipping pieces of frozen fish in batter and then coating with breading prior to deep fat frying. The table shows the degree of adhesion of batter to fish in ranges of 95 to 100%, 90 to 95%, 80 to 90%, and less than 80%. Adhesion is determined by the test described in U.S. Patent 3,607,393.

Type of Starch	Treatment	Age of Starch When Tested, Days*	Adhesion, percent 95-100	90-95	80-90	0-80
Dent corn	None (control)	3	25	37	13	25
Dent corn	0.10 $KBrO_3$	3	50	50	0	0
Dent corn	None (control)	12	0	0	13	87
Dent corn	0.10% $KBrO_3$	12	0	10	20	70
Dent corn	0.05% $Ca(OCl)_2$	12	8	50	25	17
Dent corn	None (control)	4	0	0	0	100
Dent corn	0.10% $Ca(OCl)_2$	4	42	25	25	8
Dent corn	0.05% $Ca(OCl)_2$	4	25	50	17	8
Dent corn	None (control)	9	0	17	33	50
Dent corn	0.015% $KMnO_4$	9	25	33	25	17
Dent corn	0.10% $NaClO_3$	9	33	42	17	8
Dent corn	0.20% H_2O_2	9	0	42	8	50
Dent corn	None (control)	1	0	0	12	88
Dent corn	0.05% $NaClO_2$	1	8	42	33	17
Dent corn	0.25% $NaClO_2$	1	8	33	42	17
Dent corn	0.01% $NaClO_2$	1	8	25	42	25
Waxy maize	None (control)	1	0	0	8	92
Waxy maize	0.10% $Ca(OCl)_2$	1	0	25	42	33
Dent corn	None (control)	2	0	0	0	100
Dent corn	0.10% $Ca(OCl)_2$	2	42	42	16	0

*At room temperature.

CONVERTING STARCH

C.D. Szymanski; U.S. Patents 3,748,151; July 24, 1973 and 3,635,741; January 18, 1972 both assigned to National Starch and Chemical Corporation has developed a converting starch product comprising a starch base, aqueous dispersions of which display a high viscosity when first gelatinized but upon being cooked at retort temperatures are converted resulting in a dispersion exhibiting a substantially reduced viscosity.

The term a converting starch denotes a starch which has been admixed with a specified modifying agent so that when an aqueous dispersion of the starch product reaches its gelation temperature, it displays a normal, high viscosity, and which on being cooked at retort temperatures, the starch is converted, i.e., degraded, thereby reducing its molecular weight, as manifested by the substantially reduced viscosity of the resultant dispersion, as compared to a comparable dispersion of the identical starch which has not undergone treatment with the modifying agent.

It was found that the admixture of starch with at least 0.2%, by weight, of a modifying agent selected from ascorbic acid, araboascorbic acid and dihydroxymaleic acid leads to the production of a converting starch which fulfills the above stated objects. The conversion of these starches takes place on the cooking of an aqueous dispersion of the starch ordinarily at retort temperatures. With certain of the converting starches, temperatures which are commonly employed in cooking starch dispersions are sufficient, these temperatures being, of course, in all cases, higher than the normal gelatinization temperature of starch. As used in this process, the term retort temperatures means temperatures of about 250°F and pressures of about 15 psi.

One such feature of this starch is that it may be stored in dry form, prior to use, for extended periods without any danger of its undergoing a premature degradation. Thus, it is only upon cooking under the specified conditions that the desired degradation of the converting starch of this process will take place.

It is to be noted that while all of the modifying agents used are mono- or dicarboxylic acids, the mechanism involved in the degradation of the converting starches is not dependent upon a simple acid conversion or hydrolysis as is evidenced by the neutral and slightly basic pH levels at which these starches may be converted. Each of the modifying agents used displays an ene-diol structure, i.e.

$$\begin{array}{cc} \text{OH} & \text{OH} \\ | & | \\ -\text{C} & = \text{C}- \end{array},$$

and it is believed, that it is the presence of this structure which makes the conversion reaction possible. Thus, when a large number of related acids such as citric acid, lactic acid, gluconic acid, and 2-keto gluconic acid, none of which contained an ene-diol structure, were tested in order to determine their degradative action on starch, it was found that all of these acids were completely ineffective. The possibility that ascorbic acid which exists as a lactone was undergoing hydrolysis to free the acid group was likewise considered.

Glucuronolactone was thereupon used in the test systems and also found to be without effect. It can be seen therefore, that the conversions which take place with the converting starches of this process do not result solely from the presence of carboxyl groups on the specified modifying agents.

Example 1: This example illustrates the preparation of a typical converting starch of this process as well as the viscosity characteristics of the aqueous dispersions derived therefrom.

A total of **996** parts of a waxy maize starch was thoroughly blended with four parts of finely divided ascorbic acid until the mixture was homogeneous. An aqueous dispersion comprising six parts of the starch-ascorbic acid mixture and 94 parts of water, was cooked for 15 minutes at about 200°F, displaying an expected high viscosity, whereupon the dispersion was retorted at 250°F for a period of 40 minutes. A control comprising an aqueous dispersion containing 6%, by weight, of the waxy maize starch which, in this case, was devoid of any ascorbic acid, was also prepared and cooked for a period of 15 minutes at about 200°F and likewise displayed a comparable high viscosity. The dispersion was then retorted at 250°F for a period of 40 minutes.

Subsequent to the retorting period, each dispersion was cooled to about 167°F and its viscosity was determined using a Brookfield RVF Viscometer employing a #2 spindle at 20 rpm. It was found that the dispersion of the converting starch had a viscosity of 77 cp while the control had a viscosity of 280 cp clearly indicating that only the converting starch of this process had the ability to undergo a substantial reduction in its viscosity upon being subjected to high temperature conditions. In a repetition of the above procedure, dihydroxymaleic acid in a concentration of 0.4% by weight, and araboascorbic acid in a concentration of 0.6%, by weight were each, in turn, used to prepare a converting starch product.

It was found that the dispersion prepared with the converting starch containing the dihydroxymaleic acid had a viscosity of 115 cp while the dispersion prepared with the converting starch containing the araboascorbic acid had a viscosity of only 74 cp. Food products, such as chicken gravy containing this converted starch are also included in this process.

Example 2: A total of 10 parts of an uncooked aqueous dispersion containing 6%, by weight, of a converting starch based on cornstarch which had been admixed with 0.5%, by weight, of ascorbic acid was combined with 90 parts of semicooked chicken gravy. The homogeneous blend was heated at 195°F for 20 minutes and was subsequently cooled to 165°F. When the viscosity of the homogeneous starch-gravy blend was measured at this point by means of a Bostwick Viscometer, an instrument which measures viscosity in terms of distance flowed within a specified period of time, it was found that it flowed 19.5 centimeters in one minute, indicating that it had a rather high viscosity. The viscous gravy was poured into conventional #2 cans which were then sealed in the usual manner.

The filled cans were then retorted at 250°F at about 15 psi for a period of 30 minutes so as to thereby effect a substantial reduction in the viscosity of the blend. When the viscosity of the retorted product, after being cooled to 165°F was measured by means of the Bostwick Viscometer, it was found that the material flowed 24 centimeters in 1.5 seconds, indicating that a substantial decrease in the viscosity of the chicken gravy had, in fact, occurred.

APPARATUS FOR PROCESSING STARCH

Dry Roasting Using Turbulized Suspension

G.G. Taylor and J.A. Hay; U.S. Patent 3,527,606; September 8, 1970; assigned to American Maize-Products Company describe a method for modifying starch by either the dry roasting or dry reaction technique. Particulate starch along with any desired reactant is placed in a cooker and mechanically agitated at a rate sufficient to maintain the particles in a turbulized suspension, thereby yielding rate coefficients of heat transfer substantially higher than in prior processes and assuring a uniformity of roast or reaction throughout the entire mass of product.

Processes to which this method is applicable are those for manufacturing dextrins, British gums, starch phosphate esters, organic esters, ethers and graft polymers of starch by the dry reaction technique.

Referring to Figure 2.1, the cooker **10** as shown is a horizontal cooker having a partitioned jacket **11** for zoned heating. The cooker **10** is positioned at a slight angle to the horizontal plane to tilt down toward the outlet end so that gravity will move the starch particles toward the outlet for discharge.

Positioned above the partitioned jacket **11** is a fluid pipe **12** which is tapped at four places to receive feeder stems **13**, the number of stems and tapped holes in the fluid pipe preferably being equal to the number of partitions in the jacket. This number can vary from one to as many as the user desires, and is not limited to four.

In each stem there is a control valve **14** to control the amount of heating fluid that is allowed to enter each partition in the jacket. Each control valve **14** is controllable independently of the others. This permits one to partially cool the reacted starch particles by selectively closing one or more of the control valves **14** before the starch product leaves the cooker. Similarly, the starch product may also be fully cooled while in the cooker chamber simply by supplying a cooling fluid to fluid pipe **12** after the reaction has been completed as in a batch operation.

Underneath the cooker are four outlet stems **15**, one for each partition in the jacket. The outlet stems will be as many in number as there are partitions in the jacket and should be equal to the number of inlet pipes **13**. However, it is not necessary to provide one outlet stem for each inlet stem as it is completely feasible to use two or more outlet stems for each inlet stem. Each outlet stem has a condensation trap **16** to remove condensate from the existing steam and can be connected back to the source of heating so that the steam can be recirculated after reheating.

The cooker is fed by feed hopper **17**, the feed rate being controlled by a conventional variable speed air lock **18**. A second feeder **19** is also provided for feeding fines and dust collected in a conventional dust collection **19a**. The feed rate of feeder **19** is regulated by a variable speed air lock **20**. Feeder **19** is connected to the cooker chamber through an exhaust pipe **21** which also serves as an air flow duct.

Feed hopper **17** and feeder **19** feed their charge through variable speed air locks **18** and **20** into the cooker chamber **22** through pipes **23** and **24**. Longitudinally displaced from the point at which pipes **23** and **24** enter the cooker chamber **22** and disposed at an end of the cooker chamber is an outlet **25** which has a variable speed air lock **25a** which serves to controllably remove reacted product from within the cooker chamber. Disposed above the output feeder **25** is a fluid

FIGURE 2.1: DRY ROASTING PROCESS FOR THE MODIFICATION OF STARCH

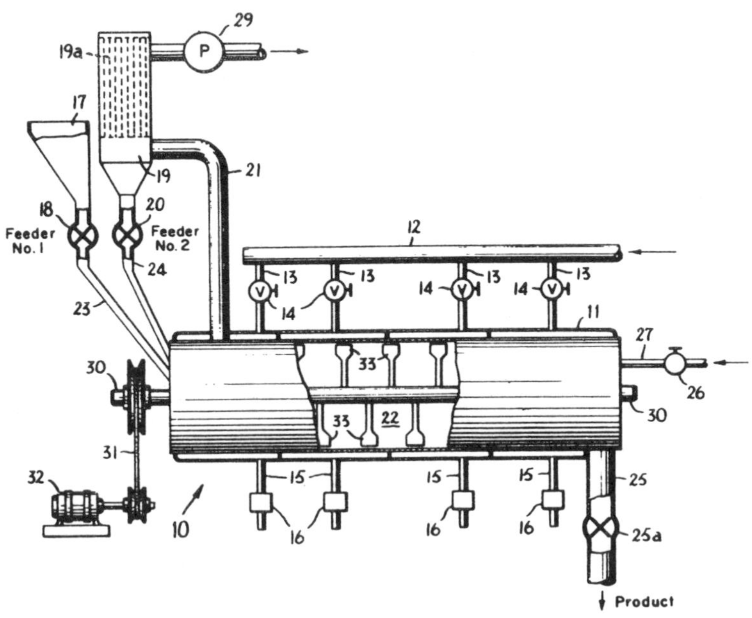

Source: G.G. Taylor and J.A. Hay; U.S. Patent 3,527,606; September 8, 1970

control valve **26** in the feed line **27** connected to the cooker chamber **22**. The fluid control valve **26** regulates the amount of fluid, preferably air, or liquid or gaseous reactant fed into cooker chamber **22**. Connected to exhaust pipe **21** is a vapor pump **29** which pulls fluid admitted to cooker chamber **22** through fluid control valve **26**, through cooker chamber **22** and out of it via pipe **21**, then through dust collector **19a** to remove entrapped fines and dust, finally through vapor pump **29** to exhaust to the atmosphere.

The speed at which the vapor pump **29** operates will determine the pressure or vacuum inside the cooker, and preferably the vapor pump is capable of sustaining a vacuum of at least 15 inches mercury within the cooker chamber **22**. When operating under pressure, the fluid may be fed under superatmospheric pressure to control valve **26** or pulled into the cooker by exhaust pump **29**.

Disposed within the cooker chamber **22** is a spindle **30** which is coaxial with the longitudinal axis of the cooker chamber **22** and is of a length greater than the axial length of the cooker chamber. Opposite ends **30** of the spindle extend past the ends of the cooker chamber and ride in bearing supports. Connected to one end of the spindle is a belt **31** connected to a drive motor **32** which drives the belt and spindle at the desired speed to maintain the starch in turbulized suspension.

Attached to the spindle are a plurality of paddles **33** which can be integral with the spindle or separately made and individually connected. The paddles are preferably oriented in a manner such that the projected length of each paddle face multiplied by the number of paddles is substantially equal to the greatest measured longitudinal distance between the first and last paddle. This assures that the paddles will have sufficient face area to turbulize the starch granules within the cooker.

The paddles may be mounted with the faces at an angle to the longitudinal axis to give a directional impetus to the starch granules in addition to the induced centrifugal force, thereby tending to force the starch granules towards the outlet end of the cooker. As shown, there are two rows of paddles set 180° apart, but this is not critical, and the paddles may be arranged in any desired manner on the spindle.

In the example, the cooker of Figure 2.1 had a 16" inside diameter and a void volume of 14 cubic feet. The sweep diameter of the paddles was approximately 15½". The cooker chamber was half filled with about 245 pounds of starch granules giving a starch to void volume ratio of 1:2.

Heat was supplied by high pressure steam in the jacket, and the temperature of the starch granules inside the cooker was recorded by means of a conventional thermocouple. In terms of peripheral velocity of the paddles, for the equipment and dimensions given, the practical upper limit was found to be about 2,000 ft per minute, which corresponds to a theoretical g level of about 52 for ordinary commercial starch granules.

Example: Commercial cornstarch granules containing approximately 10% moisture were fed into the cooker as described above and the spindle was rotated at 160 rpm to maintain the granules in turbulized suspension for 2¼ hours while maintaining the temperature of the granules at 355°F.

The dry roasting converted the starch to a British gum having the following solubility and viscosity measured with a conventional cold funnel viscosity test on a paste containing one part of roasted granules to two parts of water.

Time Hours	Solubility, percent	Viscosity, (ml.)
1.0	30	8
1.5	59	21
2.0	72	34
2.25	77	38

Vibration Ball Milling

A process described by *J.W. Knight; U.S. Patent 3,630,774; December 28, 1971; assigned to Corn Products Company* involves treating a starch-bearing material, in which the starch is present in granular form, with a source of energy of sufficient magnitude and for a sufficient length of time to disrupt the normal sequence of the molecules of the starch. The preferred treatment is vibration ball milling. The energy to which the starch-bearing material is subjected at least partially breaks down the bonding of the so-treated starch and creates a multitude of sites of increased reactivity. The energy source should not be of such magnitude or duration of application as to destroy the granular structure of the starch.

A product made in accordance with the process is particularly characterized as having a lower gelatinization temperature and a higher reactivity toward various chemical reagents than the original starch-bearing material. In particular, it has a measurably greater reactivity toward dilute alkali solutions, and greatly enhanced susceptibility to enzymatic attack.

The amount of time necessary to effect the aims of the process and achieve a suitable disrupted product will greatly vary depending on the magnitude of the energy applied to the initial starch-bearing material. As a guideline, when a vibration ball mill is utilized, a treatment time of from 5 to 180 minutes is sufficient. In the more usual case, the starch is subjected to action of the vibration mill for 10 to 60 minutes. This is sufficient partially to break down the hydrogen bonding of the starch present, and to create desired sites of increased reactivity without destroying the granular structure.

A series of vibrating mills suitable for use in the process, particularly for laboratory or pilot plant work, are marketed under different model numbers by Allis-Chalmers. These mills are formed from spring-mounted cylinders with dual eccentric mechanisms running horizontally on each side of the mill. The grinding chamber liners act as spring boards for the grinding charge.

In addition to their use in industrial adhesives, the disrupted starch granules obtained by the process have a surprisingly high viscosity, particularly in mild alkali solutions, and therefore find particular utility as a bulking, bodying or dispersing agent in a variety of foodstuffs.

Other uses for the disrupted starches can be found in converting these materials to alcohol in the brewing industy, as a feed for various animals such as calves, and for the processing of wheat flour which is intended for use in the preparation of sugars.

STARCH-FAT COMBINATIONS FOR FOOD USE

Lipid Modified Starches as Dusting Powders

Flour is commonly used as a dusting powder in the baking industry. Efforts have been made to use powdered starch, because it possesses the advantage over flour that it will not support insect life, but the results have not been entirely satisfactory. One reason is that ordinary dried and powdered starches are not sufficiently mobile or free-flowing and consequently they have a tendency toward bridging. Moreover, they flow too quickly through the screens of the mechanical dusting equipment commonly used in bakeries with flour and are too dusty, i.e., they generate a great deal of dust in the air surrounding the dusting equipment.

It is known that powdered starches can be made mobile by treatment or coating with a variety of metal salts and other compounds in a finely divided state. Such processes are described, for instance, in Canadian Patent 526,173, which employs tricalcium phosphate, calcium pyrophosphate, calcium phytate, and magnesium phosphate and in U.S. Patent 2,614,945 which employs a large group of water-insoluble salts and hydroxides.

The mobile starches produced in this manner have a flow rate which is higher than that of common powdered starches and they are substantially devoid of the tendency to cake or bridge. Such starches suffer, however, from the disadvantage that they generate too much dust. For this reason they have not found wide use in bakeries. Moreover, their flow rate is too high for use in mechanical dusting equipment of the type commonly employed in bakeries for dusting with flour.

A.D. Campbell; U.S. Patent 3,467,543; September 16, 1969; assigned to Standard Brands Incorporated has disclosed a process wherein starch is treated or coated in a slurry with both a small amount of a metal salt or other chemical compound capable of increasing its mobility substantially and a small amount of an oil or fat or equivalent agent effective to reduce the flow rate of the product, the treated product is separated from the slurry, dried and powdered. The product obtained offers advantages over dusting flour in that it generates much less dust, will not support insect life and less of it is required for effective dusting. The result obtained is surprising in view of the fact that the coating of starch with the oil or fat alone results in a product which tends to bridge and has a flow rate which is erratic and usually too low for satisfactory use as a dusting powder. The product is actually less mobile than common powdered starch yet surprisingly possesses the nonbridging characteristic commonly associated with highly mobile starches.

The starch may be treated or coated first with one of the agents used in this process and then with the other or with both simultaneously. When these agents are applied separately, it makes no appreciable difference which is applied first.

The metal salts should be used in an amount effective to substantially increase the mobility of powdered starch. Very small amounts, usually less than 1.0% are sufficient for this purpose. Greater amounts may be added if desired, although usually a relatively small further increase in mobility is imparted by increasing the amount substantially beyond about 1.25%. Of course, the amount required to impart a given mobility will vary depending upon the salt used. In the case of tricalcium phosphate satisfactory results can be obtained, for instance, with amounts varying from 0.1 to 1.0%. About 0.4 to 0.8% is preferred.

In general the amount of oil or fat added in this process should be such as to give a product wherein the amount of oil and fat extractable by CCl_4 does not exceed 0.5% because the product will not flow satisfactorily if this amount is exceeded substantially. The total extractable oil and fat will consist of extractable fatty material already present on the granular starch starting material and that absorbed by the starch granules in the coating step of this process. In the case of corn oil and olive oil the total extractable oil and fat preferably amount to 0.15 to 0.30% and advantageously about 0.22%.

It is essential that the relative amounts of the two coating agents employed be chosen to give a coated starch product having a Kerr Mobility ratio between 0.02 and less than 1.0 and thus make the starch suitable for use in mechanical dusting machines of the type of the Day Duster.

In the following example no actual determination was made of the content of
fatty material in the cornstarch used as the starting material. Since powdered
cornstarch generally contains a small amount of fatty material extractable with
CCl_4, this amount should be added to the percentage of oil added in the example to obtain the total amount of fatty material extractable with CCl_4 from
the final product.

Example: Seventy-one and one-half gallons of 23.7°Bé starch slurry, containing
300 lb dry substance cornstarch, was diluted to 19.3°Bé and heated while under
agitation to 120°F. 0.6 lb (0.2%) of refined corn oil, stabilized with TENOX-S-1,
an antioxidant composed of 20% propyl gallate, 10% citric acid and 70% propylene
glycol, was then added and the mixture was agitated for 5 minutes. One and one-
half pounds (0.5%) of tricalcium phosphate was added and the mixture was agitated
for one hour while maintaining the temperature at 120°F.

The slurry was then diluted with water to 15.0°Bé, filtered without washing and
dried in a current of air heated to 220°F to a moisture content of 10.4%. The
dried product was ground in a hammer mill so that 98% would pass a #200
U.S. Standard Sieve. The product had a Kerr Mobility of 7.8% and was considered satisfactory.

Kerr Mobility is determined by taking 200 grams of the starch being tested,
placing it on a #200 U.S. Standard Mesh Sieve and shaking the sieve on a Ro-Tap
shaker for 30 seconds. The percentage by weight of starch passing through the
sieve is taken as an index of its mobility. Common powdered starches normally
have Kerr Mobilities of the order of about 30 to 70%.

Similar products and products are also disclosed by *W.J. Ryan and L. Kogan;
U.S. Patent 3,377,171; April 9, 1968; assigned to Standard Brands Incorporated.*

Fat Treated Starch for Instant Cream Soups

Attempts have been made to produce a heavy-bodied, lump-free instant soup
having a creamy consistency from a blend of dried food ingredients. Workers
have achieved products offering a measure of convenience to the housewife, but
all of the prepared cream-style products have suffered from one or more disadvantages.

Dry soup mixes have been used but these mixes have not been able to produce
a product having the degree of creaminess achieved by the conventional household cooked soup. Moreover, dry mixes have achieved viscosity or thick-like
consistency by merely increasing the amount of solids in the reconstituted soup.
This does not achieve a true creaminess in the reconstituted soup and presents
problems of clumping, graininess, and other related problems on reconstitution.

*H.W. Block and P.B. Touher; U.S. Patent 3,433,650; March 18, 1969; assigned
to General Foods Corporation* have discovered that a dry soup mix capable of
reconstituting to a cream-like soup in 10 to 60 seconds in water above 180°F
can be produced using a modified starch. The soup composition comprises
a free-flowing mixture of dehydrated food solids and between 5 to 35% by
weight of an ungelatinized starch having a gelatinization range of 135° to 165°F
a particle size of less than 200 mesh U.S. Standard Sieve and being coated with
a fatty hydrophobic substance which delays hydration of the starch on recon-

stitution while imparting flavor to the reconstituted solids, the remaining food solids having a particle size wherein at least 50% of the particles are between 40 and 100 mesh U.S. Standard Sieve.

While the amount of starch used will vary with the type of soup, at least 5% by weight of ungelatinized starch must be present to give sufficient thickening of the soup but above 35% by weight of starch thickener will give a soup which is too thick or creamy. The starch must thus be rendered sufficiently dispersable to prevent clumping. This is accomplished by uniformly surface coating the starch with a fatty substance having sufficient immiscibility with water to delay starch hydration and gelatinization until the starch has been effectively dispersed in the reconstituting liquid and available for rehydration without clumping.

The coating material is taken from a fat composition such as butter, oleomargarine, vegetable oil and the like which will contribute desired flavor qualities to the reconstituted soup. Hydrophobic action is accomplished by only surface coating of the starch powder with fat.

Preferred among the raw starch thickeners is bland, ungelatinized potato starch. This type of starch gives the desired cream-like character to the soup while not imparting any undesirable taste to the reconstituted product.

Other starches which can be used are rye starch, sweet potato starch, water chestnut starch and lotus root starch. The remaining starches, such as tapioca starch, cornstarch, waxy maize starch, rice starch, hydrolyzed starch and pregelatinized starches will not rehydrate properly to give the desired creaminess to the soup.

A majority of the food solids (excluding the starch thickener) have a critical particle size of between 40 and 100 mesh U.S. Standard Sieve. At sizes smaller than 100 mesh, the fat coating is ineffective in preventing clumping and texture of the reconstituted product tends to be slimy.

Example 1:

Instant Chicken Soup Mix
(Cream Style)

Ingredients	Grams	Percent by Weight
Raw Idaho potato starch	4.39	27.29
Nonfat dry milk solids	2.04	18.89
Salt	2.25	13.99
Hydrolyzed starch	1.91	11.89
Freeze-dried chicken	1.40	8.74
Chicken fat	1.39	8.67
Monosodium glutamate	0.56	3.50
Sugar	0.40	2.46
Hydrolyzed vegetable protein	0.25	1.57
Butter	0.19	1.19
Powdered onion	0.13	0.79
Dried celery powder in sugar	0.067	0.42
Vanilla color	0.056	0.35
Parsley flakes	0.032	0.20
White pepper	0.005	0.03
Turmeric	0.0003	0.02
	15.0703	100.00

The dry mix ingredients with the exception of the parsley flakes, freeze-dried chicken pieces, chicken fat, and butter, were ground and sifted through a #40 mesh U.S. Standard Sieve Screen. The sifted soup ingredients were then blended in a Hobart mixer at a paddle stirrer speed of 64 rpm for 20 minutes. The chicken fat and butter ingredients were then added to the mixer and a wire whip was used at low stirrer speed (64 rpm) for 5 minutes. Mixing was continued for 5 more minutes at a higher speed (128 rpm) to assure complete dispersion of the fat with the blended ingredients.

The ingredients which had a bulk density of 0.45 g/cc were then agglomerated to achieve a density of 0.35 g/cc. The parsley flakes and freeze-dried chicken pieces were then added to the soup ingredients and mixed in the Hobart mixer at low speed (64 rpm) for 5 minutes. The blended dry soup mix was then removed from the Hobart in a free-flowing and powdered form, except for the chunks of freeze-dried chicken and parsley flakes.

About 3 teaspoons (14.3 grams) of the dry soup mix were placed in a bowl, ⅔ cup (158 cc) of boiling water was added, and the mixture stirred for several seconds. The dry soup mix reconstituted in less than one minute to give a smooth, nongrainy, cream-like consistency similar in all respects to a cream of chicken soup prepared by the conventional household cooking method. The soup did not loose it viscosity on standing. Recipe type examples are also included in the complete patent for cream style vegetable, tomato, green pea and mushroom soup.

Fat Holding Starches

J.J. Magic and R.F. Frost; U.S. Patent 3,372,034; March 5, 1968; assigned to American Maize-Products Company have discovered certain starch compositions comprising admixtures of starch and small amounts of finely divided, synthetic sodium silico aluminate particles which compositions have the capacity to maintain fats and oils in uniform suspension in foods in a superior manner not possible with starch alone. As a matter of fact, the fat suspending capacity of these compositions is some form of synergistic phenomenon since by actual tests it was found that this capacity exceeds the sum of the separate fat-suspending powers of each of the two components of the compositions.

Furthermore, when the mixture of starch and sodium silico aluminate is cooked in water under gelatinizing conditions and then cooled, a substantial increase in viscosity is achieved without any tendency on the part of the thickened liquid to set up or become immobilized in a gel.

For this process, synthetic sodium silico aluminate particles containing oxides of sodium, aluminum and silicon in such proportions that the mol ratio of Na_2O to Al_2O_3 is from 0.8:1 to 1.3:1 and the mol ratio of SiO_2 to Na_2O is from 4:1 to 16:1, substantially all of the particles being less than one micron in diameter and preferably averaging less than one-half micron in diameter, may be used. If the pH of such particles in water is not already neutral (7.0), it is preferred to adjust the pH to the neutral point with nontoxic edible acids such as citric or phosphoric acids.

The requirement that the sodium silico aluminate particles be of synthetic origin is extremely important. Attempts to duplicate the unique results achieved with the mixtures of starch and synthetic sodium silico aluminate particles by using certain naturally occurring clays which also contain oxides of sodium, aluminum, and silicon, in one or two instances even in comparable mol ratios, but the natural materials are not satisfactory.

Only a small amount of the synthetic sodium silico aluminate particles need be mixed with the starch to achieve the desired properties. As little as 0.1% of the particles based on the weight of the starch will be effective and the amount of particles can be progessively increased up to about, say, 5% based on the weight of the starch. As a matter of fact, there is no actual limit on the maximum amount of the synthetic particles; as high as 15% has been used with good results. However, use of more than 5% of the particles does not give any significant additional advantages to speak of and is ordinarily wasteful.

Further details of the process are illustrated in the following example. Unless otherwise indicated, all proportions and parts specified are by weight.

Example: In this example, the fat-suspending capacity of a variety of starch compositions was compared with that of similar compositions to which were added various amounts of synthetic sodium silico aluminate particles.

Substantially, all of the particles were less than one micron in diameter and contained oxides of sodium, aluminum and silicon in the proportions of about 0.83 mol Na_2O for each mol of Al_2O_3 and about 12 mols SiO_2 for each mol of Na_2O.

The fat suspending capacity of each composition was determined by means of a fat-folding test which consists of slurrying 6.25 grams of the composition to be tested in 198 ml of water and heating the slurry to 190°F with stirring until the starch is gelatinized. Then 30 ml of liquid hot rendered fat is added to the starch suspension with agitation and the mixture is permitted to stand. Visual observations will then show whether the composition has good fat-suspending capacity, in which case the added fat will remain uniformly dispersed throughout the aqueous starch suspension, or poor fat-suspending capacity, in which event the added fat will separate and rise to the top to form a separate layer.

The following specific compositions were tested and the results are noted in the table appearing below:

(A) 5.0 grams of waxy cornstarch crosslinked with phosphorous oxychloride and 1.25 grams of unmodified cornstarch were blended together and a fat-folding test was run. 0.1 grams of the unmodified cornstarch of a similar blend was replaced with sodium silico aluminate particles described and a fat-folding test was run.

(B) The blend first described in (A) was made up again and subjected to the fat-folding test. 0.05 gram of the unmodified cornstarch of a similar blend was replaced with sodium silico aluminate particles and a fat-folding test was then run.

(C) The blend first described in (A) was made up again and subjected to the fat-folding test. This time 0.016

gram of a similar blend was replaced with the sodium silico aluminate particles and a fat-folding test was then run.

(D) A fat-folding test was run on 6.25 grams of the waxy cornstarch crosslinked with phosphorus oxychloride. 0.125 gram of another 6.25 gram batch of this starch was replaced with the sodium silico aluminate particles and another fat-folding test was run.

| | Fat-Folding Test | |
Composition	Without sodium silico aluminate	With sodium silico aluminate
A	Poor	Good
B	Poor	Good
C	Poor	Good
D	Poor	Good

As will be noted, in every instance where the sodium silico aluminate particles were included, the starch compositions were capable of holding the added rendered fat in uniform and stable dispersion. On the other hand, the added fat could not be maintained dispersed and separated as a layer in the same compositions which lacked the sodium silico aluminate particles.

GELATINIZED STARCHES

Most native starch granules are birefringent and show the characteristic cross-shaped shadow when viewed under polarized light. Starch granules gelatinize in water when the temperature is raised to 60° to 70°C. As the temperature goes above 70°C, the granules will normally swell to form a paste or sol and the shorter molecules dissolve. A loss of the birefringence occurs during gelatinization and the starch granule may disintegrate into molecules and fragments.

Pregelatinized starch refers to commercial starch products that are used in the preparation of instant puddings and similar products. These are prepared by gelatinization and drying the starch using drying drums, spray drying and the like to prepare a cold water dispersible starch.

Depending on the process used, various degrees of gelatinization can occur such as:

> *Granular Preswollen Starch* — a cold-water-swelling starch with intact granules. When produced under preferred optimum conditions, granular preswollen starch will be free of birefringent granules, will exhibit little or no granule fragmentation, and will produce a high cold water viscosity. It is useful as a quickly soluble thickener in making such food items, for example, as puddings, gravies, and soups.
>
> *Texturized Starch* — a cold-water-swellable, nonbirefringent, substantially completely fragmented starch. When slurried in water at a solids content of as little as 5%, the product exhibits texturizing characteristics by yielding a pulpy mixture. Texturized starch is useful in preparing food products such as, for example, applesauce, spaghetti sauce and the like.

GELATINIZATION PROCESSES

Pregelatinized Waxy Starch

A process for treating waxy (i.e., high amylopectin) starches to produce cold-water-swelling starches having remarkable and unexpected viscosity characteristics has been developed by *F.J. Germino, F.E. Kite and E.H. Christensen; U.S. Patent 3,563,798; February 16, 1971; assigned to CPC International Inc.*

Briefly, the process is as follows. A waxy starch, in granular, nongelatinized form, is slurried with a liquid component which comprises a solvent for starch and an organic liquid which is itself not a solvent for starch and which is miscible with the solvent for starch. The relative proportions of the ingredients in the slurry are critical, as will be explained later.

The slurry is subjected to gelatinizing conditions for a time sufficient to result in the desired product. Immediately following the treatment it is essential that an additional quantity of the organic liquid, in an amount at least equal, by weight, to the weight of the total slurry, be combined with the slurry, preferably with agitation. The greater the amount of starch solids present in the slurry, the more vigorous agitation that is required.

The products are different in structure and in properties from treated nonwaxy starches. The starches are characterized by having 25%, up to 100%, intact granules, which exhibit complete (i.e., 100%) loss of birefringence. In many respects the behavior of these products is remarkably similar to that of the costly natural gums, such as guar, galactomannan, carrageenan, and the synthetic gums such as carboxymethylcellulose.

Like most of the natural and synthetic gums maximum paste viscosity is achieved if the starch is dispersed in water at 25°C, lower viscosities resulting as the dispersion temperature increases. Paste viscosity also varies with concentration, pH and shear in much the same manner as that of the gums. Also, like many of the natural and synthetic gums, the starch pastes exhibit excellent thermal stability, freeze-thaw stability, and storage stability, which properties make them excellent thickening agents for use in prepared food products.

By the term solvent for starch is meant any liquid in which starch will solvate, or gelatinize. Water is, of course, by far the most economical and commonly used starch solvent or gelatinizing agent, and is therefore preferred in this process. Other starch solvents, e.g., dimethyl sulfoxide, 2-aminoethanol, N-methyl-pyrrolidone, can also be used however. In the following discussion the solvent for starch component of the liquid portion of the slurry is referred to as the solvent.

The second component of the liquid portion of the slurry, the organic liquid, can be any organic liquid which is miscible with the solvent and which is not itself a solvent for starch. If the starch is to be used in food, then all of the organic liquid must be removed after treatment; for this reason it is desirable to use a liquid which is not excessively retained within the starch granules. It is also desirable, for economic reasons, to select a liquid which can be readily recovered from the solvent for reuse. Lower alcohols (e.g., methanol, ethanol, isopropanol, tertiary butanol), ketones (e.g., acetone, methyl ethyl ketone), dioxane, etc. are particularly suitalbe in the process.

Example: This example illustrates the process as applied to the following starches: native (i.e., noncrosslinked unmodified and underivatized) white milo; native waxy maize; a waxy maize starch acetate crosslinked to a Scott viscosity of 5 grams/60 seconds, 100 ml delivered paste; a white milo starch sulfate crosslinked to a Scott viscosity of 3 grams/50 seconds, 100 ml delivered paste; a waxy maize starch sulfate crosslinked to a Scott viscosity of 4 grams/55 seconds, 100 ml delivered paste; a waxy maize starch phosphate crosslinked to a Scott viscosity of 4 grams/60 seconds, 100 ml delivered paste; a cationic waxy maize starch crosslinked to a Scott viscosity of 5 grams/60 seconds, 100 ml delivered paste; and a hydroxyethyl white milo starch crosslinked to a Scott viscosity of 4 grams/55 seconds, 100 ml delivered paste.

The slurries were treated in a Votator. The treated slurries were discharged from the Votator directly into a tank containing methanol in an amount which resulted in a dilution of 1.5 parts methanol per part of slurry. The slurry was agitated during the dilution step. The treated starches were then recovered from the slurries, washed with additional methanol, and dried. In all cases the starches rapidly dispersed in cold (i.e., about 25°C) water to form smooth viscous pastes. Viscosities were determined, and reported as the percent starch concentration required to produce a 50 poise aqueous paste when measured by a model RVT Brookfield viscometer.

Starch	Liquid Composition, % Methanol in Water	Solids Level, % Starch	Processing Temp, °F	Product Viscosity, % Concentration of 50-poise Paste
Native white milo	55	20	230	6.5
Native white milo	35	20	190	6.2
Native waxy maize	55	20	230	6.4
Crosslinked waxy maize starch acetate	55	25	260	1.8
Crosslinked white milo starch sulfate	55	25	240	1.6
Crosslinked waxy maize starch sulfate	55	25	240	2.5
Crosslinked waxy maize starch phosphate	55	25	240	1.6
Crosslinked cationic waxy maize starch	55	25	240	1.9
Crosslinked hydroxyethyl white milo starch	55	25	240	2.5

As can be seen from the data in the above table, the viscosities of pastes prepared from the treated slightly crosslinked waxy starches were exceptionally high; in most cases they were actually higher than pastes prepared from the untreated starches.

Granular Gelatinized Starch

R.A. Thurston and R.E. McConiga; U.S. Patent 3,617,383; November 2, 1971; assigned to CPC International Inc. describe a method for preparing a cold-water-swelling starch, the granules of which are completely intact, are slightly swollen, and which exhibit from 40 to 100% loss of birefringence under a polarizing microscope.

Briefly, the process is as follows. Ungelatinized starch is first slurried in a mixture of (1) a solvent for starch, and (2) an organic liquid which is itself not a solvent for starch and which is miscible with the solvent for starch. The slurry

is passed through a suitable confined heat exchange zone, in which it is subjected
to gelatinizing conditions. The specific conditions of time of treatment, temperature of treatment, and proportions of the ingredients in the slurry are so selected as to permit from 40 to 100% of the starch granules to lose birefringence.
The most commonly used and most economical solvent for starch is, of course,
water, and this is the preferred solvent. Other common solvents, however, such
as dimethyl sulfoxide, dimethylformamide, N-methylpyrrolidone, or 2-amino-
ethanol, are just as suitable.

The second liquid component of the slurry can be any organic liquid which is
miscible with the solvent for starch and which is not itself a solvent for starch.
If the starch is to be used in food, then substantially all of the organic liquid
must be removed from the starch after treatment; for this reason it is desirable
to use a liquid which is not excessively retained within the starch granules. It
is also desirable, for economic reasons, to select a liquid which can be readily
recovered from the solvent for reuse. Lower alcohols, (e.g., methanol, ethanol,
isopropanol, tertiary butanol), ketones, (e.g., acetone, methyl ethyl ketone), dioxane, etc. are particularly suitable in the process.

FIGURE 3.1: PROCESS FOR PREPARING GRANULAR COLD-WATER-SWELLING STARCHES

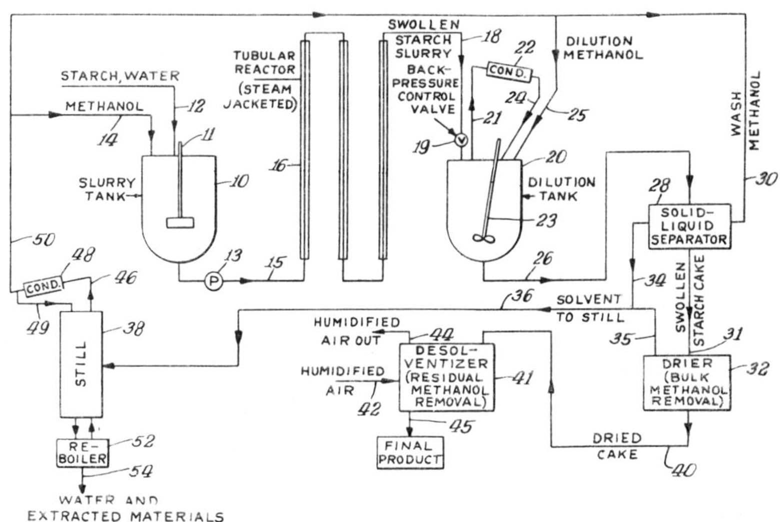

Source: R.A. Thurston and R.E. McConiga; U.S. Patent 3,617,383; November 2,
1971

Referring to Figure 3.1 by numerals of reference, the number **10** denotes a tank
that is equipped with a mixer **11**, for making up slurry. A supply line **12** feeds
the starch water slurry into the tank. A supply line **14** feeds methanol into the
tank. Since the process is intended to be continuous, ordinarily a plurality of

slurry tanks would be mounted in parallel, to permit alternate use and continuous discharge of slurry from at least one of the tanks, or alternatively, flow controllers could be installed on the lines 12 and 14 to permit continuous operation from a single tank. The slurry tank 10 is connected to a pump 13, that discharges through a line 15 into one end of a steam jacketed tubular reactor 16. The internal diameter of the heat exchange tube is selected for efficient heat exchange, preferably under turbulent flow conditions. The velocity of the slurry must be sufficiently high to prevent settling.

A velocity of one foot per second, or greater, is adequate for this purpose. This tubular reactor discharges through a line 18, that is provided with a control valve 19, into a receiving or dilution tank 20. The tank is equipped with a mixer. The tank is also equipped with a superposed reflux condenser. This includes a vapor uptake line 21, and a condenser 22, and a liquid return line 24, that drains the condenser back into the tank. A methanol supply line 25 is disposed to feed methanol into the tank.

A discharge line 26 interconnects the lower end of the dilution tank 20 with a solid liquid separator 28. This separator may be, for example, a filter, a centrifuge, etc. In practice, several similar items of separating equipment might be mounted in parallel, to permit continuous operation. A methanol supply line 30 is connected to the separator, to permit washing the separated solid material with methanol, if desired.

From the separator 28, the solid cake is transferred by the line 31 to a drier 32. Any suitable form of drying apparatus may be employed. The preferred ways of drying are by contact of the starch with hot humid air or with live superheated steam, in a tray drier, or in a fluidized bed. The liquid material from the filter or other separator is withdrawn through a line 34. The methanol that is removed in the drier is condensed and the condensate is passed through a line 35 into a line 36 that feeds a still 38.

The dried cake from the drier is transferred, as is indicated by the line 40, to a desolventizer 41, which removes residual traces of methanol from the dry cake. This can be accomplished in any desired manner, but preferably is accomplished by contacting the dried cake with live, superheated steam. Alternatively, hot humidified air can be used as shown. The humidified air is supplied to the desolventizer through a line 42, and is discharged through a line 44. The solid product is removed from the desolventizer as final product, as indicated by the line 45.

The still 38 permits purification, recovery, and reuse of the organic liquid within the system, which for economic reasons is a closed system. A reflux condenser is disposed above the still, and includes a vapor line 46, a condenser 48, and a recycle return line 49. The condenser is also connected to a manifold 50, that communicates with the methanol supply lines 14, 25 and 30 respectively, to supply fresh methanol to the several points in the process where it is used. A reboiler 52 is disposed beneath the still, and water and extracted materials are discharged from the reboiler through a line 54.

Example: This example illustrates the process conditions for the treatment of cornstarch with a water-methanol liquid system. Regular, unmodified cornstarch was slurried with several different mixtures of water and methanol, after which the slurries were rapidly heated for 1 to 2 minutes, during passage through the tubular heat exchanger. Upon discharge from the tubular reactor, additional methanol was added to the slurry, in an amount to bring the water content of the slurry to about 15%. The slurry was then permitted to cool in the dilution tank, after which it was filtered. The treated starch was washed with methanol and dried. The treatment conditions for the several runs are reported in the following table. In each case, sufficient pressure was maintained on the slurry, within the tubular reactor, to maintain the liquid vehicle in the liquid phase.

| | Composition of Slurry (%) | | | | Loss of | Brookfield Viscosity at ** | |
No.	Starch*/ H_2O/CH_3OH	Liquid/ Starch*	Liquid Phase, H_2O/CH_3OH	Temp. (°F.)	Birefrin- gence, %	25°C.	60°C.
1	16.7/41.65/41.65	83.3/16.7	50/50	230	100	9.2	8.7
2	16.7/41.65/41.65	83.3/16.7	50/50	280	100	7.5	6.9
3	20/40/40	80/20	50/50	225	100	10.3	9.5
4	20/40/40	80/20	50/50	245	100	9.4	8.7
5	16.7/20.8/62.5	83.3/16.7	25/75	280	100	8.5	7.6
6	25/18.75/56.25	75/25	25/75	285	100	10.5	9.4
7	20/30/50	80/20	37.5/62.5	260	100	8.85	8.4
8	20/30/50	80/20	37.5/62.5	280	100	8.3	7.9
9	20/30/50	80/20	37.5/62.5	285	100	7.7	7.4
10	16.7/20.8/62.5	83.3/16.7	25/75	265	50–60	25	
11	20/24/56	80/20	30/70	340	100	6.9	

*Starch reported on dry substance basis.
**Concentration of processed starch in water required to produce a paste of 50-poise viscosity.

Solutions of the processed starches, in cold water, were then prepared; in all cases the starches hydrated rapidly and completely in the water to form smooth, viscous pastes. These pastes were examined under a polarizing microscope; it was observed that all of the granules were intact, and were uniformly swollen. The loss of birefringence was determined by means of the microscopic examination; this is reported in the above table.

Brookfield viscosities were determined in the following manner. A weighed amount of processed starch was added to a weighed amount of water and stirred for 2 minutes. The paste was allowed to stand for 2 hours, after which an LVF Brookfield viscometer was used to measure the viscosity of the paste. The concentration of processed starch in water (weight percent) which yielded a 50-poise viscosity was recorded; these values are reported in the above table.

As can be seen from the data in the above table, the principal variables to be considered for a particular starch-solvent-organic liquid system are (1) the relative proportions of starch solvent and organic liquid in the liquid phase and (2) the temperature of treatment. These variables are interdependent, the lower the ratio of starch solvent to organic liquid, the higher the temperature required to produce a specific product.

This can be illustrated by comparing samples 5, 7, and 10. At a ratio of water to methanol of 25/75 (samples 5 and 10), a temperature of 265°F resulted in only 50 to 60% loss of birefringence, while a temperature of 280°F resulted in 100% loss. Run No. 7, with a water/methanol ratio of 37.5/62.5, resulted in complete loss of birefringence using a temperature of 260°F. It will be noted, however, that the temperature and composition of the liquid phase can be adjusted within relatively wide ranges, without significant differences in the finished product. Because of this, the process is very flexible, easy to control, and therefore readily adaptable to large-scale operations.

Fragmented Gelatinized Starch

Using the same apparatus and solvent mixture of U.S. Patent 3,617,383, *F.J. Germino, F.E. Kite and E.H. Christensen; U.S. Patent 3,620,842; November 16, 1971* have developed the process for producing a gelatinized starch with completely disrupted granules. This process gives a texturized starch which when dispersed in water yields a pulpy paste having excellent mouthing properties as exhibited by the complete absence of grainy, gritty or sandy characteristics.

Briefly, the process may be carried out as follows. Ungelatinized starch is mixed with a liquid medium in a proportion such that the starch is present in an amount less than 35% by weight. The slurry is preferably placed in a confined zone and heated to a temperature between 240° and 380°F. The preferred temperature is 260° to 310°F and the most preferred 280° to 300°F. The retention time at the elevated temperature is interdependent with and inversely proportional to the temperature at which the reaction is being carried out.

The time may be as short as one minute and as long as 60 minutes, or under low-temperature reaction conditions the reaction may go an unlimited amount of time without deleterious effects upon the final product. The basic premise for establishing the optimum reaction time is to continue the reaction until substantially all granules are completely disrupted or fragmentized thereby exhibiting substantially 100% gelatinization. It is also desirable to maintain the reaction under sufficient pressure to prevent vaporization of the liquid medium. In this manner, reaction time may be kept to a minimum and substantially complete recovery of the liquid medium constituents is easily achieved.

Example 1: This example illustrates the process conditions for the treatment of cornstarch with a water-methanol liquid system. Crossbonded cornstarch having a 15-gram Scott viscosity was slurried with water and methanol at different solids contents and in combination with different water-methanol liquid systems. The slurries were heated for times in the range from 1 to 3.5 minutes. In each case, after the treatment, additional methanol was added to the slurry in an amount of 10 parts of methanol for every 8 parts of slurry.

The addition of methanol reduced the water content of the slurry to about 15%. The slurry was permitted to cool, after which it was wet milled and filtered. The starch was then washed with methanol and dried. All runs in this example were performed in the continuous tubular reactor previously described. The product was slurried in water in an amount of about 5% and the texture and thickness visually observed. The observations are reported in the table below.

As can be seen from the table, the principal variables for a particular starch-solvent-organic liquid system are (1) the relative proportions of solvent and organic liquid in the liquid phase, and (2) the temperature of treatment. These variables are interdependent; the lower the ratio of solvent to organic liquid, the higher the temperature required to produce a specific product. It will be noted, however, that the temperature and composition of the liquid phase can be adjusted within relatively wide ranges without significant differences in the finished product. Because of this, the process is very flexible, easy to control, and therefore readily adaptable to large scale commercial operations.

| Sample No. | Percent | | | Reaction | | Rating |
	Starch	Methanol	Water	Temp, °F	Time, min	
1	25	41.25	33.75	230	1.5	F
2	25	41.25	33.75	250	1.5	D
3	25	41.25	33.75	270	1.5	B
4	25	41.25	33.75	290	1.5	A
5	25	41.25	33.75	310	1.5	C
6	25	41.25	33.75	330	1.5	E
7	25	41.25	33.75	350	1.5	E
8	25	41.25	33.75	370	1.5	E
9	30	31.50	38.50	240	1.5	D
10	30	31.50	38.50	260	1.5	C
11	30	31.50	38.50	280	1.5	A
12	30	31.50	38.50	300	1.5	A
13	30	31.50	38.50	290	3.5	B
14	30	31.50	38.50	290	2.5	B
15	30	31.50	38.50	290	1.5	A
16	30	31.50	38.50	290	1	B
17	30	31.50	38.50	290	30	E
18	30	31.50	38.50	290	15	D
19	10	49.5	40.5	250	1.5	C
20	20	44.0	36.0	270	1.5	A

The samples were all rated according to the following scale:

A   Excellent. Pulpy texture, high stability, resistant to shear, product uniform.

B   Very good. Pulpy texture, high stability, shear resistant, product uniformity good.

C   Good. Pulpy texture, satisfactory stability, satisfactory shear resistance, product uniformity acceptable.

D   Fair. Pulpy texture, some shear resistance.

E   Pulpy texture, no viscosity.

F   Product unsatisfactory.

Example 2: This example illustrates use of other solvents and organic liquids in accordance with the process. The procedure and rating scale of Example 1 were used.

| Sample No.: | | Percent | | Reaction | | |
	Starch	Solvent	Organic liquid	Temperature, °F.	Time, minutes	Rating
21	25	33.75 water	41.25 ethanol	280	1.5	B
22	25	do	41.25 acetone	270	1.5	B
23	25	37.50 DMSO	37.50 methanol	260	2–3	B–C
24	25	37.50 2-amino ethanol	do	260	2–3	B–C

Granular and Texturized Starch by Direct Heating

The essence of the process by *F.J. Germino, V.D. Harms and E.H. Christensen; U.S. Patent 3,607,396; September 21, 1971; assigned to CPC International Inc.* is the discovery that, by subjecting a slurry of starch material to direct (as opposed to indirect) heating as previously described in U.S. Patents 3,617,383 and 3,620,842, the process can be so controlled as to permit the preparation of products of desired characteristics, while process advantages unattainable with indirect heating are realized.

One important process advantage afforded by direct heating is the relatively high, i.e., in excess of 25%, starch solids levels that can be used. In addition, direct heating provides a substantially instantaneous process and significantly reduces equipment requirements inherent in indirect heating processes.

In this process, the slurry is passed through a direct heater, preferably a steam injection heater, in which the slurry is subjected to gelatinizing conditions. By the term gelatinizing conditions is meant the type of conditions which would be necessary to gelatinize the starch material in the particular starch solvent being used. For example, if water is the solvent, steam is introduced directly into the slurry to achieve the desired degree of birefringence loss.

After the direct heating, the slurry is rapidly cooled and the granular starch material is recovered from the liquid phase of the cooled slurry, as by filtration, and dried. With most liquid systems it is desirable to add excess organic liquid to the slurry, thereby reducing the overall solvent content, prior to recovering the starch from the slurry. The low proportionate water content effects a de-swelling of the granules, by releasing a part of the imbibed water. This step is particularly important when water is used as the solvent for the following reasons: (1) it greatly facilitates filtration of the starch material from the slurry, and (2) unless the water content is reduced, if the processed starch material is dried with the application of heat, a horny product results.

The continuous preparation of granular preswollen starch, in accordance with this process, involves subjecting the slurry to gelatinizing conditions while minimizing granule fragmentation, so that the product exhibits little or no granule fragmentation.

Example 1: The starch material employed was a thick-boiling cornstarch, having a moisture content of about 12 weight percent. The slurry contained 35 weight percent starch solids (dry basis). The balance (i.e., the other 65 weight percent) of the slurry was a liquid medium comprising 20 weight percent water and 80 weight percent methanol. A stream of this slurry was continuously supplied to the heater, at a flow rate of about 2.5 gallons per minute. Steam was introduced into the heater to maintain the heater at a temperature of about 265°F.

Conventional steam injection heaters, known in art, and sometimes referred to as continuous starch cookers or pasters, are suitable for use in the process. The heater used in this example comprised a block provided with a plenum chamber which had one inlet for the slurry stream, one inlet for the steam, and one outlet for product discharge. The slurry stream and the steam are introduced tangentially in the same plane at relatively high velocities into the plenum chamber where intimate contact and uniform heating results.

The heated slurry was immediately discharged from the heater through a suitable orifice into agitated dilution methanol. About ten parts of dilution methanol per seven parts of heated slurry was used. Thus, the direct heating is a substantially instantaneous process. The product was recovered by filtration and drying. The product recovered had highly desirable properties. The product comprised swollen, substantially intact starch granules, and produced a paste of approximately 50-poise Brookfield viscosity, at a starch concentration of 7.5%

when reconstituted in water at 25°C (2 hours age). The paste exhibited excellent shear stability having no loss in viscosity after shearing at a constant force at 1,800 rpm for 20 minutes. This example also indicates that relatively high (e.g., about 35%) starch solids levels can be processed substantially instantaneously, using relatively simple equipment. In contrast, when indirect heating is employed, although satisfactory products can be produced, operating difficulties are frequently encountered.

For example, when processing starch solids levels in excess of about 25%, using an indirect continuous heat exchanger, e.g., a Votator, plugging is frequently a problem. In addition, this latter process requires relatively higher processing temperatures and more elaborate equipment.

Example 2: A texturized starch was obtained by passing a starch material-methanol-water slurry, of a composition described below through a direct steam injection heater of the type used in Example 1, under the conditions indicated below.

The starch material employed in this example was a cross-bonded cornstarch. The slurry contained 35 weight percent starch solids. The balance of the slurry was a liquid medium comprising 30 weight percent water and 70 weight percent methanol. A stream of this slurry was continuously supplied to the heater, at a flow rate of about 2.5 gallons per minute. Steam was introduced into the heater to maintain the heater at a temperature of about 275°F. The heated slurry was discharged into dilution methanol and the product recovered.

The product recovered had highly desirable properties and the process afforded the same advantages illustrated in Example 1. When the product was added to tomato juice at a 5 weight percent level, and then heated under retort conditions of 240°F for 10 minutes, a thick, pulpy sauce resembling a natural paste was obtained.

Gelling Low Fat Starches

Starch gels which approximate gels from gelatin have prepared by the process disclosed by *E.R. Jensen, J.E. Long and L.D. Williams; U.S. Patent 3,666,557; May 30, 1972; assigned to CPC International Inc.* These starch gels are prepared by subjecting an aqueous slurry of an amylose-containing starch which has a bound fat content of less than 0.3% to a relatively high temperature (between 158°F and 220°F) and a high degree of shear. The resulting starch gels can be used as extenders or complete replacements for gelatin in many applications in which gelatin is customarily employed, e.g., in food products, in the manufacture of hard capsules, and as encapsulating agents in the microencapsulation of sensitive materials.

As a starting material, any amylose-containing starch may be used, including the high amylose starches, or the pure amylose fraction obtained by fractionating an amylose-containing starch. Amylopectin, or the waxy starches, do not form gels upon pasting, and therefore are not suitable for the process. It is essential that the starch be defatted, i.e., that the bound fat content be less than about 0.3%. The temperature range used in the process will depend upon the type of starch used; it is essential that the temperature be sufficiently high to completely gelatinize the starch; on the other hand, it must be sufficiently low so as to avoid degradation of the starch molecules themselves. In the case of cornstarch, the temperature can range from 158° to 220°F, the preferred range being between 175° and 195°F. The cooked paste, either during or preferably after the cooking, must be subjected to a high degree of shear, on the order of that obtained using a milk homogenizer, which forces the liquid through a spring loaded orifice. A colloid mill would also be suitable, but not as effective as a homogenizer.

Example: A series of 12 slurries were prepared in the following manner. Six slurries (1 through 6) used an acid modified thin boiling (67 fluidity) cornstarch, which had not been defatted. The slurries contained 8.5% solids, by weight. The remaining slurries (7 through 12) used an acid modified thin boiling (67 fluidity) cornstarch defatted to a bound fat content of between 0.1% and 0.15%. These slurries contained 6.5% solids, by weight.

Each slurry was placed in a steam-jacketed kettle and cooked at a specified temperature for a period of 25 minutes. Samples were removed at intervals of 5 minutes, 15 minutes and 25 minutes after the slurries had reached cooking temperature. After cooking, the samples taken from slurries 4, 5, 6, 10, 11 and 12 were homogenized in a Creamery Package Homogenizer at 2,000 psi at the first stage and 500 psi at the second stage.

Each sample was collected in a beaker, permitted to stand for 20 to 24 hours at 40°F, after which gel strengths were determined using a Bloom Gelometer. Conditions of treatment, and the Bloom gel strengths in grams are set forth in the table below.

Sample Number	Defatted	Cook temp., °F.	Time (min.)	Homogenized	Bloom gel strength (gms.)
1	No	158	5	No	0
			15		0
			25		20
2	No	176	5	No	96
			15		100
			25		108
3	No	194	5	No	112
			15		120
			25		118
4	No	158	5	Yes	60
			15		72
			25		67
5	No	176	5	Yes	88
			15		98
			25		101
6	No	194	5	Yes	88
			15		103
			25		89
7	Yes	158	5	No	52
			15		60
			25		75
8	Yes	176	5	No	60
			15		69
			25		94
9	Yes	194	5	No	72
			15		82
			25		110
10	Yes	158	5	Yes	108
			15		128
			25		123
11	Yes	176	5	Yes	152
			15		167
			25		162
12	Yes	194	5	Yes	139
			15		152
			25		158

As can be seen from the above table, Samples 10, 11 and 12 prepared by this process developed substantially greater gel strengths than did the remaining samples, which were not prepared in accordance with this process. In the case of the nondefatted starch, homogenization of the pastes cooked at the higher temperature (176° F and 194°F) actually resulted in decreased gel strength. In addition to measuring the gel strength, the samples were visually examined and tasted. The physical appearance and mouthing characteristics of samples 10, 11 and 12 were very similar to those of gelatin based gels, the principal difference being that the gels were hazy, rather than clear and transparent as is the case of gelatin based gels.

Gelatinization of Potato Starch

Potato starch when gelatinized in water produces pastes which are highly transparent, glossy, viscoelastic, and noncongealing or gelling on cooling. In higher concentration, or in food formulation, potato starch is apt to be very gummy or even slimy, and, in the untreated state, has not been particularly suitable for use in starch puddings, salad dressings, canned food thickeners, etc.

Attempts to overcome the pronounced gumminess or sliminess of potato starch pastes by crosslinking with chemical crosslinking agents, for example, phosphorus oxychloride, epichlorhydrin, acrolein, divinylsulfone, di-epoxy reagents, aldehydes, glycidaldehyde, thermosetting resins or their monomers, etc, have not been particularly effective.

E.T. Hjermstad; U.S. Patent 3,578,497; May 11, 1971; assigned to Penick & Ford, Limited has found that improved paste and gel properties and a higher swelling temperature are imparted to potato starch by heating an aqueous suspension of potato starch at temperatures which are at least initially below the natural gelatinization temperature of the starch, but always below the higher gelatinization temperature attained by the starch due to the heating.

According to the process, a concentrated, neutral suspension of potato starch, usually up to 40% dry substance, is first heated for a period of time at a temperature below the incipient swelling temperature of the particular batch of starch being treated. The temperature is then gradually raised until a temperature well above the original swelling temperature is attained. During subjection to these elevated temperatures the potato starch undergoes progressive changes in properties, including raising of its swelling temperature. By this means, swelling temperatures are raised as much as 20° to 30°F. The product after the steeping treatment has a high degree of granule stability.

It resists rapid gelatinization and produces a rising or fairly flat viscosity curve on cooling. The pastes of the product are very short textured, nongummy, nonslimy, cloudy and noncohesive. They form firm gels on cooling and ageing. The changes imparted are produced entirely without addition of chemical reagents and the products are, therefore, ideally suited for uses in food formulation which require short, noncohesive texture, paste viscosity stability to heat and shearing action, and in foods which require formation of a gel on cooling or ageing.

It has been discovered that potato starch can be given these desirable properties by either a long heating time at a relatively low temperature or by heating short periods of time at successively higher temperatures. Any suspension concentration can be used, providing the suspension is stirrable. Concentrations 20 to 40% dry substance are preferred. The pH should be in the neutral range, i.e., 5.5 to 8.0 especially if acid degradation or alkaline swelling or oxidation is to be avoided. A pH in the range of 6.0 to 7.5 is preferred. It is essential that swelling be avoided during the different heating periods; otherwise, higher temperatures will cause the suspensions to gelatinize to an unstirrable, sticky, nondewaterable state, especially in higher solids suspensions.

Example 1: This example shows the effect of prolonged heating of the starch suspensions at different temperature levels in raising the swelling temperature. Unmodified potato starch (commercial starch from Idaho potatoes) was suspended in water in 40% dry substance concentration and the pH adjusted to 6.5. Several such suspensions were heated and agitated gently at different temperature levels for at least 20 hours. Before and after the heating period the swelling temperature was determined by agitating a 40% dry substance suspension with a propeller, raising the temperature at around 1°F per minute and noting the temperature at which the vortex disappears and the suspension becomes semisolid and unstirrable.

Test	Temp. heated, ° F.	Time heated, hrs.	Swelling temp., ° F.
a	None	None	134
b	110	23	140
c	115	23	141
d	120	23	146
e	125	20	147
f	130	20	149
g	130	42	152

Example 2: This example shows the effect of increasing time of heating at a single temperature level. Commercial unmodified potato starch was suspended in water in a 40% dry substance concentration at pH 6.5 and agitated at 125°F for varying periods of time. Swelling temperatures were determined before heating and after each time period by the method in Example 1.

Hours Heated at 125°F	Swelling Temperature, °F
0	130
1	140
3.5	142
5	143
6	145
17	147

GELATINIZED STARCH BLENDS

The following processes are concerned with increasing the water solubility of pregelatinized starch for use in food products. This is done by blending nongelatinized starch with the second component and heating to form gelatinized starch products.

Fatty Acid Glycerides-Starch Blends

In a process described by *C. Decnop; U.S. Patent 3,443,990; May 13, 1969; assigned to Amylo Chemie NV, Netherlands* a cold-water-dispersible starch product is made by intimately mixing an aqueous suspension of a starch material with an effective amount up to 4% of an antiagglomeration and antidegradation additive consisting essentially of a finely divided mono- or diglyceride of a saturated higher fatty acid, (e.g., glycerol mono- or distearate), and then subjecting the mixture to a normal heat treatment to gelatinize and dry the starch. The starch material may have a special addition of carboxymethylcellulose or natural gums while the glyceride additive may include up to 3% by weight of triglycerides. The resulting end product shows no lump formation on addition with water and no undesired gel formation or retrogradation on aging and cooling of pastes made therefrom.

Example 1: A glyceride emulsion is prepared by dissolving 80 kg of a distilled mixture of glycerol mono- and distearates (containing about 90% of monostearate) in 700 liters of tap water and agitating the mixture as obtained for about 6 hours. After microscopic examination of the emulsion, the emulsion is diluted with water to give a volume of 1,100 liters. On the other hand, a starch suspension is prepared by introducing 900 kg of cornstarch (containing 12% moisture) into 1,250 liters of water and adjusting the concentration of the suspension to a value of 20° Baumé by addition of water.

Thereafter, 85 liters of the glyceride emulsion are added to the thoroughly agitated starch suspension and agitation is continued for one hour. The glyceride-starch suspension as obtained is then passed over steam-heated rolls of a roller drier in order to gelatinize the starch, to promote the reaction and to dry the suspension.

The dry product emerging from the rolls is ground to the desired particle size, viz, 200 mesh. There results an end product which has a neutral taste and does not manifest lump formation on addition of water. Gel formation and retrogradation on aging and cooling of pastes made from this product are practically absent.

Example 2: A glyceride emulsion is prepared by dissolving 80 kg of a distilled mixture of glycerol mono- and distearate (containing about 90% of monostearate) in 700 liters of tap water and stirring the mixture as obtained for about 6 hours. After microexamination of the emulsion, the whole is completed with water to give a volume of 1,100 liters. 320 kg of arachis oil are heated to about 50°C and added to the stirred glyceride emulsion. Stirring is continued for some time until the oil has been properly emulsified. Thereafter, the mixture is cooled to give a stable emulsion.

A starch suspension is prepared by introducing 900 kg of rice flour into 1,250 liters of water and adjusting the concentration of the suspension to a value of 20° Baumé by addition of water. Thereafter, 85 liters of the combined glycerides emulsion are added to the thoroughly stirred starch suspension and stirring is continued for one hour. The glyceride-starch suspension as obtained is then passed over steam-heated rolls of a roller drier in order to gelatinize the starch, to cause the reaction to take place and to dry the suspension. The dry product emerging from the rolls is ground to a particle size of 200 mesh. This results in the end product which has excellent qualities with regard to lump formation, gel formation and retrogradation.

Gelatinized Starch-Shortening Blends

S. Werbin, D. Weinstein and I. Rubenstein; U.S. Patent 3,582,350; June 1, 1971;

assigned to Maryland Cup Corporation have also prepared nonlumping pregelatinized starch products which are blends with a hydrophobic shortening material. The process for preparing these products comprises the steps of (a) forming an aqueous slurry of the flour or starch in water at ambient temperature; (b) mixing with the slurry a hydrophobic shortening material or a blend of a shortening material and an emulsifier; and (c) subjecting the mixture to a brief heat treatment at a temperature between 250° and 500°F.

The flour or starch product which results from this three-step treatment, when dried and comminuted to desired particle size, provides a dry, substantially completely dispersible, at least partially gelatinized flour or starch, which will absorb moisture without the application of heat. It forms a product which is capable of being used with other suitable ingredients, for the preparation of instant pudding mixes, pie fillings, ice cream mixes, fudge, chocolate and coffee flavorings, and the like.

The dispersible heat treated pregelatinized flours and starches of this process are also intermediate products suitable for further process, for example, as breading materials for meat, poultry and fish, or in baked goods in wide variety. In such applications, the products may be admixed with additional ingredients and the mixture subjected to further finishing by baking, frying and similar supplemental heat treatments. As a flour there may be used any conventional flour, for example, commercial wheat flour, and as a starch any conventionally manufactured starches, such as potato starch, tapioca, and the like.

The shortening material used may be, for example, lecithin, or a fatty-type material, such as a vegetable oil. Suitable emulsifiers include those commonly used in food processing, such as glycerol monostearate, Tween 60 (polysorbate 60), sorbitan diglycerides, and the like. The shortening material is used in an amount ranging from 4% to 8% by weight of the flour or starch, and the emulsifier in an amount ranging from 0.2% to 1%, but these proportions are not critical. However, far less shortening is required for this process than for previously known processes.

Example 1: An aqueous slurry was prepared using 200 pounds of commercial wheat flour and 100 pounds of water at 60°F. To this slurry there was added a mixture of 12 pounds of a hydrogenated shortening and 6 ounces of lecithin, and the mass was agitated until homogeneous. The slurry was pumped onto a drum dryer and heat treated at 350°F for 60 seconds. The product was substantially completely dispersible in water.

Example 2: The formulation of Example 1 was employed with further inclusion of 26 pounds of sugar, and 26 ounces of baking soda, and water sufficient to make 525 pounds. The heat treated product was suitable for baked goods having a foam structure. The baking soda is an optional ingredient, having as its primary purpose, however, to produce a more friable product having a porous cellular structure.

Example 3: An instant pudding can be prepared by adding to 100 grams of cold homogenized whole milk: 15.0 grams cane sugar, 2.2 grams dextrose, 0.2 grams salt, 0.2 grams anhydrous disodium phosphate, 0.3 grams anhydrous tetrasodium pyrophosphate and 6.0 grams finely ground pregelatinized wheat flour of Example 2.

Other examples in the complete patent included the use of these gelatinized starch blends in breading mixes, ice cream, sauces and pie fillings.

Starch-Carrageenan Blends

H. Kragen; U.S. Patent 3,524,767; August 18, 1970; assigned to Société de Produits Chimiques d'Auby, France describes a method of rendering starches soluble in the cold state, in accordance with which the starch, normally insoluble when cold, is mixed while hot in a solution comprising at least one colloidal substance. The method is mainly characterized in that the heating temperature is less than 100°C and in that the colloidal substance is advantageously anionic and of high molecular weight.

The colloidal substance is preferably selected from the family of polysaccharide esters, preferably sulfatic, and more particularly from the group of salified acid esters of polysaccharides, especially of natural carrageenans (associated or not with other colloids), the colloidal substance corresponding chemically to the general formula:

$$\left[SO_4 \underset{M}{\overset{C_6H_9O_4}{<}} \right]_n$$

in which n is a polymerization index and M is, as is well-known, a cation formed by one or more alkali, alkaline earth or magnetic metals. As natural colloidal substances possessing ester-sulfate groups, there may be cited various red and brown seaweeds, together with their extracts, such as carrageenans, furcellarans, agar-agar, fucoidin, etc.

Example 1: To 100 liters of a 1% aqueous solution of lambda-carrageenan there is added while stirring, 1 to 5 kg of cassava starch. In order to obtain a more rapid dispersion, it is possible to wet the starch previously with cold water. The mixture is then heated up to optimum solubilization of the starch, that is to say to about 95°C for 15 minutes. The starch is then coagulated in isopropanol. The coagulum is separated from the alcohol by sieving, rewashed in pure alcohol in order to obtain more complete dehydration, centrifuged and then dried under vacuum. After grinding, the finished product is presented in the form of pale beige powder which is rapidly soluble by stirring in cold water.

Example 2: To 100 liters of an aqueous solution of carrageenans having a concentration of between 1 and 3%, there are added, under the same conditions as in Example 1, quantities of maize starch ranging from 1 to 9 kg. The mixture is heated up to solubilization and is then heated to 95°C for one hour, after which it is coagulated in a polar organic solvent or is dried by spraying.

Example 3: 100 liters of an aqueous solution of potato starch flour having a concentration comprised between 1 and 4% are heated to 95°C and mixed with 100 liters of a 1% aqueous solution of carrageenan, the temperature of which has been previously brought up to 60°C. After homogenization and without intermediate heating, the mixture can be immediately coagulated. It is also possible to employ the lyophilization technique.

When used in foods, those complexes made from natural colloids have the advantage of containing no chemical additives. The complexes also have improved organoleptic properties, lack of odor, and improved texture.

STARCH-PREGELATINIZED STARCH MIXTURES IN FOODS

Nonfriable Extruded Products

A process of producing nonfriable starch forms has been developed by *C.H. Pelton; U.S. Patent 3,528,853; September 15, 1970; assigned to Standard Brands Incorporated*. These nonfriable forms after drying are significantly less dusty than dried common starch produced by prior starch drying methods. This is accomplished by providing an intimate mixture of wet common starch and a small amount of gelatinized starch, extruding the mixture and drying the extruded product. The amount of gelatinized starch present in the mixture is effective to provide a nonfriable extruded product. The extruded product when dried will be less dusty than dried common starch produced by prior art methods.

As used here, common starch is defined as starch which is non-heat-treated, ungelatinized, unmodified or otherwise altered by heating or chemical treatment. The common and gelatinized starches used in this process include, for instance, corn, milo, sorghum, wheat, potato, tapioca, yucca and the like.

The gelatinized starch may be provided in the common starch by separately producing a gelatinized starch and mixing it with common starch to provide the necessary mixture or it may be provided by subjecting common starch to controlled gelatinization conditions so that the common starch is gelatinized to only a slight degree.

When the gelatinized starch is incorporated into partially dewatered common starch, i.e., starch after filtration, it is preferred that the gelatinized starch be of a particle size such that substantially all of it will pass through a 200 mesh U.S. standard screen. The amount of moisture present in the partially dewatered common starch is not critical but starch having a moisture content between 34 and 44% may be used, since these are the approximate moisture levels that are normally obtained when starch is partially dewatered on vacuum filters on in centrifuges commonly used in starch processing.

Preferably, the amount of gelatinized starch needed to provide sufficient extrudability to the mixture is between a trace and 5%, based on the dry substance weight of both the gelatinized starch and common starch present in the mixture. Typically, the amount of gelatinized starch present is from 0.25 to 1.6% on the same weight basis.

Example: A common cornstarch slurry (21.4 Bé) at 110°F was vacuum filtered on a Büchner filter, yielding a starch cake containing 42.3% moisture. Portions of this starch cake were mixed with sufficient gelatinized cornstarch, having a moisture content of 9.1%, to achieve in one portion 0.54% and in another 1.62% gelatinized starch by weight. The particle size of the gelatinized starch was such that 0% remained on 40 mesh screen, 16% on 100 mesh screen, 20% on 140 mesh screen, 24% on 200 mesh screen, 20% on 325 screen and 20% passed through a 325 mesh screen.

These starch samples were extruded using a screw type extruder through flat plate dies having 3/16 inch and 3/8 inch diameter holes respectively. A control sample was prepared by extruding a portion of the starch cake containing no gelatinized starch through a die having 3/16 inch holes. Another control sample was prepared by air drying a portion of the starch cake containing no gelatinized starch to a moisture level of 13%.

The samples containing the gelatinized starch extruded without difficulty. However, the sample containing no gelatinized starch was difficult to extrude, and localized heating of the sample occurred due to the excessive pressures necessary to extrude the same. This sample was considered unacceptable. The extruded samples containing gelatinized starch were air dried to moisture levels shown below. These samples and the air dried starch cakes were evaluated by the following procedures.

One hundred grams of the air dried samples were gently hand screened on a 10 mesh screen. The weight of material which passed through the screen was considered as starch fines and was taken as an indication of the dustiness of the samples. Fifty-gram portions of the samples which did not pass through the 10 mesh screen above were placed into a one-pint container of a Patterson-Kelly twin shell blender, Model No. LB-2180 with six one-inch glass balls and rotated for 15 minutes at a speed of 23.5 rpm. At the end of the rotation, the samples were gently hand screened on a 10 mesh screen and the weight of the material which passed through the screen was taken as an indication of the friability of the samples and was measured as percent attrition. The results of these tests are shown in the following table.

Sample	Die opening	Percent moisture	Added gelatinized starch (Percent dry basis)	Percent fines	Friability (percent attrition)
Common starch cake....	Nonextruded ...	13	0	46	97
Common starch	3/16″ diameter ..	12.5	0	Unacceptable product (no test)	
Common starch and gelatinized starch.	3/16″ diameter ..	12.7	0.54	4	3
Do. ...	3/8″ diameter....	12.9	0.54	8	20
Do. ...	3/16″ diameter ..	13.1	1.62	3	2
Do. ...	3/8″ diameter....	13.3	1.62	3	5

Protein-Free Starch Pastes

S. Salza; U.S. Patent 3,836,680; September 17, 1974 discloses the production of alimentary pastes which are substantially free of glutens and therefore of proteins, and which are especially suitable for the nourishment of persons suffering from chronic uraemia and other dysfunctions. The problem connected with the manufacture of alimentary pastes in general and of protein-free pastes in particular, is that of their stability to cooking, which means that, when cooked, the paste does not turn sticky and does not agglomerate into a gluey mass.

A substitution of gluten by cellulose derivatives and/or natural polysaccharide gums has not solved all the problems; further, the presence of these rather costly substances in the food may in some cases be unadvisable, owing to their effects upon some physiological functions.

This process gives a starch composition capable of being formed, as by extrusion, into any possible shape of alimentary paste or pasta, such as spaghetti, macaroni, rigatoni, etc. This composition is mainly formed of a mixture of ungelatinized starch with gelatinized starch, the latter acting as a binder for the former. To promote the mutual dispersion of these two constituents, small amounts of monoglycerides of the alimentary fatty acids of C_{14} to C_{18} are added as emulsifiers, together with suitable dyes to confer a desired color to the resulting alimentary paste. The procedure can be carried out either in two stages or in a single stage, the resulting products being identical in their properties.

In the two-stage procedure, merely one part of the starch is premixed and gelatinized in the presence of the necessary amount of water and emulsifiers, by heating it, under continuous mixing, to the temperature where its gelatinization starts, which, according to the type of starch used, is in the range of 60° to 75°C. Subsequently, the remaining starch is added and the resulting mass is cooked until the mixture is obtained, which is immediately extruded to the desired shapes, the extrusion causing the quick cooling of the mass and consequently the rapid interruption of the cooking process.

In the single stage procedure, the total amount of starch is kneaded, in the presence of the monoglycerides as emulsifiers and the necessary amount of water. This first portion of the stage is performed at a temperature which does not reach the temperature where the gelatinization of the starch begins until the mixture has been kneaded to a completely homogeneous mass. From this point on the mass is cooked at a high temperature, up to a point where only part of the starch is converted to starch paste, and then it is passed through the extruder and cooled.

The proportion of monoglycerides ranges from 0.8 to 1.2 (preferably 1.0) parts by weight for 100 parts by weight of the total starch used. The proportion of water depends on the consistency which is to be conferred to the mass during its extrusion, in view of the type of alimentary paste which has to be produced. Also the evaporation of the water during the process must be taken into account, unless the whole procedure is carried out in a closed apparatus, such as a closed combination mixer-extruder unit. For conventional pastes, 52 to 64, preferably 57 to 59 parts by weight of water are used for 100 parts by weight of the total starch. Any type of starch may be used for this process, such as from wheat, potatoes, rice, corn, etc. However, a mixture of various starches is preferred for best results.

Independently of whether the procedure is performed in a single or in two stages, the cooking proper is regulated so that the mixture reaches a temperature of 80° to 90°C within 12 minutes when using an open apparatus, that is when cooking under atmospheric pressure, and a temperature 110° to 120°C, preferably between 110° and 115°C within 6 to 10 minutes, preferably within 8 minutes, when using a closed apparatus, that is under a higher pressure. This cooking stage is of a fundamental importance for the process, because tests have shown that the chemical and physical interactions which confer to the final product its stability to cooking develop during this stage.

The following examples may serve for a better understanding of the process. 12 kg of potato starch, 30 kg of cornstarch and 18 kg of rice starch are kneaded in a closed mixer with 35 liters of water in the presence of 0.600 kg of fatty acid monoglycerides of C_{14} to C_{16} as emulsifiers and a suitable, officially permitted dye. The mixture is heated so that its temperature reaches 65°C when it has already been kneaded into a homogeneous mass. Thereafter begins the cooking proper during which the temperature of the mass is raised linearly, under continuous mixing, within 8 minutes to 110°C, at the end of which period the heat supply is interrupted, the mass passed through the extruder to form spaghetti and conveyed into a drying chamber. Within approximately 15 seconds from its issue from the extruder, the temperature of the spaghetti has dropped below 65°C and when removed from the conveyor, it is already almost dry and its temperature has sunk to 35° to 40°C.

The same process may be carried out in two stages. In the first stage, part of the 60 kg of the above mixture of starches is kneaded, together with 35 liters of water and, in the presence of 0.600 kg of the monoglycerides in a closed mixer-extruder, while the temperature of the mixer is raised so that 65°C is reached when the mixture has become completely homogeneous. Subsequently, the remaining starch is added and the temperature of the mixture is raised, under continuous kneading, to reach 110° to 115°C at the end of 8 minutes. The remaining operations as well as the results are identical in both examples.

Mixed Starches for Instant Puddings

F. Germino and R.S. Golik; U.S. Patent 3,583,874; June 8, 1971; assigned to CPC International Inc. have developed a starch composition for use as a base for instant puddings which will produce a finished pudding having the texture, appearance, gel structure and flavor of a cooked starch pudding. This starch composition comprises a blend of two starches, designated as (1) a pregelatinized defatted starch having a fluidity of from 5 to 16, and (2) a granular preswollen starch having a fluidity of from 15 to 25 and preferably about 20.

By defatted starch is meant an amylose-containing starch, the bound fat content of which has been reduced to less than 0.3% and preferably not more than 0.15% (percentages by weight, based on the total weight of the starch). Any amylose-containing starch can be employed, although the amylose-containing cereal starches such as cornstarch, are preferred. The particular method employed for defatting the starch is immaterial, the only necessary qualification being that the resulting starch have a bound fat content of less than 0.3%, and preferably not more than 0.15%.

In addition to the reduction of the fat content, the starch should be pregelatinized, i.e., rendered cold-water-dispersible, by first pasting in water and then instantaneously removing the water from the starch paste. Any drying apparatus capable of instantaneously drying a starch paste, such as a roll-dryer, spray-dryer, foam mat-dryer or belt-dryer, can be employed. Another necessary requirement of the defatted starch component is that it have a fluidity of from 5 to 16, and preferably about 8. The fluidity can be adjusted by any known means, as by treatment with acid or the like.

By granular preswollen starch is meant a starch the granules of which are (1) substantially completely intact and (2) substantially nonbirefringent. A starch having these structural characteristics will have the property of rapidly dispersing in cold water to form a paste, which paste will increase in viscosity when subjected to shear. A suitable method for the preparation of a granular preswollen starch is as follows. Granular (i.e., nongelatinized) starch is slurried with (1) a solvent for starch, i.e., a liquid in which starch will gelatinize, and (2) an organic liquid which is not a solvent for starch, and which is miscible with the solvent for starch. The resultant slurry is then placed in a confined zone, wherein it is subjected to gelatinizing conditions. The process conditions are so selected as to produce the desired end product.

For reasons of economy and availability, water is the preferred solvent for starch, although other known starch solvents such as dimethyl sulfoxide may be used. As the organic liquid low molecular weight alcohols, ketones, dioxane, or the like are preferred. Methanol is particularly suitable because of its low cost and because it can be readily removed from the starch after treatment.

The granular preswollen starch component should have a fluidity of 15 to 25 and preferably about 20. As in the case of the defatted starch component, the fluidity adjustment can be performed by any known method, as by treatment with acid, enzymes, or the like. The granular preswollen starch should be present in an amount of 20 to 90%, by weight, based on the total weight of the starch blend. The defatted starch should be present in an amount of from 80 to 10%. The preferred portions are 60% granular preswollen starch and 40% defatted starch.

Example: Blends of defatted corn starches (all having a bound fat content of about 0.1%) and granular preswollen cornstarches, of various fluidities, were prepared. The granular preswollen starch was prepared by treating cornstarch with water and methanol (the slurry comprised 20% starch, 30% water and 50% methanol) at 280°F. The resultant starches contained about 0.1% bound fat. Pudding bases were then prepared using the following formulation:

		Grams
Starch		12.0
Sucrose		20.6
Vanilla		0.9
Color		0.5
Salt		0.5
	Total	34.5

The pudding bases were poured into 118 milliliters of milk (at 4°C) and the mixture was stirred vigorously for one minute and then allowed to set at room temperature for 15 minutes. Also, an instant pudding and a cooked starch pudding were prepared with commercially available products according to package directions. In addition, several puddings were prepared using gelling and thickening agents other than blends of the two starch components of this process. In the following table, where the various compositions are tabulated, these puddings and the two commercial products are listed under the heading "A. Control Samples." The various granular preswollen starch-defatted starch blends are listed under the heading "B. Experimental Samples."

The puddings were compared with the commercial products and rated according to appearance, texture, gel strength, and mouthing characteristics. These ratings appear in the following table under the subheading "Comments on Puddings."

A. Control Samples

Sample Number	Starch Components and Amounts				Comments on Puddings
	Component (a)	Component (a), grams	Component (b)	Component (b), grams	
1	Cooked pudding (commercial brand)	--	--	--	Firm gel structure, very smooth texture, excellent overall appearance.
2	Instant pudding (commercial brand)	--	--	--	Thick, no firm gel structure, grainy texture, good overall appearance.
3	Guar gum	4.0	--	--	Very thick, gummy texture.
4	Pregelatinized thick-boiling regular corn-starch	7.2	Pregelatinized defatted cornstarch, 8-F	4.8	Too runny, would not set.
5	Pregelatinized thick-boiling regular corn-starch	6.0	Pregelatinized defatted cornstarch, 8-F	6.0	Too runny, would not set.
6	Pregelatinized defatted thick-boiling corn-starch	12.0	--	--	Too runny, would not set.
7	Pregelatinized defatted 8-F cornstarch	12.0	--	--	Too runny, would not set.
8	Granular preswollen cornstarch, 20-F	12.0	--	--	Gummy texture.
9	Granular preswollen cornstarch, 20-F	11.0	--	--	Gummy texture.
10	Granular preswollen cornstarch, 20-F	10.0	--	--	Gummy texture.
11	Granular preswollen cornstarch, 20-F	3.6	Pregelatinized thick-boiling cornstarch	8.4	Too runny, would not set.
12	Granular preswollen cornstarch, 20-F	6.0	Pregelatinized modified white milo	6.0	Too runny, would not set.
13	Granular preswollen cornstarch, 20-F	6.0	Pregelatinized modified red milo	6.0	Too runny, would not set.

B. Experimental Samples

Sample Number	Fluidity of GPS*	Fluidity of DS**	GPS, g per DS, g	Wt % GPS per Wt % DS	Comments on Puddings
14	30-F	5-F	6.0/6.0	50/50	Too runny, would not set.
15	30-F	8-F	7.2/4.8	60/40	Too runny, would not set.
16	30-F	10-F	6.0/6.0	50/50	Too runny, would not set.
17	30-F	16-F	6.0/6.0	50/50	Too runny, would not set.
18	30-F	30-F	6.0/6.0	50/50	Too runny, would not set.
19	25-F	5-F	7.2/4.8	60/40	Fairly good gel strength, smooth texture.
20	25-F	8-F	7.2/4.8	60/40	Fairly good gel strength, smooth texture.
21	25-F	10-F	7.2/4.8	60/40	Fairly good gel strength, smooth texture.
22	25-F	16-F	7.2/4.8	60/40	Slightly runny, smooth texture.
23	25-F	30-F	7.2/4.8	60/40	Too runny, would not set.
24	20-F	5-F	7.2/4.8	60/40	Firm gel structure, very smooth texture, very good overall appearance.
25	20-F	8-F	4.8/7.2	40/60	Firm gel structure, very smooth texture, very good overall appearance.
26	20-F	8-F	6.0/6.0	50/50	Firm gel structure, very smooth texture, very good overall appearance.
27	20-F	8-F	7.2/4.8	60/40	Firm gel structure, very smooth texture, excellent overall appearance, practically indistinguishable from cooked pudding.
28	20-F	8-F	3.6/8.4	30/70	Fairly good gel strength, slightly runny, very smooth texture, good overall appearance.
29	20-F	8-F	2.4/9.6	20/80	Somewhat weak gel strength, fairly runny, very smooth texture, fair overall appearance.
30	20-F	8-F	1.2/10.8	10/90	Weak gel strength, runny.
31	20-F	10-F	7.2/4.8	60/40	Firm gel structure, very smooth texture, very good overall appearance.
32	20-F	16-F	7.2/4.8	60/40	Firm gel structure, very smooth texture, very good overall appearance.
33	16-F	5-F	6.0/6.0	50/50	Weak gel structure, smooth texture.
34	16-F	8-F	7.2/4.8	60/40	Weak gel structure, smooth texture.
35	16-F	10-F	6.0/6.0	50/50	Weak gel structure, smooth texture.
36	16-F	10-F	7.2/4.8	60/40	Weak gel structure, smooth texture.
37	16-F	16-F	6.0/6.0	50/50	Weak gel structure, smooth texture.
38	16-F	30-F	6.0/6.0	50/50	Too runny, would not set.
39	10-F	5-F	6.0/6.0	50/50	Too runny, would not set.
40	10-F	10-F	6.0/6.0	50/50	Too runny, would not set.
41	10-F	16-F	6.0/6.0	50/50	Too runny, would not set.
42	10-F	16-F	7.2/4.8	60/40	Too runny, would not set.
43	10-F	30-F	6.0/6.0	50/50	Too runny, would not set.
44	5-F	5-F	6.0/6.0	50/50	Too runny, would not set.
45	5-F	10-F	6.0/6.0	50/50	Too runny, would not set.
46	5-F	16-F	6.0/6.0	50/50	Too runny, would not set.
47	5-F	30-F	6.0/6.0	50/50	Too runny, would not set.
48	20-F	8-F	10.8/1.2	90/10	Slightly runny, smooth texture.

* Granular preswollen starch   ** Defatted starch

As can be seen from the above table, starch blends comprising 20 to 90% granular preswollen starch having a fluidity of 15 to 25 and 80% to 10% pregelatinized defatted starch having a fluidity of 5 to 16, (e.g., samples 19-22, 24-29, 31-37, and 48) produced very good instant puddings. These puddings were further characterized by excellent flavor and, because of their smooth texture, mouthing characteristics very similar to those of a cooked pudding.

In addition, these puddings exhibited no syneresis after standing 24 hours, as contrasted with the commercial instant pudding which showed decided water loss after this period of time. As will also be noted from the above table, Sample 27, which exemplifies the preferred blend of 60%, 20 fluidity granular preswollen starch and 40%, 8 fluidity pregelatinized defatted starch, produced a pudding which was practically identical with a cooked pudding.

USE OF PREGELATINIZED STARCHES

In addition to the following nine processes, pregelatinized starches are also used in the preparation of dehydrated alcohol products in U.S. Patent 3,786,159, page 233, used in combination with amylose in U.S. Patent 3,515,591, page 195 and pregelatinized amylose starch is used as a thickening agent in U.S. Patent 3,650,770, page 197.

With Surfactants for Whipped Products

Previously powdered, hydratable dessert mixes for use in whipped products such as fillings, custards, ice cream, sherbet, puddings, toppings and the like used powdered fat and/or protein materials. In order to enhance the whipping effect, it has been necessary to use protein materials in amounts ranging from 1 to 15% by weight of the total dry mix. Also, it has been necessary to use various gums such as carageenan, sodium alginate, gum tragacanth, gum acacia, gum karaya, locust bean and the like to provide the necessary stability and stiffness.

According to the process developed by *M.H. Katz; U.S. Patent 3,434,848; March 25, 1969; assigned to The Pillsbury Company*, an edible dry mix is prepared containing less than 8% by weight triglyceride fat and protein, and which upon hydration and mixing provides an aerated food product having a smooth, creamy consistency. The mix comprises a gelatinized starch and a surfactant composition, the surfactant being present in the mix in an amount of 2.0 to about 50 weight percent of the gelatinized starch dry weight, the surfactant comprising at least one glyceryl monoester of a fatty acid and at least one propylene glycol monoester of a fatty acid in a weight ratio of glyceryl monoester to propylene glycol monoester of 1:10 to 10:1.

This mix, upon hydration and aeration with an aqueous liquid such as water or milk, provide a fine, smooth and creamy product. Unlike previous hydrated and aerated dry mixes, the dry mixes of this process rely principally upon the gelatinized starch and surfactants to provide the ultimate whipped product. These dry mixes thus provide a whipped dessert product without the use of fats, oils and proteins previously deemed essential.

Suitable edible gelatinized starches that may be used include corn, high amylose corn, wheat, oat, potato, waxy maize, tapioca, sorghum, sago, rice, arrowroot starches, mixtures thereof and the like. Typical gelatinized starches are raw starches which have been modified via oxidation, acid hydrolysis or esterification and then cooked. Exceptional results being achieved with a gelatinized tapioca starch. A gelatinized starch having a particle size finer than #80 U.S. Standard mesh is generally employed since starches of such a particle size disperse more rapidly in an aqueous medium. The following examples will illustrate a typical product made by this process.

Example 1: Artifically sweetened lemon flavored frozen dessert mix was prepared by dry blending the following ingredients in a ribbon blender.

Ingredients	Pounds
Imitation lemon flavoring	2.5
Citric acid (powdered)	8.0
Salt	3.5
Gelatinized modified tapioca starch	117.5

(continued)

Ingredients	Pounds
Sodium carboxymethylcellulose	2.5
45:55 weight percent fused blend of glyceryl monostearate and propylene glycol mono-stearate	8.0
Dextrin	250.0
Sodium cyclamate	5.5
Sodium saccharin	0.5
Artificial coloring	25.0

To 59-g portions which had been sifted and dry blended there was added 74 ml of hot water (95°C). The resulting mixes were then blended with a standard household mixer at low speed for 2 min. An additional 74 ml of cold water (20°C) was added and the products whipped at high speed for 2 min. The resulting dessert product having 0.48 SG and short gel properties was then frozen at −20°C for 4 hr. The frozen product was very similar to ice cream in taste and texture.

Example 2: Cold water premixes were prepared by respectively dissolving in 485 pounds of water at 165°F, the following ingredients: sugar and gum arabic premix (sugar, 20 lb; gum arabic, 20 lb), gum arabic premix (gum arabic, 40 lb), and arabinogalactan premix (arabinogalactan, 40 lb). Then 10 lb of a 45:55 weight percent fused blend of glyceryl monostearate to propylene glycol monostearate was added to the above hot aqueous hydrocolloid containing solutions (165°F), allowed to melt, thoroughly mixed for 10 minutes and then spray dried.

The following mix ingredients were employed in preparing whipped toppings:

Ingredients	Ounces
Premix of Example 2	
Gum arabic	11.8
Blend of glyceryl monostearate* and pro-pylene glycol monostearate	2.9
Sugar	65.7
Gelatinized starch	11.6
Gelatin	5.2
Citric acid	1.7
Sodium citrate	0.8
Flavoring and coloring additives	0.3

> *A fused mixture of glyceryl monostearate and propylene glycol monostearate having a weight ratio of glyceryl monostearate to propylene glycol monostearate from about 35-50 of glyceryl monostearate to 50−65 propylene glycol monostearate.

A two-phase fruit flavored gelled dessert consisting of a cold set unwhipped portion (about 50%) and a fully aerated top phase was prepared in the following manner:

(1) Add ¾ cup of 95°C water to 3 ounces of the above mix;
(2) Blend for 1 minute at 350 rpm with a standard household mixer;

(continued)

(3)   Add ½ cup of cold water (5°C) and partially aerate with
a standard household mixer operated at 600 rpm for
2 minutes; and,

(4)   Refrigerate for 3 hours.

The whipped product consisted of an upper layer of whipped topping with a
nonaerated cold set lower layer. Density of the upper layer was about the same
as the fully aerated gelatin product. Both layers possessed a short gel character.

Dry Blend for Whipped Food Products

Pregelatinized starch has also been used by *L.A. Powell; U.S. Patent 3,843,805;
October 22, 1974; assigned to Wellman-Power Gas Incorporated* to prepare an
improved base composition useful in the preparation of whipped products. It
can also be used as a preservative in meat and milk products. The mixture that
is useful as a milk or meat preservative contains for each 100 parts by weight
of starch, 3 to 300 parts of sodium caseinate and 3 to 75 parts of calcium acetate. When used to prepare stable whipped products, the mixture contains
about 20 parts of sodium caseinate per 100 parts by weight of the starch.

It has been found that when an aqueous system containing a pregelatinized
starch, calcium acetate, sodium caseinate, and a suitable filler material such as
vegetable fat or skimmed or whole milk solids, is whipped under certain conditions to provide a whipped food product, the resulting product is stable against
bleeding and collapse at room temperature over long periods and can be repeatedly subjected to refrigeration and/or freezing and brought back to room temperature without adversely affecting the stability of the product. It has also
been discovered that the calcium acetate, sodium caseinate and the gelatinized
starch component cooperate in accentuating the flavor of foods and of flavoring substances added to the system and further provide the whipped products
with an illustory sense of richness.

The starch component must be hydrated or in a state of being readily hydratable
at the time the ingredients are subjected to the whipping procedure. In the dry
compositions discussed here, a pregelatinized starch is used. However, it is within
the scope of the process to start from raw starch, gelatinize it in water, and then
use the aqueous suspension of hydrated starch as a component of the aqueous
system subjected to the whipping step.

Example: As a specific example of a dry base composition that may be packaged and marketed with simple instructions for its use in preparing whipped
food products that are stable at room temperature and which may be repeatedly
subjected to alternating freezing and thawing conditions, the following dry ingredients, in the proportions indicated, were stirred together in a mixing bowl
to form a simple dry mixture.

	Grams
Dry synthetic milk solids (Richnin)	150
Cane sugar	68.1
Pregelatinized cornstarch	30.0
Calcium acetate (monohydrate)	2.0
Cerelose (corn syrup solids)	45.0

The dry mixture was then stirred into 474 grams of a 1:1 mixture of crushed ice and water, and subjected to the whipping action of a conventional household electric beater before any appreciable quantity of the ice had melted. As the whipping process transpired, a slight foam initially appeared and the volume of the mixture gradually increased as air became entrapped in the mixture until a maximum volume of 300% of the initial volume of the aqueous mixture was reached. The whipping process was continued until the whipped product thickened and set.

Samples of the whipped product were removed and exposed to room temperature conditions for a period of one month. At that time, it was ascertained that there was no collapse in the structure of the whipped product although there was a slight deterioration at the exterior surface of the product due to simple dehydration at the surface. The product was also edible at this point and showed no evidence of mold or other bacterial growth. Samples of the whipped product were also frozen in a conventional home freezer and repeatedly thawed and refrozen over a similar period without deterioration in the whip structure.

The complete patent includes numerous other examples of the use of such a base in the formulation of pudding, orange juice, cream soups, salad dressings, meat loaf, etc.

Baked Puffed Food Products

Puffed, snack type foods have been produced by *N.G. Marotta, G.A. Zwiercan and R.M. Boettger; U.S. Patent 3,652,294; March 28, 1972; assigned to National Starch and Chemical Corporation* using a pregelatinized starch in the formulation. As examples of the specialty food products which may be prepared by this process, one may list crackers, chips, cereal puffs, and other so-called snack items; the latter being available in a multiplicity of flavors and shapes. These products may also be ground and used as coatings for fried and/or baked food products.

In this process, the steps of precooking and drying the shaped product prior to final cooking can be eliminated. It is important that the pregelatinized starch used in the product must be characterized by its ability to swell in water to the extent that one gram of the starch, when mixed in water at a temperature of 25°C, will absorb at least about 10 grams of water. In preparing the product, the pregelatinized starch is moistened with 18 to 30% of water, by weight of the starch; then pressed into the desired shape without the need for heat or pressure (beyond the obvious pressure required to shape the material), and the shaped pieces are then oven-baked without any predrying of the shaped pieces prior to baking, and without any frying in fat or oil.

Since it is essential that the shaped pieces undergo a substantial increase in volume upon oven baking, and that the texture of the baked product be tender and crisp, rather than tough, there must be nothing mixed with the starch that will cause toughening or loss of volume. Specifically, the pregelatinized starch must not contain more than 5% by weight of vital gluten. The starches which must be used in this process comprise pregelatinized starches derived only from either root or root-like starch bases. As examples of such starch bases, tapioca starch may be viewed as a typical root starch, while waxy maize and waxy milo are typical root-like starches. Also included are the pregelatinized conversion products derived from any of the latter starch bases including, for example, dextrins, oxidized starches, and fluidity or thin boiling starches prepared by enzyme conversion or by mild acid hydrolysis. Additionally, any of the above starch base sources may also be inhibited, i.e. crosslinked, and the inhibited starches should have a granule swelling power value greater than 14.

Example: This example illustrates the preparation of a typical ready-to-eat food product by this process. Thus, 775 parts of a tapioca starch which had been pregelatinized by being subjected to a drum drying procedure for 10 to 12 seconds at a drum temperature of 300° to 350°F were thoroughly mixed with 225 parts of water. The resulting mix was pumped into an extruder which was maintained at room temperature and forced, by means of a three-fourths quarter inch conveying screw, into a shaping die. The solid, rope-like product which emerged from the die exhibited excellent shape-retention properties and could be sliced immediately after the shaping procedure.

The cut slices were then inserted, for 10 minutes, in a conventional oven maintained at 350°F. The resulting baked products exhibited a desirable puffed appearance and were found to be exceedingly crisp and light textured. When the above example was repeated, except that the cut slices were fried in deep fat, the resulting products were unsatisfactory because of the absence of uniform puffing. Such puffing as did occur, resulted in a large hollow bubble in the center of the fried mass, the remainder being quite dense and unpuffed. Other examples in the complete patent show the product of cheese, bacon, potato, peanut butter and chocolate flavored snacks.

Dry Starch-Fat Sponge

A method has been disclosed by *W.A. Mitchell and W.C. Seidel; U.S. Patent 3,769,038; October 30, 1973; assigned to General Foods Corporation* for preparing a fat-containing starch compound which will hold a very high proportion of fat in a dry powdered form, and which will readily release the fat when added to foodstuffs. This compound, a starch-fat sponge, has many advantages over the known fat-starch emulsions. For example, this product gives better results in preparing clear salad dressings than known starch-fat emulsions, since it introduces less starch into the salad dressing and consequently the salad oil is less cloudy.

This product has the further advantage that once the sponge is formed all or any portion of the fat can be expressed by mechanical means and replaced with a different type fat. Because of this property this fat sponge is also suited for introducing volatile flavor oils such as lemon oil, coffee oil, etc. into dry foodstuffs in the powdered form. A high fat-containing seasoning can also be prepared in the powdered form for use in basting meats and poultry. Since the starch-fat sponge contains up to 92% fat, very little starch would be added to the meat with the oil, thereby avoiding the formation of a slimy surface.

This starch-fat sponge can also be combined with seasonings such as acetic acid, powdered onions and garlic to produce a salad seasoning which can be sprinkled directly onto moistened vegetables.

The starch-fat sponge is preferably prepared by adding pregelatinized starch and fat to water and blending for a time sufficient to form an emulsion. The emulsion is then frozen and freeze-dried, yielding a product that is dry to the touch and which can be ground into a free-flowing powder. The emulsion can also be prepared by adding fat and ungelatinized starch or flour to water, cooking the mixture to gelatinize the starch and then intimately mixing the mixture to form an emulsion.

However, when using this method there is little control over the gelatinization and poor emulsions might be obtained, resulting in an inferior starch-fat sponge. Therefore, it is preferable to gelatinize the starch or flour under controlled conditions to obtain the correct degree of swelling prior to preparing the fat emulsion. The dried fat sponge can contain as much as 92% fat, and preferably contains from 75 to 92% fatty material, and can be prepared in any general shape. This process can be shown specifically in the following examples.

Example 1: A fat-containing starch sponge was prepared in the following manner. A 14-gram portion of pregelatinized cornstarch was added to 200 ml of water in a bowl of a Waring Blendor and mixed at high speed for one minute. To the mixture 100 ml of Wesson oil was added and mixed for one minute at high speed. The resulting emulsion was poured into trays and frozen with dry ice. The trays were placed in a laboratory freeze-dryer and dried, yielding a low density chunky product that had a dry appearance and feel and contained about 87% fat.

Example 2: A portion of 5 grams of the sponge product obtained in Example 1 was pressed firmly between two glass plates causing a considerable amount of the fat to be expressed. To the compressed starch sponge was added 5 cc of lemon oil. The resulting product was dry to the touch, but had not expanded to the original size of the sponge. The sample was stored for several months in a glass jar. From time to time it was examined and found to contain good quality lemon oil flavoring which could be released by dropping pieces of the product into water.

Example 3: A salad seasoning was prepared in the following manner. A fat-containing starch sponge was made and ground in the bowl of a Waring Blendor so that it passed through a 20 mesh screen. To 5 grams of this material 2 cc of glacial acetic acid was added, producing a dry powder with a strong acetic acid odor. The dry powder containing the acid was mixed with an equal weight of dry cheese garlic dressing and the mixture was placed in a capped shaker bottle. The resulting coarse, powdered material flowed easily and when dispensed onto moistened lettuce released oil and acid to produce a good-tasting salad dressing.

Dry Coatings for Foods

A method of coating foods with an edible amorphous film containing a pregelatinized starch as the essential ingredient has been disclosed by *K.A. Scheick, L. Jokay and G.E. Nelson; U.S. Patent 3,527,646; September 8, 1970; assigned to American Maize-Products Company.* The method uses very limited quantities of moisture or liquid in its operating steps and thus is essentially a dry method which gives the advantages of cost economy, high rate of production and accurate control of physical characteristics of the final product.

Briefly, the process includes the following four steps: (1) a small quantity of moisture is provided on the surface of the food to be coated if dry starch powder will not adhere; (2) a dry powder containing one or more pregelatinized starch materials as a major ingredient is applied to the premoistened surface of the food; (3) the layer of powder applied to the food is moistened in a limited manner to form a continuous colloidal suspension on the food surface; and (4) the suspension of powder is dried to form an amorphous flexible film on the surface of the food.

It is necessary that the dry powder used in this process contain one or more pregelatinized starches as the major ingredient, which means that at least 50% of the weight of the powder will comprise pregelatinized starch. In general, any pregelatinized starch may be used since the moistening step following application of the powder will cause a pregelatinized starch in the form of fine particles to form a continuous colloidal suspension to cover the food surface.

Accordingly, unmodified starch such as ordinary cornstarch and modified starch such as oxidized, acid-treated, crosslinked, partially-hydrolyzed or enzyme-converted starch which has been pregelatinized and dried to a powder may be used alone or in mixtures as the major essential ingredient in the powder to be applied to the food surface. In addition, starch derivatives such as ethers and esters, and in particular hydroxyalkyl ethers such as hydroxyethyl and hydroxypropyl starch ethers, in pregelatinized powder form may also be used.

Of particular interest are derivatives of high amylose content starches which are special genetic species of starch containing from 55 to 70% of amylose, the linear fraction of all natural starches. Thus, hydroxyethyl and hydroxypropyl ethers of high amylose starch, pregelatinized and powdered, are excellent base ingredients for producing coatings on food surfaces.

While any one of the starch ingredients described above may be used alone, the best results are achieved with powders containing mixtures of different starches. Thus, a mixture of hydroxyalkyl ether of high amylose or ordinary starch with oxidized and/or enzyme-converted waxy maize starch, in relative weight proportions of 5 to 60% hydroxyalkyl ether, 0 to 20% oxidized waxy maize and 10 to 60% enzyme-converted waxy maize, is more advantageous for practical commercial operations than use of any one of such ingredients alone. The hydroxyalkyl ether derivative gives strength and flexibility to the final coating and the waxy maize starches contribute gloss and transparency.

Examples 1 through 6: Finely divided powders were prepared containing the following ingredients in the specified relative proportions of weight percentages based on the total weight of the powder:

	\-\-\-\-\-\-\-\-\-\-\-\-\-\-\- Example \-\-\-\-\-\-\-\-\-\-\-\-\-\-\-					
	1	2	3	4	5	6
Pregelatinized:						
Hydroxypropyl ether of high amylose starch	5.6	15.0	50.0	45.6	38.1	0.0
Hydroxypropyl ether of cornstarch	0.0	0.0	0.0	0.0	0.0	38.25
Enzyme converted waxy maize starch	55.6	50.0	15.0	13.6	38.1	38.25
Oxidized waxy maize starch	16.6	15.0	15.0	13.6	0.0	0.0
Thin boiling waxy maize starch	0.0	0.0	0.0	0.6	14.3	4.75
Gelatin	16.6	15.0	15.0	13.6	0.0	9.5
Sucrose	5.6	5	5	13.6	9.5	9.5

Each of the powders were used to form coatings, by the method described below, upon various foods such as caramel cubes, semidried prunes, dates, figs, dried peaches, and an assortment of cubed fruits used in fruit cake. The food pieces were first exposed to steam to deposit a film of moisture upon these surfaces. Then the powder mixture was placed in a revolving inclined pan. The food pieces were slowly fed into the pans and particles of the powder mixture adhered to the moistened surface of the food to form substantially uniform coatings of them. The dry coated food pieces were then again exposed to steam whereby the particles of the powder mixture fused into a continuous layer having the nature of a colloidal suspension.

Following this, the food pieces were dried in ambient air of low relative humidity. This resulted in the formation of an amorphous flexible film around the entire surface of each piece of food. The film was nontacky, transparent, with varying degrees of loss, and in all cases firmly adhered to the surface of the food pieces.

Coatings for Malt Kernels

Malt is an essential ingredient in the production of alcoholic beverages, such as beer, ale, malt liquor, etc. One of the most important requirements of the brewer is that the amount of extract in the malt be high. Malt extract is usually referred to as the percentage of extractable dry matter in the malt which passes into solution during mashing in the brewing process. Other factors being equal, the malt extract will vary according to the thinness or plumpness of the malt kernels, with plump kernels giving the higher extract.

The size of the malt kernels, whether they be plump, thin or of intermediate size, depends on the size of the barley kernels which are malted. The size of the resulting malt kernel increases proportionately with increase in size of the barley kernels which are malted. Accordingly, to produce plump malt kernels, one should use plump barley kernels. Unfortunately, the available supply of plump grades of barley varies year-to-year, with perhaps only one out of four barley crop years in the United States giving large amounts of high quality plump kernels.

A method has been developed by *C.M. Hollenbeck and N.W. Kochorosky; U.S. Patent 3,446,707; May 27, 1969; assigned to North American Corporation* for increasing the size of malt kernels made from the thinner grades of barley having an extract below 76.5%. The size of the kernels is increased by applying a coating of gelatinized starch to the thin kernels prior to kilning.

Any suitable starchy material can be used in this process, providing it is acceptable in beer, forms a smooth slurry when gelatinized and the starch and other components are converted by the malt enzymes during mashing. Flours or meals of corn, barley, wheat, rye, rice, sorghum and potato are suitable for use in this process. Soluble sugars, on the other hand, are not suitable additives in this process because they do not increase the berry or kernel size and because they tend to give too much color to the malt by their carmelization during kilning.

Example 1: Samples of barley (3,000 grams of each) were steeped and germinated with the usual malting procedure. (This usually comprises steeping in water at 50° to 55°F for 30 to 50 hours until the moisture content of the barley reaches 39 to 44%. The barley is then germinated 4 to 5 days at 60°F with aeration.) On the fourth day of germination the following were added to individual samples of germinating barley (green malt):

A   Control — 650 ml of water;

B   650 ml of a slurry of pregelled barley flour containing 299 grams (about 10% of the barley now in green malt form) of a barley flour gelled by heating the slurry to 80°C, and cooling;

C   650 ml of a slurry of pregelled barley flour containing 464 grams (about 15.4% of the barley now in green malt form) of a barley flour gelled by heating as in B;

D   Same as C except 5% of the barley flour in the slurry was replaced by a diastatic malt flour.

The barley flour slurry of D was much easier dispersed throughout the green malt mass than either the slurries in B or C, due to the lower viscosity of the slurry as the result of the liquefying effect of the malt flour enzymes. The treated green malt samples were germinated with aeration for 24 hours more, then dried to 4% moisture by the usual kilning (12 hours at 120°F, 4 hours at 140°F, 4 hours at 180°F, 2 hours at 190°F). The extracts (dry) on the samples of malt were as follows:

Sample	Percent Extract
A	75.4
B	76.8
C	77.8
D	78.3

The appearance and chemical analyses of the malts were essentially normal. Sample C was less desirable appearance-wise than Sample B, and the heavier flour coating tended to flake off the malt. Sample D was higher in extract than Sample C due primarily to more efficient application of the slurry with its lower viscosity. The increased extract values, due to the added barley flour, were illustrated by Sample B being higher in extract than Sample A, and Samples C and D being higher in extract than Samples B, as well as A. A colloidal gum, such as guar gum, may be included with the gelatinized starch to improve adhesion of the starch to the barley.

Example 2: A series of malts were made with barley starch, wheat flour, and wheat starch as the starchy material additives. Wheat flour contains the normal protein content while wheat starch has had the protein removed. The starchy material was pregelled before its application to the green malt. This was done by heating the slurry to 90°C and then cooling it to room temperature. The water used in the slurries was an amount equivalent to two gallons per bushel of barley, and the starch was equal to 10% of the dry weight barley (176 grams).

A 1% amount of malt flour based on the weight of the starchy material additive was also used along with 500 ppm of sodium metabisulfite to help thin the starch slurries for easier application. The pregelled slurries were applied to green malt in equal amounts on the last two days of a five day germination period. The green malt was kilned, after germination, according to normal commercial practice (see Example 1). The following malts were made in this series:

(1)   Regular control — normal malting procedure.
(2)   Watered control — watered on the last two days of germination
        with an equivalent of two gallons of water per bushel of barley.
(3)   10% pregelled barley starch.
(4)   10% pregelled wheat flour.
(5)   10% pregelled wheat starch.

Laboratory analyses of these sample malts were as follows:

| | Malt identity | | | | |
	1	2	3	4	5
Moisture (percent)	4.3	4.4	5.1	5.2	5.1
Extract (dry) (percent)	75.1	75.1	75.7	75.3	76.0
Color (° L.)	2.26	3.28	1.64	1.68	1.68
Diastase (° L.)	119	104	165	165	173
Alpha amylase (units)	60	63	65.3	63.3	65.3
Total protein	13.51	13.78	11.45	13.70	13.80
Soluble protein	6.30	6.96	5.03	5.88	5.67
Assortment:					
7/64 screen (percent)	3.1	4.1	9.0	8.4	7.3
6/64 screen (percent)	25.9	38.3	55.5	66.4	53.4
5/64 screen (percent)	70.4	56.9	34.9	24.0	38.8
Thru (percent)	0.6	0.7	0.5	1.2	.05

The pregelled starch, wheat flour, and wheat starch additives increased the extract and increased the average kernel size of the malts in these samples.

Easily Digested Baby Food

A baby food is prepared by *B.Y.-L. Ho; U.S. Patent 3,803,311; April 9, 1974;* from cornstarch that is particularly suitable for infants in reducing digestion problems while simultaneously meeting major nutritional requirements. The above objects are obtained by a food composition comprising water, colloidal cornstarch, and milk, in admixture, the colloidal cornstarch being in homogeneous suspension in the water, the milk ranging from 1 part of milk per 3 parts of water for an infant preparation to 2 parts of milk per 1 part of water for a preparation for a baby of at least 3 months of age, the cornstarch ranging from 1 to 3 level tablespoons per quart of water, and the composition being characterized by properties dependent upon the homogeneous suspension preceding admixture of the milk.

Accordingly, the process requires that the water and cornstarch be admixed sufficiently to form a substantially homogeneous colloidal suspension prior to the addition of and mixing with the milk.

A typical process is as follows. Two level tablespoons of cornstarch are mixed with a quart of water, and the mixture is first boiled (i.e., heated to about 212°F for a period sufficient to homogenize the starch) and then allowed to simmer until the cornstarch has become homogeneously suspended in colloidal suspension (i.e., a homogenized colloidal suspension) in the nature of a soup. This soup is either stored, such as in a refrigerator, or used in the next step of mixing with milk — either already prepared whole or skim milk, or alternatively an equivalent amount of milk prepared from powdered milk. The proportion of milk for a baby bottle — for example of about eight ounces, is to be decided or determined by the physician according to the infant's age, weight, maturity (for example whether or not premature) and the like, and the soup is added to the bottle to finish filling the bottle, such as for the remaining five ounces of an eight ounce bottle. Accordingly, the final food composition includes the wholesome cornstarch food in a palatable form as well as containing the milk component in a very dilute form easily digestible. Also, to the bottle is added about two level teaspoons of sugar, this amount of sugar thereby falling within the above stated range relative to the initial water employed to produce the soup. Preferably the sugar is added before addition of the soup to the bottle, thereby facilitating the admixing of the sugar by the soup addition.

Dry Mix Thickening Agent

It was found by *D.J. Yoder and M.W. Bugg; U.S. Patent 3,554,764; January 12, 1971; assigned to General Foods Corporation* that a food product having improved freeze-thaw stability can be prepared using a thickening mixture containing pregelatinized wheat flour, and/or starches, an edible gum and sodium stearyl fumarate. The product is prepared by simple, dry blending of the ingredients, eliminating the need to chemically modify the starch or flour.

The thickener of this process has a high absorption capacity and when mixed with liquids and cooked results in a product with a desirable, smooth texture. There is little tendency for the product to break down at high temperatures, thus, when used as a thickener for precooked foods it offers improved eating characteristics. The stringiness associated with gelatinized starches is reduced and the tendency toward gel formation and water separation are eliminated. The reduction of heat sensitivity may be particularly useful in preparing a thickener for canned goods which are to be retorted. The product when used as a thickener for frozen gravies or as a thickener for frozen foods imparts freeze-thaw resistance to the foods. Weeping and syneresis after thawing are eliminated.

The thickener of this process is basically a blend of farinaceous material, an edible gum and sodium stearyl fumarate. While other ingredients may be added when preparing specific blends of the product for specific uses, such as constituents for flavoring or coloring, these other ingredients do not add to or subtract from the basic discovery. The farinaceous material is the major component of the dry mix and the moisture content of the mix is less than 10% and preferably less than 7%.

An edible gum at a level of 0.5 to 10.0% is necessary. While the use of gums as stabilizers and thickeners has long been known, and the fact that gums are known to impart a degree of freeze-thaw stability to some products, the degree of stability achieved in the product of this process is far greater than that which can be attributed to a gum alone or a gum in combination with a farinaceous material. While all edible gums have not been tested, results of experiments show that basically all edible gums can be useful in the thickener and it is to be understood that the term edible gum is further limited to hydrophylic gums.

The preferred gums are generally vegetable gums and the most preferred of these is guar gum. However, it has been found that carboxymethylcellulose (CMC) among other nonvegetable gums is also a preferred gum. It has been found that the gums are most effective when used at a level of about 3.0 to 8.0%. The use of sodium stearyl fumarate (NaSF) is essential. Sodium stearyl fumarate sold as Pruv has been described in two publications, the March 1966 issue of *Food Engineering* and the February 1966 issue of *Cereal Science Today,* Vol II, No. 2.

It has been found that NaSF is useful at levels of about 0.5 to 6.0% by weight of the dry product. The preferred level of NaSF being from 1.0 to 4.0% and the most preferred level is 2.0%. Unexpectedly, it was discovered that the synergistic effect of a gum and NaSF in combination with a major percentage of a farinaceous material is most effective when the farinaceous material is precooked and dried before the gum and NaSF are added.

Example 1: Wheat flour was precooked in a Wenger type cooker such that the flour was highly gelatinized on discharge from the cooker. The gelatinized flour was then dried conventionally and ground into a powder. A dry blend of 93% precooked wheat flour, 5% guar gum and 2% Pruv was then made by mixing the ingredients in a ribbon blender for about 5 minutes, or until a homogeneous mix was obtained.

The resultant mix was used to thicken a commercially available beef broth. It was added to the broth at a 6% concentration and allowed to cook in the broth at about 190°F for 10 minutes. The resulting gravy was then cooled until frozen and held in a frozen state overnight. The gravy was then allowed to thaw and was tested by visual inspection and by centrifugation.

Visually, the texture was found smooth and an absence of gelling was observed. There was no sign of liquid or fat separation. A spin test was run on an International (Type SB-100) centrifuge. After 20 minutes at 2,000 rpm the centrifuge tubes were inspected and it was found that 0% of water settled out. No separation of water on spin test (0%) is the most preferred rating as shown by the following table.

Spin Test Results

Percent Water Separation	Rating of Freeze-Thaw Stability
0%	Excellent
Less than 25%	Acceptable
25% to 50%	Some freeze-thaw stability, but not acceptable
Greater than 50%	Poor — unacceptable

Example 2: The flour mix of Example 1 was used to make a tomato sauce having the following ingredients:

Item	% by Weight in Mix
Water	60.7
Tomato paste	21.9
Onions	8.8
Flour mix	3.0
Soy oil	2.9
Salt	1.5
Sugar	0.9
Seasoning	0.3

The dry ingredients were dry blended. Then the tomato paste, onions in ground state, soy oil and half of the water were mixed together and brought to a boil, stirring constantly. Upon reaching a boil the rest of the water was added and then the dry ingredients were mixed into the liquid. The resultant mix was brought to a boil and allowed to simmer at about 190°F for 5 minutes. The tomato sauce was then poured into containers and frozen. Upon thawing the sauce was found to have a very desirable texture and exhibited no syneresis.

Multilayered Pudding Mixtures

Powdered whippable dessert compositions for use in making differently textured, multilayered pudding-like desserts and methods of preparing them are disclosed by *R.R. Cassanelli, R.P. Wauters, A.C. Wirchansky and C.R. Wyss; U.S. Patent 3,702,254; November 7, 1972; assigned to General Foods Corporation.*

Katz, in U.S. Patent 3,434,848 discloses a food mix and process for obtaining a two-phase, fruit-flavored gelled dessert, comprising a pregelatinized starch and surfactant setting system where the surfactant is a blend of glyceryl and propylene glycol monostearates. This process, however, produces a two- or three-layered pudding dessert of superior taste and texture, using a settling system of pregelatinized starch and gelatin to form the bottom pudding-like layer, and an emulsified fat system for forming the upper aerated layer.

The Katz patent specifically keeps the fat and protein level to an absolute minimum, below 8% of the mix weight and preferably below 5%. The high quality of the dessert of this process depends greatly on the presence of a greater amount of emulsified fat of the type to be described, and a lower level of pregelatinized starch.

The particular edible fats to be incorporated in the dessert composition according to this process may be edible oils, such as coconut oil, etc., and semisolid or solid fats, i.e., the usual shortenings. In addition to the fat itself, if the fat is not preemulsified, the fat portion can include one or more emulsifying agents for fat, such as lecithin and polyglycerol esters of fatty acids.

The fat portion should exhibit a plastic range after whipping with the gelatin-starch portion according to the recipe, such that it will melt readily at or about mouth temperature. More particularly, the fat portion should show a rapid decrease in solids content over a range between room temperature and body temperature. Thus, it is desirable that between 70° and 100°F there be a sharp drop in the solids content index of the fat portion. The fat portion employed in this process will preferably have a solids content of 64.4% at 50°F, and a solids content of 4.2% at 110°F. The Wiley melting point range of the fat portion is between 99°F and 120°F. Any pregelatinized starch can be used such as potato starch, and preferably tapioca starch.

Example 1: The range of ingredients used in the multilayered, multitextured pudding dessert mixes are given on the following page. The fat, emulsifiers and flavors are heated to about 140°F in a mixing tank. Next, about one part of the melted fat portion and about five parts of granulated sugar are simultaneously added to a liquid/solid continuous mixing device to form a fat/sugar paste. The resulting paste is then fed into a 5-roll refiner where it is crushed to a relatively dry fat/sugar powder, in which the fat is plated on the sugar. The fat/sugar powder is cooled to about 45°F, and then ground into a homogeneous mixture in a Fitz Mill. Next, the homogeneous fat/sugar powder is blended in a solids mixture with about one part of additional granulated sugar and the other ingredients above, and packaged for storage.

	Percent of Total Mix Weight
Sugar	40.0 – 85.0
Gelatin-starch portion:	
Nonfat milk solids	0 – 30
Gelatin	3.0 – 7.0
Pregelatinized starch	1.0 – 7.0
Flavor	0.2 – 4
Salt	0.1 – 0.2
Coloring	0.02 – 0.05
Fat portion:	
Hydrogenated vegetable oil	6 – 13
Polyglycerol esters of fatty acids	
(diglycerol monostearate)	0.2 – 3
Flavor	0.03 – 0.1
Lecithin	0.02 – 0.5

Example 2: A dessert mix was prepared according to the process of Example 1, containing the following ingredients:

	Grams	Percent of Total Mix Weight
Sugar	74.17	59.78
Gelatin-starch portion:		
Nonfat milk solids	25.0	20.15
Pregelatinized starch	5.65	4.63
Gelatin	5.4	4.35
Vanilla flavor	0.27	0.22
Acetaldehyde	0.20	0.16
Salt	0.133	0.11
Coloring	0.033	0.03
Fat portion:		
Hydrogenated vegetable oil	9.64	7.78
Polyglycerol esters of fatty acids		
(diglycerol monostearate)	1.48	1.19
Vanilla flavor	0.05	0.04
Lecithin	0.04	0.03

One-half cup of boiling water is added to the mix and blended at low speed with an electric mixer for about 30 seconds in order to melt the fat portion, solubilize the gelatin and disperse the starch. The mixture is then whipped at medium to high speed for 3 minutes with an upright electric mixer or 4 minutes with a portable electric mixer or a rotary hand beater, lowering the temperature of the emulsion, solidifying and aerating the fats within the range for best whipping. The speed is reduced to low and 1¼ cups of cold milk is blended into the mixture, breaking the emulsion and allowing the fat and gelatin foam to rise to the surface.

Additionally, the milk adds to the creaminess of the mixture and the taste of the end product. The mixture is poured into glasses and chilled until set. The mixture gradually separates into a two-layered vanilla-flavored dessert having a lower pudding-like layer, and an upper foamy creamy fat layer.

STARCH ESTERS

The starch molecule contains on the average, one primary and two secondary alcohol groups per D-glucose unit. These groups provide the sites for formation of esters and the ethers covered in the following chapter. One of the first starch esters to be prepared commercially was a low DS acetate sold under the trade name Feculose which was used as a substitute for gelatin and natural gums. Beginning in 1905, this ester was produced in Britain by a semidry process from acetic anhydride. Starch esters are permitted in limited amounts in food products by the Federal Drug Administration. In addition to use in food, starch esters are also used in the paper and textile industries.

STARCH CARBOXYLATES

In addition to the following six processes, starch esters can be used to prepare dehydrated alcohol products as disclosed in U.S. Patent 3,786,159, page 233.

Magnesium Hydroxide or Oxide Control of Esterification

C.D. Bauer; U.S. Patent 3,839,320; October 1, 1974; assigned to Anheuser-Busch Incorporated has disclosed a process for producing starch esters from mono- and dibasic organic acid anhydrides using magnesium oxide or magnesium hydroxide to control the formation of excess acid. The oxide or hydroxide of magnesium is used as a pH control agent in starch esterification reactions, particularly in the reaction of starch with an anhydride of a monobasic or dibasic organic acid. It maintains the pH in the desired range for the esterification reaction or acetylation reaction if the reactant is acetic anhydride. The magnesium oxide or hydroxide acts to regulate the pH without the problems associated with the use of sodium hydroxide.

The magnesium oxide or hydroxide preferably is added at one time, thus neutralizing any acid initially present and also neutralizing acid formed by the reaction of starch or water with the subsequently added acid anhydride. The acid anhydride is metered into the reaction vessel after the magnesium oxide or hydroxide has been added. This avoids loss of anhydride and avoids formation of excess acid. The conventional alkaline additives, e.g., NaOH, may cause a rapid localized rise in pH upon their addition to the starch slurry if insufficient amounts of acid are present to prevent the rise in pH. This can cause the starch to gelatinize and lose its granular structure. It is advantageous to crosslink the starch prior to esterification when the final product is used in foods.

Example 1: A starch slurry is formed containing 1,000 parts waxy maize starch and is adjusted to a Baumé of 20.5 and held in a water bath at 105°F with continuous stirring. To the slurry are added 3 parts of calcium hydroxide and 2.5 parts of epichlorohydrin. After allowing to react for 6 hours the pH is adjusted to 9.0 with dilute hydrochloric acid. To the reaction mixture is added 30 parts of a reactive form of magnesium hydroxide. Stirring of the 105°F starch slurry is continued while 120 parts of acetic anhydride are added at a uniform rate over a period of 20 minutes. At the conclusion of the acetic anhydride addition, 2 volumes of the original amount of water are added; the slurry is filtered and dried in the usual manner. The resulting starch, when pasted, was found to have a high viscosity, short texture and good clarity suitable for use in such food products as puddings, sauces, or fruit pies.

Example 2: 100 parts of waxy maize starch is slurried in water to a Baumé of 20.5 and held in a bath at 95°F. 1.5 parts of a powdered reactive form of magnesium oxide is added and the slurry kept well stirred with a mechanical stirrer. 6 parts of acetic anhydride are slowly added at a rate controlled to maintain the pH of the reaction mixture between 8.0 to 9.0. At the conclusion of the acetic anhydride addition, 2 volumes of water are added (2 times the original water used) and the pH is adjusted to 5.0 to 5.5. The slurry is filtered and dried in the conventional manner. The resulting starch was found by analysis to have an acetyl content of 1.5 to 2.5%.

Example 3: 500 parts waxy maize starch are suspended in 800 parts of water and the slurry maintained at a temperature of 100° to 105°F with agitation. 2.0 parts of magnesium oxide are added. 6.6 parts of succinic anhydride are added in small portions over a period of 10 minutes followed by 15 minutes additional mixing. The pH of the starch slurry is adjusted to 5.5 with hydrochloric acid, and the slurry is filtered, washed and dried.

Levulinate Esters Using α-Angelica Lactone

In a process described by *E. Bernasek and C.N. Eaton; U.S. Patent 3,580,906; May 25, 1971; assigned to R.J. Reynolds Tobacco Company* starch levulinates are prepared by reaction of starch in either aqueous or nonaqueous medium with α-angelica lactone. These starch levulinates are produced according to the following reaction sequence:

$$HC\!\!-\!\!-\!\!-\!\!CH_2 + HO\text{-}Starch \longrightarrow CH_3\overset{O}{\overset{\|}{C}}CH_2CH_2\overset{O}{\overset{\|}{C}}\!-\!O\text{-}Starch$$
$$H_3C\!-\!\underset{\|}{C} \quad \underset{C=O}{|}$$
$$\diagdown O \diagup$$

where HO-starch represents the starch molecule. It has been further found that the resulting starch levulinates can be readily crosslinked, as required, for various applications where these derivatives can be used to advantage. It has been observed that crosslinking occurs in cornstarch levulinate when the degree of substitution (DS) is greater than about 0.1. When the DS is below about 0.1, crosslinking is not apparent for cornstarch levulinate but does occur with potato and waxy maize starch levulinates. The following examples illustrate the process.

Example 1: For the reaction between cornstarch and α-angelica lactone, the principal variables, the pH of the reaction, the temperature of the reaction, and the length of time of the reaction were varied. The following preferred procedure was used. Cornstarch (1.0 mol) was suspended in 250 ml of water. The temperature of the slurry was raised to 93°F, and the pH was adjusted to 8.7. α-Angelica lactone (0.1 mol) was added dropwise maintaining the pH at 8.7 by the simultaneous addition of a 7:1 mixture of saturated sodium chloride solution and 30% sodium hydroxide solution. The suspension was stirred vigorously for 1.0 hour at 93°F. The product was filtered, slurried successively in three 500 ml portions of water and two 500 ml portions of ethanol, filtered and air dried. The results of these tests are outlined in the table.

Reaction of Cornstarch with α-Angelica Lactone

pH	Temperature, °F	Mols Lactone per Mol Starch	Reaction Time, hours	- - - -Percent- - - -		DS
				Levulinyl	Yield	
8.7	93	0.1	0.5	4.23	73.4	0.07
8.7	93	0.1	1.0	3.76	65.3	0.06
8.7	93	0.1	1.0	4.71	81.8	0.08
8.7	93	0.1	1.0	5.08	88.2	0.09
8.7	93	0.1	1.0	5.00	86.8	0.09
8.7	93	0.1	3.0	3.78	65.6	0.09
8.7	125	0.1	1.0	4.64	80.6	0.08
9.0	93	0.1	2.0	3.96	68.8	0.07
9.2	82	0.1	3.5	2.98	51.7	0.05
9.2	93	0.1	0.5	2.95	51.2	0.05
9.2	93	0.1	1.0	4.23	73.4	0.07
10.2	96	0.1	1.0	0.92	15.9	0.015
10.8	93	0.1	1.0	0.31	5.2	0.01
10.8	93	0.1	2.5	—	—	—
8.0	93	0.2	2.5	4.19	38.5	0.07
9.2	82	0.2	2.5	6.88	63.2	0.12
9.2	93	0.2	0.5	6.48	59.5	0.11
9.2	93	0.2	2.5	7.59	69.7	0.13
9.2	93	0.2	7.0	7.26	66.7	0.13
9.2	97	0.2	2.5	6.94	63.7	0.12
9.2	104	0.2	1.0	7.59	69.7	0.13
9-10	75	0.4	4.5	1.93	9.6	0.03
9-10	120	0.4	61.0	4.50	22.5	0.08

Example 2: This example uses commercially dry starch in the levulinylation process. A reaction vessel equipped with mechanical agitation was charged with 162 parts of cornstarch (13.2% moisture), which previously had been treated with base in an aqueous slurry, filtered and allowed to air dry. Into the starch was sprayed 9.8 parts of α-angelica lactone. The contents of the vessel were then heated to a temperature of 120°F and agitation was continued for 6.5 hours. The starch was then slurried in 1 liter of water, and the pH of the system was adjusted to a pH level of 8.3 by the addition of a 5% sodium hydroxide solution.

The resulting product was filtered, washed with ethanol and water, and air dried. This derivative contained 3.94%, by weight, of levulinyl groups. The latter concentration corresponds to a reaction efficiency of 68.4%.

Use of Mixed Carbonic Carboxylic Anhydrides

In the process described by *M.M. Tessler and M.W. Rutenberg; U.S. Patent 3,720,662; March 13, 1973; assigned to National Starch and Chemical Corporation* aqueous slurries or dispersions of starch are reacted with mixed carbonic-carboxylic anhydrides of monocarboxylic acids under alkaline conditions to yield ester derivatives of starch. These noncrosslinked starch products can also be prepared by a dry reaction process. According to this process starch or a starch derivative is reacted in an aqueous suspension with a mixed anhydride of carbonic acid and a monocarboxylic acid.

The reaction may be carried out at temperatures ranging from somewhat below room temperature to somewhat above room temperature and at a pH between 6.5 and 12.5. By a suitable choice of starting materials, reagents and reaction conditions, very useful modified starches may be prepared conveniently and easily which are useful in foods. The following examples illustrate the process in aqueous suspension using various mixed anhydrides.

Example 1: In preparing these derivatives (see the table below), the procedure used comprised the suspension of the respective starch bases in 1.25 to 1.50 parts of water per each part of starch whereupon the indicated amounts of the selected mixed carbonic-carboxylic anhydride were introduced. The pH was controlled at the indicated value by periodic addition of 3.0% aqueous sodium hydroxide solution during the entire reaction. The reaction was allowed to proceed, under agitation, at the desired temperature until there was no further change in pH. The resulting starch ester derivatives were then acidified with dilute sulfuric acid, recovered by filtration and then washed with water to remove residual salts.

The table below presents the pertinent data relating to various derivatives which were prepared. In the case of derivative 18, the pH was controlled by addition of solid calcium hydroxide; in the case of derivative 19, by addition of 3.0% aqueous sodium carbonate solution. Each reacted starch was examined for acyl content, calculated from the saponification number.

| | | | | Reaction conditions | | | |
No. Starch base	Esterification reagent name	Percent on Starch	Controlled pH	Temp., °F.	Time, min.	Percent acyl
1 Corn starch	Ethylcarbonic acrylic anhydride	5.0	8.0	73	75	0.75
2 do	Ethylcarbonic methacrylic anhydride	5.0	8.0	73	40	0.97
3 Hydrolized corn starch (75 fluidity)	do	10.0	8.0	73	60	2.14
4 Corn starch	Benzylcarbonic propionicanhydride	20.0	8.0	73	180	1.16
5 do	Ethylcarbonic propionic anhydride	5.0	7.0	73	60	0.45
6 do	do	5.0	8.0	73	70	0.67
7 do	do	5.0	9.0	73	70	0.86
8 do	do	5.0	10.0	73	70	0.60
9 do	do	5.0	8.0	104	65	0.71
10 do	do	5.0	8.0	50	60	0.77
11 Hydrolyzed waxy maize starch (85 fluidity)	do	7.0	8.0	73	50	1.10
12 Potato starch	do	7.0	8.0	73	50	1.05
13 Tapioca starch inhibited with 0.02% phosphorus oxychloride and hydroxypropylated with 7.5% propylene oxide.	do	10.0	8.0	73	90	1.69
14 Corn starch	Ethylcarbonic lauric anhydride	10.0	8.0	73	480	2.35
15 do	Ethylcarbonic benzoic anhydride	7.5	8.0	73	125	2.61
16 Waxy maize	Ethylcarbonic crotonic anhydride	10.0	8.0	73	180	3.26
17 High amylose corn starch (55% amylose by weight).	do	5.0	8.0	73	120	1.99
18 Corn starch	Ethylcarbonic propionic anhydride	10.0	8.5	73	30	0.96
19 do	do	10.0	8.5	73	45	1.76
20 Waxy maize	Ethylcarbonic crotonic anhydride	2.5	8.0	73	120	1.09
21 Corn starch	Ethylcarbonic stearic anhydride	30.0	8.0	73	960	5.01

Example 2: This example illustrates the preparation of a starch ester by means of a dry reaction. About 200 parts of waxy maize starch was pretreated by suspending the starch in 300 parts of water containing 1.6 parts of sodium hydroxide and stirring for a period of 15 minutes. The pH of the suspension was found to be 11.8. The suspension was thereafter filtered and the starch was air dried to have a moisture content of 17%. A solution of 5 parts ethylcarbonic propionic anhydride in about 15 ml acetone was sprayed onto 50 parts of the previously treated waxy maize starch. The sprayed starch was then placed in an oven set at a temperature of 45°C (113°F) for a period of 5 hours, after which the starch was cooled and poured into 100 parts of water. The pH of this suspension was adjusted to 6.5 with dilute sulfuric acid, and the starch recovered by filtration, washed three times with water and air dried. The starch ester product contained 1.53% propionyl groups.

Imidazolides of Carboxylic and Sulfonic Acids as Acylating Agents

In a process described by *M.M. Tessler; U.S. Patent 3,720,663; March 13, 1973; assigned to National Starch and Chemical Corporation* aqueous slurries or dispersions of starch are reacted with imidazolides of carboxylic or sulfonic acids to yield crosslinked and uncrosslinked starch ester derivatives. These starch products can also be prepared in nonaqueous solvents or by a dry reaction process. For purposes of this process the term imidazolides of carboxylic or sulfonic acids means compounds corresponding to the formulas

$$(1) \quad
\begin{array}{c}
CH=CH \\
|\qquad\; \diagdown \\
|\qquad\qquad N-\overset{\displaystyle O}{\overset{\|}{C}}-N \\
|\qquad\; \diagup \\
N=CH
\end{array}
\begin{array}{c}
CH=CH \\
\diagup \qquad\; | \\
\qquad\qquad | \\
\diagdown \qquad\; | \\
CH=N
\end{array}
\qquad
(2) \quad
R_1-\overset{\displaystyle O}{\overset{\|}{C}}-N
\begin{array}{c}
CH=CH \\
\diagup\quad| \\
\quad\;| \\
\diagdown\quad| \\
CH=N
\end{array}$$

$$(3) \quad
\begin{array}{c}
CH=CH \\
|\qquad\; \diagdown \\
|\qquad\qquad N-\overset{\displaystyle O}{\overset{\|}{C}}-R_2-\overset{\displaystyle O}{\overset{\|}{C}}-N \\
|\qquad\; \diagup \\
N=CH
\end{array}
\begin{array}{c}
CH=CH \\
\diagup\quad| \\
\quad\;| \\
\diagdown\quad| \\
CH=N
\end{array}
\qquad
(4) \quad
R_1-SO_2-N
\begin{array}{c}
CH=CH \\
\diagup\quad| \\
\quad\;| \\
\diagdown\quad| \\
CH=N
\end{array}$$

where R_1 is selected from alkyl, substituted alkyl, unsaturated alkyl, cycloalkyl, aryl, substituted aryl, and arylalkyl, and R_2 is selected from alkylene, substituted alkylene, bis-alkylene ether, cycloalkylene, arylene, and substituted arylene. R_1 and R_2 may each contain between 1 and 20 carbons.

Suitable imidazolides of carboxylic or sulfonic acids corresponding to Formulas 2 through 4 may be prepared using acids such, e.g., as acetic acid, propionic acid, stearic acid, trimethylacetic acid, phenylacetic acid, p-dimethylaminobenzoic acid, succinic acid, glutaric acid, adipic acid, dimethylmalonic acid, sebacic acid, terephthalic acid, diglycolic acid, 3,3'-oxydipropionic acid, benzenesulfonic acid, p-toluenesulfonic acid and methanesulfonic acid. The preparation of imidazolides of carboxylic and sulfonic acids is well described in the literature and is ordinarily carried out by reacting selected sulfonic or carboxylic acids such as those listed above with 1,1'-carbonyldiimidazole or 1,1'-thionyldiimidazole. The following examples illustrate the starch esterification process.

Example 1: This example illustrates the use of various imidazolides of carboxylic and sulfonic acids in preparing starch esters where the resulting products are not inhibited (crosslinked) and display an intact granule structure. In preparing the starch derivatives, listed in the table below, 1.00 part of cornstarch were suspended in 1.25 parts of water, and the indicated amounts of the selected imidazolide of a carboxylic or sulfonic acid were added to the dispersion. The pH was controlled at the indicated value by periodic addition of 3.0% aqueous sodium hydroxide solution during the course of the reaction.

The reaction was allowed to proceed at room temperature (RT) until there was no further change in pH. The resulting starch ester derivatives were then acidified with dilute sulfuric acid, recovered by filtration, and subsequently washed with water to remove residual salts. The acyl content of each of the reacted starches calculated from the saponification number, was determined and is listed in the table below.

Derivative	Esterification Reagent	% on Starch	Reaction Conditions Controlled pH	Time, hours	% Acyl
1	N-acetylimidazole	6.0	8.0	2.0	1.57
2	N-benzoylimidazole	7.0	8.0	2.0	3.29
3	N-benzoylimidazole	7.0	9.0	2.0	3.08
4	N-acryloylimidazole	7.0	8.0	1.75	0.48
5	N-methanesulfonylimidazole	10.0	8.0	16.0	0.53
6	N-(p-toluenesulfonyl)imidazole	7.0	10.0	8.0	2.18
7	N-stearylimidazole	20.0	9.0	18.0	4.43

Example 2: This example illustrates the use of 1,1'-carbonyldiimidazole in preparing inhibited (crosslinked) starch esters. In preparing these derivatives, listed in the table below, 1.00 part of the respective starch bases was suspended in 1.25 to 1.50 parts of water and the indicated amounts of 1,1'-carbonyldiimidazole were added to suspension.

Derivative Number	Starch base	Parts by wt. 1,1'-carbonyldiimidazole per 100 parts starch	Reaction conditions Controlled pH	Reaction temp. (°C.)	Sediment volume (ml.) Reaction product	Base
10	Corn starch	1.5	8.0	RT	13.0	62.0.
11	Waxy maize (acid converted to a degree known in the trade, as 85 fluidity).	1.5	8.0	RT	29.0	None.
12	Oxidized corn starch (converted by reaction with NaOCl to a degree known, in the trade, as 75 fluidity).	1.5	8.0	RT	16.0	Do.
13	Corn starch which was previously hydroxypropylated with 5% propylene oxide.	1.5	8.0	RT	23.0	91.0.
14	Potato starch	1.5	8.0	RT	21.0	75.0.
15	Waxy maize	1.5	4.0	RT	56.0	None.
16	do	1.5	5.0	RT	33.0	Do.
17	do	1.5	6.0	RT	18.5	Do.
18	do	1.5	7.0	RT	19.2	Do.
19	do	1.5	8.0	RT	24.5	Do.
20	do	1.5	9.0	RT	23.5	Do.
21	do	1.5	10.0	RT	45.0	Do.
22	do	1.5	8.0	2-3	15.5	Do.
23	do	1.5	7.5	48	63.0	Do.
24	do	75.0	8.0	RT	8.0	Do.
25	do	0.1	8.0	RT	40.5	Do.
26	do	0.2	8.0	RT	35.0	Do.

The pH was controlled at the indicated value by periodic addition of 3.0% aqueous sodium hydroxide solution during the course of the reaction. The reaction was allowed to proceed with agitation, at the desired temperature until there was no further change in pH. Most of the reactions were completed in about 1 hour. The resulting starch ester derivatives were then acidified with dilute sulfuric acid, recovered by filtration, and washed with water to remove residual salts. The degree of inhibition was determined by cooking an aqueous suspension of the resulting starch product having a concentration of 1%, by weight, solids in a boiling water bath for a period of 30 minutes.

The cooked dispersion was then allowed to stand at room temperature in a 100 ml graduated cylinder for a period of approximately 16 hours. In order to show comparative values, the sediment volume of the base starch was also determined by this method.

Acyl Guanidines as Esterification Agents

In a similar process described by *M.M. Tessler and M.W. Rutenberg; U.S. Patent 3,728,332; April 17, 1973; assigned to National Starch and Chemical Corporation* starch dispersions in aqueous or nonaqueous media are reacted with acyl guanidines to produce starch esters. The acyl guanidines used in the process are compounds corresponding to the formulas

$$(1) \quad R_1-\underset{\underset{O}{\|}}{C}-NH-\underset{\underset{NH}{\|}}{C}-NH_2 \qquad \text{and} \qquad (2) \quad NH_2-\underset{\underset{NH}{\|}}{C}-NH-\underset{\underset{O}{\|}}{C}-R_2-\underset{\underset{O}{\|}}{C}-NH-\underset{\underset{NH}{\|}}{C}-NH_2$$

where R_1 is selected from alkyl, substituted alkyl, unsaturated alkyl, cycloalkyl, aryl, substituted aryl, and aralkyl, and R_2 is selected from alkylene, substituted alkylene, bisalkylene ether, cycloalkylene, arylene, and substituted arylene. R_1 and R_2 may each contain between 1 and 20 carbon atoms. Suitable guanidine derivatives of carboxylic acids corresponding to structures (1) and (2) may be prepared using acids such as acetic acid, propionic acid, benzoic acid, phenylacetic acid, n-butyric acid, oleic acid, lauric acid, stearic acid, terephthalic acid, 1,4-cyclohexanedicarboxylic acid, succinic acid and adipic acid.

The preparation of acyl guanidines of monocarboxylic acids is described in U.S. Patent 2,408,694. The same procedure may be used to prepare guanidine derivatives of polycarboxylic acids. In general, esters of the selected carboxylic acids are reacted with guanidine in ethanol solution, and the insoluble acyl guanidines are recovered by filtration.

The process comprises reacting a selected acyl guanidine, such as described above, with a selected starch base which is ordinarily suspended in water. The reaction of the acyl guanidine with the suspended starch is carried out at temperatures from 70° to 100°F and for a period of 1 to 24 hours. The pH of the reaction mixture is ordinarily controlled so as to be above 4.0 and below 11.0 with the preferred range being from 7.0 to 10.0. The pH is conveniently controlled by the periodic addition of a dilute aqueous solution of hydroboric acid, but other common acids, such as sulfuric or acetic, may also be used with equal success. The starch products resulting from this process are starch esters, with the reactions using acyl guanidines (1) and (2) being represented as shown below.

$$R_1-\overset{\overset{\textstyle O}{\|}}{C}-NH-\overset{\overset{\textstyle NH}{\|}}{C}-NH_2 + StOH \longrightarrow R_1-\overset{\overset{\textstyle O}{\|}}{C}-O-St + NH_2-\overset{\overset{\textstyle NH}{\|}}{C}-NH_2$$

(1)

$$NH_2-\overset{\overset{\textstyle NH}{\|}}{C}-NH-\overset{\overset{\textstyle O}{\|}}{C}-R_2-\overset{\overset{\textstyle O}{\|}}{C}-NH-\overset{\overset{\textstyle NH}{\|}}{C}-NH_2 + 2StOH \longrightarrow St-O-\overset{\overset{\textstyle O}{\|}}{C}-R_2-\overset{\overset{\textstyle O}{\|}}{C}-O-St + 2NH_2-\overset{\overset{\textstyle NH}{\|}}{C}-NH_2$$

(2)

where StOH represents the starch molecule and R_1 and R_2 are as defined above. The following examples illustrate the process.

Example 1: This example illustrates the use of various acyl guanidines in preparing starch esters according to this process where the resulting products are not inhibited and display an intact granule structure. In preparing the derivatives listed in the table below, 1.0 part of cornstarch was suspended in 1.25 parts of water and the indicated amounts of the selected acyl guanidine were introduced. The pH was controlled at the indicated value by periodic addition of dilute hydrochloric acid during the entire course of the reaction.

The reaction was allowed to proceed with agitation at room temperature until there was no further change in pH. The resulting starch ester derivatives were then acidified with dilute sulfuric or hydrochloric acid, recovered by filtration, and subsequently washed with water to remove residual salts. The table below gives the pertinent data for the various derivatives which were prepared. Each reacted starch was examined for acyl content, calculated from the saponification number.

	Esterification reagent		Reaction conditions		
Derivative number	Guanidine derivative of—	Percent on starch	Controlled pH	Time, hrs.	Percent acyl
1	Stearic acid	10.0	8.0	16	1.38
2	Phenylacetic acid	10.0	8.0	16	0.90
3	Acetic acid	8.0	8.0	16	1.32
4	Benzoic acid	6.7	8.0	24	0.23

Example 2: This example illustrates the use of acyl guanidines in preparing inhibited starch esters according to this process. In preparing the derivatives listed in the table below, 1.0 part of each of the listed starch bases was suspended in 1.25 to 1.50 parts of water and the indicated amounts of acyl guanidines were introduced.

The pH was controlled at the indicated value by the periodic addition of dilute hydrochloric acid during the reaction. The reaction was allowed to proceed with agitation at the desired temperature until there was no further change in pH. This usually required about 1 hour. Stirring for an additional 15 hours gives a slight improvement in reaction efficiency. The resulting starch ester derivatives were then acidified with dilute sulfuric or hydrochloric acid, recovered by filtration, and washed with water to remove residual salts. The table below presents the pertinent data relating to the various starch derivatives of this example and their preparation.

No.	Starch base	Esterification reagent		Reaction conditions			Sediment volume	
		Guanidine derivative	Percent on starch	pH	Temp., °C.	Time, hrs.	Product	Base
10	Waxy maize	Terephthalic acid	10.0	8.0	RT	16	31.0	None
11	do	1,4-cyclohexane dicarboxylic acid	10.0	8.0	RT	16	16.5	None
12	do	Succinic acid	1.0	8.0	RT	1	25.0	None
13	do	do	1.0	8.0	RT	16	18.0	None
14	do	Adipic acid	10.0	8.0	RT	16	7.5	None
15	do	do	1.5	4.0	RT	2	80.0	None
16	do	do	1.5	5.0	RT	2	57.0	None
17	do	do	1.5	6.0	RT	2	52.0	None
18	do	do	1.5	7.0	RT	2	35.0	None
19	do	do	1.5	8.0	RT	2	21.0	None
20	do	do	1.5	9.0	RT	2	22.0	None
21	do	do	1.5	10.0	RT	2	20.0	None
22	do	do	1.5	8.0	2–3	2	45.0	None
23	do	do	1.5	8.0	48	2	17.0	None
24	Waxy maize (acid converted to 85 fluidity)	Succinic acid	1.5	8.0	RT	2	19.0	None
25	Oxidized corn starch (75 fluidity)	do	1.5	8.0	RT	2	28.0	None
26	Potato starch	do	1.5	8.0	RT	2	10.0	88.0
27	Corn starch (previously treated with 3% diethylaminoethyl chloride hydrochloride).	Adipic acid	1.5	8.0	RT	2	42.0	None
28	Waxy maize	Succinic acid	0.1	8.0	RT	2	56.0	None
29	do	Adipic acid	70.0	8.0	RT	2	10.0	None

The degree of inhibition was determined by cooking an aqueous suspension of the resulting starch product, having a concentration of 1% solids, by weight, in a boiling water bath for a period of 30 minutes. The cooked dispersion was then allowed to stand, at room temperature, in a 100 ml graduated cylinder for a period of approximately 16 hours. In order to show comparative values, the cookability of the base starch was also determined by this method.

Starch Fatty Acid Esters in Chewing Gum

For a long time, natural water-insoluble gums of vegetable origin, i.e., chicle, were used for the major portion of chewing gum base. Unfortunately, the natural gums are subject to substantial price fluctuations as well as unpredictability of supply. Because of these reasons, manufacturers have in recent years tried with varying degrees of success to duplicate the desirable properties of natural gum by means of snythetic resins, rubber and other polymers.

J. Teng; U.S. Patent 3,666,492; May 30, 1972; assigned to Anheuser-Busch, Inc. has developed a chewing gum made from a fatty acid ester of starch such as starch laurate which meets the industry criteria for functional and esthetic qualities. Starch laurate is prepared by reacting lauroyl chloride with starch suspended in dioxane. The substitution reaction is catalyzed by pyridine. Generally about 50% excess of the theoretical amount of lauroyl chloride is used. The reaction is facilitated by heating at higher temperature. Generally, the reaction is maintained at a temperature range of 95° to 100°C and a reaction time of 6 to 8 hours.

After the reaction, laurate gum is precipitated from the reaction mixture by the addition of methanol. The gum is then purified by dissolving in chloroform, followed by repeated water washes and the precipitation of the gum from chloroform by methanol addition. The gum thus collected is then dried at 80°C under reduced pressure. If necessary, the purification by dissolving in chloroform may be repeated until the gum is free of odor. In order to be acceptable as chewing gum base, the starch laurate must have a degree of substitution higher than 2 (theoretical maximum 3). Starch laurate with DS lower than 2 usually does not possess the gummy plasticity. As the DS value decreases, the laurate becomes increasingly less hydrophobic and more water soluble. The starch laurate suitable for making chewing gum has high oil solubility and the starch laurate does not separate out from oil solution upon cooling.

Starch esters of other saturated fatty acids having a chain length of C_8 to C_{14}, also produce gum suitable for chewing gum. Starch esters having a chain length less than 8 carbons are generally amorphous solids. Gummy properties are exhibited by those with a chain length of 8 to 14 carbons. Above the chain length of 14 the esters tend to be brittle. Unsaturated esters often appear as greasy pastes.

Example 1: 405 grams of lauroyl chloride is reacted with 72 grams of cornstarch suspended in 925 ml dioxane. 220 grams pyridine is used to catalyze the reaction. The reactants are maintained at 100°C for 6 hours. 2,000 ml methanol is added to the mixture at 50°C. 475 grams starch laurate precipitates from the mixture and is removed therefrom. The starch laurate is dissolved in 700 ml chloroform, washed with water, and the starch laurate gum is precipitated from the chloroform by the addition of 2,000 ml methanol. The starch gum precipitate is dried at 80°C for 5 hours. The starch laurate has a degree of substitution of 2.9 and a melting point of 85°C.

5 grams of the foregoing starch laurate was mixed with 0.8 ml dipentyl phthalate, 0.3 ml polyglycerol oleate, and 0.3 ml of glycerol. The mixture was placed in an 80°C oven and mixed until uniform. This chewing gum base heated to 80°C was mixed with 50 grams of corn syrup at 80°C. When they were mixed thoroughly, 4.0 grams of powdered sugar was added in small portions and mixed. When the mixture became uniform, flavoring agent was added. The flavoring agent was Perma-Stabil Orange Flavor No. 6033.

Example 2: 10 grams of the above starch derivative is mixed with 1.8 grams of glycerol, 0.2 gram of propylene glycol, and 3.0 grams of infusorial earth. 0.5 gram polyethylene glycol also is added to the mixture. The mixture is heated in a 100°C oven and mixed further to form a gum base. Into the gum base 33 grams of powdered sugar, 15.8 grams of corn syrup, and 1.32 grams of starch are added. The mixture is heated in an oven (100°C) until melted and mixed again. The mixture is kneaded, shaped and rolled into slabs of chewing gum. The following table compiles the results when the foregoing gum is tested by the chewing criteria described below in Numerical Evaluation. The numeral 10 is optimum and both above and below 10 are less than optimum. Brands A through C are commercially purchased chewing gums.

Chewing Test

Sample	Brand A	Brand B	Brand C	Starch laurate chewing gum
Initial stage				
Flexibility	10	10	10	10
Stiffness	10	11	11	10
Adhesion	10	10	9	10
Intermediate stage				
Firmness of low point	9	8	7	10
Duration (mins.)	3	4	4	4

(continued)

Sample	Initial stage			
	Brand A	Brand B	Brand C	Starch laurate chewing gum
	Final stage (base only)			
Smoothness	9	10	7	10
Stiffness	8	9	10	10
Lift	9	8	10	10
Freeness	11	10	10	10
Cohesion	13	11	10	10
Stretch	11	8	8	12
Taste	10	10	10	10

Numerical Evaluation

		0	10	20
Initial	Flexibility	Brittle	Good	Floppy.
	Stiffness	Raglike	Medium	Leathery.
	Adhesion	Crumbly	Gummy	Tacky.
Intermediate	Firmness of low point.	Sloppy	Firm as final	
Final	Smoothness	Rough	Silky	Too slippery.
	Stiffness	Soupy	Medium	Leathery.
	Lift	Plastic	Round	Rubbery.
	Freeness	Tacky	Free	
	Cohesion	Waxy	Slight wet paper	Too wet papery.
	Stretch	Stringy	Full and wide	None (tears).
	Taste	Objectionable	None	

The term "medium" as used above defines a chew which is between softness and firmness.

CROSSLINKED STARCH CARBOXYLATES

Starch may be inhibited, while in granular form, by reaction with polyfunctional reagents, such as diepoxides and dianhydrides, etc., which crosslink the starch molecules within the granule. This reaction results in the formation of covalent chemical linkages between the molecules, thereby adding to the bonding forces which hold the granule together. This increase in the strength of the cohesive forces in the granule thus results in the need for a greater energy requirement in order to disrupt the granule upon cooking such inhibited starch in water.

Inhibited granular starches are desired for various industrial uses. They are particularly useful in applications where the cohesiveness and stringiness of native starches are found to be objectionable. Thus, for example, in the food industry starch products yielding short, smooth pastes on cooking are used as thickening agents in pies, sauces and soups, etc.

It is often advantageous that the crosslinkages which are introduced into inhibited starch products should be labile, i.e., they should be readily destructible. Among the methods for destroying or eliminating these crosslinkages are treatments of the inhibited starches with heat, acids or bases, etc. Crosslinked starch esters are also produced by the processes disclosed in U.S. Patents 3,728,332 page 105, 3,720,662 page 102, 3,839,320 page 99, and 3,720,663 page 103.

Crosslinked Starch Acylates

J.V. Tuschhoff and C.E. Smith; U.S. Patent 3,238,193; March 1, 1966; assigned to A.E. Staley Manufacturing Company have developed a method of preparing crosslinked starch acylates having paste properties which make them particularly suitable for use as an ingredient in foods, such as pie fillings, to which the crosslinked starch acylates impart body and consistency.

These crosslinked starch acylates are prepared by reacting granular starch with a monofunctional acylating agent and a polyfunctional crosslinking agent. However, not all crosslinked starch acylates have the paste properties necessary for use in fruit pie starches. The crosslinked cereal starch acylates of this method have an alkaline fluidity of 50 to 90 cc. The alkaline fluidity is a measurement of the viscosity of the crosslinked starch having essentially no monofunctional acylate groups.

In other words, the alkaline fluidity is purely a measurement of the viscosity of the crosslinked starch. The monoacylate groups are apparently saponified during the alkaline fluidity test. If the alkaline fluidity of the crosslinked cereal starch is below 50 cc the baked pie filling is watery and runs. In effect, the starch granule has not been toughened sufficiently to resist the acid pH. On the other hand, if the alkaline fluidity of the crosslinked cereal starch is above 90 cc, the starch is inhibited to such an extent that the fruit settles out of the pie filling prior to baking.

The acyl groups of the crosslinked cereal starch acylates not only give the pie starch improved clarity, texture and freeze-thaw resistance, but also lower the pasting temperature of the starch. In the absence of these acyl groups, crosslinked cereal starches having an alkaline fluidity of from 50 to 90 cc may be too inhibited to paste properly.

The polyfunctional crosslinking agents of this process are etherifying and esterifying agents having at least two functional groups that react with hydroxyl groups of the starch. The term polyfunctional crosslinking agent refers to compounds having at least two groups capable of reacting with the hydroxyl groups of starch, such as, aldehyde groups, ethylenically unsaturated groups, epoxy groups, halo groups, keto groups, etc. Starches crosslinked with monofunctional crosslinking agents, such as acetaldehyde, are subject to partial hydrolysis when cooked for extended periods at an acid pH.

Representative of the monofunctional acylating agents which may be used in this process are vinyl esters of monocarboxylic aliphatic acids having up to 18 carbon atoms, such as vinyl acetate, vinyl hexoate and vinyl stearate, and anhydrides of aliphatic monocarboxylic acids, such as acetic anhydride and propionic anhydride. The vinyl esters are the preferrred acylating agents since their reaction with starch is more efficient than the reaction of the anhydrides with starch. This effect is illustrated by the following examples.

The CIV cooking test used in the examples approximates the conditions employed in the preparation of fruit pie fillings. This test comprises suspending 60 grams of crosslinked starch acylate in 1,100 grams of a 35% (by weight) aqueous solution of sucrose and then lowering the pH of the suspension to 3.5 with citric acid. The suspension is then placed in a Corn Industries Research Foundation Viscometer equipped with a heating jacket which is maintained at 201°F. The viscosity is recorded at its peak and at 10, 15 and 40 minutes after the suspension has been placed in the viscometer.

Example 1: 30 grams of sodium carbonate in an 8% by weight aqueous solution was added to 1,000 grams of cornstarch (dry solids basis) suspended in 1,150 ml of water. After the alkaline starch suspension (pH 10.2) was heated to 100°F, 2 grams of acrolein was added to the stirred starch suspension. The alkaline fluidity of the starch suspension reached 71 cc in 1 hour and 20 minutes. Four grams of sodium bisulfite was then added to terminate the crosslinking reaction. After mixing for 10 minutes, 70 grams of vinyl acetate was added to the suspension of crosslinked starch. After 45 minutes, the reaction was terminated by adjusting the pH to 6 with aqueous HCl.

The crosslinked starch acylate was then filtered, washed with water and then reslurried in 1,000 ml of water. After the pH was again adjusted to 6 with HCl, the crosslinked starch acylate was filtered, washed with water and then dried to a 12% by weight moisture content. The starch, on a dry solids basis, had an acetyl content of 2.6% by weight. The CIV data on this sample is set forth in the table below.

Example 2: Example 1 was repeated except that the sodium bisulfite was added to the reaction mixture after the crosslinked starch had an alkaline fluidity of 68 cc.

Example 3: Example 1 was repeated except that the sodium bisulfite was added to the reaction mixture after the crosslinked starch had an alkaline fluidity of 74.5 cc.

Example 4: Example 1 was repeated except that the sodium bisulfite was added to the reaction mixture after the crosslinked starch had an alkaline fluidity of 55 cc.

Example 5: Example 1 was repeated except that the sodium bisulfite was added to the reaction mixture after the crosslinked starch had an alkaline fluidity of 63 cc.

Starting Starch	% Acetyl*	Pasting Temperature, °F	Alkaline Fluidity, cc	Viscosity in g-cm 10 min	15 min	Peak	40 min
Unmodified cornstarch	0	184	36	173	175	177	157
Example 1	2.6	177	71	792	880	880	744
Example 2	2.8	179	68	735	856	864	768
Example 3	2.6	178	74.5	776	880	880	776
Example 4	2.3	180	55	704	840	840	720
Example 5	2.3	180	63	672	760	760	696

*By weight on dry solids basis

Starch Carbonates with Labile Crosslinks

W. Jarowenko and M.W. Rutenberg; U.S. Patent 3,376,287; April 2, 1968; assigned to National Starch and Chemical Corporation have prepared inhibited, granular starch derivatives crosslinked by labile, carbonate ester linkages characterized by their instability in the presence of acids, bases or heat. These derivatives have use as thickening agents for food products and in various sizing, coating and adhesive applications. It was found that by reacting granular starch bases with phosgene at controlled temperatures and pH levels in either aqueous or nonaqueous media, inhibited granular starch products, presumably crosslinked by carbonate ester linkages were obtained. These products are produced according to the equation shown on the following page.

$$StOH \ + \ Cl-\overset{\overset{\textstyle O}{\|}}{C}-Cl \ + \ StOH \ \longrightarrow \ StO\overset{\overset{\textstyle O}{\|}}{C}OSt \ + \ 2HCl$$

where StOH represents the starch molecule. The resulting carbonate ester linkages are labile, thereby providing the inhibited granular starch products with the unique ability to be subsequently converted into completely swollen products. In the following examples which illustrate the process all parts given are by weight unless otherwise noted.

Example 1: 100 parts by weight cornstarch (in intact granule form); 125 parts by weight water and 5 parts by weight phosgene were introduced into a reaction vessel fitted with a condenser and a means for mechanical agitation. The phosgene was introduced into the agitated starch dispersion in a gaseous state. While agitation continued, the pH of the system was maintained at approximately 7 by the continued slow addition of a 3% by weight, aqueous sodium hydroxide solution. The temperature of the system was kept at 5°C during the entire reaction procedure. After agitation for 2 hours, at these conditions, the slurry was adjusted to a pH level of 5, filtered, washed and dried. The resulting product had maintained its original granular form.

The degree of inhibition was determined by cooking an aqueous suspension of the resulting starch product, having a concentration of 1%, by weight, solids, in a boiling water bath for 20 minutes. The cooked dispersion was then allowed to stand, at room temperature, in a 100 ml graduated cylinder for approximately 16 hours. In order to show comparative values, the cookability of the raw cornstarch base was also determined by this method. The cornstarch base showed a sediment volume of 46 ml in contrast to the 15 ml sediment volume of the inhibited starch reaction product. This comparison clearly illustrates the effectiveness of the process as a means of inhibiting granular starches.

Example 2: This example illustrates the use of a variety of granular starch bases in this process. It also illustrates the high degree of inhibition of the resulting starch reaction products. The general procedure used to prepare the products was identical to that described in Example 1. In all cases, an aqeous system was utilized and the pH was adjusted to the desired level by the addition of a 3%, by weight, aqueous sodium hydroxide solution. The following table lists the starch products, the components of the respective reaction systems, and the reaction conditions utilized.

Starch Base	Parts by Weight Phosgene per 100 parts of Starch	pH Level of Reaction	Reaction Temperature, °C	Reaction Time, hours	Sediment Volume, ml	
					Reaction Product	Base
Cornstarch	5.0	7-8	RT	16	14	46
	5.0	9	<8	2	15	46
	5.0	11	<8	2	20	46
Potato starch	5.0	7	<10	2	20	†
High amylose cornstarch (55% amylose)	5.0	7	<10	2	5	13
Waxy maize	0.1	7	RT	2	85	90
	0.2	7	RT	2	51	90
	0.4	7	RT	2	39	90
Oxidized cornstarch*	5.0	7	<10	2	53	†
Diethylamino ethyl ether of starch**	5.0	5	8	2	75	†
Cornstarch octenyl succinate***	5.0	6	8	2	25	45

†No sediment
*Converted by reaction with NaOCl to 75° fluidity
**Prepared as in Examples 1 and 2 of U.S. Patent 2,813,093
***Prepared as in Example 9 of U.S. Patent 2,613,206

The table further presents a comparison of the degree of inhibition of the resulting reaction products as compared with the starch bases from which they were respectively derived.

Labile Crosslinks Using Mixed Anhydrides

M.M. Tessler and M.W. Rutenberg; U.S. Patent 3,699,095; October 17, 1972; assigned to National Starch and Chemical Corporation have also found that one can prepare inhibited starch products containing labile ester crosslinkages by reacting starch bases with mixed carbonic-carboxylic anhydrides of polycarboxylic acids.

The inhibited starch products of this process are characterized by their ability upon cooking to initially yield thin low viscosity dispersions which can be subsequently thickened to high viscosities by the destruction of their labile crosslinkages, thereby leading to the further swelling and possible rupture of the starch granules.

These inhibited starch products are particularly useful in the food industry where they may be employed as thickeners for pies, sauces and soups, baby foods, etc. They are also useful in the canning industry as a result of their unique behavior during retorting of the canned food products. In this procedure, the crosslinkages of these inhibited starch products are initially intact and the starch dispension is in a thin state, thereby enabling the heat utilized for sterilization of the food product to rapidly penetrate the can and its contents. The continued application of heat, however, serves to subsequently destroy the crosslinkages of the inhibited starch thickeners thereby activating their delayed thickening properties to produce desirable high viscosity dispersions.

Suitable mixed carbonic-carboxylic anhydrides of a polycarboxylic acid may be prepared using dicarboxylic acids such, for example, as glutaric acid, adipic acid, pimelic acid, suberic acid, azelaic acid, sebacic acid, fumaric acid, 2,5-dimethyl adipic acid and cis-cyclohexane-1,2-dicarboxylic acid. Other mixed carbonic-carboxylic anhydrides may be prepared using such tricarboxylic acids as pentane-1, 3,5-tricarboxylic acid, benzene-1,3,5-tricarboxylic acid and citric acid.

Example 1: This example illustrates the preparation of inhibited starch products of this process. A total of 1,000 parts of waxy maize starch was suspended in 1,500 parts of water at room temperature and the pH adjusted to 8.0 with a 3% aqueous solution of sodium hydroxide. The starch suspension was stirred and 6.4 parts of bis(ethylcarbonic) adipic anhydride was added slowly over a 30 minute period. A pH of 8.0 was maintained by the addition of 3% sodium hydroxide at periodic intervals during the entire reaction.

The reaction was complete after 3 hours, at which point there was no further change in pH. The pH was then lowered to 6.5 with 6 N sulfuric acid and the product recovered by filtration. The product was washed 3 times with water to remove salts, and dried. In this manner, an inhibited waxy maize was also prepared using 10.2 parts of bis(ethylcarbonic) adipic anhydride.

Example 2: This example illustrates the preparation of additional inhibited starch products where potato starch is treated with 2.0% of bis(ethylcarbonic) fumaric anhydride, based on the weight of the starch. The procedure used to inhibit the starch was identical to that described in Example 1 except that the starch base was potato starch and the mixed anhydride was bis(ethylcarbonic) fumaric anhydride. A total of 15 parts of the thus-prepared starch derivative and 270 parts of a pH 6.07 maleic acid buffer solution were placed in 8 ounce cans.

The filled cans were then heated in a water bath at 212°F for 20 minutes, after which the viscosity of the starch solution was measured. The cans were retorted in a glycol bath maintained at 255°F while being mechanically tumbled. The cans were removed after 15 minutes, rapidly cooled to room temperature, and the viscosities of their contents were measured with a Brookfield RVF viscometer employing the indicated spindle at 10 rpm. The above procedure was repeated except that the temperature of the glycol bath was lowered to 240°F. The table below summarizes the data obtained.

	Brookfield Viscosity, cp	Spindle Number
Starch dispersion cooked at 212°F for 20 minutes	1,950	4
Starch dispersion retorted at 240°F for 15 minutes	28,000	5
Starch dispersion retorted at 255°F for 15 minutes	36,000	5

The above data likewise show the crosslinkages of these inhibited starch products to withstand cooking at temperatures of 212°F and unable to withstand temperatures of 240° to 255°F.

STARCH PHOSPHATES

Starch phosphates are anionic polyelectrolytes which give pastes with increased clarity over the parent starch. The products are used mainly as thickeners in the food industry and show good freeze-thaw stability. Commercial starch phosphate containing 0.35 to 2.5% bound phosphorus is prepared by heating a dry mixture of starch and sodium tripolyphosphate or an orthophosphate. In addition to the following six processes, see U.S. Patent 3,553,194 page 143, describing starch ether phosphonates.

Continuous Phosphorylation Apparatus

In a process described by *J.A. Hay; U.S. Patent 3,399,200; August 27, 1968; assigned to American Maize-Products Company* starch granules, free of excess unabsorbed water and containing an absorbed alkali metal phosphate salt, are passed through a phosphorylation zone which is disposed horizontally lengthwise and has a plurality of parallel spaced heated surfaces coextensive with its length. These surfaces are rotated through the granules as they traverse the zone to effect a rapid and uniform reaction between the starch and the phosphate salt.

One of the most important steps in this method is that of causing the starch particles to tumble into contact with heated surfaces within the zone of phosphorylation. This tumbling movement, which takes place continuously, maintains direct contact between a maximum proportion of the total surface area of the starch granules and the heated surfaces in the phosphorylation zone. As a result, a large amount of heat is quickly and uniformly absorbed by the starch granules and their temperature rapidly raised to reaction temperatures. Thus, the residence time of each granule in the phosphorylation zone is relatively short, usually not more than 1 hour, so that the granules are quickly and uniformly reacted without excessive coloration or deterioration as might be the case with more prolonged exposures to the high reaction temperatures.

One preferred apparatus for carrying out the process is a horizontally positioned rotatable drum into one end of which starch granules containing the absorbed alkali metal phosphate salt are introduced. The drum is provided with heating means which comprises a plurality of tubes positioned within the drum in spaced relationship around the interior wall. A heating medium such as steam or high temperature liquids or gases is introduced into these tubes and, as the drum rotates, the starch granules are tumbled into direct contact with the tubes.

This heats the starch granules to reaction temperature of 120° to 175°C and the phosphorylation reaction will be completed by the time the starch granules reach the discharge end of the drum. A flow of air through the drum, countercurrent to the direction of travel of the starch, removes moisture and dust as well as any gases which may be evolved during reaction.

As shown in Figure 4.1, a cylindrical drum **10** is connected for rotation through the chain drive **12** to motor **14**. A hopper **16** leads into one end **18** of the drum. From the end **18** extends a vent line **20** through which dust and vapors may exit from within the drum **10**. Positioned adjacent the opposite end **22** of the drum **10** is a discharge chute **24**. Leading into the end **22** is a pipe **26** through which air may be injected into the drum **10**. Also leading into end **22** is a steam line **28** which is connected to the tubes **30** which are positioned parallel to the longitudinal axis of the drum **10** and concentrically around the interior wall.

FIGURE 4.1: CONTINUOUS PRODUCTION OF STARCH PHOSPHATE PRODUCTS

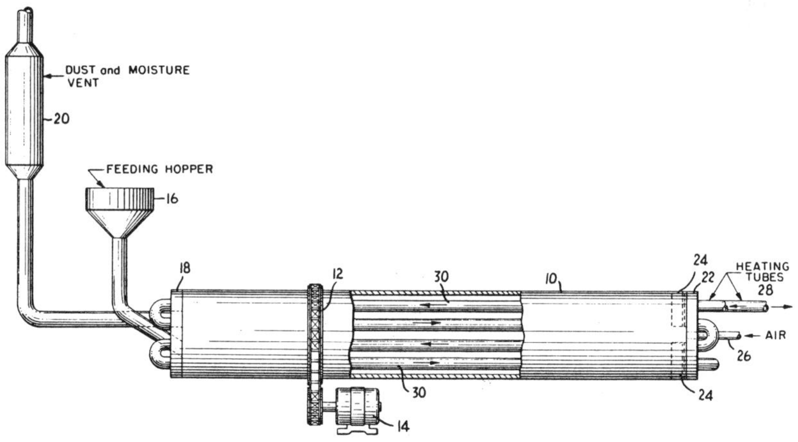

Source: J.A. Hay; U.S. Patent 3,399,200; August 27, 1968

The operation and function of the apparatus is as follows. High pressure steam is fed into tubes **30** through the line **28** and preferably dry air is injected into the interior of drum **10**, and out through vent line **20** as drum **10** is rotated. Starch granules containing absorbed alkali metal phosphate salt are then introduced into the end **18** of drum **10** from the feed hopper **16**. The starch granules are continuously tumbled into contact with the hot tubes **30** as they travel through the drum **10** toward the opposite end **22**.

This rapidly raises the temperature of the starch granules to 120° to 175°C where the phosphorylation reaction takes place. Any inherent moisture of the starch is volatilized and carried away by the dry air moving in countercurrent flow through the starch from line **26** and out through vent line **20**. This air stream also removes fine dust and any gases which may be evolved during the reaction. Since the starch granules are continuously being injected into the drum **10** from hopper **16**, the granules are forced to gradually move towards the opposite end **22** of the drum **10** until they are discharged through chute **24**.

The total time of travel through the drum **10** is relatively short, usually not more than 1 hour, and the starch granules are uniformly reacted with alkali metal phosphate salt by the time they reach the chute **24**. Moisture is substantially completely removed from the starch granules so that the final product is usually ready for end use or packaging immediately after discharge. The examples illustrate the production of starch phosphate reaction products using the process.

Example 1: Commercial grade cornstarch granules were mixed with an aqueous solution of mono- and disodium phosphates and the mixture was gently agitated for a sufficient period of time to allow the granules to absorb some of the solution. Thereafter, excess unabsorbed solution was removed and the granules were air dried to a moisture content below 20% by weight.

The cornstarch granules containing the absorbed alkali metal phosphate salts were fed into the apparatus shown in Figure 4.1 at the rate of ¼ pound per minute. Oil at a temperature of about 176°C was circulated through the heating tubes **28** and a continuous discharge of starch phosphate reaction product was collected at chutes **24** with the average residence time of each granule in the drum **10** being about 70 minutes. A 5% aqueous solution of the product had a Brookfield viscosity of 50,000 cp.

Example 2: Example 1 was repeated except that the feed rate was doubled to ½ pound per minute and average residence time consequently halved. Under these conditions a starch phosphate reaction product was continuously collected having a Brookfield viscosity of 18,000 cp. Because of the shorter residence time, the product was also somewhat lighter in color.

Mixed Phosphate Esters

N.E. Lloyd; U.S. Patent 3,539,551; November 10, 1970; assigned to Standard Brands Incorporated describes starch mixed esters containing orthophosphate groups and orthophosphate groups esterified with a nonstarch organic group. These starch mixed esters may contain from 0.05 to 1 mol of nonstarch organic group per mol of bound phosphorus. Little or no crosslinked esters are found in the products. The starch mixed esters containing orthophosphate groups and orthophosphate groups esterified with a nonstarch group are prepared by reacting starch with a substantially water-soluble pyrophosphate diester at a temperature of 100° to 170°C. The starch mixed esters are useful as industrial starches and as thickeners for foods. The following examples illustrate the process.

Examples 1 through 4: These examples illustrate the effects of the initial pH of the slurries on the efficiency of the reaction, and on the viscosity of aqueous pastes of the starch esters produced. In Example 1, 41.2 grams of dimethyl acid pyrophosphate was added slowly to 223 ml of a 1.53 N sodium hydroxide solution. The addition was made slowly with cooling. Then, a wet filter cake composed of 324 grams of cornstarch and 277 grams of water was added, and the resulting slurry stirred for 30 minutes at 50°C. The pH of the slurry was 3.

In Example 2, a slurry was prepared at a pH of 5 in the same manner as in Example 1, except that 221 ml of a 1.65 N sodium hydroxide solution was used. In Example 3, the dimethyl acid pyrophosphate was added to 500 ml of a 0.8 N sodium hydroxide solution. This solution was cooled, and 324 grams (dry basis) of cornstarch added. The pH of the slurry was 8.5. In Example 4, a slurry was prepared at a pH of 10 in the same manner as shown in Example 1, except that 259 ml of a 2 N sodium hydroxide solution was used.

The slurries were stirred for 30 minutes at 50°C, and then filtered on a Buchner funnel with vacuum to remove excess liquid. The resulting filter cakes were air-dried. The air-dried filter cakes were heated for 2 hours at 130°C in a forced air oven. During the heating step, the pH of all of the products decreased indicating that acidic groups had been generated. After the heating step, the products were purified by washing repeatedly with water or with aqueous methanol solutions.

As seen from the following table, the pH of the slurries affected the paste viscosity of the starch esters markedly. The slurry prepared at a pH of 3 gave a starch ester which was depolymerized as indicated by a viscosity considerably lower than that of a paste of unmodified cornstarch. Starch esters prepared at higher slurry pH's gave higher viscosities. At a slurry pH of 10, the paste viscosity of the starch ester was about 4 times the viscosity of a paste of unmodified cornstarch. Pastes of all of the starch esters were considerably clearer than pastes of unmodified cornstarch. Reaction efficiencies varied from 26 to 46% and generally increased with decreasing slurry pH due chiefly to increased formation of orthophosphate groups.

Examples 5 through 8: These examples illustrate the effects of heating at different temperatures. The starch esters were prepared in the manner described in Example 4, with the exception that the air-dried filter cakes were heated at temperatures from 140° to 170°C. The table shows that reaction efficiency increased with increasing temperatures and the starch esters contained lower mol ratios of organic group to phosphorus group (OR/P).

The starch esters were extremely thick-boiling. The starch esters of Examples 7 and 8 were so viscous that final viscosities of 6% pastes prepared in the Brabender Visco/Amylo/Graph were beyond the range of the instrument (i.e., above 2,500 Brabender units). This necessitated that their paste viscosities be determined at a concentration of 4 grams of starch ester per 100 ml of slurry.

Examples 9 through 13: These examples illustrate the preparation of starch esters containing different organic groups by the procedure in Example 3. In general, the greater the number of carbon atoms in the organic group, the lower the reaction efficiency. This was probably due to steric hindrance. Owing to the low degrees of substitution, it was not possible to ascertain the second inflection points in the titration curves of the starch esters of Examples 11, 12 and 13. Consequently, the mols of organic groups could not be calculated. All of the starch esters tested gave pastes more viscous and clearer than an unmodified cornstarch paste.

Ex. No.	Organic group	Moles pyrophosphate diester per mole starch in the air-dried filter cake	pH			Heating conditions		Titration analysis			Total reaction efficiency,[1] percent	Final Brabender viscosity (Brabender units)
			Slurry	After air-drying	After heating	Time (hrs)	Temp. (° C.)	P/AGU	OR/AGU	OR/P		
1	Methyl	0.048	3.0	3.1	3.0	2.0	130	0.0213	0.0027	0.13	44	65
2	do	0.042	5.0	5.3	4.3	2.0	130	0.0195	0.0060	0.31	46	460
3	do	0.028	8.5	7.8	6.8	2.0	130	0.0095	0.0030	0.33	34	1,285
4	do	0.036	10.0	10.1	9.3	2.0	130	0.0083	0.0051	0.61	26	2,120
5	do	0.037	10.0	8.8	6.9	2.0	140	0.0137	0.0066	0.48	37	1,560
6	do	0.037	10.0	8.8	6.9	2.0	150	0.0164	0.0081	0.49	44	1,800
7	do	0.037	10.0	8.8	6.8	2.0	160	0.0233	0.0102	0.44	63	[2]1,210
8	do	0.037	10.0	8.8	6.7	2.0	170	0.0300	0.0060	0.20	81	[2]1,190
9	Ethyl	0.032	8.5	8.0	7.2	2.0	130	0.0085	0.0024	0.28	27	1,340
10	Butyl	0.023	8.5	7.9	7.1	2.0	130	0.0057	0.0039	0.67	27	950
11	Iso-amyl	0.022	8.5	8.0	7.4	2.0	130	0.0039			18	1,090
12	2-ethylhexyl	0.036	8.5	7.4	7.1	2.0	130	0.0051			14	930
13	Octyl	0.025	8.5	8.1	7.8	2.0	130	0.0020			8	

[1] Mole ratio of P to AGU in the mixed ester expressed as a percentage of the mole ratio of pyrophosphate to AGU in the dried filter cake just prior to the heating step.

[2] Concentration, 4%.

The ratio of mols of phosphorus per mol of anhydroglucose unit (P/AGU), and the mols of organic group per mol of anhydroglucose unit (OR/AGU) were calculated as follows:

$$P/AGU = \frac{162V_1 N}{1,000W}$$

162 = molecular weight of anhydroglucose unit. V_1 = volume (ml) of NaOH solution required to reach the first inflection point on the titration curve (at pH 6) corresponding to complete neutralization of the first titratable hydrogen (strong acid group). N = normality of NaOH solution (0.1 N). W = weight of starch ester (8 grams).

$$OR/AGU = \frac{162N(3V_1 - 2V_2)}{1,000W}$$

V_2 = volume (ml) of NaOH solution required to reach the second inflection point on the titration curve (at about pH of 8.2) corresponding to complete neutralization of first titratable hydrogen (strong acid group) and half neutralization of the second titratable hydrogen (weak acid group).

Ammonium Phosphate Derivatives

A process described by *N.E. Lloyd; U.S. Patent 3,539,553; November 10, 1970; assigned to Standard Brands Incorporated* relates to ammonium monostarch phosphates having a molar ratio of nitrogen to phosphorus of 0.01 to 2. The ammonium monostarch phosphates, depending upon the reagents used and the conditions of preparation, have properties which make them suitable for a large number of industrial applications and as thickeners for foods. The ammonium monostarch phosphates may be produced by reacting an ammonium metaphosphate or an ammonium polyphosphate with starch. It has been found that the pH of the slurry containing the ammonium metaphosphate or ammonium polyphosphate, water and starch plays an important part in the properties of the ammonium monostarch phosphates produced. The following examples illustrate the process.

Examples 1 through 7: These examples illustrate the effects of the initial pH of the slurries on the efficiency of the reaction and on the viscosity and fluidity of aqueous pastes of the ammonium monostarch phosphates. 350 ml of deionized water was placed in a 1,000 ml beaker and the electrodes of a pH meter immersed therein. 60 grams of metaphosphoric acid pellets (0.75 mol of phosphorus per mol of starch) were added with stirring and while they were dissolving, concentrated ammonium hydroxide (29% NH_3) was added at a rate just sufficient to maintain the pH of the solution as close as possible to the selected pH indicated in the table below under the column marked Initial mixture.

After dissolution of the acid was complete, a final adjustment of the pH was made to exactly that shown in the table under the column marked Initial mixture. The volume of solution was then brought to 500 ml with water and 180 grams of cornstarch containing 10% moisture were added. The mixture was stirred for 20 minutes at a temperature of 25° to 30°C and then filtered with suction. The filter cake was crumbled and air-dried at room temperature to a moisture content of 9 to 16%. A small portion of the sample was suspended in water and the pH measured as shown in the table, under the column marked After air-drying.

A total of seven starch-ammonium metaphosphate mixtures were prepared in this manner with the pH of the mixture adjusted to the values indicated in the table under the column marked Initial mixture. The air-dried mixtures were heated in a vacuum oven for 2 hours at 120°C and 24 inches of mercury negative pressure. A small amount of dry air was passed through the oven continuously to aid in the removal of moisture.

After cooling, the products were suspended in 500 ml of water. The pH's of the products were determined and are shown in the table, under the column marked After heating. The pH of the slurries was adjusted to 5.0 with dilute ammonium hydroxide or phosphoric acid. The slurries were then filtered and the filter cakes washed with distilled water and then twice with water or with 50% aqueous methanol. The resulting filter cakes were air-dried. The table shows the results of analysis and testing of the products prepared at the different pH levels. In Example 7, a strong ammoniacal odor was noted during the air-drying step indicating that excess ammonium hydroxide was present.

Aqueous pastes of these ammonium monostarch phosphates were compared with aqueous pastes of unmodified cornstarch in respect to fluidity and viscosity. Pastes of the ammonium monostarch phosphates were very stable, showing no appreciable change in fluidity on standing at room temperature for 24 hr (see column marked 24-hr paste in the table.) There was no discernible change in clarity or appearance of the pastes on standing. Unmodified cornstarch paste became much less fluid on standing for the 24-hr period (see column marked 24-hr paste), with the formation of lumps, and showed a very noticeable increase in opacity. The viscosities of pastes of the products of Examples 1, 2 and 3 were lower than that of a paste of unmodified starch. This was due to the acidic conditions that prevailed during their preparation and resulted in depolymerization of these products.

Examples 8 through 11: These examples illustrate the use of ammonium polyphosphate as the phosphorylating agent and the effect of 4 different initial pH levels on the efficiency of the reaction and the viscosity of the ammonium monostarch phosphates produced. The polyphosphoric acid used in this example and other examples was a commercial product having an average P_2O_5 content of 83% and approximately the following composition:

Anion Classification	No. of Phosphorus Atoms in Chain	Percent
Orthophosphate	1	5
Pyrophosphate	2	18
Tripolyphosphate	3	18
Tetrapolyphosphate	4	16
Pentapolyphosphate	5	13
Hexapolyphosphate	6	9
Heptapolyphosphate	7	7
Higher homologues	>7	14

To a mixture containing 486 grams of conrstarch (3 mols) and 824 grams of water was added 63.8 grams of polyphosphoric acid (0.75 mol of phosphorus per mol of starch) plus a sufficient amount of ammonium hydroxide (29% NH_3) to achieve the selected pH indicated in the table under the column marked Initial mixture. This mixture was stirred for 30 minutes at 50°C and then filtered with suction. The mixtures were stirred for 20 minutes at a temperature of 25° to 30°C and filtered with suction. The filter cakes were crumbled and air-dried at room temperature to a moisture content of about 9 to 16%. Small portions of the dried filter cakes were suspended in water and the pH measured (see column marked After air-drying in the table).

The mixtures were placed in an air-oven for 30 minutes at 130°C. The pH of the mixtures was measured upon a small portion of the mixtures (see column marked After heating in the table). The mixtures were suspended in water and the pH of the suspensions was adjusted to 5.0 with either ammonium hydroxide or phosphoric acid. The slurries were then filtered and the filter cakes washed with water and methanol. The viscosities and fluidities of the products are shown in the table.

Ammonium Metaphosphate as the Phosphorylating Agent

Example No.	pH Initial mixture	pH After air-drying	pH After heating	Degree substitution	Moles of nitrogen per mole of phosphorus	Aqueous paste fluidity Fresh paste	Aqueous paste fluidity 24-hr. paste	Aqueous paste fluidity Conc. (g./300 ml.)	Viscosity (Brabender units)
1	3.1	3.6	3.7	0.021	0.8	80	79	6	184
2	4.0	4.7	4.5	0.017	0.8	68	67	6	380
3	5.0	5.7	5.6	0.020	0.7	59	56	6	565
4	6.0	6.6	6.2	0.010	1.0	75	75	6	650
5	7.1	7.5	6.7	0.006	1.5	55	59	8	720
6	8.1	7.7	6.4	0.007	1.1	60	59	8	760
7	9.1	7.8	6.8	0.005	1.5	57	58	8	745

Ammonium Polyphosphate as the Phosphorylating Agent

Example No.	pH Initial mixture	pH After air-drying	pH After heating	Degree substitution	Moles of nitrogen per mole of phosphorus	Aqueous paste fluidity Fresh paste	Aqueous paste fluidity 24-hr. paste	Aqueous paste fluidity Conc. (g./300 ml.)	Viscosity (Brabender units)
8	3.0	4.6	4.1	0.021	1.2				425
9	5.0	5.9	5.0	0.016	1.5				605
10	8.0	7.1	5.4	0.022	1.3				715
11	10.0	7.4	5.6	0.016	1.4				620
Untreated cornstarch						46	31	9.5	580

o-Carboxyaryl Phosphates as Phosphorylating Agents

Three processes have been developed at the National Starch and Chemical Co. using phosphorylating agents in preparing starch esters. The starch phosphate esters prepared by these methods may be used as food thickeners in such products as soups, and canned vegetables and fruits, where a thick-bodied, creamy, relatively clear thickening agent is preferable to an opaque, gel-like agent obtained with many untreated cereal starches. Cationic starches which in addition have also been phosphorylated by these processes are useful in the manufacture of paper.

The first process described by *M.M. Tessler; U.S. Patent 3,719,662; March 6, 1973; assigned to National Starch and Chemical Corporation* relates to the preparation of starch phosphate esters by the reaction of selected o-carboxyaryl phosphates with a starch base under specified reaction conditions which includes an aqueous reaction medium.

The reaction is carried out at temperatures from 95° to 125°F for periods of 1 to 24 hours. The pH of the reaction medium may be between 3.0 to 8.0. It is not ordinarily necessary to control the pH of the medium during the reaction. The addition of sodium sulfate to the reaction medium in a concentration of 40 to 60%, based on the weight of the starch, has been found to increase reaction efficiency. The o-carboxyaryl phosphates useful in preparing the starch phosphate esters correspond to the following formula:

$$
\begin{array}{c}
\quad\quad\quad\;\; O \quad OX \\
\quad\quad\quad\;\; \parallel \;\; / \\
\quad\quad O\!-\!P \\
\quad\; / \quad\quad\; \backslash \\
R \quad\quad\quad\;\; OX \\
\;\; \backslash \\
\quad\;\; C\!-\!OX \\
\quad\;\; \parallel \\
\quad\;\; O
\end{array}
$$

where R is selected from phenyl, naphthyl, alkyl substituted phenyl, alkyl substituted naphthyl, the alkyl substituent being up to 6 carbon atoms, halogen substituted phenyl, halogen substituted naphthyl, and X is hydrogen or other cation such as sodium, potassium, lithium, calcium, ammonium and the like. In all instances, the carboxylic group attached to R is in the adjacent position with respect to the attachment of the phosphate group.

The preparation of the various o-carboxyaryl phosphates is well described in the literature. For example, the preparation of salicyl phosphate by the reaction of salicylic acid and phosphorus pentachloride and subsequent hydrolysis of the reaction product has been described by R. Anschuetz and W.O. Emery in *Ann.* 228, 308 (1885). An improved procedure for the isolation and purification of salicyl phosphate has been reported by J.D. Chanley, E.M. Gindler and H. Sobotka in *JACS* 74, 4347 (1952). The preparation of o-carboxynaphthyl phosphates is described by J.D. Chanley and E.M. Gindler in *JACS* 75, 4035 (1953).

Example 1: This example illustrates a typical preparation of a starch phosphate ester in accordance with the process. A total of 2 parts of salicyl phosphate was added to a solution of 12 parts of sodium sulfate in 30 parts of water. The pH of the solution was adjusted to 6.0 with aqueous sodium hydroxide (50%, by

weight) and 20 parts of waxy maize was added. The starch slurry was allowed
to react with agitation for a period of about 16 hours. The temperature of the
starch slurry was maintained at about 113°F throughout the entire reaction period.
The starch product was recovered by filtration and washed 4 times with distilled
water, and air-dried. On analysis, the starch phosphate ester was found to contain 0.11% of phosphorus, by weight. In a repetition of the above procedure,
waxy maize was reacted with salicyl phosphate in similar manner, but the sodium
sulfate was purposely omitted. Analysis of this starch product showed it to contain 0.07% phosphorus, by weight.

The absence of inhibition was determined by cooking an aqueous dispersion of
the starch phosphate ester (0.1%, by weight) in a boiling water bath for 20 minutes. The cooked dispersion was then cooled and allowed to stand at room temperature in a 100 ml graduated cylinder for a period of about 16 hours. A sample of untreated waxy maize was carried through the identical procedure for comparison purposes. Neither one of the two starch phosphate esters formed any
sediment and the resulting dispersions of each ester were very clear. The dispersion produced with the untreated waxy maize was cloudy but no layers or sediment were observed. The above results clearly illustrate that phosphorylation
of the starch had taken place without producing any inhibition.

Example 2: This example illustrates additional representative preparations of
starch phosphate esters in accordance with the method. A total of 2 parts of
5-methyl-salicyl phosphate, 5-chlorosalicyl phosphate and 2-carboxy-1-naphthyl
phosphate, respectively, was added to a solution of 12 parts of sodium sulfate
in 30 parts of water. The pH of each solution was adjusted to 6.2 with aqueous
sodium hydroxide. About 20 parts of a cornstarch which had been previously
hydroxypropylated with 5%, based on the weight of starch, of propylene oxide,
cornstarch which had been acid converted to a degree known in the trade as 75
fluidity, and tapioca starch was added to each of the above solutions, respectively.

Each of the three resulting starch slurries was agitated for 16 hours at 113°F.
At the completion of the reaction period, each of the starch products was recovered by filtration, washed 3 times with distilled water, and air dried. On
analysis, the starch phosphate esters were found to contain 0.13, 0.09, and 0.22%
of phosphorus, respectively.

N-Phosphoryl Imidazole Salts as Phosphorylating Agents

In the second process, *M.M. Tessler; U.S. Patent 3,838,149; September 24, 1974;
assigned to National Starch and Chemical Corporation* prepared starch phosphate
esters by the reaction of N-phosphoryl-N'-alkyl or N-phosphoryl-N'-alkenyl or
N-phosphoryl-N'-aralkyl substituted imidazole salts with a starch base under similar reaction conditions. The N'-substituted N-phosphoryl imidazole salts useful
in preparing these starch phosphate esters correspond to the following formula:

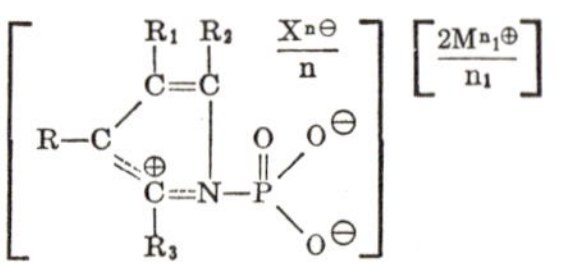

$$\left[\begin{array}{c} \overset{\displaystyle R_1 \quad R_2}{\underset{\displaystyle R_3}{}} \\ \text{structure} \end{array} \right] \left[\frac{2M^{n_1\oplus}}{n_1} \right]$$

The N'-substituted N-phosphoryl imidazole salt has the ring structure with substituents R_1, R_2 on the $C{=}C$, $R{-}C$, R_3 on the $C{=}N{-}P$ bearing O, $O^{\ominus}$ and $O^{\ominus}$, with counterion $\dfrac{X^{n\ominus}}{n}$ and $\left[\dfrac{2M^{n_1\oplus}}{n_1}\right]$.

where R may be an alkyl, an alkenyl, or aralkyl group, the groups having 1 to 8 carbons. R_1, R_2, and R_3 may be hydrogens or, at times, one of these designations may be an alkyl group having 1 to 3 carbons, and the remaining two are hydrogens; X is an anion, e.g., chloride, bromide, sulfate, etc; M is a cation, e.g., sodium, calcium, potassium, ammonium, etc or hydrogen at relatively low pH levels; and n and n_1 represent the valence number of X and M, respectively.

The preparation of the above N'-alkyl or N'-alkenyl or N'-aralkyl substituted N-phosphoryl imidazole salts is well described in the literature. For example, the preparation of the barium salt of N-phosphoryl-N'-methylimidazole by the reaction of N-methylimidazole with phosphorus oxychloride in an aqueous, alkaline medium at a low temperature followed by the addition of a barium chloride solution is described by E. Jampel et al in *Tetrahedron Letters,* No. 31, pp 3535-6 (1968). This reaction which involves N-methylimidazole and phosphorus oxychloride is typical of those occurring in the preparation of any of the useful N-substituted imidazole phosphates described above.

A typical preparation of a starch phosphate ester is carried out by mixing a selected N'-alkyl or alkenyl or aralkyl substituted N-phosphoryl imidazole salt and water. The selected starch base is then added to the resultant mixture. If preferred, sodium sulfate may be added to increase reaction efficiency. The amount of phosphorylating reagent used may vary from 0.5 to 100%, based on the weight of starch depending on the selected starch base employed, the degree of modification desired in the end product, the phosphorylating reagent used, the reaction temperature, and the pH.

The reaction temperature may vary from 65° to 205°F (18.3° to 96.1°C), depending on the ingredients used, particularly the starch base. For example, a granular starch phosphorylation is carried out at 95° to 125°F (35° to 51.7°C), and a nongranular starch phosphorylation at 95° to 155°F (35° to 68.3°C). The pH level of the reaction may be from 9.5 to 13 and preferably from 11 to 12.5.

Reaction time will vary from 1 to 24 hours depending on the reactivity of the reagent used, the amount of reagent used, the temperature employed, etc. After completion of the reaction, the pH of the reaction is preferably adjusted to 5.0 to 7.0 using any common acid such as hydrochloric acid, sulfuric acid, acetic acid, etc. The resultant granular starch product is recovered by filtration and washed free of residual salts with water, and then dried.

Example 1: A total of 4.0 parts of the insoluble barium salt of N-phosphoryl-N'-methylimidazolium chloride was added to a suspension comprising 50.0 parts of corn starch in 65.0 parts of water. The pH of the resultant mixture was adjusted to 11.0 with 3% aqueous sodium hydroxide. Then the mixture was stirred for 6 hours at 104°F (40°C) while the pH was controlled at 11.0 by the addition of 3% sodium hydroxide as required. The reaction mixture was cooled to about 77°F (25°C), whereupon the pH was adjusted to 3.0. The starch product was then recovered by filtration, washed three times with water and air dried. A portion of the starch product was washed with dilute nitric acid to remove any residual inorganic phosphates and the starch was then analyzed for phosphorus.

The analytical method consists of combusting the starch in an oxygen filled Schoniger flask to convert all phosphorus into orthophosphate. The amount of orthophosphate is then determined colorimetrically based on the formation of reduced heteropolyphosphomolybdic acid (molybdenum blue). Upon analysis, it was determined that the starch phosphate ester contained 0.32% of phosphorus, by weight.

Example 2: This example illustrates the preparation of starch phosphate esters using varied starch bases. Into each four beakers, A, B, C, and D, were placed 1.5 parts of the barium salt of N-phosphoryl-N'-methylimidazole, 9.0 parts sodium sulfate, 0.9 parts sodium hydroxide and 38.0 parts of water. The solutions were stirred and to each there was added 30.0 parts of a particular starch base as follows:

> A—potato starch
> B—cornstarch which contained 5.1%, by weight, dry basis, of propylene oxide
> C—cornstarch which contained 0.30%, by weight, dry basis, of nitrogen obtained
>   by treating the starch base with the reaction product of epichlorohydrin
>   and trimethylamine
> D—cornstarch which contained 0.31%, by weight, dry basis, of nitrogen obtained
>   by treating the starch with 4.0% diethylaminoethylchloride hydrochloride

In each case, the resulting slurry was agitated at 104°F (40°C) for 16 hours and cooled to about 68°F (20°C). The pH was lowered to 3.0 by the addition of 10% aqueous hydrochloric acid solution, and the starch products were then recovered by filtration, washed three times with distilled water and dried. A small portion of the starch product was washed with dilute nitric acid and analyzed for phosphorus as described in Example 1. The results are given in the table below.

Material Tested	Percent Phosphorus
Sample A	0.24
Sample B	0.21
Sample C	0.20
Sample D	0.20

N-Acylphosphoramidic Acid Salts as Phosphorylating Agents

In the third process *M.M. Tessler; U.S. Patent 3,842,071; October 15, 1974; assigned to National Starch and Chemical Corporation* uses an N-acylphosphoramidic acid salt with a starch base to produce starch phosphate esters. The N-acylphosphoramidic acid salts used in preparing these esters correspond to the following formula:

$$R-\overset{\overset{O}{\|}}{C}-\overset{\overset{H}{|}}{N}-\overset{\overset{O}{\|}}{P}\overset{O^{\ominus}}{\underset{O^{\ominus}}{<}}\left[\frac{2M^{n\oplus}}{n}\right]$$

where R is selected from a halogen substituted methyl group, a halogen substituted phenyl group in which the halogen, in any group, is fluorine or chlorine or bromine; a phenyl group; an alkyl substituted phenyl group; and an alkoxy substituted phenyl group, where the alkyl and alkoxy groups contain 1 to 6 carbons; and M is a cation, e.g., sodium, potassium, calcium, ammonium, hydrogen, etc.; and n represents the valence of M.

In general, the preparation of the various N-acylphosphoramidic acid salts used in this process is well described in the literature. For example, the N-benzoylphosphoramidic acid may be synthesized by reacting benzamide with phosphorus pentachloride, followed by the addition of water, as discussed by A.W. Titherley and E. Warrall in the *Journal of the Chemical Society,* 95, 1143 (1909). Additional information concerning the preparation of the various N-acylphosphoramidic acid salts may also be obtained by referring to C.Z. Zioudron, *Tetrahedron,* 18, 197 (1962), and A. Lapidot and D. Samuel, *Journal of Chemical Society,* 2110 (1962).

The amount of phosphorylating reagent used to react with the starch base may vary from 1 to 100%, based on the weight of starch. The reaction temperature may vary from 68° to 203°F (20° to 95°C), depending on the ingredients used, particularly the starch base. For example, the preferred temperature for a granular starch phosphorylation is 86° to 125°F (30° to 51.7°C), and that for a nongranular starch phosphorylation is 86° to 155°F (30° to 68.3°C). The pH level at which the reaction is normally carried out may be from 2.5 to 13 and preferably 3.0 to 5.5.

Reaction time will vary from 1 to 24 hours depending on the reactivity of the reagent used, the amount of reagent used, the temperature employed, etc. After completion of the reaction, the pH of the reaction mixture is preferably adjusted to 5.0 to 7.0 using any common acid or common base. The resultant starch product is recovered by filtration and washed free of residual salts with water and dried.

Example 1: To a slurry of 50 parts of cornstarch suspended in 65 parts of water, 3.75 parts of N-benzoylphosphoramidic acid were added. This mixture was stirred, and the pH adjusted to 4.5 by the addition of a 3.0% aqueous, sodium hydroxide solution. This mixture was then agitated for 16 hours at 40°C, and cooled to about 23°C, whereupon the pH level was raised to 5.0. The starch product was then recovered by filtration, washed three times with distilled water, and air dried. On analysis, the resulting starch phosphate ester was found to contain 0.04%, by weight, of phosphorus.

Example 2: A total of 30 parts of waxy maize which had been previously acid converted to a degree known in the trade as 85 fluidity was added to 300 parts water, and the resulting suspension was heated in a boiling water bath for a period of 20 minutes. The resulting starch dispersion was cooled to room temperature and a mixture of 20.3 parts disodium N-benzoylphosphoramidate and 9.7 parts sodium chloride was added. The pH was adjusted to 4.5 with 10% hydrochloric acid and the dispersion was then agitated for 20 hours at 40°C. The reaction product was dialyzed against distilled water for a period of 24 hours and the starch was isolated by precipitation from ethyl alcohol. On analysis the resultant starch phosphate ester contained 0.21% phosphorus, by weight.

CROSSLINKED PHOSPHATE ESTERS

Distarch phosphates may be prepared by treating starch with sodium trimetaphosphate or phosphorus oxychloride. These crosslinked starches have stability to shear and acid thinning needed in some food products. Reference should also be made to U.S. Patent 3,751,410, page 159, on ethers of crosslinked starch phosphates.

Use of Phosphoric Anhydride

A process described by *S. Suzuki, K. Kainuma, K. Tanida and T. Oda; U.S. Patent 3,555,009; January 12, 1971; assigned to Food Research Institute* relates to improvements in the synthesis of starch derivatives. This process involves the synthesis of inhibited-swelling crosslinked starch phosphate by using phosphoric anhydride. The process comprises the steps of suspending starch in water, adjusting the pH value of the aqueous starch suspension to an alkaline state by adding an alkali metal hydroxide or an alkaline earth metal hydroxide, hydrating phosphoric anhydride in the aqueous starch suspension so as to form several kinds of phosphoric acids of different degree of hydration with which the starch is reacted at a temperature lower than the gelatinization temperature of the starch and at a pH value of greater than 8.0, and recovering an inhibited-swelling crosslinking type starch phosphate from the suspension.

An alkali carbonate may be added into the suspension as the buffering agent for keeping the suspension in the alkaline state for a long time. The reacted alkaline suspension is neutralized by adding an acid and is allowed to stand until a sufficient quantity of orthophosphate is formed, then the aqueous phase is filtered to obtain a cake which has absorbed thereon a sufficient amount of phosphate. The cake is dried and heated up to a temperature of from 110° to 170°C for proceeding with esterification, whereby a starch derivative is produced. In the search for a single phosphoric acid or a single salt that may form mono- and diesters in excellent yields, it has been found that phosphoric anhydride and a fresh aqueous solution thereof have remarkable crosslinking effect.

It is possible to control the properties of these products within an appreciable range depending upon the temperature and the duration of time that this temperature is maintained, the temperature being from 110° to 170°C. Thus with a lower temperature and shorter duration, the properties of the diesterified starch predominate so that the products are superior in the acid resistant properties and the electrolyte resistance and have the low temperature stability. When the heating temperature is elevated and the heating period is extended, the diester combination once formed is slowly thermally decomposed and, at the same time, a monoesterifying reaction is promoted between the phosphoric acid absorbed and starch. By virtue of these reactions, near monoester derivatives are produced, which gives transparent liquid paste having the acid proof property, electrolyte resistance, and excellent low temperature stability.

The starch derivatives obtained as above, which have superior acid resistant properties and electrolyte resistance as well as excellent resistance in the freeze-thawing tests, are excellent as the viscosity increasing agent and the moistening agent for sauce, mayonnaise, ice cream, etc. and also as the stabilizing agent for canned foods and the refrigerated foods. It is also possible to use these starch derivatives as the molding agent for medicinal tablets, paste for printing, separator for dry cells, and sizing agent for paper and textiles.

Example: 12 mols of cornstarch (2,260 grams of the material containing 13.9% of water) were suspended in 1.2 mols of aqueous phosphoric acid solution prepared by dissolving 174 grams of phosphoric anhydride into water. The pH value of the suspension was adjusted to 11.0 by adding sodium hydroxide. The suspension was left to proceed reacting at a temperature of 60°C for 3 hours, yielding a crosslinking type starch suspension containing a very small quantity of phosphorus but having a superior swelling inhibition and showing less breakdown in the amylogram. The crosslinking type starch suspension was adjusted in pH value by orthophosphoric acid. After standing over one night, the suspension was filtered, of which a cake obtained on the filter paper was air-dried in a draft at a temperature of from 50° to 60°C. When the water content had become less than 20%, the cake was esterified by heating up to a temperature of 150°C. The variations in water content and in phosphoric acid content obtained by means of sampling every 2 hours during the reaction period are illustrated in the table below.

	Reaction time for esterification, hours				
	0	2	4	6	8
Water content, percent	12. 4	2. 4	0. 6	0. 4	0. 2
Combined phosphorus content, percent	0. 023	0. 23	0. 50	0. 60	0. 63
Ratio of esterification of absorbed phosphorus, per-cent [1]		16. 1	35. 0	42. 3	43. 8

[1] The ratio of esterification of absorbed phosphorus represents the ratio of phosphorus combined with starch by means of the esterifying reaction to that absorbed in the cake on the filter paper.

Use of Tripolyphosphate Salts

W. Jarowenko; U.S. Patent 3,553,195; January 5, 1971; assigned to National Starch and Chemical Corporation describes a process for the preparation of inhibited, intact granule starch products which comprises the reaction, in aqueous medium of a starch base and a tripolyphosphate salt. These inhibited starches have use as thickening agents in foods and in various sizing, coating and adhesive applications.

The inhibited, granular starch products resulting from the process may show varying degrees of inhibition depending upon the extent of the reaction and the consequent number of resulting crosslinkages. The amount of granule inhibition may be determined by performing a sediment volume test. In the latter procedure, an aqueous dispersion of the inhibited product, having a concentration of 0.5% by weight solids, is cooked on a boiling water bath for about 20 minutes. The cooked dispersion is then allowed to stand in a graduated vessel, such as a 100 ml graduated cylinder, at room temperature for a period of about 16 hours. The cooked product will proceed to separate into layers on the basis of its relative inhibition.

In extreme cases, it will completely settle out with the sediment so formed occupying different volumes, depending upon the degree of inhibition of the reaction product. These sediments constitute insoluble granules of the starch derivative whose swollen volumes are relative to the degree of inhibition of the derivatives. Thus, because of their lower swelling and hydration capacity, the more inhibited, i.e., the more crosslinked, products will yield smaller sediment volumes than correspondingly less inhibited products.

However, where the original starch base, e.g., potato starch, exhibits no sediment formation because of the completely swollen highly hydrated and/or disrupted nature of its granules, inhibition in the crosslinked product will be evidenced by the subsequent formation of sediment. This result is directly attributable to the toughened state of the crosslinked granules. In the following examples which illustrate the process all parts are by weight unless otherwise noted.

Example 1: This example illustrates the preparation of inhibited granular starch products by means of the process. A reaction vessel with means for mechanical agitation was charged with 125 parts of water, 1.5 parts of sodium tripolyphosphate and 15 parts of sodium sulfate. While the resulting solution was under agitation, 100 parts of a waxy maize starch were suspended therein.

The pH of the starch suspension was then adjusted to 11.0 by the addition of 0.8 parts of sodium hydroxide. The temperature was raised to 37°C and the inhibition reaction was allowed to proceed for 17 hours. The resulting slurry was adjusted to a pH of 5.0, filtered, washed and dried. The inhibited product obtained was found to have maintained its original granular form. The sediment volume test, as previously described, was used to determine the degree of inhibition of the resulting starch product. Thus, the untreated waxy maize starch base exhibited a sediment volume of 90 ml in contrast to the 12 ml sediment volume of the inhibited starch product of Example 1.

The degree of inhibition of the resulting inhibited starch product was further evaluated by means of a viscosity determination conducted with a Brabender Viscograph. Thus, 23 parts of the above-prepared inhibited starch were mixed with 50 parts of a citrate buffer (a 3:2 mixture of 1 molar citric acid and 0.29 molar sodium citrate) and sufficient distilled water so as to have a total of 460 parts in the system. The temperature of the system was then increased at a rate of 1.5°C per minute until a temperature of 95°C was attained; the latter level then being maintained for a period of 10 minutes. The viscosity, in Brabender Units (BU), was recorded on a continuous chart and was compared with another chart which reflected the viscosity characteristics of the base starch which, in this instance, was a waxy maize starch which had been submitted to a similar heat treatment.

Inhibition of the starch products of this process, under the stated conditions, is generally reflected by a moderate peak viscosity and minimum degradation during the heating cycle, as contrasted with the higher peak viscosities and the substantially greater degradation during the heating cycle which is exhibited by the base starches. Thus, the following viscosity readings were obtained by means of the above-described procedure. The data summarized below clearly illustrate the effectiveness of the process as a means for inhibiting granular starches.

Starch Product	Peak Viscosity (BU)	Viscosity (BU) After 10 min. at 95°C.
Inhibited waxy maize	750	750
Waxy maize base	1,000	300

Example 2: This example further illustrates the inhibition procedure, utilizing a variety of reactants at different concentration levels and under varying reaction conditions. The procedure used to prepare these products was identical to the procedure in Example 1. In all instances, 0.8 part of sodium hydroxide were added to maintain the pH level within the specified range.

The table below lists the starch bases, the components of the respective reaction systems, and the reaction conditions utilized. It further presents a comparison of the degree of inhibition of the resulting reaction products as contrasted with that of the starch bases from which they were respectively derived.

	Parts							
Formulation Number	1	2	3	4	5	6	7	8
Corn starch	100	100	100					
Waxy maize starch				100	100		100	100
A high amylose corn starch containing 70%, by seight of amylose						100		
Sodium tripolyphosphate	3	3	3	1	10	1	0.5	0.5
Sodium sulfate	5	10	20	15	40	30		
Sodium nitrate							10	
Sodium chloride								10
Water	125	125	125	125	125	125	125	125
Reaction temperature (° C.)	40	40	40	37	40	23	23	23
Total reaction time (hrs.)	16	16	16	17	7	16	16	16
pH of reaction system	10.9	10.8	10.5	11.0	10.5	10.5	9.6	9.6
Sediment volume of inhibited product (ml.)	14	9	9	12	13	5	13	14
Sediment volume of starch base (ml.)	35	35	35	90	90	9	90	90
Peak viscosity (B U) (inhibited)				1,740			980	
Viscosity (B U) after 10 min. at 95° C. (product)				1,460			920	
Peak viscosity (B U) (starch)				1,000			1,000	
Viscosity (B U) after 10 min. at 95° C. (base)				300			300	

The results summarized above clearly indicate the effectiveness of the inhibition
process in the utilization of a wide range of reagents and reaction conditions.

Use in Puffed Snack Foods

*L.D. Williams and E.R. Jensen; U.S. Patent 3,666,511; May 30, 1972; assigned
to CPC International Inc.* have developed an improved starch material suitable
for making puffed, deep fat fried snack products comprising a combination of
30 to 70% of a crosslinked waxy starch with 30 to 70% of a crosslinked nonwaxy
starch. The nonwaxy starch preferably is milo starch, cornstarch, or mixtures
thereof.

It is essential that the starch material contain waxy starch and a nonwaxy starch
in the proportion as specified. When these proportions are adhered to, the resulting product is puffy and, in the absence of added flavors, bland tasting. The
nonwaxy starch contains less than 50% amylose and preferably is milo starch,
cornstarch, tapioca or mixtures thereof.

Bland taste is an extremely desirable characteristic, since the natural flavor of
the material will not interfere with an added flavor. When the starch material of
this process is used, puffed snacks can be produced that have any desired flavor,
by adding a selected flavor to the initial raw batch material.

Preferably, the starch material used has a Scott viscosity of from 10 to 23 grams.
That is, a 100 ml portion of a solution of 280 ml volume containing from 10 to
23 grams of the starch material, dry basis, will flow from a standard Scott apparatus in from 20 to 70 seconds. This Scott viscosity is imparted to the material
by crosslinking both the waxy starch and the nonwaxy starch with a crosslinking
agent. Typical crosslinking agents which may be used to attain this Scott viscosity include alkali metal trimetaphosphates, such as sodium trimetaphosphate,
and potassium trimetaphosphate, and anhydrides of organic acids such as acetic
anhydride, succinic anhydride, maleic anhydride, and the like. The preferred
crosslinking agent is sodium trimetaphosphate. The process can be illustrated
by the following example.

Example: Formulation and Evaluation of Puffed Rings — A series of ten starch
blends were formulated, using the proportions of different starches shown below

in the table. Each of the blends was then converted into fat fried snacks using the following procedure. Each blend of starch was separately mixed, respectively, with 3.76% liquid shortening (a highly stable hydrogenated and winterized cottonseed oil-soybean oil blend) and 0.26% of glyceryl monostearate. Water, 15.59% based on the total blend weight, was added.

The resulting wet dough was then fed into a Wenger Model X-25 Continuous Extruder Cooker where it was heated to 300°F while under a pressure of from about 200 to 500 psi. It was simultaneously sheared and agitated by passage through this screw-type extruder, which has a 5 inch diameter shaft and screw. The resulting gelatinized starch paste left the extruder in the form of a rope.

The rope was fed into a Wenger Forming Extruder. The forming extruder was maintained at a temperature of approximately 200°F. In the forming extruder the gelatinized starch paste was passed through a 5 inch diameter screw type extruder section, to remove entrapped air and to compact the gelatinized paste. At the extruder discharge nozzle, the gelatinized paste was forced through shaping dies and was cut into the shape of rings. The rings were dried at room temperature (70° to 120°F) for a measured time period and then were deep fat fried at 390°F for 45 seconds.

The percentages of moisture in the rings after 1 hour and 7 hours of drying, respectively, but prior to cooking, was determined. In addition, the specific volumes of the fried rings and the percentages of fat abosrbed by the rings on frying was measured. The surface smoothness of the fried rings was rated by visual observation. The table below summarizes the results of this testing.

Starch blend, 50/50 mixture	Hours dried	Moisture, percent	Fat absorbed, percent	Specific volume, ml./g.	Surface evaluation*
14 g. Scott, waxy and non-waxy milo	1	15.98	36.7	2.72	2–3
	7	10.67	21.7	3.40	1
17 g. Scott, waxy and non-waxy milo	1	16.83	40.6	3.02	2–3
	7	10.84	26.3	3.23	1
20 g. Scott, waxy and non-waxy milo	1	17.60	38.2	3.56	4–3
	7	12.25	35.0	3.77	1
22 g. Scott, waxy and non-waxy milo	1	19.66	24.4	3.33	4
	7	12.50	33.8	3.70	1–2
13 g. Scott waxy milo and 14 g. Scott non-waxy milo	1	16.60	37.9	4.31	3
	7	11.66	33.3	4.27	1
12 g. Scott waxy milo and 17 g. Scott non-waxy milo	1	15.56	39.3	4.00	2–3
	7	11.47	28.0	4.00	1
12 g. Scott waxy milo and 23 g. Scott non-waxy milo	1	17.47	27.6	3.54	4–3
	7	12.02	35.4	3.94	1
14 g. Scott, waxy and non-waxy milo, and 4% potato	1	17.05	40.5	3.56	4–3
	7	11.00	28.3	3.21	1
22 g. Scott, waxy and non-waxy milo, and 4% potato	1	16.27	34.0	3.66	3
	7	11.83	28.2	3.72	1
12 g. Scott waxy milo, 17 g. Scott non-waxy milo, and 4% potato	1	18.22	34.1	2.90	3
	7	11.70	34.2	3.95	1

* 1=smooth; 2=moderately smooth; 3=moderately rough; 4=rough.

Blends of waxy cornstarch and nonwaxy cornstarch, and of waxy cornstarch and nonwaxy milo starch give similarly good fried snacks. The data indicate that puffed deep fat fried snacks can be produced by the method of this process from blends of waxy and nonwaxy starches of different Scott viscosities. Also indicated is the advisability of drying the extruded shaped snack items prior to deep fat frying to obtain better surface characteristics in the final deep fat fried product. The waxy starches and nonwaxy starches of different viscosities were obtained by crosslinking the respective starches with sodium trimetaphosphate.

The deep fat fried snack products produced as described above were very bland in flavor. The effect of using a second and different drying procedure upon each of the ten blends was also determined. The second drying procedure used was as follows.

Rings produced from each of the ten blends were dried to a moisture content of from 17 to 21% at room temperature (70° to 120°F). The rings from each blend were then further dried at 250°F for 4 minutes. The moisture contents of the rings from each blend both before and after the 250°F treatment were measured. The percentages of fat absorbed on frying and the specific volume after frying of each set of rings were determined. The table below summarizes the results of this testing.

	Moisture, percent		Fat absorbed percent	Specific volume, ml./g.
Starch blend, 50/50 mixture	Before 250° F. drying*	After 250° F. drying		
14 g. Scott, waxy and non-waxy milo	18.1	15.4	25.9	2.98
17 g. Scott, waxy and non-waxy milo	20.7	18.0	28.1	3.00
20 g. Scott, waxy and non-waxy milo	18.3	16.1	19.7	4.62
22 g. Scott, waxy and non-waxy milo	19.3	18.0	17.4	3.42
13 g. Scott, waxy milo and 14 g. Scott non-waxy milo	18.2	16.9	25.0	5.32
12 g. Scott waxy milo and 17 g. Scott non-waxy milo	17.7	15.0	21.9	4.07
12 g. waxy milo and 23 g. Scott non-waxy milo	17.1	15.3	13.6	3.62
14 g. Scott waxy and non-waxy milo and 4% potato	18.7	16.5	29.6	4.32
22 g. Scott, waxy and non-waxy milo and 4% potato	18.8	16.4	25.9	3.92
12 g. Scott waxy milo, 17 g. Scott non-waxy milo, and 4% potato	16.8	14.7	17.4	4.51

*1 hour drying at room temperature (70° F.–120 F.°).

The data indicate that short time drying at 250°F is effective in reducing the amount of absorbed fat and usually increasing the specific volume of the puffed deep fat fried snack products.

Use in Artificial Rice

A method of producing artificial rice enriched with a large amount of amino acids has been developed by *N. Katsuya, T. Sagara, R. Takahaski, T. Yoshida and T. Ojima; U.S. Patent 3,628,966; December 21, 1971; assigned to Ajinomoto Co., Inc., Japan.* Formerly amino acids could not be formed into grains when mixed with water. When they amount to 6 to 50% by weight in the grains formed by mixing amino acids with starch or protein, the obtained grain was rather brittle and its hardness was merely 1 to 2 kg by hardness meter. It has been found that it is essential to mix the raw materials in two steps.

The first step is to mix amino acids with starch, knead the mixture with water and heat with steam; and the second step is to mix suitable binder such as gluten and starch (with vitamins and minerals, if necessary) with the product obtained in the first step and knead it with water. The product of the second step is shaped in a granulator.

In the first step of mixing amino acids with starch, kneading the mixture with water and heating it with steam, the crystals of the amino acids are covered and solidified by starch, which makes the mixing in the second step satisfactory and brings about good hardness and elasticity and prevents breaking of the arti-ficial rice grains. Amino acids used in the first step are L-lysine hydrochloride, L-threonine, L- or DL-methionine and L- or DL-tryptophan. The starch used in the method, should be waterproof and have a low expansion rate such as is found in crosslinked starch (e.g., distarch phosphate, and starch crosslinked with epichlorohydrin and amylose, for example, corn amylose). The ratio of amino acids and starch may be varied according to the desired enrichment. The amount of water used for kneading the amino acids and starch should be 20 to 50% based on the weight of starch, which is enough to gelatinize the starch partly during heating with steam.

The heat treatment is performed for partly gelatinizing the starch at 90° to 120°C for 45 minutes to 1 minute. If the heat treatment is more severe, the starch granules are expanded and broken. The viscosity and elasticity of the starch and its ability to form grains are reduced as its tackiness increases, so that uniform grains of artificial rice cannot be obtained. If the heat treatment with steam is performed at lower temperature for a shorter time, the starch granules do not gelatinize and do not adhere to nutrients.

In the second step, gluten and/or starch, and optional vitamins and minerals are added. Glutens are especially effective for maintaining the mechanical strength of the grains during boiling. Water is added as needed and the mixture is kneaded. The water content of the mixture at the time of forming grains is preferably 28 to 40%. Wheat gluten, corn gluten, and purified soybean protein are most suitable, and preferably amount to about 5 to 50% of the product. After mixing in the second step, the mixture is formed into grains and dried.

The grains formed are dried in warm air of less than 80°C for a suitable time, until the water content is reduced to about 5 to 15%. It is necessary to coat the surface of the grains in order to avoid leaching of the nutrients during prolonged washing and soaking. It is known to gelatinize the surface of rice with formaldehyde and to coat the surface of rice with vinyl resin. It has been found that a coating of shellac or ethylcellulose greatly improves the water resistance of the artificial rice. When crosslinked starch or amylose is used in the first step, dissolution of amino acids during washing and soaking can be avoided adequately even if a pinhole happens to be in the coating.

Example: To a mixture of 400 grams of distarch phosphate, 140 grams of L-lysine hydrochloride and 70 grams of L-threonine, 300 ml of water was added and the mixture was heated at 100°C for 35 minutes with steam, whereby starch granules were semigelatinized. Then 90 grams of cornstarch, 200 grams of wheat gluten, 10 grams of calcium carbonate and 150 ml of water were added, and the mixture so obtained was rolled into sheets (water content 33%) which were converted to grains in a double roller granulator.

The grains were screened and only rice-shaped grains were put into a dryer to reduce their water content to 13%. An alcohol solution containing 34% of shellac was sprinkled (30 ml at one time) on the grains in a pancoating machine under warm air of 40° to 60°C until the added amount of shellac reached 80 grams. The artificial rice thus produced was white, contained 11% L-lysine hydrochloride, 5.5% L-threonine and 4% water, had a hardness of 10 kg by hardness meter, and could hardly be distinguished from natural rice in appearance and taste.

For comparison purposes, 490 grams of cornstarch, 140 grams of L-lysine hydrochloride, 70 grams of L-threonine, 200 grams of wheat gluten, 20 grams of calcium chloride and 300 ml of water were mixed and put into a granulator. The grains obtained were heated with steam at 100°C for 3 minutes. The grains were not uniform and adhered to one another so that their surface became rough when they were separated. Although the grains were coated according to the method mentioned above, they broke and lost their shape during cooking after soaking in water.

ORTHOPHOSPHATE MODIFICATION OF CROSSLINKED STARCH

A process is described in U.S. Patent 2,328,537 where an aqueous suspension of starch is treated with a reactive chloride such as phosphorus oxychloride in an alkaline environment having a pH of about 8 to 12. Although the starch product obtained has proved to be of commercial value, it has a decided disadvantage. When used in canned and frozen foods the starch does not have sufficient thickening power at the temperature at which the food is sold or served. As a result the food is watery and a great amount of thickening agent has to be added to improve the consistency and appearance.

E.M. van Patten and E.L. Powell; U.S. Patent 3,422,089; January 14, 1969; assigned to American Maize Products Company have found that modified starch products having exceptional thickening power at room temperatures are obtained when a starch slurry is treated with a reactive chloride in the presence of a small amount of orthophosphate salts of an alkali metal.

In forming the modified starch by this process, raw starch granules such as, but not limited to the waxy maize starch, waxy sorghum starch, tapioca, cornstarch and wheat starch are suspended in water. The specific gravity of the suspension preferably does not exceed 25°Bé. If the starch granular slurry exceeds 25°Bé, it becomes a puddled cake and can no longer be readily stirred or pumped about. The optimum specific gravity of the starch slurry for reasons of economy and handling was found to be between 12° and 20°Bé. A reactive chloride such as phosphorus oxychloride, phosphorus pentachloride, thio phosphoryl chloride, antimony pentachloride, antimony oxychloride and epichlorohydrin is added to the starch slurry.

Of the various chlorides, phosphorus oxychloride is best. The amount of reactive chloride to be added to the slurry is about 0.01 to 0.3% by weight based on the dry weight of starch in the slurry but other concentrations may be employed if desired. To increase the thickening power of the starch the reaction must be carried out in the presence of small amounts of orthophosphate salts. The quantity of the alkali metal phosphate salts to be added to the starch slurry depends on the pH of the slurry. It is important to maintain a pH of at least 8 and preferably between 10 and 12. Therefore the quantity of salt used should be sufficient to raise the pH of the suspension to at least 8.

It has been found that when anhydrous trisodium phosphate salt was used, a range from about 0.7 to 2% by weight based on the dry weight of starch in the slurry resulted in a pH between 10 and 12 respectively. After the slurry has been prepared the crosslinking reaction will take place. However, since at cold or room temperatures the reaction is very slow the slurry is heated preferably to a temperature from 90° to 120°F, to increase the rate of reaction. It has been found that temperatures about 130°F result in the starch becoming pasty and very difficult to handle. After the process has proceeded sufficiently to obtain the

desired amount of crosslinking in the starch the reaction is stopped by adjusting
the pH below 7 preferably between 5 and 6. This is done by adding an acid
such as hydrochloric acid to the suspension. The modified starch is then removed
from the slurry by filtration or centrifuge, and washed to remove any excess salt
in the product. The modified starch is then dried by conventional means.

Example 1:

	Grams
Waxy maize starch (dry basis)	500
Water	900
Trisodium phosphate, anhydrous	6.5
Phosphorus oxychloride	0.75

The starch was mixed in the water to form a slurry having a specific gravity of
20°Bé. Trisodium phosphate was then added to adjust the pH of the slurry to
10.0. The phosphorus oxychloride was then added and the mixture was heated
to 110°F and agitated for 2 hours. The $POCl_3$ starch reaction was killed by acidifying the mixture with an 18% solution of HCl to establish a pH of 5.5. The
modified starch was then filtered, resuspended in 1 liter of water, refiltered and
dried in an oven at about 180°F for 17 hours. A 5% autoclaved paste of the
dried sample in water was tested with a Bostwick apparatus and recorded a
reading of 5.5 at room temperature.

Example 2:

	Grams
Waxy sorghum starch (dry basis)	500
Water	900
Trilithium phosphate	4.6
Phosphorus pentachloride	1.0

The method of Example 1 was repeated except the pH of the slurry was adjusted
to 11.8, the temperature of the slurry was 110°F and the reaction was allowed
to proceed for 2 hours. A 5% autoclaved paste of the dried sample was tested
with a Bostwick apparatus and recorded a reading of 5.6.

STARCH ETHERS

Like starch esters, starch ethers or hydroxyalkyl starches have improved properties over unmodified starches. In addition to lower gelatinization temperatures, faster granule swelling, gels with improved clarity and cohesiveness which are also properties of starch esters, the ether derivatives are more stable to cleavage by acids. In addition to many industrial uses, hydroxy propyl ethers and cross-linked ethers are useful in food processing.

PREPARATION OF NON-CROSSLINKED ETHERS

The preparation of alkyl ethers of starch has been disclosed in scientific and patent literature for many years. Starch can be alkylated by processes similar to those used to alkylate alcohols. In these reactions the hydrogen of a hydroxyl group is replaced by an alkyl or substituted alkyl radical and an ether group is formed. Thus, starch will react with alkyl halides to form alkyl ethers, and with certain unsaturated compounds by 1 to 4 addition to form substituted alkyl ethers.

Processes for the preparation of starch ethers having either low or high degrees of alkyl group substitution have been disclosed in the patent literature. Low-substituted starch ethers having a substitution range up to around 0.15 alkyl group per anhydroglucose unit have been prepared by reacting etherifying reagents with ungelatinized starch in alkaline aqueous suspensions. These processes generally result in cold-water-insoluble starch ethers which can be readily purified by dewatering on a filter and washing the filter cake.

In the United States starch industry, the process described by Kesler and Hjermstad, U.S. Patent 2,516,633 has been used for many years to monofunctionally etherify starch with alkylene oxides, particularly ethylene and propylene oxides, while obtaining a filterable product which is essentially in native granule form. An alkaline catalyst is used to promote the etherification reaction. Such catalysts include inorganic alkalies, namely the alkali metal or alkaline earth metal hydroxides, and also organic bases, such as amines or quaternary ammonium compounds, including quaternary ammonium hydroxides, trialkyl amines and tri(hydroxyalkyl) amines. In commercial practice, the preferred alkaline catalyst has been sodium hydroxide because of its cheapness and efficiency.

The use of an alkaline catalyst tends to promote swelling or gelatinization of the starch, but with suitable temperature control and/or the presence of a swelling inhibitor, a useful degree of modification can be obtained without swelling the starch to a nonfilterable state. Heretofore, the extent of the hydroxyalkyl substitution has been regarded as sharply limited if a filterable product is to be obtained.

In etherifications with ethylene oxide or propylene oxide where the starch is suspended in water for the etherification reaction, the maximum practical level for obtaining filterable products has been accecpted as not over 0.15 DS, that is, 0.15 mol of ethylene oxide or propylene oxide reacted per anhydroglucose unit of the starch. In commercial practice, even lower degrees of substitution have been accepted to avoid difficulties in filtration and washing, most modifications being with ethylene oxide in the range of 0.05 to 0.1 DS.

High DS Hydroxypropyl Starch

In a process described by *T. Tsuzuki; U.S. Patent 3,378,546; April 16, 1968; assigned to American Maize Products Company* an aqueous mixture of starch granules, alkali catalyst and propylene oxide is held for at least 15 minutes at a temperature of up to about 94°F, prior to reacting the starch and propylene oxide at higher temperatures. The holding step enhances the reaction efficiency to produce granular hydroxypropyl starch ether having an increased quantity of hydroxypropyl substituents.

Example 1: An aqueous slurry containing 500 grams (with 10% moisture) of ordinary cornstarch, 800 ml of water, 5 grams of sodium hydroxide and 70 grams of sodium sulfate was prepared and held at a temperature of 65°F. While the slurry was being stirred, 50 ml of propylene oxide was introduced to the slurry and mixed for approximately one-half hour. The temperature of the mixture was increased to 120°F and maintained at that level for 8 hours before the slurry was neutralized with hydrochloric acid to pH 5.5.

The product hydroxypropyl starch ether, which was still in the granular state, was filtered, washed thoroughly and dried. The DS of this product was 0.050. Another batch of ordinary cornstarch was reacted with propylene oxide in the manner described above, except that the slurry was heated to 120°F prior to the addition of propylene oxide, without use of a one-half hour holding period at 65°F. The resulting product had a DS of 0.035.

Example 2: An aqueous slurry containing 500 grams (with 10% moisture) of ordinary common starch, 800 ml of water, 5 grams of sodium hydroxide and 58 grams of sodium chloride was prepared and held at 70°F. To this slurry 50 milliliters of propylene oxide was added and mixed for approximately 15 minutes. The temperature of the mixture was raised to 120°F and maintained at that level for 8 hours before the slurry was neutralized to pH 5.5. The resulting hydroxypropyl starch ether in the granular state was filtered, washed thoroughly and dried. The DS of this product was 0.047. Another batch or ordinary cornstarch was reacted with propylene oxide in the manner described above, except that the slurry was heated to 120°F prior to the addition of propylene oxide whereby the holding period of 70°F was omitted. The product had a DS of 0.030.

Highly Hydroxypropylated Potato Starch

A hydroxypropyl starch which remains in the granule form even when alkylated to a high degree of substitution and the processes for making the same are disclosed by *E.T. Hjermstad; U.S. Patent 3,577,407; May 4, 1971; assigned to Penick & Ford, Limited.* Potato starch in granule form is given a pretreatment in aqueous suspension at elevated temperature to markedly increase the stability of the granules against swelling or gelatinization. The pretreated starch is then reacted with an hydroxypropylating agent in an alkaline aqueous system to yield a high substituted hydroxypropyl starch in granule form which can be dewatered and washed with water to a high purity with little or no swelling.

The highly substituted hydroxypropylated starch remains in the ungelatinized granule form and can be readily dewatered and washed free of salts and side reaction products. The product dries to a white, nongritty state which disperses readily in water without lumping. When untreated potato starch is hydroxypropylated under identical conditions to the extent obtained by this process, it swells and becomes unfilterable and unwashable with water, due to forming a water-impervious filter cake.

Such products, even if purified by extreme methods, such as pressure filtration, centrifugation or dialysis, are very sticky or gelatinous and when dried form a very hard, horny mass which is extremely difficult to grind and which does not readily redissolve in water. Warm water stabilized potato starch can be hydroxypropylated to levels as high as 20% hydroxypropyl content or 0.7 mol of hydroxypropyl per $C_6H_{10}O_5$ mol of starch and remain readily filterable and water washable without swelling.

One property of higher substituted starch ethers, for example, 4 to 10% hydroxyethyl or hydroxypropyl content, is the very greatly increased resistance to freeze-thaw effects as compared with underivatized commercial starches and low substituted ether derivatives. When normal starch pastes are subjected to slow freezing and thawing or repeated freezing and thawing, their pastes undergo syneresis, become spongy and insoluble, and lose their smooth, cohesive texture. Highly substituted starch ethers form pastes which are practically unchanged on repeated freezing and thawing. An edible starch ether, for example, highly substituted, purified hydroxypropyl starch prepared by the process, is therefore ideally suited for use as thickeners and stabilizers in frozen foods such as frozen soup, gravies, fillings, etc. The process is illustrated in the following example.

Example: The effects of warm-water treatment on hydroxypropylation of various starches are shown below. Potato, corn, tapioca and waxy maize starches were suspended in water in 35% dry substance concentrations at pH 6.5 and the suspensions agitated at 132°F for 20 hours. The suspensions were then reacted with propylene oxide at 100°F using 1.5% NaOH on dry starch, 20% Na_2SO_4 on total water, and a starch to water ratio of 30:70. The results are given in the table on the following page. The data indicate that corn, tapioca and waxy maize starch are not benefited by prior heating of their suspensions at 132°F, while potato starch treated this way was substituted to a high level without swelling. Tests 5 and 6 showed that much higher degrees of hydroxypropyl substitution can be obtained on cornstarch which has not been treated by heating in aqueous suspension.

| Test | Starch | Heat—H_2O treatment | | | | Total P.O. used starch, percent | No. of P.O. addi-tions | Suspen-sion charac-ter | Wash-ability | Purification | Filter cake | Ash, per-cent | H.P.[1] content, percent |
		Concen-tration, percent	pH	Temp., °F.	Time, hrs.								
1	Potato	35	6.6	132	24	30	2	Fluid	Fast	Dil. and washed 3 times.	Normal	.12	17.6
2	Corn	35	6.5	132	24	10	1	Thick	Nil	Swollen, discarded			
3	Tapioca	35	6.5	132	24	10	1	...do	Nil	...do			
4	Waxy Maize	35	6.5	132	24	10	1	...do	Nil	...do			
5	Corn			None		16	1	Fluid	Fair	Dil. and washed 3 times.	Normal	.01	7.33
6	do			None		20	1	Thick	Slow	Dil. and washed 2 times.	Soft	.54	9.35
7	Potato	35		None		30	2	...do	Nil	Swollen, discarded			

[1] Hydroxypropyl.

High DS Hydroxyalkyl Starch Ethers

In a different process *E.T. Hjermstad; U.S. Patent 3,706,731; December 19, 1972; assigned to Kcinep & Drof, Inc.* has prepared uninhibited granule starches with high levels of alkylene oxides by a water suspension etherification process using an alkaline catalyst in combination with an alkali metal sulfate. The alkali metal sulfate is essential, and it is used at a concentration of at least 3 to 5 parts per 100 parts of water.

Although the reaction is carried to a high substitution level measured as hydroxy-alkyl (DS), the starch, although not crosslinked, remains in native granule form and is both filterable and washable. This permits substantially complete removal of the salts and side reaction products by ordinary filtration and washing tech-niques. The method can be used to produce starch products, which are believed to contain monofunctionally linked oxyalkyl polymer chains, and the products are free of bifunctional crosslinks.

The process involves suspending starch in water containing sufficient alkali, such as NaOH or KOH, to promote the etherification, and a concentration of an alkali metal sulfate sufficient to prevent swelling of the substituted reaction product by the alkaline water. Ungelatinized and uninhibited starches which have been etherified to a high degree with ethylene or propylene oxide in an alkali metal sulfate solution can be washed on a filter with little or no swelling, even though degrees of ether group substitution (measured as hydroxyalkyl content) are obtained at levels above those which heretofore have been found sufficient to make the starch cold-water swelling.

The products are different than any heretofore known, since they require heat for gelatinization although having hydroxyalkyl contents which would have made prior art products cold-water swelling, unless the products were inhibited by bifunctional crosslinks between ether groups. The process is advantageous for obtaining a substitution level (measured as hydroxyalkyl content) within the range from 0.2 to 1.5 DS.

The products, when dried, are similar to normal commercial starches in their physical form, except that they are much more resistant to swelling and gelati-nization than would be predicted based on measured DS. Their grits disintegrate readily in water and when the temperature is raised their water suspensions gelati-nize. However, their gelatinization properties are very different from those of starch ethers having equal or even lower monofunctional molar ether contents.

Such prior art ethers of starch gelatinize or swell very suddenly in cold water and are usually difficult to disperse to a lump-free suspension because the rapid swelling of the surface of their grits produces a somewhat water-impervious barrier which retards further dispersion. The highly substituted ethers produced by the process have no tendency to lump when added to water at normal or low temperatures. They are also bland tasting and their films clear and glossy and non-hygroscopic, due to the absence of salts and side reaction products. The experimental evidence indicates that the process promotes polymerization of the alkylene oxides to form graft polymers of starch with chains of oxyalkyl groups, such as oxyethyl or oxypropyl groups.

Example 1: Unmodified cornstarch was made up into a suspension having a starch to water ratio of 30:70 and containing 5 parts by weight of Na_2SO_4 per 100 parts by weight of water and 1.5 parts by weight of NaOH per 100 parts by weight of starch solids. The salt and alkali were added together slowly with agitation sufficient to give uniform distribution without local swelling of the starch at the point of addition. Five parts by weight of ethylene oxide per 100 parts of starch solids were added and the suspension agitated in a closed vessel at a constant 100°F temperature for 24 hours. This was repeated with another 5 parts by weight of ethylene oxide for another 24 hours.

The suspension was then neutralized with acid, diluted to around 20% starch concentration and dewatered in a suction filter. The filter cake was resuspended in about 20% starch solids concentration and dewatered 2 more times. The filter cake was similar to that of normal starch and was not soft and sticky. The dried filter cake was nongritty and dispersed readily in water. The dried starch contained 0.19% ash and analyzed 6.3% by weight of hydroxyethyl by the analytical method given in *Anal. Chem. 28:* 892 (1956), which shows that 0.24 mol of the ethylene oxide reacted per mol of starch.

The gelatinization temperature range of the product was determined by slowly raising the temperature of a 5% suspension of the product and placing a drop of the suspension in an Abbé Refractometer at intervals and reading the Brix scale. The temperature at which a Brix reading is first obtained and the temperature at which a 5% reading is obtained was considered to be the gelatinization temperature range of the starch.

Example 2: Unmodified potato starch was made up into a suspension having a starch to water ratio of 35:65 and containing 10 parts by weight of Na_2SO_4 per 100 parts by weight of starch solids. The salt and alkali were added together slowly with agitation sufficient to give uniform distribution without local swelling of the starch at the point of addition. Fifteen parts by weight of ethylene oxide per 100 parts by weight of starch were added in third portions at 24 hour intervals and the suspension was agitated in a closed vessel at a constant temperature of 100°F.

After 72 hours the suspension was neutralized with acid, diluted to a 20% starch concentration with water, dewatered and washed. The filter cake was resuspended, dewatered and washed 2 more times, and dried. The filter cake dried to a non-gritty, white powder which analyzed 11% by weight of hydroxyethyl (0.446 DS) and 0.24% ash. The product dispersed readily in water and gelatinized in a temperature range of 137° to 140°F by the test given in Example 1.

Hydroxypropylation Using Phosphite Catalyst

J.V. Tuschhoff and C.E. Hanson; U.S. Patent 3,705,891; December 12, 1972; assigned to A.E. Staley Manufacturing Company describe ungelatinized cold-water-swelling hydroxypropylated starch materials produced by reacting a dry granular starch material directly with propylene oxide in the presence of a polybasic, water-soluble salt of phosphoric acid. The starch derivatives produced by this method are particularly useful as food additives.

Polybasic water-soluble salts of phosphoric acid which are particularly good catalysts for the process include the water-soluble secondary and tertiary, metal ortho- and pyrophosphates, and especially those in which the metal cation is an alkali or alkaline earth metal. Examples of such catalysts include disodium phosphate, trisodium phosphate, dipotassium phosphate and tripotassium phosphate. The following examples illustrate the process.

Example 1: To a 21°Bé starch slurry, solid disodium phosphate catalyst was added in an amount sufficient to provide a catalyst concentration of 1 to 1.5% based on the weight of the dry starch. The catalyst starch slurry was then filtered and the starch cake dried to a moisture level of between 8 and 11%. The catalyst treated starch was then placed in a one quart beverage bottle, and propylene oxide added in an amount of 25% based on the weight of the starch dsb. The bottle was capped and placed in a hot water bath maintained at a temperature of 165° to 180°F for a period of 8 to 9 hours.

The product was analyzed and was found to be 11 to 15% hydroxypropylated and substantially completely (100%) cold-water-swelling. The above example was repeated with the exception that the disodium phosphate was replaced with sodium sulfate. It was found that to obtain a product having cold-water-swelling properties comparable to that obtained with the disodium phosphate catalyst, it was necessary to carry out the reaction under essentially the same reaction conditions for a period of 16 to 18 hours, about twice that required when using the disodium phosphate catalyst.

Examples 2 through 10: Example 1 was repeated with the exception that an acrolein crosslinked starch having an alkali fluidity of 49 cc (0.25 N NaOH), was used to demonstrate the catalytic effect that various polybasic water-soluble metal salts of phosphoric acid had on the preparation of hydroxypropyl ethers of crosslinked starches. The reaction was conducted at a temperature of about 180°F for periods of from 4 to 24 hours. A Brookfield viscosity for each of the products was determined, as reported in the table below. The Brookfield viscosity reflects the degree of substitution of the hydroxypropyl starch.

Example	Catalyst	Reaction time (hours)	Starch slurry concentration, percent	Brookfield viscosity data Reading (cps.) 5 min.	Flavor and color properties
2	NaCl	4	10	135	Unsatisfactory.
3	Na_2HPO_4	4	10	2,760	Satisfactory.
4	Na_2SO_4	4	10	7	Unsatisfactory.
5	NaCl	8	5	9,800	Do.
6	Na_2HPO_4	8	5	10,000	Satisfactory.
7	Na_2SO_4	8	5	3,700	Unsatisfactory.
8	NaCl	15	5	8,280	Do.
9	Na_2HPO_4	15	5	8,280	Satisfactory.
10	Na_2SO_4	15	5	8,260	Unsatisfactory.

The catalytic advantage of disodium phosphate over sodium chloride and disodium sulfate is clearly shown in the preceding table. After 15 hours, the degree of substitution on the starch material was essentially the same for all three catalysts; however, the flavor and color properties of products obtained with nonphosphate catalysts were generally unsatisfactory as a food additive.

Example 11: This example shows that the process can be used for producing granular hydroxypropyl starches on a semicommercial basis. A 1,000 gallon granular cornstarch slurry was prepared by adding 3,500 to 4,000 pounds of granular cornstarch to 720 gallons of water. To the starch slurry 3 pounds of anhydrous disodium phosphate per 100 pounds of starch dry substance basis was added. The phosphate catalyst and granular cornstarch were mixed for 15 minutes with a high-speed, six-blade turbine mixer (120 rpm) at a temperature of 90° to 110°F. The catalyst-starch mixture was shakered through a Roball 10XX silk screen, filtered and then dried without washing at a temperature of 225° to 250°F to a starch moisture content of 8 to 11%.

Tricalcium phosphate was mixed with the catalyst-starch mixture in an amount of ½ lb/100 lb of starch dsb. Six hundred pounds of the above catalyst-starch mixture were added to a reaction vessel which was evacuated to a 25 inch vacuum. To remove air, the vessel was flushed by adding 50 psig of nitrogen and then venting to the atmosphere. The reaction vessel containing the starch mixture was then heated to 175°F and evacuated to a vacuum of 25 inches (2 psi absolute).

Propylene oxide was added over a 2-hr period to a pressure of 35 psig. The reaction was conducted for about 7 hr at 175° ± 5°F. After the reaction had run for 6 hr, samples of the reacted starch were removed every ½ hr and the viscosity determined on a Brookfield Viscometer Model RVF by taking 20 g of the reacted starch and adding 380 ml of 75°F tap water. The reacted starch was stirred for about 10 min before the viscosity was taken using a No. 3 spindle at 20 rpm. When a viscosity of between 3,000 to 4,000 cp was obtained (generally in 6 to 8 hr), the reaction was stopped by venting the propylene oxide from the reaction vessel. It was pressurized with nitrogen, evacuated, repressurized with nitrogen and once again evacuated before it was opened and the starch removed.

The reaction product was screened through a U.S. No. 60 screen and reground to the desired particle size. On analysis, it was found that substantially all of the hydroxypropyl starch material produced was cold-water-swelling. When the above process was repeated with a disodium sulfate catalyst rather than disodium phosphate catalyst, the reaction time required to obtain a completely cold-water-swelling product was 12 to 16 hr. When conducted in the absence of tricalcium phosphate, the product did not appear to be as uniformly substituted or possessed of the free-flowing, fluidized characteristics obtained when the tricalcium phosphate was present.

Purification of Hydroxypropyl Starch

A method for purifying crude cold-water-soluble hydroxypropyl starch derivatives resulting from the dry granular reaction of propylene oxide with starch in the presence of a polybasic, water-soluble salt of phosphoric acid has also been developed by *C.E. Hanson and J.V. Tuschhoff; U.S. Patent 3,725,386; April 3, 1973; assigned to A.E. Staley Manufacturing Company.* The purification is accomplished by contacting the crude granular hydroxypropyl starch product with a water-alcohol mixture having a water-to-alcohol weight ratio of 0.1:1 to 0.7:1. It has been found that when the crude granular hydroxypropyl starch is at least 98% cold-water-soluble, the water-alcohol mixture should then have a water-to-alcohol weight ratio of 0.1:1 to 0.5:1. If the amount of water in the mixture is substantially above the 0.5:1 ratio, swelling of the granular hydroxypropyl starch is likely to occur with loss of the granular structure. However, if the amount of water present in the mixture is substantially below the 0.1:1 ratio, only small amounts of the undesirable odors, flavors and other contaminants are removed from the crude starch.

It was also found that a wash solution of 85 to 90% by weight alcohol, in an 0.18:1 to 0.11:1 water-alcohol ratio will not only remove the highly substituted starch fraction from the purified starch product but will also remove substantially all of the other dimers and trimers of propylene glycol and other contaminants, so that the purified starch product is substantially free of undesirable odors and flavors, and is food acceptable.

Ethanol is the preferred alcohol used in the process because it is most easily removed and is most efficient in removing the by-products from the crude starch product. Other monohydric alcohols such as methanol, propanol, isopropanol, butanol, isobutanol, the pentanols, hexanols can be used, but are not as easily removed, or are not as effective as ethanol in removing the by-products. Excellent results have been obtained when 1.0 to 2.0 parts by weight of solvent is used per part of crude starch material.

Although temperatures of 35° to 150°F are in most instances satisfactory, elevated temperatures of 80° to 120°F are generally preferred when a cold-water-soluble hydroxypropyl starch is being washed with a water-ethanol mixture. Effective purification of the crude hydroxypropyl starch is achieved when the water-alcohol mixture is adjusted and maintained at a pH of 7 or below, preferably at a pH of 3.5 to 7.0. If the water-alcohol wash is maintained at a pH of more than 7, there is a good possibility that the granular characteristics of the starch material will be destroyed. If the pH of the water-alcohol mixture is substantially below 3.5, hydrolysis of the starch material can occur.

Example 1: To a 5,000-milliliter, 3-necked, round-bottom flask, 500 grams of a crude, cold-water-soluble granular hydroxypropyl starch was obtained by directly reacting granular starch with propylene oxide without any liquid reaction medium and using a phosphate salt catalyst. This crude product was suspended in 750 ml of a water-alcohol mixture having a weight ratio of water to alcohol of 0.3:1. The round-bottom flask was equipped with a mechanical stirrer having a glass shaft fitted with a Teflon paddle.

The water-alcohol-starch suspension was adjusted to a pH between 5.5 and 6.5 and heated to 100°F. After the suspension had been stirred for 30 minutes, the granular starch was separated from the water-alcohol mixture by filtration and examined. It was found that the washed product had good flavor and relatively low (5 to 7) Gardner color (based on a Gardner white-plate secondary standard).

An analysis of the purified filtrate further showed that the purification process was capable of removing from the granular cold-water-soluble hydroxypropyl starch essentially all (98 to 100%) of those materials which were responsible for the bad taste in the starch material. This filtrate also contained highly substituted hydroxypropyl starch. The granular starch remaining was more uniformly substituted than was the crude hydroxypropyl starch starting material.

Examples 2 through 7: These examples show that the ratio of water to alcohol is extremely important if substantially all of the interfering materials are to be removed while at the same time retaining the granular form of the hydroxypropyl starch materials. With the exception of varying the water-to-alcohol ratio, the procedure in Example 1 was followed. The results obtained are reported in the following table. Percent Purification Efficiency is a measure of the amount of contaminants, particularly dimers and trimers of propylene glycol, and other by-products in the crude hydroxypropyl starch which were removed by water-alcohol washing. The flavor of the starch was determined by a panel of five experienced judges using standard testing procedures.

Example	Weight Ratio, Water-Alcohol	Purification Efficiency, Percent	Starch Form	Flavor of Starch
2	1:1	- - - -	Swollen	Poor
3	0.7:1	87.0	Starting to swell	Satisfactory
4	0.5:1	93.0	Granular	Good
5	0.3:1	99.0	Granular	Good
6	0.1:1	83.0	Granular	Satisfactory
7	0.05:1	10.0	Granular	Poor

Starch Ether Phosphonates

In a process described by *F. Verbanac and K.B. Moser; U.S. Patent 3,553,194; January 5, 1971; assigned to A.E. Staley Manufacturing Company* starch ether phosphonates are prepared by reacting starch with phosphonating agents, where the organo group bonded directly to the phosphorus atom is a vinyl group, an alkane substituted with a vicinal halohydroxy group, a vicinal epoxy group or an omega-halo group. The starch ether phosphonates are those having the structure:

$$St-O-R_1-\overset{\displaystyle O}{\underset{}{\overset{\|}{P}}}\overset{OR_2}{\underset{OM}{}}$$

where St is starch; R_1 is methylene, alkylene of 2 to 18 carbons, or hydroxyalkylene of 2 to 18 carbons; R_2 is alkyl or substituted alkyl of 1 to 18 carbons, alkenyl of 2 to 18 carbons, R_1OSt, St or M; and M is a cation.

The starch ether phosphonates have the same food uses as the direct starch phosphate esters. The phsophonating agents can be used in appropriate concentrations to produce starch ether phosphonates having a DS of 0.0001 to 3. Generally, it is preferred to react 0.01 to 100 parts by weight phosphonating agent per 200 parts by weight starch (dry substance).

In the preferred method of preparing the starch derivatives of this process, granular starch is suspended in a polar solvent, preferably water, in such manner that the starch comprises 2 to 60% by weight of the composition and then an alkaline catalyst is added. Alternatively, the granular starch may be suspended in a polar solvent containing the alkaline catalyst and/or the starch pasted in

water. The phosphonating reaction is then carried out at a pH of 9 to 13 with
one of the preferred phosphonating agents (vicinal halohydroxyalkanephospho-
nate, vicinal epoxyalkanephosphonate or vinylphosphonate) and sufficient alkali
to establish the pH. The reaction proceeds rapidly at moderate temperature,
and in some cases, may require external cooling to prevent gelatinization of the
starch derivative where a granular product is desired. Further details of the
process are shown in the following examples.

Example 1: 1,180 grams granular cornstarch (1,036 grams dry solids basis, 6.4
mols) was slurried in 2,105 grams water containing 13 grams calcium hydroxide
at 45°C. Sixty and seven-tenths grams of diethyl chlorohydroxyethanephosphonate (0.28 mol) was added dropwise over a 5-hour period while maintaining the
reaction mixture at about 50°C with heating and at a pH of about 9 to 11.1 by
adding 35.36 grams of solid calcium hydroxide as needed.

Nineteen hours after the addition of diethyl chlorohydroxyethanephosphonate
was ended, the reaction mass was neutralized wth 6 N hydrochloric acid, filtered,
washed with water and air dried. The granular starch product contained 0.29%
by weight phosphorus and had a DS of 0.015.

The diethyl chlorohydroxyethanephosphonate was prepared by sparging 164.2
grams diethyl vinylphosphonate (1 mol) dissolved in 640 ml of distilled water
with about 72 grams chlorine for 1½ hours while cooling the reaction mass at
about 14° to 30°C. The reaction was terminated when the solution turned yellow
and then sparged with air until colorless. The solution was adjusted to pH 5.0
with aqueous sodium hydroxide, extracted once with 200 ml of methylene
chloride and four times with 100 ml portions of methylene chloride. The extracts were combined, dried over sodium sulfate, concentrated by distilling off
methylene chloride, and fractionally vacuum distilled.

Vacuum distillation of the residue at 105°C/0.34 mm to 105°C/0.35 mm yielded
118.7 grams of diethyl vicinal chlorohydroxyethanephosphonate containing
33.04% C (33.27% theoretical), 6.08% H (6.51% theoretical), 14.09% P (14.30%
theoretical) and 16.56%·Cl (16.37% theoretical).

Example 2: Twenty-seven and seventy-five hundredths grams granular cornstarch
(24.4 grams dry solids basis, 0.15 mol) was slurried in 38.6 ml water containing
1.25 grams sodium hydroxide and 17.7 grams sodium sulfate. After the starch
slurry was sparged with nitrogen, 1.63 grams diethyl vicinal chlorohydroxyethanephosphonate (0.0075 mol) was added. The bottle was capped and rotated in a
polymerization bath at 45°C for about 18 hours. The contents of the bottle
were poured into a beaker, adjusted to pH 7 with 6 N HCl, filtered, washed
with distilled water, reslurried in 80 ml water, filtered, washed and allowed to
air dry. The granular starch phosphonate had a DS of 0.0138 as determined by
phosphorus analysis.

Example 3: This example illustrates the reaction of granular starch with diethyl
2,3-chlorohydroxypropanephosphonate. Example 2 was repeated except that
the diethyl chlorohydroxyethanephosphonate was replaced with 1.23 grams of
diethyl 2,3-chlorohydroxypropanephosphonate (0.0053 mol). The starch phosphonate had a DS of 0.0090 as determined by phosphorus analysis. The diethyl
2,3-chlorohydroxypropanephosphonate was prepared in the same manner as the
diethyl chlorohydroxyethanephosphonate of Example 1 by replacing the diethyl

vicinal phosphonate with an equivalent amount of diethyl 2-propenephosphonate.

α-Hydroxycarboxy Starch Ethers

In a process described by *R.J. Hathaway; U.S. Patent 3,702,847; November 14, 1972; assigned to A.E. Staley Manufacturing Company* chlorohydroxy acidic reagents, such as chlorohydroxypropionic acid (CHP) and chlorohydroxysuccinic (CHS) acids produced by chlorination of acrylic acid and maleic acids, respectively, are used to introduce carboxyl groups into starch in the presence of an alkaline catalyst.

It is expected that the modified starches of the process will be useful in foods. Cleavage of the ether linkage could take place during metabolism, and the reaction by-products should include glyceric and tartaric acid, both of which are naturally occurring metabolites. A nontoxic starch derivative can be most valuable for use as a constituent of various food products, for example, as a binder, filler, or coating material for frozen foods which are ingested into the human system. The chlorohydroxypropionic acid (CHP)-starch slurry reaction, as presently understood, is believed to proceed generally according to the following steps.

$$(1) \quad CH_2{=}CHCOOH + Cl_2 \xrightarrow{H_2O} \underset{\underset{(CHP)}{Cl \quad OH}}{CH_2CHCOOH}$$

$$(2) \quad \underset{\substack{[Ca(OH)_2\ can\ be\ used \\ instead\ of\ NaOH]}}{Starch + CHP} \xrightarrow{NaOH} \left[\underset{(epoxy\ intermediate)}{CH_2\underset{O}{\diagdown\diagup}CHCOONa} \right] + NaCl$$

$$(3) \quad \longrightarrow \underset{\underset{OH}{\big|}}{Starch{-}O{-}CH_2{-}CH{-}COONa}$$

$$Product = glyceric\ acid\ starch\ ether\ (GAE)$$

The above reaction mixture is then neutralized with HCl and isolated with an excellent yield of hydroxycarboxy modified starch at a pH near neutrality. The following examples illustrate the process.

Example 1: CHP-Cornstarch — Chlorohydroxypropionic acid (CHP) reagent was prepared by dissolving 288 grams acrylic acid in 5 liters water and adding Cl_2 gas below the surface while stirring rapidly at 15° to 25°C. The pH was controlled at 1 to 1.5 during the addition by continuous addition of 318 ml of 10 N sodium hydroxide. Control of pH allows more rapid additon of Cl_2, but does not affect the final yield of CHP. The yellow color of excess Cl_2 signalled the end of the reaction, after which the solution was aerated to remove Cl_2 and neutralized to pH 5 with another 465 ml of 10 N NaOH.

The CHP reagent was mixed with 3,000 grams (dry basis) cornstarch, 500 grams NaCl, and 150 grams calcium hydroxide to form a slurry, and stirred at 24°C under a nitrogen atmosphere for 22 hours. The pH was 11.2. The mixture was cooled, neutralized with hydrochloric acid, filtered and washed with 80-to-20 water-methanol mixture. The product contained 3.4% carboxyl, dry basis, which was 60% of the theory based on acrylic acid. The degree of substitution (DS) for carboxyl is about 0.13. Reaction of similarly prepared CHP reagents with waxy maize and tapioca starches gave 3.85% carboxyl in both cases, which was 70% of theoretical yield based on acrylic acid, and a DS of about 0.15.

Example 2: CHS from Maleic Acid-Tapioca Starch — 49.0 grams (0.5 mol)
maleic anhydride was reacted with 700 ml hot H_2O, cooled and neutralized
partially with 17.0 grams sodium hydroxide in 100 ml H_2O. The volume was
adjusted to 1 liter having pH 2.8, and cooled to 30° to 35°C. Sodium maleate
will precipitate if the solution is cooled further. Chlorine gas was added at 30°
to 35°C while stirring rapidly. The pH was maintained at 1.5 to 2.0 by addition
of 30 ml 40% NaOH during the reaction. The temperature was maintained at
30° to 35°C.

After 5 hours a permanent yellow color developed indicating excess chlorine.
An analysis for inorganic chloride produced during the reaction indicated the
yield of chlorohydroxysuccinic acid (CHS) was 90% of theoretical. One part
tapioca starch was suspended in 1.8 parts of the CHS reagent prepared above
to form a slurry and treated while stirring with 0.2 part of calcium hydroxide.
The mixture was stirred for 22 hours at 42°C (pH 11.1), then neutralized to pH
5 with HCl, filtered, washed with water and air dried. The product contained
2.1% carboxyl (dry basis), which is a carboxyl DS of about 0.04.

Derivatives of Peroxide Thinned Starches

The thinning of derivatized starches is generally accomplished with the use of a
mineral acid such as sulfuric acid or hydrochloric acid. The use of acidic materials generally requires that special acid resistant equipment be employed in the
manufacture of the derivatized starches. This, of course, adds to the cost of the
manufacturing process.

*H.W. Durand; U.S. Patent 3,655,644; April 11, 1972; assigned to Grain Processing
Corporation* has disclosed an improved process for thinning derivatized starch
using hydrogen peroxide under alkaline conditions required for the derivatization.
The temperature at which thinning of derivatized starches is accomplished with
the use of hydrogen peroxide is 80° to 130°F and at a pH of 7.0 to 12.0. A
catalytic material such as copper ion is used in the amount of 5 to 100 parts per
million based on the weight of starch to assist the thinning action of the hydrogen
peroxide. The process is illustrated by the following example. Here "line starch"
refers to the refined starch slurry (~40% solids) resulting from a typical corn
wet-milling operation and taken just before the final filtration and drying steps.

Example: Line starch (4,900 grams at 40.8% solids, equivalent to 2,000 grams
dry starch), under continuous stirring at 110°F, was alkalinized to pH ~11.5
(from an initial value of ~6.0) by slow addition of a mixture of 100 grams of
30% aqueous sodium hydroxide and 448 grams of 26% aqueous sodium chloride
(corresponding to 1.5% of anhydrous sodium hydroxide based on the dry starch
and 4% of anhydrous sodium chloride based on the water in the starting line
starch). Ethylene oxide (50 grams or 2.5% based on the dry starch) was then
introduced and the mixture allowed to react for approximately 16 hours.

At the end of this time 50 parts per million (starch basis) of copper ion was
added in the form of aqueous copper sulfate, and the mixture (still at pH ~11.5)
was split into 6 equal parts. To these were now added sufficient amounts of
35% aqueous hydrogen peroxide to provide a series of reaction systems containing 0.0, 0.2, 0.4, 0.6, 0.8 and 1.0% levels of anhydrous hydrogen peroxide based

on the dry starch solids. After another 8 hours of stirring, still at 110°F, sufficient sodium bisulfite was added in each case to eliminate any residual hydrogen peroxide followed by adjustment to pH 5.5 to 6.0 by means of 20% aqueous sulfuric acid. Finally, the various mixtures were vacuum filtered, the filter cakes washed thoroughly on the filter with distilled water and air dried at 50° to 60°C.

For control purposes, an additional preparation was carried out applying the above hydroxyethylation procedure to line starch, but conducting the subsequent thinning step (to a final fluidity of 80+ Buel) under acidic rather than peroxidic conditions, in the following manner: On completing the reaction with the ethylene oxide the mixture was taken to a filtrate acidity of 0.216 normal by addition of sulfuric acid and heated at 125° to 130°F for 17 hours. [Shortly after heat-up sodium chlorite (0.05% starch solids basis) was added as a bleaching adjunct.]

At the end of this time the alkali fluidity value obtained was 161 (20 grams) and so the process was terminated by first adding a small amount of sodium bisulfite to get rid of the trace excess of chlorite, then adjusting the pH to 5.5 by introduction of 3% aqueous sodium hydroxide. The product was vacuum-filtered, washed and dried in the usual manner. The properties of hydroxyethyl starches (applied ethylene oxide level ~2.5%) as affected by postreaction with various levels of hydrogen peroxide (Cu^{++}-catalyzed) are shown in the following table.

Assay	H_2O_2 level (percent as anyhdrous, starch basis)						
	0.0 (Unthinned)	0.2	0.4	0.6	0.8	1.0	0.0 (acid-thinned to 80+ Buel)
1... Moisture, percent	8.6	6.8	7.8	6.3	7.5	10.3	8.6
2... Ash, percent	0.09	0.11	0.30	0.16	0.18	0.19	0.19
3... Alkali fluidity (x g.)	38 (10 g.)	92 (20 g.)	142 (20 g.)	152 (20 g.)	158 (20 g.)	{168 (20 g.) / 122 (40 g.)	166 (20 g.) / 122 (40 g.)
a. Buel rating	(1)	20	55–60	65–70	70–75	80+	80+
4... Carboxyl percent [corrected] [2]	~0	~0	0.06	0.05	0.08	0.09	(3)
5... Brabender findings at percent solids indicated (pH preadjusted to 6.5).	8.0	10.0	15	20	20	25	25
a. pH before adjustment	6.85	7.3	6.8	7.3	7.1	7.2	6.9
b. Gelatinization temp. (° C.)	62	61	61	58	58	58	58
c. Peak consistency (B.U.) [4] at T. ° C.	~1,000 (71)	510 (68.5)	600 (69.5)	1,130 (68)	635 (67)	820 (66.5)	>1,000
d. B.U. at 95° C.:							
Initial	580	60	40	50	30	35	40
After 30 minutes	370	35	20	40	25	30	35
e. B.U. at 50° C.:							
Initial	870	200	205	320	220	215	300
After 30 minutes	725	185	205	335	230	230	310
Setback at 50° C, (ΔB.U., 30 min.)	−145	−15	~0	+15	+10	+15	+10

[1] Thick boil.
[2] Carboxyl content determined by titration with 0.1 N NaOH and corrected by subtracting the value for a line starch control.
[3] Negligible.
[4] Brabender Units.

The data of the table point up the strong thinning action of hydrogen peroxide and the fact that the resulting derivatives maintain their excellence in pasting properties and relative freedom from paste setback and retrogradative change throughout the range of fluidity increase covered. That the mode of degradation in the thinning process is hydrolytic rather than oxidative is indicated by the only slight increase of carboxyl content obtained and by the fact (determined in a separate experiment) that similar treatment with as much as 1.0% of chlorine (well recognized as a means of oxidatively modifying starch and derivatives thereof) gives a product of only 5 to 10 Buel fluidity, considerably less than the value gained using hydrogen peroxide at the 0.2% level. Other examples in the complete patent show the thinning of line starch followed by organoethylation and by acetylation.

STARCH ETHERS IN PROCESSED FOODS

Dip Coatings for Protective Films

A great deal of research work has been done to develop an edible water-soluble film coating on edible foods. The Army Quartermaster Corps is particularly interested in coating foods for use by astronauts in space flights and the food industry desires to coat foods to improve packaging techniques and preservation of packaged foods.

F.J. Mitan and L. Jokay; U.S. Patent 3,427,951; February 18, 1969; assigned to American Maize-Products Company have developed a coating using a high amylose starch (over 50% amylose) modified by ethylene or propylene oxide as shown in U.S. Patent 3,378,546 on page 136.

Briefly, protective coatings are formed on the surface of foods and other articles by forming an aqueous slurry containing a hydroxyalkyl ether of starch having a natural amylose content of at least about 50% by weight, cooking the slurry under super-atmospheric pressure to at least partially gelatinize the starch ether granules, releasing the cooked slurry to atmospheric pressure and then applying it as a film on the surface of the food or article to be coated.

Further details of the process may be readily understood from the examples which have been selected for the purpose of illustration. In the following examples starch granules which contained 70% amylose were reacted with about 10% propylene oxide by weight and then compositions were formed with the stated parts by weight of water, plasticizer and humectant. All values are parts by weight.

Composition	\- \- \- \- \- \- \- \- \- \- \- \- \- \- Example \- \- \- \- \- \- \- \- \- \- \- \- \- \-					
	1	2	3	4	5	6
Starch granules	30.0	28.5	19.0	14.25	28.5	30.0
Plasticizer*	6.0	6.0	5.0	3.0	6.0	- - - -
Humectant**	- - - -	1.5	1.0	0.75	1.5	- - - -
Water	64.0	164.0	175.0	182.0	64.0	70.0

    *Glycerine
  **Humectant was gelatine in Examples 2, 3 and 4; GMC in Example 5

The above compositions were heated to a temperature at which the granules gelatinize which in the examples was 190°F and cooked preferably with stirring under 15 psi pressure for about 30 minutes to complete the cooking. The water content of the composition of Example 1 was adjusted to the solid content set forth on the following page to form liquids which were held at the indicated temperature while caramel candy cubes, fruit cake and Graham crackers were submerged in the liquid by dipping to form a water-soluble edible amylosic starch ether film on the food. The viscosity of the dip liquids was between about 30 to 3,000 cp.

Percent solids by weight:

A	8 - 10
B	10 - 20
C	20 - 36

Temperature, °F:

A	60 - 100
B	100 - 160
C	160 - 190

The starch film coated foods were placed in an oven where hot air at a temperature of about 100° to 150°F was circulated to dry the film. Compositions B and C were also applied with a doctor blade and in other cases with a brush to form starch films on the foods specified in the examples.

Films formed with Composition A were effective to resist shattering of the Graham crackers. Films made with Compositions B and C were continuous solid films without pin holes or other discontinuities and these films enveloped the food as a protective shield against atmospheric air. The films were not tacky or sticky and there was no tendency to chip or flake off. When eaten, the films readily dissolved in the mouth without objectionable odor, taste, or texture.

The compositions of Examples 2 through 5 were used as a liquid dip to form a water-soluble edible amylosic starch ether film on cereal fruit cubes, cheese sandwiches and peanut butter sandwiches. In all cases the films formed a satisfactory protective envelope on the food. The composition of Example 6 containing 20% solids by weight was heated to 190°F and poured over fresh eggs to form a protective coating. Coated and uncoated eggs taken from the same batch were stored at room temperature for 34 days. When placed in water the coated eggs sank while the uncoated control eggs floated showing that the uncoated eggs were no longer edible. A continuous process is also outlined in the complete patent for applying the coating to food products.

Freeze-Thaw Stable Hydroxypropyl Starch Thickeners

In both the cooking and freeze-thawing of food products containing starch as a thickener, serious difficulties are encountered. One difficulty is that liquid separates out of the food, giving distinct liquid and solid portions instead of a homogeneous mixture. Another difficulty is that foods set up to a cloudy gel upon cooking and cooling. Still another difficulty is that the viscosity of the food decreases too much during cooking and before cooling; that is sometimes called "cookout."

The process disclosed by *A.J. Ganz and G.C. Harris; U.S. Patent 3,369,910; February 20, 1968; assigned to Hercules Incorporated* either substantially eliminates or minimizes these difficulties by using hydroxypropyl starch as the thickener. Although not certain why, thus far this improvement is restricted to hydroxypropyl starch. Even hydroxyethyl starch did not give satisfactory results.

Preparation of the hydroxypropyl starch is not critical nor per se a part of the process. The hydroxypropyl starch may be either of the granular (ungelatinized) or gelatinized type. That is, the starch granule may have been substantially maintained or substantially destroyed during preparation of the hydroxypropyl starch. Likewise, crosslinking of the hydroxypropyl starch is not critical.

Examples 1 through 9: A conventional chicken broth containing 5% suspended thickener was heated to the boiling point and boiled 5 minutes in an open vessel. The water which had evaporated was replaced. The resulting broth was quick frozen at –10°C and stored at that temperature. Once each 24 hours the broth was removed from storage, allowed to come to room temperature and then examined for the presence of separated water, curdling, graininess, lumping, etc. The freezing at –10°C, storing at that temperature and allowing to come to room temperature completed a cycle. Further details appear in the following table.

Evaluation of Thickeners in Gravy

Example No.	Thickener	Stability, No. of Cycles
1	Wheat Starch	0
2	Hydroxypropyl Wheat Starch M.S. 0.065	25
3	Hydroxypropyl Wheat Starch of Example 2 cross-linked with 100 p.p.m. $POCl_3$.	18
4	Corn Starch	0
5	Hydroxypropyl Corn Starch M.S. 0.07	8
6	Waxy Maize Starch	8
7	Hydroxypropyl Waxy Maize Starch M.S. 0.07.	25
8	Potato Starch	0
9	Tapioca Starch	2

If a material failed on the first cycle, it was given a rating of 0.
If a material had not failed after 25 cycles, the test was discontinued.
All values between 0 and 25 indicate that the material failed in the following cycle.

The stability of the hydroxypropyl starch when used in white sauce, fruit pie fillings and pudding is also illustrated in the complete patent.

Hydroxyalkyl Starch plus Carrageenan for Stabilized Puddings

Preparation of high quality starch and milk-containing food products such as canned puddings, cream soups, sauces and gravies has been complicated by problems arising as a result of the retorting process. The term "retorting" refers to the heating, under pressure, of the food products to temperatures above 212°F to sterilize the products. After being retorted these products have poor texture and viscosity stability, discolorization and off-flavors due to nonenzymatic browning and carmelization of available sugars, and incipient curdling of the milk proteins.

R.P. Vilim and H. Bell; U.S. Patent 3,628,969; December 21, 1971; assigned to National Starch and Chemical Corporation have found that the use of a starch which has been etherified with sufficient alkylene oxide so as to have a minimum degree of substitution of 0.056, in conjunction with a carrageenan, in a starch-milk system, yields a finished product which is not only stable at normal cooking temperatures, but even after retorting is characterized by the absence of discoloration and off-flavors due to carmelization as well as nonenzymatic browning, and a lack of incipient curdling of the milk protein. It is believed that the relatively high degree substitution of alkylene oxide on the starch leads to these improvements over the prior art.

By alkylene oxide is meant any of the epoxides of lower aliphatic hydrocarbons
in which the oxygen is linked to adjacent carbon-hydrogen groups. After etherification with the alkylene oxide, the starch is referred to as a hydroxyalkyl
starch. The hydroxyalkyl starches preferred for use are the hydroxyethyl and
hydroxypropyl starches. Any type of starch may be used in the preparation
of these hydroxyalkyl derivatives. It is preferable, however, to use previously
crosslinked starches from corn, waxy maize, tapioca, or wheat. The hydroxyalkyl starch should be present at a weight concentration of 2 to 7% of the
total food product.

Besides the hydroxyalkylated starch, the other essential ingredient necessary in
the product is carrageenan. It is preferred that the carrageenan be the relatively
pure natural gum and not one of the blends with other natural gums known
under various trade names. Based on a 1 to 10% weight concentration of milk
solids in the total formula, the preferred range for carrageenan is between 3.5
and 7.1%, by weight, of the total hydroxyalkyl starch present. At less than
3.5%, a significant increase in undesirable features such as off-colors and curdling
may occur. At greater than 7.1%, resultant viscosities are excessively gelled and
product body has been found to be rubbery in texture. The following example
illustrates the process; all parts are by weight unless otherwise specified.

Example: This example illustrates the preparation of a pudding typical of the
process. The following ingredients were used in the preparation of the product.

Ingredients	Parts
Water	53.05
Emulsion system (as described hereinbelow)	25.00
Vanilla powder	2.20
Sugar	13.45
Nonfat dry milk solids	2.50
Salt	0.20
Hydroxyalkyl starch (waxy maize crosslinked with 0.03 part by weight of phosphorus oxychloride with propylene oxide to the DS listed in the table on the following page)	3.40
Carrageenan	0.12-0.28*

    *See table on following page. This range of 0.12 to 0.28 part of carrageenan
corresponds to the range of 3.5 to 7.5% of carrageenan based on the weight
of the starch.

In adding the fats, a previously formed emulsion system yields optimum results
for long term product stability. This is not a necessary step in this process but
a recommended step in obtaining optimum quality. The emulsion system was
formulated as follows.

Ingredients	Parts
Water	69.35
Fat (coconut oil with lecithin)	20.00
Corn syrup solids	6.50
Nonfat dry milk solids	3.00
Purity gum 539 (pregelatinized waxy maize starch)	0.75
Emulsifier (blend of mono- and diglycerides)	0.40

This fat emulsion system was mixed and homogenized before being added to the final formulation. The total mixture was brought to 195°F, canned in 211 x 300 millimeter cans and retorted for 60 minutes at 245°F. The following table illustrates the result of varying the DS of propylene oxide onto starch, and of varying the amount of carrageenan in the above preparation.

Formulation	#1	#2	#3	#4	#5	#6
Starch	0.040 D.S.	0.056 D.S.	0.056 D.S.	0.056 D.S.	0.14 D.S.	0.14 D.S.
Carrageenan	0.24 parts.	0.12 parts.	0.24 parts.	0.28 parts.	0.12 parts.	0.24 parts.
Results observed	Sl. browning, sl. curdling.	Smooth, no off-color or flavor.	Smooth, no off-color or flavor.	Heavy, cohesive, sl. incipient curdling.	Smooth, no off-color or flavor.	Smooth, no off-color or flavor.

No. 1 - below minimum DS of hydroxyalkyl on starch, maximum amount of carrageenan
No. 2 - amount of carrageenan
No. 3 - minimum DS, maximum amount of carrageenan
No. 4 - minimum DS, greater than maximum amount of carrageenan
No. 5 & 6 - higher levels of DS with minimum and maximum amounts of carrageenan, respectively

The desired results are obtained in Samples 2, 3, 5 and 6 where the DS is 0.056 or higher and the carrageenan to starch ratio is between 3.5% (0.12 part of carrageenan) as in Sample 2 and 7.1% (0.24 part carrageenan) as in Sample 3.

Etherified Tapioca Starch in Pudding

Frozen puddings having improved freeze-thaw stability have been developed by *A.D. D'Ercole; U.S. Patent 3,669,687; June 13, 1972; assigned to General Foods Corporation.* Broadly, this process gives a frozen pudding containing a modified food starch by cooking and cooling the pudding during its preparation under carefully maintained and critical temperature conditions. Typically, the frozen pudding is prepared by cooking the ingredients including modified food starch at temperatures of 230° to 260°F and cooling the cooked pudding to 100°F. Following the cooling, the pudding is packaged in suitable containers and is then further cooled quiescently and frozen.

While modified food starches have been suggested for use in pudding compositions of various kinds and even for frozen puddings, such starches, because of their crosslinks and substituent groups, generally exhibit a not-too-well-defined gelatinization temperature. Moreover, while individual starch granules may gelatinize quite sharply, not all the granules in a given quantity of starch gelatinize at the same temperatures but instead gelatinize over a range of say, 10° to 20°F.

It has now been found that by cooking a pudding composition containing a modified starch within the range of 230° to 260°F, gelatinization of the starch is able to be controlled in a unique manner which is not completely understood. It appears that cooking at this range of temperatures causes swelling of the granules and cells of the modified food starch so as to obtain the desired gel strength and viscosity. However, rupturing or breaking of the granules and cells is limited and may not occur at all thereby resulting in a frozen pudding characterized by freezer and refrigerator storage stability and freeze-thaw cycling stability.

On the other hand, conventional puddings are usually prepared by cooking until the granules and cells are ruptured and broken so as to develop the requisite gel strength and viscosity. Yet, such conventional puddings do not exhibit desired freezer and refrigerator stability nor freeze-thaw cycling stability.

The modified food starches which may be used to prepare the frozen puddings include those derived from wheat starch, cornstarch, waxy maize starch, potato starch, tapioca starch, and the like, all of which may be modified. The amount of modified food starch in the frozen pudding generally is from 3 to 10% by weight. An especially suitable modified food starch is an etherified tapioca starch. A preferred range of ingredients for the frozen pudding is as follows.

	Percent by Weight
Water	45 - 70
Sweetening agent	12 - 18
Milk and/or milk derivative	3 - 10
Modified food starch	3 - 10
Fat or oil	2 - 10
Protein	1 - 3
Emulsifiers	0.1 - 0.5
Stabilizers (gums)	0.01 - 1.0
Salt	Less than 1
Flavor and color	0.002 - 0.02

In order to illustrate the process, the following example is given.

Example: Dry pudding base ingredients comprising, by weight, 8 parts of sugar, 5 parts of nonfat dry milk solids, 4 parts of modified food starch (etherified tapioca starch) and 0.17 part of emulsifier (sodium stearoyl 2-lactylate) are thoroughly mixed in a ribbon blender. The blend is added with agitation to 53 parts, by weight, of water in a suitable vessel and 0.001 part of suitable flavor and color is then incorporated. A nondairy fatty emulsion as shown and described in Example 1 of U.S. Patent 3,431,117, which has been homogenized but not whipped nor frozen, is then added to the slurry in a weight ratio of 1 part of emulsion to 3 parts of slurry.

The resultant mixture is then passed into a scraped surface heat exchanger where the temperature of the pudding mix is increased until it reaches 245°F. The mix is held at this temperature for 60 seconds and it is then circulated through the cooling section of the scraped surface heat exchanger where it is cooled to 100°F. The pudding is then suitably packaged and while in containers is quiescently cooled and frozen at 0° to –10°F. The pudding obtained is then available for distribution through frozen food outlets.

Examination of the pudding indicated no breakdown or structural change in the pudding when it was stored at 0° to –10°F for more than 6 months. Moreover, the pudding was found to be resistant to syneresis and separation during repeated freeze-thaw cycling. Further, the pudding when thawed and stored at refrigerator temperatures was stable for periods in excess of 2 weeks. Finally, the frozen pudding was noted to be syneresis- and separation-resistant for 8 to 12 weeks when stored at 15°F, a temperature frequently found to cause structural breakdown in conventional puddings.

Photomicrographs of the pudding indicated that at the cooking temperature used, swelling of the starch granules and cells resulted but rupture and breakage were not evident. Such controlled cooking and cooling conditions appear to bring about stability of the pudding under a wide variety of storage conditions.

Cooked Pudding Mixes

In conventional cooked pudding mix compositions, raw cornstarch is widely used as the sole setting or gelling agent and during cooking the starch is converted, by gelatinization, to the sought--after firmness and texture associated with cooked puddings. However, unless care in preparation is exercised, a conventional cornstarch cooked pudding may be either thin and runny or firm and pasty.

Attempts have been made to use potato starch, tapioca starch, rye flour, wheat flour, and combinations for all or part of the cornstarch to provide pudding mixes which do not have to be cooked to a full rolling boil during preparation. Yet, the resultant puddings lack, to a certain extent, desired texture and taste. Moreover, when such "no boil" mixes are cooked to a full boil or beyond, the puddings are excessively firm and/or pasty and/or gummy.

A pudding mix has been developed by *B.J. Bahoshy, R.R. Ferguson and J.L. Hegadorn; U.S. Patent 3,619,208; November 9, 1971; assigned to General Foods Corporation* which gives a fully acceptable pudding regardless of whether the product is cooked to less than a boil, the first sign of a boil, full boil, or even beyond. This is realized by including in the pudding mix an amylaceous ingredient and a modified amylaceous ingredient.

The modified amylaceous ingredient is one characterized by its gel point being lower than that of an unmodified, raw starch of the same type. On cooking the pudding mix composition, a pudding of controlled texture is obtained. Although raw starches may be modified by any of a number of methods, it appears that a raw starch which has been etherified and/or esterified is especially suitable in the pudding mix composition of this process. More particularly, it would seem that a starch which has been etherified and/or esterified so as to introduce epichlorohydrin and/or succinyl groups is most advantageously used.

A modified cornstarch found especially useful in the pudding mix composition of this process is a modified cornstarch identified as Delta Food Starch 07444 (Anheuser-Busch, Inc.). In a preferred embodiment, the amylaceous ingredient is cornstarch and the modified amylaceous ingredient is a modified cornstarch.

The ratio of amylaceous ingredient to modified amylaceous ingredient may be from 2 to 1 to 8 to 1, by weight. The pudding mix composition may also comprise other amylaceous ingredients such as potato starch, tapioca starch or flour and/or a thin boiling colloidal or cellulose gum system as provided for by including carrageenan, pectin, algin, or a cellulose derivative and combinations thereof. Suitable levels in grams of the various ingredients for use with one pint of milk are, for example:

Sugar	40 - 80
Raw starch	10 - 30
Modified starch	3 - 15
Flours or other raw starches	0 - 15

(continued)

Colloid or gum	0-3
Salt	¼-3
Flavor/color	As needed
Surface active agent	0-2

Example 1: A dry vanilla pudding mix is made by blending 55 parts of sucrose, 20 parts of dextrose, 20 parts of raw cornstarch, 4 parts of modified cornstarch, 1.5 parts of salt, and color and flavor as desired in a ribbon mixer for 20 minutes. Three to three and one-half ounces of the blended mix are added to two cups of whole milk. The admixture is then cooked to a boil over medium heat. The cooked pudding is then allowed to cool to develop its set.

Example 2: The procedure of Example 1 is repeated in all essential respects except that 0.15 part of calcium carrageenan is included in the formulation. A summary of observations regarding texture and appearance appears in the following table.

| | Pudding mix composition | | Commercially available pudding mix | |
| | | | "A" (cornstarch | "B" (cornstarch |
Cook condition	Example 1	Example 2	only)	and carrageenan)
Thickening stage	Watery, not gelled	Slightly soft creamy	Watery, not gelled	Very soft, creamy.
First sign of boil	Slightly soft, very creamy.	Good, creamy	Very soft, watery	Soft, creamy.
Partial boil	Good, creamy	do	Slightly soft, creamy	Good, creamy.
Full boil	do	do	Good, creamy	Do.
Extended boil (30 seconds).	do	do	Fair, slightly pasty	Fair, slightly tacky, off taste.
Extended boil (60 seconds).	do	do	Pasty, gummy	Slightly tacky, caramelized off taste, gummy.

PREPARATION OF CROSSLINKED STARCH ETHERS

The term "inhibited starch" refers to a crosslinked starch in which the starch granules have been toughened so that they are more resistant to rupturing during cooking than ordinary starch granules. Inhibited starches may exhibit a markedly reduced tendency to swell or gelatinize and generally display a comparatively short, noncohesive consistency after cooking. The degree of inhibition can be controlled and varied over a wide range so as to produce starches in which the tendency of the swollen granules to rupture is decreased through successive stages to starch products in which the swelling of the granules is so highly restrained that they will not swell noticeably when cooked in boiling water.

It is well known that starch may be inhibited, while in granular form, by reaction with polyfunctional reagents, such as epichlorohydrin, phosphorus oxychloride, divinyl sulfone, sodium trimetaphosphate, etc., which crosslink the starch molecules within the granule. Such inhibition is disclosed in U.S. Patents 2,328,537; 2,500,950; 2,524,400 and 2,884,413. This reaction results in the formation of covalent chemical linkages which reinforce the normal hydrogen bonds between starch molecules which hold the granule together. As a result, when the treated starch is cooked under conditions which normally weaken or destroy the hydrogen bonds, the granules are swollen but remain intact because of these covalent linkages.

Inhibited granular starches are desired for various industrial uses. They are particularly useful in applications where the cohesiveness or stringiness of native starches is found to be objectionable, for example, in the food industry where starch products yielding short, smooth pastes on cooking are used as thickening agents in pies, sauces and soups, etc.

Epichlorohydrin Inhibited Hydroxypropyl Starch

Previously commercial preparation of high quality starch-containing food products such as canned puddings, cream soups, sauces and gravies has been complicated by problems arising as a result of the retorting process. The term "retorting" refers to the heating of food products under pressure to temperatures above 212°F to cook or sterilize the products. After being retorted these products often exhibit poor texture and viscosity stability.

In the commercial preparation of food products incorporating a starch thickener, it is necessary that the starch display a good texture (smooth and homogeneous), viscosity stability, and, preferably, thin-thick properties. The initial thin state of the starch dispersion facilitates a more rapid penetration by the heat required for sterilization. Upon cooling, the cooked starch must thicken in order for the food product in final form to possess a desired viscosity and texture.

C.D. Szymanski, M.M. Tessler and H. Bell; U.S. Patent 3,804,828; April 16, 1974; assigned to National Starch and Chemical Corporation have described epichloro-hydrin-inhibited hydroxypropyl starch products which have a thin viscosity on normal cooking with water but display a higher viscosity upon being heated at retort temperatures. Such starch products find a particular use in the commercial preparation of canned foods. It has been determined that the starches of this process should have a degree of substitution (DS) with respect to the hydroxy-propyl group of at least 0.06 and no greater than about 0.3.

The usual process for the preparation of these modified starch products involves suspending an applicable starch (which has previously been stabilized by reaction with propylene oxide) in an aqueous medium containing epichlorohydrin to effect inhibition. The inhibition reaction is normally carried out at a pH level between 11.3 and 12.3 as the mixture is stirred over a period of 5 to 8 hours at a temperature from 35° to 45°C. The amount of epichlorohydrin necessary in the reaction medium will vary depending on the time and temperature at which the reaction is run and the necessity for having a final product which falls within a specified narrow range of inhibition as determined by its viscosity characteristics. Generally, the amount of epichlorohydrin necessary is from 0.015 to 0.023 part of epichlorohydrin to 100 parts of starch.

Subsequent to the reaction, the pH level of the suspension is adjusted to 6.0 by the addition of dilute acid. The suspension is then stirred slightly and filtered to remove the reaction product. The starch is repeatedly washed and refiltered to remove any residual salts or other contaminants. Finally the resultant starch product is dried to a moisture content of about 12% of the total weight. Any conventional means such as oven or air drying is suitable for this purpose.

Example: This example illustrates the preparation of the modified, hydroxy-propylated starch products of this process. Into a reaction vessel containing a solution comprising 15 parts of sodium hydroxide and 300 parts of sodium sulfate in 2,000 parts of water there was introduced 1,000 parts of cornstarch.

The suspension was briefly stirred, and thereafter there was added to the slurry 75 parts of propylene oxide. The vessel was sealed, and the contents therein were allowed to react for 16 hours at 40°C, while the vessel was continuously tumbled to assure uniform suspension of the starch throughout the mixture. Upon completion of this initial reaction, a sample of the product was extracted from the vessel and it was determined that the hydroxypropylated starch therein had a DS of 0.14. The pH of the reaction mixture was 11.7.

Shortly thereafter 20.0 parts of an aqueous epichlorohydrin (1.0% by weight) solution was added to the vessel. The epichlorohydrin was then reacted with the starch for 5 hours at 40°C, as the vessel was again continuously tumbled. The pH of the resultant solution was adjusted to 6.0 by the addition of 74.8 parts of a 21.2% hydrochloric acid solution. The epichlorohydrin inhibited starch was recovered by filtration, followed by repeated washings and subsequent drying.

Phosphorus Oxyhalide Crosslinked Starch Ethers

Two processes have been developed at A.E. Staley on the crosslinked starches. In the first process *J.V. Tuschhoff, G.L. Kessinger and C.E. Hanson; U.S. Patent 3,422,088; January 14, 1969; assigned to A.E. Staley Manufacturing Company* prepared a phosphorus oxyhalide crosslinked hydroxypropyl cereal starch having a hydroxypropyl DS of 0.10 to 0.30 and pH 6.5 buffered salt CIV viscosity of about 60 to 100 gram-centimeters after 10 minutes and 80 to 150 gram-centimeters after 40 minutes. The products are suitable for the preparation of thin-thick starch pastes which attain their full viscosity after cooking under pressure.

Crosslinking the hydroxypropyl cereal starches with phosphorus oxyhalide imparts freeze-thaw resistance, gives the starch paste greater viscosity stability and permits their conversion into thin-thick starches. The degree of crosslinking must be carefully controlled. If the hydroxypropyl cereal starch is reacted with too much phosphorus oxyhalide, the starch will be inhibited, i.e., incapable of pasting. On the other hand, if the hydroxypropyl crosslinked cereal starch is reacted with too little phosphorus oxyhalide, the final product will not have the necessary viscosity characteristics and/or freeze-thaw resistance.

Accordingly it is preferred to treat the hydroxypropyl cereal starches with an amount of phosphorus oxyhalide equivalent to 0.01 to 0.05 part by weight phosphorus oxychloride for each 100 parts by weight hydroxypropyl starch on a dry solids basis. As shown in the prior art of U.S. Patent 3,238,193, it is not the amount of phosphorus oxyhalide added to the reaction vessel that controls the properties of the final product, but it is the amount of phosphorus oxyhalide reacted with the hydroxypropyl cereal starch.

It has been found that the degree of crosslinking is best controlled by a 40-minute pH 6.5 buffered salt CIV viscosity cook. The details of this test are described below. This CIV viscosity cook is representative of the conditions under which canners cook starch and various foods. It is, of course, well known that salts retard the gelatinization of starches. However, it is also well known that most food preparations have various condiments added, particularly salt, and allowance must be made for the effect of salt on the gelatinization of the granular starch.

The pH 6.5 buffered salt CIV viscosity is determined in the following manner.

Fifty grams of starch (dry solids basis) is suspended in 940 grams of a pH 6.5 buffer solution. The buffered solution comprises a 1% by weight aqueous solution of disodium phosphate (Na_2HPO_4) and 0.2% by weight sodium benzoate which has been adjusted to pH 6.5 with citric acid (approximately 0.35 gram of citric acid is required by each 100 grams of solution). Ten grams of sodium chloride is added to 990 grams of starch slurry. The starch-buffered salt slurry is added to the CIV viscometer while it is running and with its temperature maintained at about 201° to 203°F. The viscosity is recorded at its peak and at 10 and 40 minutes after the suspension has been placed in the viscometer.

Example: A granular phosphorus oxychloride crosslinked hydroxypropyl cereal starch was prepared in the following manner. 300 parts by weight sodium sulfate was added to 1,000 parts by weight of granular cornstarch (dry solids basis) suspended in 1,150 parts by weight water. After the starch suspension was heated to about 110°F, 10 parts by weight sodium hydroxide (dry solids basis) was added as an aqueous 5% by weight solution to the suspension. Nitrogen gas was bubbled through the starch slurry in order to replace the air in the reaction vessel and the reaction vessel was sealed.

Then 82.5 parts by weight propylene oxide was added to the starch slurry through a dip-tube while the reaction mixture was maintained at 108° to 112°F, continuously sparged with nitrogen and stirred for 18 hours. Nitrogen sparging was discontinued; the reaction vessel was unsealed and 0.16 part by weight phosphorus oxychloride was added to the granular hydroxypropyl starch slurry. After reacting for one-half hour, the pH 6.5 buffered salt CIV viscosity of a sample of the phosphorus oxychloride crosslinked hydroxypropyl cereal starch was determined. The product had no initial peak viscosity.

The viscosity after 10 minutes was 84 gram-centimeters and after 40 minutes was 107 gram-centimeters. The starch suspension was filtered, washed with water, reslurried with water, adjusted to pH 5, filtered again, washed carefully, and dried to between 9 and 11% moisture. The granular product had a two gram alkaline fluidity of about 60 cc and 4.5% by weight hydroxypropyl groups (hydroxypropyl DS of 0.135).

A starch thickener having different CIV properties has been prepared by *J.W. Robinson, G.N. Bookwalter and J.V. Tuschhoff; U.S. Patents 3,437,493; April 8, 1969 and 3,719,661; March 6, 1973; both assigned to A.E. Staley Manufacturing Company.*

Here the phosphorus oxyhalide crosslinked hydroxypropyl cereal starch has a hydroxypropyl DS of at least 0.10 and pH 6.5 buffered salt CIV viscosity of about 200 to 400 gram-centimeters after 10 minutes and 190 to 300 gram-centimeters after 40 minutes.

Example 1: A granular phosphorus oxychloride crosslinked hydroxypropyl cereal starch was prepared in the following manner. 300 parts by weight of sodium sulfate was added to 1,000 parts by weight of granular cornstarch (dry solids basis) suspended in 1,150 parts by weight water. After the starch suspension was heated to about 110°F, 10 parts by weight sodium hydroxide (dry solids basis) was added as an aqueous 5% by weight solution to the suspension. Nitrogen gas was bubbled through the starch slurry in order to replace the air in the reaction vessel and the reaction vessel was sealed. Then 82.5 parts by weight propylene oxide was added to the starch slurry through a dip-tube while the reaction mixture was maintained at 108° to 112°F, continuously sparged with nitrogen, and stirred for 18 hours.

Nitrogen sparging was discontinued; the reaction vessel was unsealed, and 0.09 part by weight phosphorus oxychloride was added to the granular hydroxypropyl starch slurry. After reacting for 1 hour, the pH 6.5 buffered salt CIV viscosity of a sample of the phosphorus oxychloride crosslinked hydroxypropyl cereal starch had a peak viscosity of 340 gram-centimeters, 292 gram-centimeters after 10 minutes, and 245 gram-centimeters after 40 minutes.

The starch suspension was filtered, washed with water, reslurried with water, adjusted to pH 5, filtered again, washed carefully, and dried to between 9 and 11% moisture. The granular product had a two gram alkaline fluidity of about 7 cc and 4.6% by weight hydroxypropyl groups (hydroxypropyl DS of 0.14). These crosslinked starch ethers form excellent pie fillings when mixed with cross-linked starch acylates.

Example 2: 740 grams of a dry blend of 555 grams of the granular phosphorus oxychloride crosslinked hydroxypropyl starch of Example 1 and 185 grams of a granular acrolein crosslinked cornstarch acetate (prepared by the method of Example 1, U.S. Patent 3,238,193 having an alkaline fluidity of 70 cc, 2.5% by dry weight acetyl and CIV viscosity at pH 3.5 after 15 minutes of 880 gram-centimeters and after 40 minutes of 744 gram centimeters) was suspended in 3,500 grams of water. 2,350 grams of water, 2,350 grams of 43°Bé Sweetose corn syrup, 1,380 grams sucrose and 60 grams sodium chloride were added to the starch suspension.

The resulting starch suspension was cooked in a continuous starch cooker at 220°F. 700 grams of cooked starch paste was cooled to 85°F, mixed with 750 grams of drained cherries and 500 grams of cherry juice, deposited into a pie shell and baked at 450°F for 22 minutes. The pie was permitted to stand over-night at room temperature and the pie was cut. The cherry pie filling had excellent eye appeal (clarity) and mouth feel. Essentially the same results were obtained using equal parts by weight of the two starch thickeners (370 grams of each) except that the clarity of the pie filling was slightly cloudy.

When this example was repeated using 740 grams of the granular phosphorus oxychloride crosslinked hydroxypropyl starch of Example 1 in place of the starch blend, the pie filling gelled and had an undesirable lumpy texture on eating. When the example was repeated using 740 grams of the granular acrolein crosslinked cornstarch acetate used in this example in place of the starch blend, the pie filling had poor clarity.

Ethers of Crosslinked Starch Phosphates

A process has been developed by *J.R. Caracci, F.J. Germino and T.D. Yoshida; U.S. Patent 3,751,410; August 7, 1973; assigned to CPC International Inc.* for producing stable starch derivatives having use as food components, particularly as thickeners. Starch is crosslinked by reacting with sodium trimetaphosphate. After crosslinking, the starch is modified by incorporation of a hydroxypropyl functional group. The resulting product is stable and performs well in food compositions including acid systems such as fruit pie fillings or lemon puddings.

or in neutral systems such as sauces, gravies, baby foods and chocolate puddings.
The starch used may be waxy milo, potato, tapioca, corn, waxy corn, red milo,
and high amylose starches. Preferred are potato starch, waxy starch and tapioca
starch.

More particularly, it has been found that each mol of starch must be crosslinked
with 0.01 to 0.3 gram of sodium trimetaphosphate. Moreover, the molar substitution (MS) should fall within a rather narrow range of 0.03 to 0.09. Only
such defined starches possess the above discussed properties and as well have
good heat stability and proper viscosity characteristics. One specific method
for carrying out the process may be considered as being composed of four
steps: (1) crosslinking, (2) washing, (3) substitution, and (4) final product
purification stage, as typified in the following.

Crosslinking: A slurry of waxy starch is prepared so that it has a total solids
content of 35 to 36%, with a slurry temperature of between 110° and 125°F,
and is transferred to a reaction tank. Sodium chloride in an amount of 6 pounds
per 100 pounds of starch is added. 1.5 pounds of sodium hydroxide per 100
pounds of starch is added, as a 4 to 50% solution. 100 grams of sodium trimetaphosphate (STMP) per 100 pounds of starch is added, as a 1.6% solution.
The slurry is heated to 110° to 125°F and held at temperature for 5 to 8 hours.
The slurry is then neutralized to pH 5 by addition of hydrochloric acid (16 to
37%).

Washing: The slurry is washed, using conventional means, e.g., a Merco-Dorrclone
wash, to remove the salts present.

Substitution: The wet slurry is then transferred to a reaction tank and the
solids adjusted to a content of 35 to 36%, the temperature of the slurry being
maintained at 110° to 125°F. 6 pounds of sodium chloride per 100 pounds of
starch are added. There is then added 1.5 pounds of sodium hydroxide per
100 pounds of starch as a 4 to 50% solution. 2 to 7 pounds of propylene
oxide per 100 pounds of starch are then added. The slurry is heated to 110°
to 125°F and held at temperature for about 18 hours. The slurry is then neutralized to pH 5 to 6 with hydrochloric acid solution (with 16 to 30%).

Product Purification: The product is then washed and dried using conventional
wash procedures as in the washing stage described above.

Example 1: To a starch slurry containing 1 mol of waxy milo starch is added
0.163 mol NaCl and 0.053 mol NaOH. The slurry is heated to 46°F and 0.20
gram STMP is added. The mixture is allowed to react for 6 hours. The slurry
is neutralized to 6.0 pH, filtered, washed and reslurried. The slurry is heated to
50°F and 0.13 mol of propylene oxide is added. The mixture is allowed to proceed for 7 hours before being neutralized, filtered and washed. The product has
a MS of 0.065. The product forms pie fillings which do not show syneresis for
8 cycles and gives a cut pie rating of 9 for a 0 to 10 scale. The product also
gives a lemon pudding viscosity of over 16,000 cp (acceptable viscosity, 9,000
cp).

Example 2: This run is carried out the same as in Example 1 except that the
propylene oxide is increased to 0.185 mol thus giving a MS of 0.092. This
product performs well in sauces, gravies, baby foods, or chocolate puddings, i.e.,
in a neutral environment.

Starch with Labile and Nonlabile Crosslinks

M.W. Rutenberg, M.M. Tessler and L. Kruger; U.S. Patent 3,832,342; August 27, 1974; assigned to National Starch and Chemical Corporation have disclosed a process for preparing granular starch inhibited with two different crosslinking agents, one producing a labile crosslinkage and the other a nonlabile crosslinkage. These starch derivatives display utility in food products.

According to this process a starch base is usually first reacted with a reagent which will produce relatively nonlabile crosslinkages, as for example, epichlorohydrin, phosphorus oxychloride, 1,4-dichlorobutene-2, cyanuric chloride, sodium trimetaphosphate, followed by reaction with a reagent which will produce a labile crosslinkage such as an ester crosslinkage, as for example, bis(ethylcarbonic)-adipic anhydride, linear polymeric adipic anhydride, bis(ethylcarbonic)fumaric anhydride, succinyl guanidine, adipyl guanidine, phosgene, 1,1'-carbonyldiimidazole, diimidazolide of succinic acid, diimidazolide of adipic acid, divinyl adipate, mixtures of adipic acid with acetic anhydride, etc.

Labile crosslinkages can also be introduced by reacting starch with chlorine and glycine, as disclosed in U.S. Patent 3,463,668. The resulting dually inhibited starch products are characterized by their ability to be subsequently controllably and readily converted into more highly swollen products, similar to the base starch after treatment with the nonlabile crosslinking reagent, but before treatment with the labile crosslinking reagent.

The inhibited starch products of this process are characterized by their ability upon cooking in aqueous medium to initially yield thin, low viscosity dispersions which can be subsequently thickened to high viscosity by means of the swelling of the starch granules as a result of the destruction of the labile crosslinkages. The labile crosslinkages of these starch products may be readily and controllably removed or disrupted by heat at temperatures above the boiling point of water, e.g., 220° to 300°F. Destruction of the labile crosslinkages can be controlled by pH, temperature, as well as the time period of exposure to heat. The results of the process are shown in the following example.

Example: The preparation of the starches, labeled A, B, C, D, E and F used in the example, was as follows.

Starch A — The pH of a slurry of 600 parts potato starch in 900 parts water was adjusted to 8.0. The starch suspension was stirred at room temperature and 0.25 part of cyanuric chloride was added. A pH of 8.0 was maintained by the addition of 3.0% sodium hydroxide at periodic intervals during the entire reaction. The reaction was complete after 2.7 hours, at which point there was no further change in pH. The pH was then lowered to 5.0 with 10% hydrochloric acid and the product recovered by filtration. The product was washed three times with water and dried.

Starch B — A total of 200 parts of Starch A was suspended in 300 parts water at room temperature and the pH adjusted to 8.0 with 3.0% sodium hydroxide. The starch suspension was stirred and 0.60 part of bis(ethylcarbonic) adipic anhydride was added slowly over a 20-minute period. A pH of 8.0 was maintained by the addition of 3.0% sodium hydroxide at periodic intervals during the entire reaction. The reaction was complete after 2.5 hours at which point there was no further change in pH. The pH was lowered to 5.0 with dilute hydrochloric acid and the product recovered by filtration. The product was then washed three times with water and dried. An analogous dually inhibited starch was prepared by reversing the reaction sequence used to prepare Starch B, i.e., reacting first with bis(ethylcarbonic) adipic anhydride and then with cyanuric chloride.

Starch C — A total of 1,000 parts of cornstarch was suspended in a solution of 300 parts sodium sulfate and 15 parts sodium hydroxide in 1,250 parts water. After addition of 50 parts propylene oxide, the slurry was sealed in a jar and reacted at 40°C for 24 hours while being continuously tumbled to keep the starch in suspension. The reaction was cooled to room temperature and the pH lowered to 6.0 with 10% hydrochloric acid. The product was recovered by filtration, washed three times with water and dried.

A total of 200 parts of the thus prepared hydroxypropylated cornstarch was added to a solution of 1.6 parts sodium hydroxide in 250 parts water. After the addition of 0.02 part 1,4-dichlorobutene-2, the mixture was sealed in a jar and reacted at 40°C for 16 hours while being continuously tumbled to keep the starch in suspension. The reaction was then cooled to room temperature and the pH was lowered to 6.0 with 10% hydrochloric acid. The product was washed three times with 50% aqueous ethanol and dried.

Starch D — A total of 100 parts Starch C was suspended in 150 parts water and the pH adjusted to 8.0 with 3.0% sodium hydroxide. The starch suspension was stirred and 2.0 parts linear polymeric adipic anhydride (prepared by the method of J.W. Hill, *J. Am. Chem. Soc.,* 52, 4110 [1930]) added. The pH was maintained at 8.0 by the addition of 3.0% solution of sodium hydroxide at periodic intervals during the entire reaction. The reaction was complete after 24 hours, at which point there was no further change in pH. The pH was lowered to 5.5 with 10% hydrochloric acid and the starch recovered by filtration. The product was washed three times with 50% aqueous ethanol and dried.

Starch E — A total of 600 parts of tapioca starch was suspended in a solution of 4.8 grams sodium hydroxide in 900 parts water. The starch suspension was stirred and 0.12 part phosphorus oxychloride added. The reaction was stirred at room temperature for 1 hour and the pH was then lowered to 5.0 with 10% hydrochloric acid. The product was recovered by filtration, washed three times with water and dried.

Starch F — A total of 100 parts Starch E was suspended in 150 parts water and the pH adjusted to 8.0 with 3.0% sodium hydroxide. The starch suspension was stirred and 1.0 part bis(ethylcarbonic) fumaric anhydride was added slowly over a 15-minute period. A pH of 8.0 was maintained by the addition of 3.0% sodium hydroxide at periodic intervals during the entire reaction. The reaction was complete after 50 minutes at which point there was no further change in pH. The pH was lowered to 5.0 with 10% hydrochloric acid and the starch recovered by filtration. The starch was washed three times with water and dried.

A slurry of each of the inhibited starch derivatives in a pH 6.1 maleic acid buffer (see table on following page for amounts) was placed in an 8-ounce can. The filled cans were then heated in a boiling water bath for 20 minutes after which the cans were sealed and retorted in a bath maintained at 250°F while being mechanically tumbled. The cans were removed after 15 minutes, rapidly cooled to room temperature, allowed to stand overnight, and the viscosities measured. The following table summarizes the viscosity data obtained on testing aqueous dispersions of each of the above described starches. The viscosities were measured with a Brookfield RVF viscometer using the indicated spindle at 10 rpm at a temperature of 72°F.

Sample	Parts starch	Parts buffer solution	Brookfield viscosity, cps. after—			
			20 min. at 212° F.	Spindle No.	15 min. at 250° F.	Spindle No.
Starch A _	13.0	234.0	12,740	5	16,060	5
Starch B _	13.0	234.0	34	1	23,020	5
Starch C _	15.0	180.0	10,000	5	31,800	5
Starch D _	15.0	180.0	2,280	5	39,000	5
Starch E _	13.0	195.0	22,000	5	34,050	6
Starch F _ _	13.0	195.0	466	2	34,200	6

CROSSLINKED STARCH ETHERS IN PROCESSED FOODS

Grainy Starch Thickeners

The appearance and over-all consumer appeal of many processed food products are greatly enhanced by the presence of a pulpy texture. Such food products thus appear to retain much of their natural texture and, in so doing, exhibit a rich, highly concentrated appearance as opposed to the thick, pasty character which often results from the use of conventional starch thickeners. In addition, the presence of a pulpy texture is often accompanied by other improved properties of color and taste. Attempts to incorporate cracker crumbs or tapioca pearls into the food has not given the desired grainy texture.

N.G. Marotta, P.C. Trubiano and K.S. Ronai; U.S. Patent 3,443,964; May 13, 1969; assigned to National Starch and Chemical Corporation have found that starch-containing food products exhibiting a highly desirable grainy, pulpy texture can be prepared by incorporating pregelatinized, crosslinked, amylose-containing starch products into food systems, prior to cooking and sterilization.

These starch products are ideally suited for such use because they exhibit sufficient cold water swelling ability so that upon being cooked they will produce swollen, discrete particles having an appropriate particle size along with excellent resistance to heat, acidity, and agitation which will permit these swollen particles to remain intact during any subsequent processing operations.

The applicable starch bases which may be used are corn, potato, sweet potato, wheat, rice, sago, tapioca, sorghum or the like as well as the high amylose containing varieties of these sources. Also included are the conversion products derived from any of the latter bases including, for example, dextrins prepared by the hydrolytic action of acid and/or heat, oxidized starches and fluidity or thin boiling starches prepared by enzyme conversion or by mild acid hydrolysis. In addition, the amylose fraction derived from any of the above starch bases may also be used. It is also possible to use any substituted ether or ester derivative of these starch bases or their amylose fractions.

The amount of crosslinking agent needed for this process is determined by the granule swelling power (GSP) of the resulting crosslinked starch. GSP is a measure of the extent of granule inhibition, and may be defined as the amount of swollen, hydrated paste which is formed by the cooking, in water under specific conditions, of one gram of dry starch as divided by the weight of anhydrous starch in the swollen paste.

The GSP is determined, in practice, by dispersing one gram of starch (anhydrous weight) in enough distilled water to give a total weight of 100 grams. Normally, the starch is suspended in this water, stirred over a boiling water bath for 5 minutes, and then covered for the remainder of the cooking cycle. After cooking is complete, the sample is readjusted to a weight of 100 grams and transferred, quantitatively, into graduated 100-milliliter centrifuge cups.

The sample is then centrifuged at 2,000 rpm for exactly 20 minutes and the starch dispersion is removed as a clear supernate and a compacted swollen paste. The percent solids in the supernate is determined by evaporation of an aliquot. The wet weight of the swollen paste is determined directly after the decantation of the supernate and the amount of dry solids in the paste is determined by evaporation. The granule swelling power is then calculated by the formula:

$$\text{GSP} = \frac{\text{Wet weight of swollen paste}}{\text{Weight of dry starch in swollen paste}}$$

Unconverted starch will ordinarily exhibit higher granule swelling power than crosslinked, i.e., inhibited, starches. Thus, raw cornstarches have a GSP of 33 to 35. However, in order to function effectively as texture producing starches, these mildly inhibited starches should have a GSP of 8 to 32 since within this range they provide food products with an optimum degree of pulpy texture. Therefore, the quantity of crosslinking reagent used in the inhibition process may be defined as that amount required to obtain a product having a GSP of between 8 and 32.

It should be noted, however, that only inhibited high amylose-containing starches and inhibited amylose provide satisfactory products displaying a pulpy texture at the lower GSP values, i.e., 8 to 10; whereas the other inhibited starch bases provide optimum pulpy textured products when they have a GSP value in excess of 10. Size of the starch product used as the thickener is important. After forming the pregelatinized starch by known methods, the starch product is dried on a drum dryer to form thin solid sheets. These sheets are then pulverized to form particles of which no more than 25%, by weight, will be retained on a No. 12 U.S. Standard Sieve, while no more than 60%, by weight, will pass through a No. 100 U.S. Standard Sieve.

Thus, the use of starch products which contain more than 25%, by weight, of +12 material will result in the formation of undesirably thick unnatural textures in the final food products. On the other hand, starch products which contain more than 60%, by weight, of –100 material cannot provide the particles whose large size is the basis of the resulting pulpy textured effect.

Example: This example illustrates the correlation between the granule swelling power of inhibited starch products and their performance as texture imparting additives in the food products. In particular, it illustrates the necessity for inhibiting the starch bases to within the required limits in order to obtain products whose pulpy texture exhibits maximum resistance to disintegration

during various subsequent processing operations. To determine the stability of
the grain effect imparted by various starches, the following test procedure was
utilized.

Part A — A dry blend of 7.5 parts of the specified inhibited starch and 7.5 parts
of sugar was admixed with 200 parts of distilled water. The resulting slurry was
then cooked to a temperature of 190°F, immediately diluted with 200 parts of
water and poured over a tared No. 20 U.S. Standard Sieve. The material retained
on the screen was then thoroughly washed, dried and weighed.

Part B — Thereupon, a second portion of the above prepared cooked slurry was
poured into a No. 2 can, which was sealed and retorted, i.e., pressure cooked,
at a temperature of 245°F and a pressure of 15 psi for a period of 30 minutes.
The can was then cooled for 30 minutes and its contents diluted with 200 parts
water. The resulting slurry was poured over a tared No. 20 U.S. Standard Sieve;
the material retained thereon was then washed, dried and weighed.

The presence, as determined by the method of Part A, of +20 particles, i.e.,
particles which were retained on the No. 20 mesh screen, and the subsequent
retention of a predominant portion of these particles in their initial bulky form,
as shown by the results of the procedure of Part B, is indicative of the fact that
these inhibited starch products are capable of providing satisfactory pulpy textured food products which will not be deleteriously affected by retorting conditions.

The following table lists the various amylose-containing, pregelatinized, inhibited
starches which were tested and provides information relating to the pulpy textured products resulting from their use. It should be noted that these inhibited
starches were drum dried and inhibited by means of the procedure referred to
in the description of the starch product utilized in this example.

Starch	GSP	Parts of +20 material in Part A	Parts of +20 material in Part B
Corn starch crosslinked with 0.05% epichlorohydrin	17.0	3.3	2.5
Corn starch crosslinked with 0.30% epichlorohydrin	7.9	0.2	Trace
Wheat starch crosslinked with 0.05% epichlorohydrin	17.7	2.9	2.0
Wheat starch crosslinked with 0.50% epichlorohydrin	6.9	0.2	Trace

The above data clearly indicates the excellent pulpy textured products resulting
from the use of the specified amylose-containing, pregelatinized, crosslinked
starches. It also illustrates the necessity for carefully controlling the concentration
of the crosslinking agent so as to be able to maintain the GSP within the range
necessary for attaining optimum results.

Baby Food Formulations

In the preparation of wet baby food formulations containing particles of meat
and vegetables, it is necessary to incorporate a thickening agent such as starch
to obtain a product attractive to the consumer. However, during prolonged
periods of storage at low temperatures, naturally occurring (unmodified) starches

do not retain consistent gel characteristics. In conventional practice, it has been found that certain starches such as a waxy maize (amioca) starch will maintain the desired cold stability in this type of wet formulation, provided that the amylopectin molecules in the starch are modified by crosslinking and acylating prior to introduction into the wet formulation. However, waxy maize starch is considerably more expensive than other types of naturally occurring grain starches.

Naturally occurring nonwaxy grain starches such as corn, wheat, sorghum, rice and like have not been employed in the preparation of baby food formulations containing meat and/or vegetable particles as amylose molecules suffer from linear association during prolonged storage especially at low temperatures. This linear association causes syneresis, i.e., a squeezing out of water, when the starch gel is subjected to storage conditions in combination with meat and/or vegetable particles.

It has now been found by *V.J. Kelly and W.G. Fry; U.S. Patent 3,685,999; August 22, 1972; assigned to Gerber Products Company* that ungelatinized nonwaxy grain starches which have been modified by at least partial reaction with both a bifunctional and monofunctional etherifying and/or esterifying agent can be incorporated with wet formulations of meat and/or vegetables to obtain ready-to-eat baby food formulations which have excellent cold stability over prolonged periods.

In fact, when such modified grain starches, ordinarily in their nonpeptized form, are incorporated with conventional baby food formulations prior to cooking, the resulting precooked product has been found to have rheological character- istics at least as beneficial as those present in a product prepared through the use of the more expensive modified waxy maize starch. Most surprisingly, when a nonwaxy grain starch that has been modified (in an ungelatinized condition) with propylene oxide and epichlorohydrin is employed, the product has been found to be significantly superior in prolonged cold stability to wet baby food formulations prepared with modified waxy maize starch.

The term "bi- or polyfunctional agent" is intended to include those etherifying and esterifying agents that are capable of reacting with two or more hydroxyl groups of the starch molecule, and include bifunctional esterifying or etherifying (crosslinking) agents that include epihalohydrins such as epichlorohydrin wherein a single molecule of the agent can attach itself across a pair of starch molecules. The term "monofunctional agent" is intended to include well known mono- functional etherifying and esterifying agents such as halogenated fatty acids, halogen hydrins, reactive epoxyalkanes, dialkylsulfates, alkyl halides, polybasic acid anhydrides and the like. Specific materials found to be especially advan- tageous include epichlorohydrin, vinyl acetate and acetic anhydride.

To further illustrate the process, the following example is provided. Split green peas were included in the formulation because they are found to produce the most severely adverse conditions for retaining satisfactory cold stability over prolonged storage periods.

Example: A strained meat-vegetable formulation was prepared in the proportions set forth on the following page.

Ingredient	Pounds*
Carrots	80
Liver	40
Bacon	20
Tomato puree	40
Inhibited cornstarch**	20
Baked wheat flour	15
Peas, split green	10
Potato flour	10
Seasoning, etc.	10

*Per 100 gallons of formulation at 200°F
**Modified with propylene oxide and epichlorohydrin

The liver was passed through a Rietz extruder and an Autio grinder having three-sixteenth inch plates. The meat was then slurried with hot water and the potato flour and split peas added; the combined slurry was then pumped to a cone tank. The carrots were passed through a Robinson cutter and also collected in the cone tank. The mixture was adjusted to 10 gallons and pre-cooked for 5 minutes at about 210°F.

The other dry ingredients, including tomato puree, were slurried with 15 gallons of cold water and passed through a finisher, then combined with the meat-vegetable slurry. Water was added at about 200°F to adjust the volume to 100 gallons and the formulation was pumped into small glass containers and capped. The capped glass containers were then heated to about 250°F (initial temperature, 160°F) for about 40 minutes to insure adequate sterilization.

Similar packs of strained vegetables and liver were prepared employing modified waxy maize starch and unmodified cornstarch. All the packs were placed in storage at 40°F, and samples of each were tested after 8 weeks of storage. The formulations were graded according to consistency, sheen and degree of water separation. The formulations prepared with the modified cornstarch were found to be at least equal to those prepared with the modified waxy maize starch. In fact, the formulation prepared in accordance with this process displayed only trace amounts of water even after 52 weeks.

AMYLOSE AND AMYLOPECTIN

Ordinary starch is known to consist of two polymers of glucose, the linear polymer called amylose (sometimes referred to as the A-fraction), and the branch-chain polymer called amylopectin (sometimes referred to as the B-fraction). The relative content of amylose and amylopectin varies with the source of the starch. It has been estimated that tapioca contains about 17 to 21% amylose; potato starch, 22 to 25%; cornstarch, 22 to 30%; and so on. The amylose molecule is considered to be a long, linear chain of anhydroglucose units. The amylopectin molecule, on the other hand, is considered to be a larger complex branched chain of tree-like structure with many of the branches themselves having branches, and so on.

The two fractions have substantially different properties. The amylose molecule is of low molecular weight as compared to the amylopectin molecule (a few hundred anhydrogluclose units with only one nonreducing end group per molecule). On the other hand, amylopectin is a high-molecular-weight molecule (more than 1,000 anhydroglucose units with one nonreducing end group for each 20 to 30 glucose units). Amylose has a high intrinsic viscosity and a low solution stability in water at ordinary concentration, while amylopectin has a fairly high solution stability but about the same intrinsic viscosity.

Relatively recently, several different approaches have been taken with the view of producing the individual fractions. One approach has been the genetic development of waxy maize whose starch consists essentially of amylopectin. Currently, there is a major program underway to breed varieties of corn whose starch is high in amylose. Further, over the last twenty years, a substantial number of patents have issued on methods for the separation of the two fractions.

None of the methods formerly used to produce the separate starch fractions has proved really successful in an economic sense. The amylopectin derived from waxy maize corn hybrids is expensive because of the care required in growing and processing this hybrid variety. The genetic program to breed corn whose starch consists essentially of amylose has met with some success, as a mutant has been found that has of the order of twice to three times the amylose content

of normal corn. However, the costs for growing and processing such hybrids can
be high. On the other hand, the chemical methods of separation are based on
the formation of a chemical complex of the fractions, particularly of the amylose,
or on a fractional salting-out process which may also use a complexing agent.
For example, in one method, an alkaline earth hydroxide complex is used; in
another, an alkyl alcohol complex (pentanol or butanol for example) is produced;
and in another, solutions of certain inorganic sulfates are used in a kind of fractional crystallization. In another method, described in U.S. Patents 3,313,654
and 3,323,949, carbon tetrachloride was used to facilitate the precipitation of
the amylose by complexing with carbon tetrachloride and then removing the
carbon tetrachloride by a series of controlled heating steps.

U.S. Patent 3,067,067 also discloses a process whereby a mixture of starch and
water is heated above 120°C to provide a fluid solution of starch in water. The
starch solution is then carefully cooled and maintained at a temperature above
49°C to provide a stabilized, noncongealing starch solution from which a solid
fraction of starch is precipitated. The precipitate is then separated from the
fluid fraction in any suitable manner. For example, a high-speed centrifuge of
the type employed in cornstarch wet milling for separating the granular starch
from the crude may be used.

In recent years enzymes which specifically hydrolyze the α-1,6-glucosic bonds
of amylopectin have been isolated. These enzymes are used to debranch the
amylopectin molecules of starch to produce an all-amylose starch where the
amylose has different chain lengths.

ALPHA-1,6-GLUCOSIDASE PRODUCTION

Alpha-1,6-glucosidase has been applied as a generic name for the various enzymes
that hydrolyze the α-1,6 linkages of amylopectins. Other names for the enzymes
having this specific activity are isoamylase, pullanase and amylo-1,6-glucosidase,
which are used by the various workers in this field. Isoamylase production by
use of yeasts was detailed by Maruo et al, *Symposia on Enz. Chem.* (1949).
Moreover, Bender et al [Biochem. Z., 334, 79-95 (1961)] reported that the
enzyme produced by *Aerobacter aerogenes,* which decomposes polysaccharides
pullulan produced by a strain of Pullularia, has properties similar to isoamylase.

Culture Media for Large Scale Enzyme Production

An improved culture medium has been developed by *K. Masuda, M. Mitsuhashi,
M. Hirao, Y. Sato and K. Sugimoto; U.S. Patent 3,622,460; November 23, 1971;
assigned to Hayashibara Co., Japan* for the industrial production of isoamylase.
The process consists of using a culture medium, which consists of ammonium
salts as nitrogen source and liquefied starch as carbon source, to produce and
accumulate a large amount of isoamylase in the culture medium. This process
is illustrated in the following examples.

Example 1: A streak culture medium of 1.0% peptone, 0.5% yeast extract, 0.05%
$MgSO_4 \cdot 7H_2O$, 0.05% KCl, 0.001% $FeSO_4 \cdot 7H_2O$ and 1.0% maltose is inoculated with *Aerobacter aerogenes* and the strain is cultured for 20 hours. (This
culture is used in this experiment as a seed culture.) A medium composed of
0.4% $(NH_4)_2SO_4$, 0.7% K_2HPO_4, 0.3% KH_2PO_4, 0.05% $MgSO_4 \cdot 7H_2O$, 0.05% KCl,

0.001% $FeSO_4 \cdot 7H_2O$, and 1.4% maltose, soluble starch or liquefied starch which was poured into a 500-ml conical flask and sterilized as usual was inoculated with one platinum loop of the resulting seed culture and the charge was then cultured at 30°C for 40 hours. Thereafter pH, absorbency and enzyme activity of the resulting solution obtained were determined. The results are tabulated in the following table.

Carbohydrate Used as Substrate	pH	Absorbency	Enzyme Activity, ml
Soluble starch	7.80	0.500	35
Maltose	6.90	0.570	39
Liquefied starch*	7.70	0.520	138

*The liquefied starch refers to a solution wherein a commercially available liquefying enzyme (alpha-amylase) was added to a potato starch emulsion to be treated and the pH of it was adjusted to about 6 and continuously liquefied at 88°C.

As the results from the above table show it has been found that the use of liquefied starch in the culture medium is obviously successful for producing isoamylase.

Example 2: The following culture media A, B and C, each of which have different nitrogen sources, were cultured in the same manner as described in Example 1. However, the pH of culture medium C was adjusted to 3 with alkali.

Culture Medium A

Peptone	1.0%
Yeast extract	0.5%
$MgSO_4 \cdot 7H_2O$	0.05%
KCl	0.05%
$FeSO_4 \cdot 7H_2O$	0.001%
Liquefied starch	1%

Culture Medium B

Corn steep liquor	3%
$MgSO_4 \cdot 7H_2O$	0.05%
KCl	0.05%
$FeSO_4 \cdot 7H_2O$	0.001%
Liquefied starch	1.0%

Culture Medium C

NH_4NO_3	0.48%
K_2HPO_4	0.7%
KH_2PO_4	0.3%
$Mg_2SO_4 \cdot 7H_2O$	0.05%
KCl	0.05%
$FeSO_4 \cdot 7H_2O$	0.001%
Liquefied starch	1.0%

Transition of Isoamylase Activity with Different Nitrogen Sources

Nitrogen Sources in Culture Medium	pH	Absorbency	Enzyme Activity, ml
Culture medium A	7.65	0.440	40
Culture medium B	8.40	0.400	25
Culture medium C	7.00	0.365	70

Bacterial Isoamylase from Pseudomonas

Y. Yokobayashi, K. Sugimoto and Y. Sato; U.S. Patent 3,560,345; February 2, 1971; assigned to Hayashibara Co., Japan have found that *Pseudomonas amyloderamosa*, a new species of the genus Pseudomonas produced a large amount of isoamylase in the culture medium. The bacteria of the genera Aerobacter and Pseudomonas are non-spore-forming Gram-negative rods. However, there are clearcut taxonomical distinctions between them in that the flagella of the organism of the former germs are peritrichous whereas those of the latter are polar, and the former decomposes various sugars aerobically and anaerobically while the latter decomposes them only aerobically.

Pseudomonas amyloderamosa discovered by these workers at Hayashibara Co. apparently differs in various properties from the organisms of the genus Aerobacter and those of the family Enterobacteriaceae. A marked difference is also noted in the properties between *Pseudomonas isoamylase* and *Aerobacter isoamylase*-like enzyme (pullulanase). For example, the optimum pH for the action of the latter enzyme is 6, while that for the action of the former enzyme ranges from 2.5 to 3.5.

Detailed bacteriological features of the *Pseudomonas amyloderamosa* SB-15 (ATCC 21,262) used in the process are given in the complete specification. Culture processes for preparing this isoamylase are shown in the following examples.

Example 1: A culture medium composed of

	Percent
Maltose	2.0
Sodium glutamate	0.2
$(NH_4)_2HPO_4$	0.3
KH_2PO_4	0.1
$MgSO_4 \cdot 7H_2O$	0.05

Tap water adjusted to pH 7

was inoculated with *Pseudomonas amyloderamosa* strain SB-15, and the strain is cultured with shaking at 30°C for 120 hours. After the incubation, the isoamylase activity was determined giving culture fluid activities of 180 to 220 units per ml. This culture fluid was centrifuged at 10,000 rpm for 10 minutes. The cells were removed and a supernatant fluid was obtained. While this liquid was being cooled and agitated, cold acetone was added to it to a concentration of 75% to cause precipitation of the enzyme. The precipitate was centrifugally collected, freeze-dried in vacuo and then isoamylase was obtained in powdery form. The recovery of enzyme in this procedure was 80 to 90%. This enzyme preparation was highly stable when stored dry. It can be further purified by the combination of this procedure with salting out with ammonium sulfate or other treatment.

Example 2: A culture medium composed of

	Percent
Soluble starch	1
$(NH_4)_2HPO_4$	0.15
KH_2PO_4	0.1
Polypeptone	0.1
$MgSO_4 \cdot 7H_2O$	0.05

Tap water adjusted to pH 7

was inoculated with this organism and cultured with shaking at 30°C for 72 hours. The isoamylase activity of this culture filtrate was 164 units. The culture filtrate was subjected to the same treatment as described in Example 1, and iso-amylase activity was obtained from 100 ml of the culture fluid, in a yield of 82%.

The method used for determining isoamylase activity conforms essentially to the method proposed by Maruo and Kobayashi [*Journal of Agricultural Chemical Society, Japan,* 23 (1949), pp. 115-120] and was carried out in the following manner. A reaction mixture consisting of

	Milliliters
Soluble glutinous rice starch solution (1.0%)	5.0
0.5 N acetic acid buffer solution (pH 4.0 for the present organism)	1.0
Enzyme solution	1.0

was incubated at 40°C for appropriate periods of time. To 1 ml of the mixture incubated was added 1 ml of 0.01 N iodine-potassium iodide solution, and the mixture was diluted with water to 25 ml. On standing for 15 minutes, the ex-tinction coefficient of the solution at 610 mμ was determined using 1 cm cell, and the amount of enzyme which caused the increase of the extinction coefficient by 0.1 in one hour was expressed as 10 units.

Escherichia intermedia for Isoamylase Production

S. Ueda, N. Nanri and M. Hongo; U.S. Patent 3,716,455; February 13, 1973; assigned to Hayashibara Co., Ltd., Japan have found that a strain of *Escherichia intermedia,* isolated from soil, accumulates isoamylase in the culture medium and have established a method for commercial production of the enzyme. The culture medium to be used in the present process may be synthesized or of nat-ural substances. It should contain carbohydrates, other carbon sources, inorganic substances and other nutrients in appropriate amounts as shown in the following examples.

Example 1: Escherichia intermedia (ATCC 21,073) was used as a seed microorganism. This microorganism previously seed-cultured in the culture medium containing 0.5% by weight maltose, 1% by weight peptone, 1% by weight meat extract and 0.5% by weight $NaNO_3$ was inoculated in the fermentation medium at the rate of 5% (by volume). The fermentation medium of the following composition was used and the shake culture was conducted at 30°C.

Composition of the fermentation medium: 5% by weight dextrin, 1% by weight of peptone, 0.5% by weight of K_2HPO_4, 0.05% by weight of $MgSO_4 \cdot 7H_2O$, 0.05% by weight of KCl, 0.001% by weight of $FeSO_4 \cdot 7H_2O$, and 0.5% by weight of $NaNO_3$. pH was adjusted to 7.5 before sterilization. The fermentation medium cultured for 48 hours showed 11.1 units of isoamylase activity. Enzymatic activity was measured as follows: 1 ml of 0.5 N acetate buffer solution at pH 6.0 is added to 5 ml of a 1% aqueous solution of soluble glutinous rice starch prepared by Lindner's method and 1 ml of the enzyme solution is added.

The solution is left to stand at 40°C for one hour. To 1 ml of the solution are added 1 ml of a 0.01 N iodine-potassium iodide solution and water to make the total amount 25 ml. The extinction coefficient at 620 mμ is measured. As a control, the extinction coefficient of the solution at the beginning was measured. The increase of extinction coefficient was proportional to the amount of enzyme if it was in the range of about 0.1. Under these conditions, the enzyme amount to show 0.1 of the increase of extinction coefficient was provided for 10 units.

The filtrate obtained by removing cells from the fermentation medium was salted out with an aqueous ammonium sulfate solution (0.15 to 0.45 saturation) according to the conventional method. Resalting out with an aqueous ammonium sulfate solution (0.18 to 0.40 saturation) was conducted. Subsequently, the solution was adsorbed on calcium phosphate and further purified by DEAE cellulose column chromatography. The specific activity showed about 100 times increase. The activities and recovery ratios in each step are determined as follows.

In determination of the activity, if 1 ml of an enzyme solution is used for the reaction, the amount of the enzyme which shows 0.1 of the increase of extinction coefficient is represented as 10 μ/ml of enzyme solution. Therefore, the specific activity is calculated by dividing the above value with the number of milligrams of the protein in 1 ml of the enzyme solution, since a pure enzyme is protein, and the purity of the enzyme in the purification steps can be shown by this activity.

	Total Volume, ml	Total Activity, units	Specific Activity	Recovery Rate, %
Supernatant solution of broth	3,000	33,000	0.88	100
Salting out 1	230	26,730	14.2	81
Salting out 2	50	23,100	18.1	70
Ca phosphate gel	21	11,500	53.0	35
DEAE cellulose	10	6,600	*99	22

*Increased about 112 times

$$\text{Specific activity} = \frac{\text{Total activity (units)}}{\text{Total protein (mg)}}$$

This purified enzyme showed about 47°C of optimum temperature and about 6.0 of optimum pH. Moreover, the present enzyme reacted with β-limit dextrin from glycogen. It was recognized to be isoamylase.

Example 2: The same fermentation as in Example 1 was conducted, except that the shake culture was first conducted and after 10 hours of culturing, it was changed to a stationary culture with or without the addition of a 0.1% by weight of toluene to the culture medium. The fermentation medium showed 7.8 units of isoamylase activity in 24 hours after the beginning of culturing.

ENZYMATIC AMYLOSE PRODUCTION

From Stabilized Starch Solutions

M. Seidman; U.S. Patent 3,532,602; October 6, 1970; assigned to A.E. Staley Manufacturing Company describes a method for reducing the viscosity of a stabilized solution of starch by contacting the solution with a material exhibiting pullulanase activity. The process has special applicability for separating amylose from stabilized starch solutions. The separated amylose may be cast from solution as a film useful in packaging, particularly foodstuffs (e.g., as sausage casings) since the amylose is digestible by humans. The structure of amylose resembles cellulose and similarly many of its derivatives are thermoplastic.

Briefly stated, the amylose is obtained by contacting a stabilized, high-viscosity, fluid aqueous solution of starch which is capable of forming and growing separable solid fractions rich in high-molecular-weight amylose with a material exhibiting pullulanase activity for a period sufficient to reduce the solution's viscosity to a viscosity which will facilitate and accelerate the formation and separation of the solid starch fraction without requiring the addition of a diluting liquid and thereafter separating a solid fraction enriched in amylose and a fluid fraction enriched in amylopectin.

Example 1: A 10% by weight dry substance cornstarch solution was prepared by mixing 10 parts of starch in 90 parts water to form a slurry, adjusting to a pH of 6.8, and cooking the slurry at 150°C in a continuous autoclave for about 1 minute. The starch solution was then stabilized by rapidly cooling from 150° to 100°C and then slowly cooling from 100° to 45° to 50°C at a rate of about 1°C per 3 to 4 minutes over a period of about 3 hours. This controlled rate of cooling was accomplished by placing the starch solution in an oven heated to 100°C and then cooling the oven at a rate of about 1°C per 3 to 4 minutes.

The stabilized starch solution was then divided into two portions and allowed to stand for 24 hours at 45° to 50°C to continue the formation and growth of a solid starch fraction. To one of the starch portions, pullulanase preparation was added in an amount of 100 mg/1,500 ml of 10% by weight DS starch slurry. The pullulanase preparation had an activity of 320 units/g. One unit of pullulanase is defined as that amount of enzyme present in 1.0 ml of solution which, with 1% pullulan as a substrate under standard conditions of assay, raises the reducing value of the substrate in 1 hour at 45°C to a reducing value equivalent to 1 mg of maltose. To the other starch portion no enzyme was added. The starch solution containing the enzyme was then allowed to stand at 45° to 50°C for an additional 24 hours. (The pullulanase was obtained from the organism

Aerobacter aerogenes by the Bender and Wallenfels process described in *Biochem. Z.,* vol. 334, pp. 79-95 (1961). The viscosities of the two starch solutions were determined prior to centrifugation. The starch solution which was treated with pullulanase had a viscosity of 265 cp, while the untreated starch solution had a viscosity of 1,760 cp. The starch solutions were then centrifuged at various speeds for different periods of time in a Lourdes centrifuge Model No. LCA-2, having an angle head of about 30° and a radius of about 4 inches. The degree of separation of the solid starch fraction for each of the starch portions at different centrifuging speeds was noted and is shown below.

Centrifuging Speed, rpm	Centrifuging Time, min	Untreated Starch Portion	Enzyme Treated Starch Portion
2,000	10	No appreciable separation obtained	Fair separation
4,000	10	Poor separation	Good separation
8,000	30	Good separation	Excellent separation

Poor = Less than 30% separated
Fair = Between 30 and 60% separated
Good = Between 60 and 85% separated
Excellent = Above 85% separated

These results show that the starch solution treated with the enzyme required substantially lower centrifuging speeds as well as shorter centrifuging times to obtain a good separation of the amylose fraction. On the other hand, the starch which had not been treated with a pullulanase preparation required substantially higher centrifuging speeds and longer centrifuging times to separate the solid amylose fraction from the water-soluble fraction.

Examples 2 through 5: A stabilized cornstarch solution was prepared following the procedure of Example 1. The stabilized starch solution was divided into four equal portions and identified as Examples 2 through 5 respectively. To one of the portions, pullulanase in an amount equivalent to 100 mg/1,500 ml of starch solution was added. The enzyme had a potency of about 320 units/g as defined in Example 1. To another portion 0.34 g of a malt amylase was added per 100 g of starch on a dry substance basis. To the third portion 1.53 ml of concentrated hydrochloric acid per 100 g of starch on a dry substance basis was added.

After two hours at 49°C the starch solution containing the hydrochloric acid was neutralized with 5 N potassium hydroxide. The last portion was untreated and served as a control. Viscosities were determined for each of the starch solutions prior to separation of the solid fraction. The solid starch fractions enriched in amylose were separated from the starch solution by centrifuging and washed by slurrying in distilled water and again centrifuged. This washing procedure was repeated 3 times. The amylose fractions were then washed 3 times with methanol and 3 times with acetone before drying in a vacuum oven at 50°C. After the amylose fractions had been isolated and purified, a reduced viscosity was determined for each of the amylose fractions. Reduced viscosity is specific viscosity divided by the concentration of the amylose fraction. The reduced viscosities reported in the table were determined on 0.4 g of the amylose fraction per 100 ml in 1 N KOH at 30°C. The specific viscosity was determined as described in *Kerr's Chemistry and Industry of Starch*, 2nd edition, pp. 675-676 (1950). The reduced viscosity of a linear polymer is empirically related to the polymer's molecular weight. Therefore a reduction in the polymer's reduced viscosity indicates a reduction or lowering of the polymer's molecular weight. The results obtained are reported in the table below.

Example	Viscosity of Solution, cp	Type of Treatment	Reduced Viscosity of Amylose
2	265	Pullulanase	1.92*
3	240	Malt amylase	0.61
4	180	Hydrochloric acid	1.18
5**	750	Control	2.07

*The decrease in reduced viscosity of the amylose fraction is attributed to the low-molecular-weight amylose (straight chain polymer) obtained from the action of the pullulanase on the amylopectin.

**Example 5 was diluted with water to obtain a viscosity of about 240 cp prior to centrifuging.

These examples show that high-molecular-weight amylose can be recovered by treating a starch solution with pullulanase without encountering substantial degradation of the amylose fraction. These examples further show that with other thinning agents substantial degradation of the amylose fraction is encountered. This is comparatively shown by the reduced viscosity of the amylose fraction treated with either malt amylose or hydrochloric acid as compared with the amylose fraction treated with pullulanase.

From Cooked Starch Solutions

R.J. Hathaway; U.S. Patent 3,556,942; January 19, 1971; assigned to A.E. Staley Manufacturing Company describes a method for fractionating and separating amylose from a cooked starch solution by contacting the cooked starch solution with a material exhibiting pullulanase activity for a time sufficient to reduce the solution's viscosity to a viscosity of preferably less than 10,000 cp measured at 40°C, and then separating the amylose from the solution. The process differs from that reported in U.S. Patent 3,532,602 in that the starch is not stabilized prior to adding the pullulanase but is used as a solution of cooked starch. The entire process encompasses the following steps:

(1) Heating a suspension of starch in water to a temperature and for a time sufficient to transform substantially all of the suspended starch into a fluid solution without causing substantial degradation of the starch molecules;

(2) Cooling rapidly said starch solution to a temperature below that which will substantially deactivate pullulanase activity;

(3) Releasing a solid starch fraction enriched in amylose from the solution by treating the cooled starch solution with a material exhibiting pullu-

lanase activity for a time evidenced by a reduction in the viscosity of
the starch solution to a viscosity of preferably below 10,000 cp measured at 40°C; and

(4) Separating from the treated starch solution the solid starch fraction
enriched in amylose.

Example 1: A 22% by weight DS potato starch slurry was prepared by suspending 100 g (dry basis) of potato starch in 350 ml of distilled water. The starch
suspension (pH 5.4) was heated rapidly to 97°C by passing steam into the suspension and then cooked for 30 minutes at 97° to 100°C. The addition of steam
reduced the starch solids content to 17%. The cooked or pasted starch was then
cooled to 40°C over a period of 15 minutes. The cooked starch at this point
was very thick and was difficult to stir. To the cooked starch was added 67 ml
of pullulanase beer, having a pH of 8.5 and containing 2,000 units of measurable
pullulanase activity, which is equivalent to 20 units of enzyme per gram of starch.

Ten milliliters of toluene was then added to prevent yeast growth. The starch
solution was adjusted to a pH of 6.8 and maintained at a constant 40°C. Within
four hours, after the enzyme had been added, the Brookfield viscosity at 40°C
and 20 rpm had dropped from an unmeasurably high viscosity to a viscosity of
680 cp. Approximately at this point the amylose began precipitating slowly, in
amorphous form from the starch solution, accompanied by an increase in viscosity of up to about 2,000 cp after about 28 hours.

The resulting starch mixture was centrifuged in 250-ml steel bottles at 9,000 rpm
(about 10,000 g) on a Serval high-speed centrifuge for two hours at about 40°C.
The centrifuged solution was then decanted and the residue washed three times
with 250 ml of distilled water, and again centrifuged. The final wash contained
only 0.3% solids by weight. The residue was precipitated with methanol, filtered
and dried at room temperature. The weight yield was 27% based on starch. The
amylose-rich product had a reduced viscosity of 1.54 and contained 72% amylose
(dry basis). This amounted to a recovery of 70% based on the amylose originally
present (27%).

Example 2: This example shows that amylose can also be separated by this process from jet cooked cornstarch. 300 grams of pearl starch was suspended in
water to produce a slurry containing 10% solids DS and adjusted to a pH of 6.3.
The slurry was introduced into a jet cooker at 140°C with minimum hold time.
The pasted starch obtained therefrom was cooled rapidly to 45°C and treated
for 18 hours with pullulanase in an amount corresponding to 11 units of pullulanase activity per gram of starch.

The Brookfield viscosity at 20 rpm was below 100 cp. The precipitated product
was redissolved by heating to 140°C and reprecipitated by slow cooling over 24
hours. The reprecipitated amylose was then centrifuged at 9,000 rpm and washed
twice with water. After drying the residue was found to contain 77% apparent
amylose. These examples show that only a relatively short period of enzyme
treatment is required in order to obtain a solid starch enriched in amylose.

Low MW Amylose Production

A process is disclosed by *K. Sugimoto, M. Hirao, M. Yoshida, M. Shiosaka and
Y. Yokobayashi; U.S. Patent 3,730,840; May 1, 1973; assigned to Hayashibara*

Company, Japan for the preparation of straight-chain amyloses having molecular chain lengths of 20 to 30 that have never been obtainable before. Straight-chain low molecular weight amyloses where the molecules have relatively uniform chain length are obtained by liquefying a starch slurry to a relatively low extent, rapidly cooling the liquefied slurry before a substantial retrogradation can occur, and selectively decomposing the α-1,6-glucoside bonds of amylopectin contained in the starch.

For convenience, the process will be described below as applied to waxy cornstarch, a variety of starch with a simplified composition. Because this starch consists solely of amylopectin having branched structure, its decomposition product should also be simple in structure. A 5 to 10% aqueous suspension of waxy cornstarch is gradually heated with stirring. After the suspension has been gelatinized, it is heated in a pressure oven at 1.5 to 2 atm for 30 minutes to effect complete gelatinization and dispersion. The resultant is cooled, adjusted to pH 4.5 and α-1,6-glucosidase obtained from the bacteria of the genus Pseudomonas is added at a rate of 20 to 50 units per gram of the starch. At regular intervals the mixture is sampled, and the degree of reaction is determined using the color reaction of the amylose.

The blueness with the addition of a predetermined amount of an iodine-potassium iodide solution is evaluated in terms of the extinction coefficient at 570 mμ. When the enzyme is used in an amount of 50 to 100 μ/g starch, the decomposition proceeds almost to the limit in 50 to 20 hours. Where an Aerobacter enzyme is used, the pH is adjusted to 6. When a Pseudomonas enzyme is used, the amylose produced forms mycels because of the crystallinity and precipitates. Upon completion of the reaction, therefore, it can be readily separated from water by centrifuge. The yield can range from 80 to 90% of the starting material. The product when washed several times with small amounts of water and then dried will take the form of white powder.

When the polymerization degree is determined by the reducing end groups using the periodic-acid oxidation method, the precipitated part indicates a polymerization degree of 23 to 24. This agrees well with the accepted values of 20 to 30 of the branched chain parts of amylopectin.

The present process has so far been described primarily in connection with the enzyme from the genus Pseudomonas. In the case of the enzyme from Aerobacter bacteria, the enzymatic activity apparently differs somewhat. The product shows relatively high solubility and is difficult to fractionate as a crystalline precipitate. Precipitation with butanol or concentration of the whole mass is required for the purpose of fractionation. As for other amylose-containing starches than waxy cornstarch, e.g., potato starch, the amylose contents obviously cause a widespread distribution of the polymerization degrees of the amylose molecules, and the mean polymerization degree of those molecules are accordingly increased. The process and the behaviors of the products are illustrated in the example below.

Example: Decomposition of Various Starches With α-1,6-Glucosidases — Waxy cornstarch, glutinous rice starch, and potato starch are purified and are suspended in water to prepare 5 to 10% aqueous suspensions. The suspensions are adjusted to pH 4.0 to 5.0 when an enzyme produced by bacteria of the genus Pseudomonas is used, or to pH 5.8 to 6.0 when an enzyme from bacteria of the genus Aerobacter (U.S. Patent 3,622,460, page 169) is used. They are heated with stirring to gelatinize the starch. For complete gelatinization, each gelatinized solution is further heated at a pressure of 1.5 to 2.0 kg/cm^2 for 20 to 30 minutes, and then rapidly cooled to 45°C. The enzyme is added in an amount of 20 to 100 units per gram of the starch, and the mixture is reacted with stirring at 45°C. After the lapse of about 16 hours, the reaction rapidly takes place. The progress of the reaction is determined as the solution turns blue upon the addition of an iodine-potassium iodide solution which initiates a reaction characteristic of the particular amylose produced.

After one hour of reaction the viscosity is remarkably decreased. When the Pseudomonas enzyme is used, the precipitation of amylose occurs with progress of the reaction. The rate at which the reaction proceeds is determined by adding 9 ml of N/10 hydrochloric acid per milliliter of the reaction solution thereby terminating the reaction and testing the solution by the anthrone method. An amount of the solution equivalent to 10 mg of sugar is sampled and with the addition of 2 ml of an acetic acid buffer solution at pH 4.0 and 0.5 ml of a 0.3% aqueous solution of iodine and potassium iodide, the total amount is increased to 100 ml. After 30 minutes the extinction coefficient as measured with light of wavelength of 570 mμ is recorded.

Thus, when the reaction reaches a point as determined by the iodide color test, the solution is concentrated by boiling, and is then powdered. When the Pseudomonas enzyme is used, the amylose produced precipitates. Therefore, the solution is concentrated to a concentration of about 20% solids, cooled, and the precipitate is allowed to deposit overnight, and is then separated by centrifuging. Next, the precipitate is dissolved to 20% solids, reprecipitated, and the new precipitate is separated and dried at 40° to 50°C. The supernatants obtained by the centrifuging of reaction mixtures are mixed and concentrated in vacuo, and finally dried and powdered. The treatment for precipitation and separation as above described is rendered easiest when the Pseudomonas enzyme is used, by the use of waxy cornstarch or glutinous rice starch as the starting material.

K. Sugimoto and M. Yoshida; U.S. Patent 3,729,380; April 24, 1973; assigned to Hayashibara Company, Japan have also developed a commercial process for producing a relatively low molecular amylose having straight chain structure by selectively hydrolyzing with α-1,6-glucosidase only the branched parts in amylopectin molecules. The amylose obtained by this process is a somewhat crystalline powder free from hygroscopicity, and can be used as an absorbent of perfume, or medicine, filler, and stable additive. The amylose has a water solubility of a few percent, and it is suitable for the production of films or foamed products which should disappear after being used as edible or digestible products.

The process can be explained with reference to manufacturing steps. Preferable starches used as starting raw materials are glutinous starches such as rice, waxy corn and waxy milo. Others such as corn, tapioca, sago, sweet or white potato, rice and wheat can also be used. It is necessary to purify these starches as much as possible by washing, since purification after the hydrolysis is difficult. Then, the starch must be liquefied to facilitate and to uniformalize the action of enzyme. The starch slurry can be gelatinized homogeneously at 130° to 170°C and the decomposition does not proceed to the extent that glucose is formed. Cornstarch can be liquefied in a concentration of 30 to 40%. The resulting liquefied product is highly viscous, but that starch slurry can be easily gelatinized homogeneously by stirring the same under heating and with introduction of steam in a vertical liquefaction tank with a continuous stirring system.

A second method comprises heating starch slurry to a temperature of 100° to 130°C to effect gelatinization and dispersion, where the temperature is first raised to 100°C and then to 130°C to complete the gelatinization and dispersion. If the concentration is increased, the operation becomes impossible and the dispersion is somewhat insufficient.

In a third method, a liquefying enzyme (α-amylase) is used to effect dispersion and gelatinization to a low DE. In this process, α-amylase is used in a usual quantity of 10 to 20 units and stirring is effected quickly at pH 6.0 to 6.5 while heating to 85° to 96°C in the above continuous liquefying apparatus to obtain a liquefied product of DE of 0.5 to 3% within a few minutes.

The gelatinized starch dispersion is preferably cooled as quickly as possible because the dispersion is very viscous due to its low decomposition rate and since the viscosity increases and retrogradation of amylose occurs as temperature lowers. For this purpose, cooling by self-evaporation by spraying the liquor into a vacuum cooling tank is efficient. If enzyme is introduced at the same time by way of spraying, instantaneous cooling and instantaneous mixing of enzyme are possible.

The preferable enzymes are heat-resistant ones such as those from Lactobacillus (*Lactobacillus plantarum* ATCC 8008, *Lactobacillus brevis* IFO 3345) or ray fungus (*Streptomyces diastatochromogenes* IFO 3337, *Actinomyces globisporus* IFO 12208, *Nocardia asteroides* IFO 3384, *Micromonospora melanosporea* IFO 12515, and *Thermonospora viridis* IFO 12207). In case those enzymes are used, retrogradation can be prevented by treating a liquor at pH 5 to 7 at 50° to 60°C to lower the viscosity. After residence for more than one hour, in a mixing tank provided with a stirrer, mixed solution having a reduced viscosity is continuously taken out and introduced into a reaction tank in which the reaction is carried out completely at an optimum pH and temperature for 24 to 48 hours. Preferable reaction temperature is in the range of 40° to 50°C.

Degree of reaction is determined by absorbency of blue color of the liquid mixed with iodine taking advantage of the color reaction of iodine-potassium iodide solution. Preferable concentration of solids at the completion of the reaction is 10 to 20%. At a concentration higher than 20%, reaction velocity is low and complete decomposition becomes difficult. When color reaction of the liquid becomes maximum showing the completion of the reaction, the liquid is concentrated and then made into powdery form. The pulverization may be effected according to drum-drying or spray-drying.

If impurities derived from enzyme added or starch used as starting raw material are in a large quantity, reaction liquid can be purified by cooling the liquid, subjecting the resulting precipitate to centrifugation and washing the precipitate with water. Also, the precipitate can be dissolved again to form 20% solution, then reprecipitated and centrifugalized to obtain product of less than 50% water content, which is then dried. The filtrate or liquid separated by the centrifugation is concentrated to form precipitate, which is then separated and dried.

Example: Purified glutinous starch in the concentration of 30% by weight having pH 6.0 is continuously introduced under pressure in the bottom of a cylinder of

multiwing system. At the same time, steam is also introduced under pressure
pressed in the bottom of the cylinder. The starch is heated to 160° to 165°C
under vigorous stirring to effect gelatinization. The resulting gelatinized liquid
is taken out from the upper part of the cylinder. Then the liquid is introduced
into a series of several cylinders for residence time of 10 to 20 minutes to obtain
gelatinized liquid of DE 0.5 to 3%. The viscous liquefied product is quickly
cooled to 60°C by spraying into a vacuum cooling tank through a reducing valve,
and at the same time, enzyme produced by Lactobacillus is sprayed. After the
instantaneous cooling and mixing, the cooled liquid is taken out by a pump and
mixed in a mixing tank.

Then the reaction is carried out in a reaction tank in batch system while the reaction liquid is continuously taken out in such a manner that residence time is
longer than one hour. The reaction tank is controlled to pH 6.0 and temperature
of 50°C. After completion of the reaction, the reaction product is heated to inactivate the enzyme and then cooled, and the resulting precipitate is centrifuged,
dissolved again in water to form 20% solution, which is then allowed to cool to
form precipitates and again centrifuged to isolate the product. The product is
dried at 45°C and then pulverized.

High MW Amylose

A similar process whereby high MW amylose is obtained on an industrial scale
is disclosed by *M. Yoshida and M. Hirao; U.S. Patent 3,830,697; August 20,
1974; assigned to Hayashibara Company, Japan.* Amylomaize starch having a
high amylose content or amylose starch separated from sweet potato starch,
white potato starch or cornstarch is subjected to hydrolysis to selectively hydrolyze only α-1,6-glucoside bond of the amylopectin contained in an amount of
20 to 40% with an α-1,6-glucosidase, i.e., an enzyme produced by Aerobacter,
Pseudomonas, Lactobacillus or Escherichia. The Lactobacillus-produced enzyme
is the only α-1,6-glucosidase claimed.

Example: Amylomaize starch thoroughly washed with water and sieved, in the
form of 3% aqueous suspension, is heated to 100°C with stirring to obtain a
gelatinized product, which is then heated at 5 kg/cm^2 for 30 minutes to complete
the dispersion. The gelatinized liquid is quickly cooled to 60°C and pH is adjusted
to 5.5, and an aqueous solution of salted-out enzyme of Lactobacillus in an
amount of 50 units per gram of starch is added. Under vigorous stirring, the
whole is cooled to 45°C during 1.5 hours. The mixture is kept at 45°C to carry
out the reaction for 35 hours under stirring. Thereafter, the resulting precipitates
are centrifugalized.

The product is washed once with a small amount of water and centrifuged to
obtain 43%, based on the dry raw starch, of white, dry amylose precipitates.
The supernatant liquid is condensed and cooled to obtain additional 30% of
white amylose. The determination of average polymerization degree shows the
degree of 580 in the former and 160 in the latter. The determination according
to Smith's decomposition method suggests that the former has no branch whereas
the latter has about one branch per molecule. The hydrolysis with β-amylase
gives 99% yield as maltose in the former and 80% yield in the latter. Therefore,
it may be concluded that amylose of a very high purity, as a whole, may be obtained.

Separation of High and Low MW Amylose

The process disclosed by *K. Sugimoto, M. Yoshida and M. Kurimoto; U.S. Patent 3,632,475; January 4, 1972; assigned to Hayashibara Company, Japan* provides for the preparation of long-chain and short-chain amyloses from starches. This process comprises separating amyloses from a starch by decomposing the amylopectin through the agency of α-1,6-glucosidase into straight-chain dextrin of a low molecular weight (or short-chain amylose), thereby giving rise to a difference in molecular weight between the decomposition product and the high-molecular natural amylose (long-chain amylose). Efficient fractionation of the two by taking advantage of the difference in molecular weight gives long-chain and short-chain amyloses of high purity.

The first step of this process is for the dispersion of starch. Since starch is likely to be oxidized and decomposed at elevated temperatures, sufficient care must be taken in this stage. A temperature of 100°C and the use of an equipment of totally enclosed, continuously operable type or heating in an inert gas are desirable, but heat up to about 130°C may be used. The second step is for decomposition of amylopectin in the starch with the action of α-1,6-glucosidase. First, the dispersed solution must be cooled to 45°C or the optimum temperature for the enzymatic activity, which may be as high as 70°C for the Lactobacillus and Norcardin-produced enzymes.

Because the solution still remains highly viscous at 45°C and retrogrades with time, it should be instantaneously cooled to this temperature and mixed with the enzyme. One way of realizing this objective is to introduce both the gelatinized starch and enzyme solutions in an atomized state into a vacuum cooler from the top so that the two can be thoroughly mixed up and the mixture can be cooled down to 45°C all at once. The starch solution is then dropped into a large amount of decomposition liquid in a storage tank and mixed up with agitation. This results in decomposition and a decrease in the viscosity without time for retrogradation.

Thus, amylopectin alone is decomposed into a straight-chain dextrin, or low molecular weight amylose. The molecular weight of this amylose is only a fraction of that of a natural amylose. The third step consists of fractionating the natural amylose (long-chain amylose) and low molecular weight amylose (short-chain amylose) by the difference of molecular weights. This may be accomplished in any of the following four methods.

(1) One method is by fractional precipitation in an aqueous solution by precipitating the long-chain amylose at 90° to 40°C and the short-chain amylose at 5°C or less. While the solubility of long-chain amylose at normal temperature is not more than 0.1%, that of molecules with polymerization degrees of 18 to 25 of short-chain amylose, the decomposition product of amylopectin, is usually between 1 and 5% at 10°C and between 5 and 13% at 40°C.

(2) Where necessary the butanol precipitation method may be resorted to. In this case, 1-butanol is added to a starch-enzyme decomposition solution in an amount equivalent to 4 to 7% of the solution. The solution is heated to a homogeneous state, and then is allowed to cool slowly over a period of 20 hours. This leads to butanol precipitation. The precipitate can be centrifuged and recrystallized for increased purity.

(3)   Fractional precipitation with the use of magnesium sulfate is also ap-
        licable. To a decomposition solution at concentration of 10 per-
        cent, magnesium sulfate is added in an amount equivalent to 10% of
        the solution. The mixture is heated to about 90°C, cooled slowly
        down to 40°C, and precipitation of amylose is completed. The pre-
        cipitate is separated and washed several times to remove salts there-
        from. The remaining solution is concentrated to an increased concen-
        tration, cooled, precipitated, and finally the objective substance is
        separated.

(4)   This is a process wherein a reacted solution is heated to 70° to 80°C
        without the addition of any precipitant, and the crystallization and
        precipitation of the objective substance are promoted by means of
        ultrasonic energy.

Example 1: Purified cornstarch in a 15% aqueous suspension was adjusted to
pH 5 to 6, forced into a continuous multiblade agitation column, heated rapidly
to 100°C with the supply of steam, and was agitated for gelatinization. The
gelatinized solution, while being readjusted to a concentration of 10% and pH
5 to 6, was injected into a vacuum cooler and cooled quickly to 50°C. At the
same time, a solution of the α-1,6-glucosidase produced by the bacteria of the
genus Aerobacter described in U.S. Patent 3,622,460 was injected into the vacuum
cooler at a rate of 50 units per gram of the starch, and was rapidly and thoroughly mixed with the gelatinized solution. Upon cooling and mixing, the solution was immediately fed into a decomposition tank kept at 45°C and was completely mixed up with stirring.

With the average residence time in the decomposition tank set for 1 hour, the
mixture was continuously fed into the main reaction tank, where it was kept
at 45°C for 40 hours. In the cases where soluble matter precipitated during the
reaction at 45°C, the reaction mixture was heated to 80°C for 10 to 15 minutes,
and the precipitate was centrifugally removed, and this precipitate (1) was washed
with cold water. Next, the remainder of the decomposition solution at 80°C
was slowly cooled down to 50°C with stirring over a period of 20 to 40 hours.
The resulting solution was precipitated and separated through a centrifuge. This
precipitate (2) was washed twice with cold water. The precipitates (1) and (2)
thus obtained were dehydrated and dried solid.

On the other hand, the supernatant fluid left behind was cooled to downwards
of 5°C, stirred for 12 hours to precipitate short-chain amylose, centrifuged, and
washed with cold water in the similar way. The remaining supernatant fluid ·
was concentrated in a vacuum and dried. Precipitates (1) and (2) were long-chain
amylose with a polymerization degree of 750, and the combined yield was 20%.
The secondary precipitate, which was short-chain amylose, had a polymerization
degree of 30, and the yield was 65%. The remainder with a polymerization degree of 20 was obtained at a yield of 10%.

Example 2: Potato starch in the form of an aqueous suspension with a concentration of 10% was adjusted to pH 5.0, and was gelatinized and dispersed in the
same manner as described in Example 1. Then, while the pH was being adjusted
to 4.5, it was cooled and a Pseudomonas enzyme was added at a rate of 50 units
per gram of the starch. The reaction was carried out at 45°C and a precipitate
(1) formed in 30 hours was centrifugally separated and washed with water. The
remaining solution was cooled down to 5°C, allowed to stand for 12 hours, and

the resulting precipitate (2) was separated, cooled and washed with cold water in the same manner as in Example 1. The remaining solution was concentrated and dried by atomization. The total recovery rate was 90% on the basis of dry starch. The polymerization degrees of the products are shown in the table below.

	Polymerization Degree	Yield, %
Precipitate (1) (long-chain amylose)	850	15
Precipitate (2) (short-chain amylose)	35	35
Residual dry matter	25	40

Two Enzyme Process

M. Yoshida and H. Mamoru; U.S. Patent 3,721,605; March 20, 1973; assigned to Hayashibara Company, Japan; use two α-1,6-glucosidase enzymes at different temperatures to produce high yields of macromolecular amylose. Starch rich in amylose is heat-gelatinized and then rapidly cooled to 50-60°C and treated with a heat-resistant α-1,6-glucosidase. The mixture is cooled further to 40-50°C and treated with a second α-1,6-glucosidase to complete debranching of the amylopectin and permit precipitation of the amylose.

The enzyme used in the first step is produced by a strain selected from *Lactobacillus brevis* IFO 3345, *Lactobacillus plantarum* ATCC 8008, *Streptomyces diastatochromogenes* IFO 3337, *Actinomyces globisporus* IFO 12208, *Nocardia asteroides* IFO 3384, *Micromonospora melanospora* IFO 12515, and *Thermonospora viridis* IFO 12207. The second α-1,6-glucosidase is produced by a strain selected from *Pseudomonas amyloderamosa* ATCC 21262, *Escherichia intermedia* ATCC 21073, *Agrobacterium tumefaciens* IFO 3085, *Azotobacter indicus* IFO 3426, *Mycobacterium phlei* IFO 3158, *Micrococcus lysodeikticus* IFO 3333, *Pedicoccus acidilactici* IFO 3884, *Serratia indica* IFO 3759, *Streptococcus fecalis* IFO 3128, *Staphylococcus aureus* IFO 3128, and *Staphylococcus aureus* IFO 3061.

Example: In this example amylomaize starch was gradually heated at a concentration of 10% and pH of 6.0 under agitation in a nitrogen gas stream to obtain paste of high viscosity. The paste obtained was rapidly cooled to 60°C, then mixed with heat-resistant enzyme produced by Streptomyces in an amount of 20 units per gram of starch with agitation. After lowering of viscosity for one hour, the temperature was lowered to 50°C. Then the enzyme of Pediococcus was added in an amount of 20 units per gram of starch, the temperature was lowered to 45°C and reaction was carried out for 45 hours.

The reaction mixture obtained was concentrated to one-half the original volume and made to stand overnight, and the resulting precipitate was separated by centrifuging, suspended in equal amount of water, again centrifuged and dried in vacuum. The yield obtained was 80%. Further product of yield of 10% could be obtained similarly by concentrating and cooling the liquid after separation. The degrees of polymerization were 550 and 105 respectively in the former and latter products. The process which was otherwise fairly difficult to operate due to high viscosity was considerably facilitated by two-step reaction. The amylose once washed with water was spray-dried from a concentration of 20 to 30% to obtain stable porous crystalline powder.

This yield and DP of product can be compared to products listed in Table 1 where a single enzyme was used and in Table 2 where two enzymes were at a single temperature of 45°C.

TABLE 1

Enzyme used	Amount of enzyme units/g. starch	Precipitate		Soluble part	
		Yield	Degree of polymerization	Yield	Degree of polymerization
Lactobacillus	40	45	750	30	120
Streptomyces	40	41	832	32	110
Actinomyces	40	43	731	31	110
Nocardia	40	40	780	30	121
Micromonospora	40	44	832	32	105
Thermonospora	40	42	785	35	100

TABLE 2

No.	Enzyme used	Amount of enzyme units/g. starch	Time of addition	Precipitate		Soluble part	
				Yield	Degree of polymerization	Yield	Degree of polymerization
1	Pseudomonas / Escherichia	20 / 20	Simultaneous	45	770	30	110
2	Pseudomonas / Micrococcus	20 / 20	Initial stage / Middle stage	41	785	32	120
3	Pseudomonas / Azotobacter	20 / 20	Middle stage / Initial stage	43	790	30	100

Freeze-Thaw Separation of Amyloses

Normally, on cooling amylose mixtures with different DPs, the macromolecular amyloses are retrograded and separated by precipitation. On the other hand, since the moisture content of the precipitate is high (approaching 90%), the 90% moisture content dissolves the low molecular amylose within the mixture and decreases greatly the macromolecular amylose content. Therefore the separation and purification of amyloses are not attained.

An improved separation and purification of amyloses is described by *M. Kurimoto and M. Yoshida; U.S. Patent 3,766,011; October 16, 1973; assigned to Hayashibara Company, Japan*. In this process, amylopectin is debranched by the use of an α-1,6-glucosidase after first gelatinizing the starch by heat treatment. The improved separation involves freezing the debranched starch mixture, thawing the frozen product, heating the thawed material to a temperature below 70°C, eluting the low molecular amyloses and then removing the retrograded starch so formed to purify the macromolecular amyloses.

Before developing the above process, Kurimoto and Yoshida had used methods comprising cooling the aqueous starch solutions to precipitate the amylose as retrograded crystals, freezing the solutions to separate the large amount of water hydrated to the starch molecules in the form of ice, thawing to facilitate separation of retrograded crystals of amylose from water, and then removing the thawed moisture content, namely the aqueous portion of the solution, by centrifuge or filtration. Thus the spongy retrograded starch was recovered by separation. However the freezing after cooling causes the coexistence of low molecular amyloses or amylopectin in the macromolecular amylose portion.

Next they found that retrograded macromolecular amylose can be purified by heating the thawed starch solution to not more than 70°C for a short period, which does not effect gelatination and dissolution of the retrograded and crystallized macromolecular amylose portion. By simultaneous centrifuging, filtering or pressing the resultant product to remove the amylopectin portion and low molecular amylose portion in the form of aqueous solutions, purified retrograded macromolecular amyloses are obtained.

The efficiency of the process is best when a starch with a macromolecular amylose content over 50% is used as starting material. Such starches are amylomaize starch, amylose starch separated from gelatinized starch by precipitation with the use of salts, and starches which possess high contents of macromolecular amyloses obtained by fractionation. The following examples will illustrate the further details of the process. All portions given in the examples are by weight unless otherwise specified.

Example 1: A 15% aqueous suspension was prepared with amylomaize starch containing 70% of amylose, and adjusted to pH 4.0 to 4.5. The suspension was gelatinized by autoclaving with stirring to 135°C which was then cooled rapidly to 50°C. It was then hydrolyzed at 45° to 50°C after adding 30 units of enzymes, produced by *Pseudomonas amyloderamosa* ATCC 21262, per gram starch. The hydrolyzate contained 70% of amylose with DP 100 to 1,000, the remainder being of amylose with DP 15 to 50. The hydrolyzate was heated to inactivate the enzyme or was cooled intact and placed in a refrigerator at – 15°C to freeze the hydrolyzate.

The hydrolyzate was next thawed with warm water, 30° to 40°C, heated to 60° to 65°C with agitation and then the retrograded amyloses were separated by centrifuge. The precipitate was then washed once with water. Macromolecular amyloses with a moisture content of 60% were obtained. The solid portion was 90% amylose, dry substance, with DP exceeding 50. The separated aqueous portion contained amyloses of DP 15 to 50. The recovery rates of each portion in solid were 73% and 27%, respectively, dry substance. In addition, when the frozen starch solution was agitated violently after thawing, the spongy thawed starch was crushed, heated to 60°C and centrifuged, amylose with DP over 50 constituted 95% of the precipitation portion and the molecular weight distribution was found sharpened.

Example 2: A 20% suspension of potato starch was prepared after purifying the starch. The suspension was heated to 130°C and gelatinized, whereupon it was incubated for 40 hours at pH 4.0 and 45° to 50°C with the addition of α-1,6-glucosidase produced by *Pseudomonas amyloderamosa* ATCC 21262. The reaction mixture was allowed to stand intact for a day and was then centrifuged at 45°C. The precipitates obtained had a moisture content of 79% and were obtained in a yield of 75% based on the starting material, dry substance. The precipitates were dissolved by heating to 60°C and then cooled gradually and subsequently frozen at – 20°C.

After thawing at 40°C and heating to 60°C the liquid portion of dissolved low molecular amyloses was centrifuged from the precipitate and the precipitate then washed with warm water. The precipitates possessed a moisture content of 60% and recovery rate of 25%, as solid, against starting material and contained 90% of natural macromolecular amyloses exceeding DP 50.

REVERSING THE RETROGRADATION OF POTATO AMYLOSE

When potato starch is heated in the natural water content of the potato, it undergoes gelatinization. Two types of starch are found in potatoes. The amylose or straight chain fraction, which is about 22 to 26% of the total starch, is soluble when freshly gelatinized, and when liberated, as by cell rupture, imparts gumminess, stickiness, or pastiness. This undesirable characteristic is reduced in producing dehydrated instant mashed potato products by (1) reducing cell breakage, (2) tying up soluble amylose with complexing chemicals such as monoglycerides, (3) precooking and cooling before final cooking, whereby the amylose is retrograded to its insoluble form which is not sticky, gummy, or pasty, or (4) retrograding the liberated amylose after cooking and before drying by allowing to stand for a period of time at reduced temperatures.

The rate of retrogradation increases inversely with temperature. All of these techniques are used in the production of the various forms of dehydrated instant mashed potatoes where mealiness is a most important attribute. However, these products are not satisfactory without binder addition in the production of potato-base snacks where a cohesive dough is required. To make such products suitable for this use, extensive cell breakage, such as by extremely fine milling, is required to liberate sufficient free starch to serve as the cohesive agent or binder. An alternative method of making potato-base snacks is to add starches, gums, or binders to the dehydrated instant mashed potato products to create the required cohesiveness or to obtain cohesiveness by the use of extreme high temperature—high-pressure extrusion usually coupled with extensive mixing.

A process of reversing retrogradation of potato starch in dehydrated potato products has been developed by *M.A. Shatila; U.S. Patent 3,830,949; August 20, 1974; assigned to American Potato Company* to avoid the need for added binder in producing chip-type snacks. This process consists of two important steps. The first is to subject dehydrated potato products containing retrograded amylose, after adjusting to a controlled intermediate moisture content, to a temperature which seemingly reverses retrogradation, thereby converting at least part of the retrograded amylose to its original soluble form. The second important step is to dry the product to a stable moisture content quickly before the amylose fraction again retrogrades and becomes insoluble. In the dry state, the tendency to retrograde is arrested.

The moisture content at the time of heat treatment and the product temperature required for the conversion are both extremely critical. For example, heating a potato product to 212°F or below, even at optimum moisture level, would tend to promote retrogradation, (amylose insolubility) but at 220°F or above, the reverse is true—the amylose starch fraction solubility is increased significantly. Higher temperatures than 220°F, however, increase the likelihood of color damage. Although the desired conversion takes place over the wide moisture range between 25 and 70%, if the critical product temperature is reached, the preferred moisture is 25 to 45%. Samples adjusted to moisture levels below 25% and then heated, exhibited color damage and the formation of an undesirable cereal-type flavor.

Samples above 45% in moisture, when heated, are difficult to sheet and handle. When this process is done with dehydrated potato granules, the converted product when dried and ground, rehydrates quickly in 5 parts by weight of cold water to form a mix with the consistency of sticky mashed potatoes, whereas the same product before treatment forms a thin slurry-like soup under the same rehydration conditions. The rehydrated product of this process has a gummy and sticky texture, whereas the same product before treatment is friable and mealy, even when reconstituted with hot water as in making mashed potatoes. In another embodiment, the moist heat converted product can be directly sheeted, cut, and fried at 350°F for 12 to 20 seconds to form excellent potato chip-like snacks. Alternatively, the sheeted cut pieces of the converted product can be dried to the range of 12 to 15% moisture content to assure microbial stability and stored for subsequent frying to snacks.

Example 1: Potato granules which are essentially dehydrated intact cooked potato cells, were mixed with water and salt to give a damp mix with 31% moisture and about 2 to 3% salt. A ¼" thick layer of the above mix was treated in an autoclave at 15 pounds pressure (250°F) for 5 minutes. The hot treated mix was sheeted 0.03" and cut into pieces which were directly fried at 350°F for 15 to 20 seconds in deep fat to produce expanded snacks of excellent flavor and texture. When the same procedure was followed, eliminating the autoclaving but heating to boiling water temperature, the dough could not be sheeted, and formed pieces disintegrated in the frying step.

In the production of converted pieces for snack use, the thickness of the sheet was found to be critical. Pieces thinner than 0.015 inch produced snacks which were too fragile, resulting in excessive breakage. Pieces thicker than 0.04 inch were judged to be too tough and leathery in texture after frying. Pieces with a thickness in the range of 0.02 to 0.03 inch were judged excellent and 0.025 inch was judged optimum thickness prior to frying.

Example 2: Two parts by weight of potato granules were mixed with 1 part by weight of freshly cooked and riced potatoes to give a damp mix of about 30% moisture. The mix was steam treated as in Example 1 and formed into a dough which fried to an excellent snack.

Example 3: Two parts by weight of potato granules were mixed with 1 part by weight of ground raw potatoes to give a moisture content of about 30%. The mix was steam treated and handled as in Example 1. The finished snacks had a good texture and a stronger potato taste.

AMYLOSE OR HIGH AMYLOSE STARCH IN FOODS

Snack Food Formulations

D.G. Murray, N.G. Marotta and R.M. Boettger; U.S. Patent 3,407,070; Oct. 22, 1968; assigned to National Starch and Chemical Corporation have found that when high amylose starch is used as part of the starch of farinaceous food products, such products rapidly set immediately upon their emergence from the apparatus used in their manufacture. This increases the shape retention of these products and results in firm products which can better withstand further treatments such as cutting, drying and frying. Furthermore, these products exhibit prolonged storage stability, limited oil absorption upon frying and controlled texture and taste characteristics. Examples of the specialty products which may be prepared by this process are crackers, chips, cereal puffs, and other so-called snack items in a multiplicity of flavors and shapes. The term amylose or amylose product as used here refers to the amylose from the fractionation of whole starch, or to whole starch composed of at least 50% by weight of amylose. The amylose may be further treated with heat and/or acids or with oxidizing agents to form so-called thin boiling products. In addition, the amylose may be chemically derivatized, as by means of an esterification or an etherification reaction.

In this process farinaceous base refers to whole and comminuted tubers and whole or cracked cereal grains as well as flours and meals derived from the latter. The whole cereal grains may, in turn, be precooked or pretreated in order to obtain fast cooking products. In addition, conventional, i.e., low amylose, starches can be used as the farinaceous base material such as corn, tapioca, rice and waxy maize varieties.

The concentration of the amylose present in the initial dry blend should be selected so as to insure a total amylose concentration of 5 to 60%, based on the total weight of solids in the blend; the total amylose concentration referring only to the amylose content of the amylose product and not to the amylose which is present in any conventional, low amylose starches that may serve as part of the farinaceous component of the dry blend. Amylose concentrations which fall below 5% by weight, result in extruded products which tend to be gummy and of low rigidity, making the shaping, cutting and handling of the extruded product extremely difficult. On the other hand, amylose concentrations which exceed 60% by weight make the extruded product exceedingly dense, thereby precluding the preparation of a porous, puffy appearance upon completing the preparation of these products. The preparation of these snacks is shown in the following example.

Example: A dry mix comprising: (1) 3,000 parts of a waxy maize starch which had been inhibited and acetylated by treatment with 4.6% by weight of an adipic-acetic mixed anhydride containing one part of adipic acid and 50 parts of acetic anhydride, and (2) 5,000 parts of an acetate ester of a high amylose cornstarch resulting from the treatment of a high amylose cornstarch containing 55% by weight of amylose with 5% by weight of acetic anhydride, was thoroughly blended and then added, under agitation, to 10,500 parts of water. The mixture was then pumped into an extruder having a barrel length of 25" and having a screw with a diameter of 1¼" which rotated at a speed of 125 rpm. The barrel temperature was maintained within 245° to 250°F and a pressure of from 80 to 200 pounds per square inch.

The cooked material was then passed from the heated barrel into a cooled barrel consisting of a water-cooled cylinder in which a ¾" conveying screw moved the material in contact with the cold cylinder walls into a shaping die. The temperature of the material in the shaping die was about 125°F. The solid, rope-like food product which emerged from the die exhibited excellent shape-retention properties and could readily be cut immediately after extrusion. These cut slices were then dried, at room temperature, for 16 hours and deep-fried, for 30 seconds, in corn oil at 400°F. The resulting fried products were exceedingly tasty, crisp and light textured. In contrast, when the above procedures were used to prepare a food product which did not contain a high amylose starch, the resulting extruded product exhibited an undesirable gummy, tacky appearance and could not be cut immediately after its emergence from the extruder die.

Fried Potato Coatings

Amylose or high amylose starch is also used by *D.G. Murray, N.G. Marotta and R.M. Boettger; U.S. Patent 3,597,227; August 3, 1971; assigned to National Starch and Chemical Corporation* as a coating for deep frying of potato products. The products have good properties of appearance, texture and taste. The amylose or amylose product has the same usage here as in U.S. Patent 3,407,070. In conducting this process, the selected amylose product is first suspended in water in a concentration of 1 to 15% by weight. The aqueous amylose suspension is then heated at a temperature of at least 180°F for ½ to 60 minutes until the amylose product has been completely dispersed, i.e., gelatinized as a colloidal dispersion. The precise combination of time and temperature which are required will, of course, vary according to the particular amylose product whose dispersal is desired.

Prior to the actual coating of the raw potato slices which are to be used, the slices are usually washed to remove excess starch as well as to prevent their adhering to one another during the subsequent deep frying operation. Excess water may be removed with a sponge rubber roller or by subjecting the slices to a hot air blast. In all cases, the natural starch content of the potato slices or of the compressed, raw potato fragments used in this process will be completely ungelatinized prior to the time they are subjected to the deep frying operation.

The raw slices are then immersed in the amylose dispersion maintained at a temperature of 100° to 210°F. The concentration of amylose product in the dispersion is 1 to 15% by weight. The immersion of the raw potato slices is best accomplished by either passing them through a dip tank, by mechanical means, or by spraying the slices from both above and below while they are being conveyed upon a mesh belt. The slices are then allowed to drain for 1 second to 10 minutes. In some cases, the coated slices may be subjected to either a warm air blast or to a cold water dip to remove excess amylose.

The coated slices are then subjected to the conventional deep frying process in an edible cooking oil such as cottonseed, corn, coconut, soy, or any mixtures of the latter oils, and cooked for 1 to 5 minutes at a temperature of 300° to 400°F. Here again, the precise combination of time and temperature in the deep frying operation will, of course, depend upon the particular deep fried product which is being prepared as well as on the oil being used.

Example: This example illustrates the preparation of french fried potatoes. Potatoes which had been stored at 45°F to reduce formation of reducing sugars were washed, peeled, trimmed and then cut into rectangular solids whose average dimensions were about ½" x ½" x 3". The slices were then washed to remove excess starch. The slices were placed in a wire mesh basket which was then immersed for one minute in a 6% by weight aqueous dispersion of amylose acetate at a temperature of 185°F. This amylose acetate had a DS of 2.5 and had been prepared by the reaction of acetic anhydride with a high amylose cornstarch having an amylose content of 55% by weight.

After being removed from the dispersion of amylose acetate, the coated slices

were drained for one minute and then deep fried, for about two minutes, in cottonseed oil at 380°F. The fried potatoes were drained for 1½ minutes and were
then quick frozen at –30°F. Upon being reheated, the french fried potatoes
were found to have good flavor and were exceedingly crisp while displaying excellent strength without being tough. Variations in color as well as in the amount
of oil which had been absorbed were minimal.

In a repetition of the above procedure, french fries displaying comparable properties were prepared using the following amylose coating products.

(1)   A high amylose cornstarch having an amylose content of 55% by weight
(2)   A high amylose cornstarch having an amylose content of 70% by weight
(3)   Amylose derived from the fractionation of potato starch
(4)   An acetate ester of amylose having a DS of 2.0 prepared from amylose
        derived from the fractionation of potato starch
(5)   A hydroxypropyl ether of amylose having a DS of 1.5, as prepared from
        a sample of high amylose cornstarch having an amylose content of
        70% by weight
(6)   A thin boiling amylose product prepared by the treatment, with sodium
        hypochlorite, of a sample of high amylose cornstarch having an amy-
        lose content of 65% by weight; the final product having been converted
        to a degree known in the trade as 70 fluidity.

Potato chips and potato products from reconstituted potato scraps were also
prepared by this process in other examples of the complete specification.

*E.M. Van Patten and J.A. Freck; U.S. Patent 3,751,268; August 7, 1973; assigned
to American Maize-Products Company* have disclosed a related process for producing french fried potatoes. In this process, the french fries are coated with
0.1 to 1.5% of an unmodified ungelatinized high-amylose starch having an amylose content of at least 50%. The resulting product has improved texture and
absorbs less oil than french fries prepared by conventional methods.

The preferred method of coating the potato pieces is to slurry the starch in a
dip tank in its ungelatinized state. The temperature of the dip tank is maintained
below the gelatinization temperature of the starch and is preferably 120° to 180°F.
Best results are obtained when the temperature is 145° to 160°F. It has been
found that a starch concentration in the dip tank of 2.5 to 8.0% solids is sufficient to coat the potato pieces with the desired pick-up of 0.1 to 1.5% of the
starch when the potato pieces are immersed for 15 to 60 seconds, preferably
from 25 to 30 seconds.

After the potato pieces have been coated with the ungelatinized unmodified high-
amylose starch, they are immersed in an edible frying oil. The edible frying oil
has a temperature above the gelatinization temperature of the starch and thus
gelatinizes the starch on the surface of the potato pieces. This gelatinization
forms an effective barrier against oil penetration and thus reduces the oil absorbed
by the potato pieces. This is especially true at the normal deep frying temperatures of 325° to 380°F since the gelatinized coating is formed very quickly at
these temperatures.

Example: Kennebec potatoes from the Red River Valley were peeled and sec-
tioned into thin shoestrings and then blanched in water for ten minutes at 150°F.

Thereafter, the potato pieces were dipped into a standard sugar flume containing 1.0% dextrose, sugar and 1.0% sodium acid pyrophosphate. The sodium acid pyrophosphate is an antioxidant and improves the color of the finished fry. The flume temperature was held at 150°F and the potato pieces had an average dwell time in the flume of 26 seconds. Thereafter, the potato pieces were removed from the sugar flume, shaken to remove excess dip solution and then deep-fried for 1.5 minutes at 350°F in hydrogenated cottonseed oil. The finished potato pieces were classified as standard grade according to customary industry practice. Total limps and bad textured pieces amount to an average of 22 per 100 pieces. Oil pickup was 11 pounds per 100 pounds of finished fries.

The same procedure was repeated except that in accordance with the present process, the sugar flume solution contained, in addition, 6.5% solids of Amylomaize VII, an ungelatinized unmodified high-amylose starch having a pure amylose content of 70%. The suspension was 4.5° Baumé and was held at a temperature of 150°F. The potato pieces dipped into this suspension had a pickup coating of 0.5% Amylomaize VII by weight of the finished fries. The potatoes treated with the high-amylose dipped solution were classified as fancy grade according to customary industry practice. The total limps and bad textured pieces were only four per 100 pieces and the oil pickup was only seven pounds per 100 pounds of finished fries.

Pizza Dough

A new and improved dough for leavened pastries such as pizza is disclosed by *E.M. Van Patten and J.A. Freck; U.S. Patent 3,777,039; December 4, 1973; assigned to American Maize-Products Company*. The dough has from 2 to 14% of a high-amylose material containing 50% or more amylose and thereby the stickiness of the dough is materially reduced, the proofing time is reduced, and yet a tender structurally stable crust is formed. Additionally, after baking, the crust has less tendency to shrink than normal pizza pastries and the thickness of the crust varies substantially less than standard pizza pastries.

For purposes of the present process, a high-amylose material is defined as a mixture of solids which includes at least 50% by weight of pure amylose or amylose derivatives such as hydroxypropyl amylose and amylose ethers, amylose acetates and similar esters. Excellent results have been obtained from Amylomaize VII as defined in U.S. Patent 3,751,268. It has been additionally discovered that when from 2 to 14% of the high-amylose material is included in the leavened dough, the mixing time can be reduced to as little as two minutes and the proofing time can be substantially reduced and even eliminated and yet there will still result a dimensionally stable attractive looking crust.

This is especially useful since it allows pizza doughs to be continuously prepared. Up until the present time, pizza doughs have not lent themselves to continuous preparation primarily because of the long proofing time, the critical mixing time, and the sticky bucky doughs which are obtained. These disadvantages are materially reduced when a high-amylose material is added in accordance with the present process, and as a result, the pizza doughs can be continuously prepared.

Example: A control dough was formed by mixing with a mixing blade 340 g of flour, 3.4 g of salt and 6.8 g of sugar for five minutes. Over the next five minutes 13.6 g of shortening were mixed in small portions. In a separate operation, 8.4 g of yeast was hydrated in 34 ml of water for five minutes. The hydrated yeast was added to the dries in the mixing bowl and then water was added to the mixing bowl. The mixing blade was used for an additional half minute until the water had picked up the dries. Thereafter, the mixing blade was replaced with a dough hook and the dough hook was kneaded for an additional two minutes.

An additional batch of dough was prepared in the same manner except that in accordance with this process 22 g of Amylomaize VII replaced 22 g of flour, i.e., 318 g of flour were employed. The Amylomaize VII was added after the shortening and mixing was continued for an additional two minutes after the addition of the Amylomaize VII. Both the control dough and the dough of this process were divided into two equal parts by weight, formed into balls, and pinned to about ¾ the size needed to fill a 12-inch aluminum pizza pan. It was noted that doughs of this process were considerably less sticky than the control doughs and could be pinned and punched out more readily than could the controls.

Additionally, dusting flour, which was necessary to facilitate sheeting of the controls was found not to be necessary for sheeting of the doughs made by this process. Each of the doughs was pricked and allowed to stand for five minutes and each of the doughs was then baked at 425°F for five minutes to set the crust. The crusts were cooled to room temperature and then frozen. Standard tomato pizza sauce was applied to the frozen crusts and the sauced pies were baked for 14 minutes at 425°F. Measurements were made of the average diameter and the thickness of each of the pies taken 1½ inches from each edge and at the center. The results are given in the table below.

	Proof time (mins.)	Average diameter before baking, inches	Average diameter finished pizza crust, inches	Average thickness (mm.) of finished crust (edge-center-edge)
Control	0	12	11	5-2-5
Present process	0	12	$11\frac{3}{8}$	5-5-5

Puffed Snack Foods

The process disclosed by *G.T. Popel; U.S. Patent 3,800,050; March 26, 1974* produces a puffed snack food from starch-containing foodstuffs without the use of expensive high-pressure extrusion equipment and in a continuous process without the need for a holding or refrigeration step prior to shaping and drying it. This puffed snack food product is made using a relatively dilute mixture of starch and water, together with flavoring materials as desired, heating the mixture sufficiently to gelatinize a substantial part of the starch in the mixture, forming a thin sheet of the gelled material, partially drying the surface of the formed gel, subdividing the surface-dried gel into pieces of convenient size, finish drying the pieces of gel, and then frying the pieces in hot cooking oil to puff the pieces to a density less than the oil.

The gelatinized mixture prepared as described above keeps the desired thin shape when it is formed into a piece and has sufficient gel strength to support discrete particles of food such as ground wheat, potatoes, corn, and the like. Preferably, such food particles are distributed through the mixture to give the final product a fresh-food taste and improve its texture and appearance. A unique feature of the process is the inclusion in the starting mixture of a relatively high proportion of starch with a high setback or congealing power. Starches having high setback are defined as those starches which set to a firm gel when cooled after cooking, specifically, those starches in which the viscosity of cooked starch pastes cooled to 50°C is more than 2.2 times the viscosity of the cooked paste before cooling.

In general, the tendency of starch pastes to set to a firm gel is determined by the quantity of amylose or equivalent, such as acid-modified starches, in the starch. One method of defining the desired formulation of this process is by calculation of the percent amylose of the starch in the snack mixture. It has been found that if the percent of amylose in the total solids is 20% or higher, the extrusion and handling of the gel is greatly simplified. An important advantage of this process is that a wide variety of foodstuffs can be included in the final product. One major class of foodstuffs which has been used successfully is dry grain, such as wheat and corn. These can be used in the process as whole grains, cracked or partially ground, or completely ground into a flour in which all particles would pass through, say, a 60-mesh screen.

Example 1: The composition by weight on a moisture-free basis is shown in the following table.

	% of Solids	% Starch in Mixture	% Amylose in Starch	% Amylose in Mixture
Cracked Wheat	34.20	29.30	26	7.62
Whole Wheat Flour	9.38	8.00	26	2.08
Corn Starch	41.50	41.50	28	11.61
Corn Flour	13.10	11.35	28	3.18
Salt	1.94			
TOTAL				24.49

Cracked wheat having a screen analysis of 12% on U.S. No. 7 mesh, 75% on No. 10 mesh, 9% on No. 20 mesh, the balance passing under No. 20 mesh, was added to 3 times its weight of water, maintained at 170°F for 1.5 hours. The soaked wheat was then combined with the other ingredients together with sufficient cold water to increase the moisture content of the mixture to 75%. This mixture was pumped through a scraped-surface heat exchanger, which provided about one minute residence time at a speed of 88 rpm. Sufficient steam was supplied to the jacket of the exchanger to raise the temperature to about 185°F and substantially uniformly gelatinize the starch in the mixture.

The cooked, gelatinized mixture was then extruded through a flat nozzle having an opening of 0.074 x 4 inches, which was large enough to permit the wheat particles distributed throughout the gelled mixture to pass without clogging. The strength of the gel was sufficient to hold the product in a coherent ribbon as it flowed from the extruder nozzle. The ribbon of gel was deposited on an endless moving woven-wire mesh belt coated with Teflon and exposed to a current of warm air until the moisture content reduced to 70 to 75%. At this point, the surfaces of the gel were sufficiently dehydrated so that it could be cut by conventional cutters into pieces approximately 1 inch square. These pieces were then dried by conventional means at a temperature of 160°F to a final moisture content of 10 to 12%. Some of the pieces were fried immediately in hydrogenated cottonseed oil at 385°F for 20 seconds. The finished product had a uniformly puffed texture and provided a crisp, fried food snack with a fresh wheat flavor. Other pieces were stored in a sealed container for several weeks and then fried to produce a puffed snack product as just described.

Example 2: The composition of the formulation had the composition by weight on a moisture-free basis as shown in the following table.

	% of Solids	% Starch in Mixture	% Amylose in Starch	% Amylose in Mixture
Potatoes	37.60	32.30	21	6.78
Corn Starch	47.30	47.30	28	13.23
Corn Flour	11.80	10.80	28	3.02
Salt	3.23			
TOTAL				23.13

Peeled raw potatoes, Russet Burbank variety, having about 20% solids, were cut into slices about one-fourth-inch thick and blanched by placing in water at 195° to 200°F for 4.5 minutes. Under these conditions the enzymes normally associated with potato darkening were substantially inactivated. The blanched potatoes were cooled in a stream of tap water to halt the blanching step and then ground to pass through a one-eighth-inch opening in a standard meat grinder. They were then combined with the other ingredients and the solid content adjusted to 25% by adding cold water. The product was then cooked and dried substantially as described under Example 1.

The character of the extruded gel was somewhat sticky, making handling difficult in the initial drying and cutting steps. The product, when fried in hydrogenated cottonseed oil at 375°F, provided a crisp fried-food snack with a fresh potato flavor, similar to a potato chip.

Amylose-Gelatinized Starch Mixtures in Puddings

Instant starch puddings, whether made from pregelatinized starch and phosphates or from pregelatinized starch alone, generally have a coarser texture than freshly cooked starch puddings. The reason for the poor texture of instant puddings containing pregelatinized starch is that the amylose in the starch, which had been converted to a soluble form during gelatinization, retrogrades as the pregelatinized starch ages. As noted above, the amylose crystals caused by retrogradation impart a gritty texture to the prepared pudding.

According to the disclosure by *J.R. Feldman, R.E. Klose and R.V. MacAllister; U.S. Patent 3,515,591; June 2, 1970; assigned to General Foods Corporation,* a cold water-dispersible, powdered material containing a high percentage of noncrystalline amylose is prepared by solubilizing the amylose and stabilizing it while it is in soluble form. This is accomplished by mixing the solubilized amylose with a pregelatinized starch. Since the amylose molecules are now substantially separated from each other, they cannot aggregate and hence are prevented from forming crystals. This gives an improved instant pudding having a smooth texture and which does not undergo syneresis on standing. Generally speaking, any starch can be used to disperse the amylose. Tuber starches, however, are preferred to cereal starches, especially when the final product is a foodstuff, because tuber starches have a better flavor and better keeping qualities than cereal starches. Typical tuber starches which can be used are potato and tapioca. Other materials which can be satisfactorily used in place of the starch components are high amylopectin-containing starch, such as waxy maize, and amylopectin obtained as a by-product in the manufacture of amylose.

The amylose may be either a commercial amylose extracted from cereal or tuber starches, or a starch bearing a high percentage of amylose. The amylose content of these starches should preferably be at least 65%, and must not contain any components which will inhibit the solubilization of the amylose or adversely affect the resulting product. An example of a suitable high amylose containing starch is the high amylose hybrid corn which contains 70% amylose and 30% amylopectin. The advantage of using a high amylose starch instead of commercial amylose is that it is cheaper and less likely to contain degraded amylose since it has never been subjected to chemical processing.

In the preferred embodiment, the amylose or high amylose-containing starch is dissolved in water by heating an aqueous suspension of amylose or high amylose-containing starch in an autoclave or continuous heat exchanger to a temperature of 150°C for five minutes. The temperature of the solubilization step may vary from 140° to 170°C. The amylose suspension is usually held at the elevated temperature for one to five minutes, but the holding time can be longer, such as up to thirty minutes.

Following the autoclaving step, the hot solution may be cooled to a suitably lower temperature, preferably below the atmospheric boiling point of water, to enable it to be handled more conveniently, although the temperature must be maintained above that at which the dissolved amylose would precipitate. The cooled amylose solution is next mixed with an aqueous suspension of the gelatinized starch. The starch can be gelatinized by any of the known methods. Alternatively, the dispersion may be formed by adding dry pregelatinized starch to the amylose solution. After the amylose-gelatinized starch mixture has been thoroughly blended it is dried by any known means, such as spray drying, drum drying, or vacuum drying. The dried product is a fluffy white powder which dissolves easily in cold water or milk, to produce a firm gel having a smooth texture.

The ratio of solubilized amylose to gelatinized starch can be varied according to the properties desired in the final product. Typically, the amount of solubilized amylose added to the starch component will vary from 20 to 65% of the combined weight of the solubilized amylose and pregelatinized starch, and the weight of gelatinized starch may vary from 35 to 80% by weight of the composition with the preferred range of solubilized amylose being about 25 to 35% for most applications. The amylose-starch mixture is useful in the preparation of instant foods, such as instant puddings, powdered gravy mixes and the like.

Example 1: An aqueous solution of amylose was prepared by heating a 5% suspension of amylose derived from potato starch in an autoclave to 145°C and holding it at that temperature for five minutes. A 10% aqueous suspension of potato starch was gelatinized by heating the suspension to a temperature of 100°C. A sufficient quantity of the dissolved amylose was added to the potato starch suspension at 100°C to constitute a 1:3 solubilized amylose-dispersed starch mixture. The mixture was blended thoroughly and dried on a double drum drier heated with 60 psig steam. The dried amylose-starch mixture was then ground in a Homoloid Mill until it passed through a 200 mesh screen (U.S. Standard).

A pudding was prepared by adding 17.5 g of the above-prepared amylose-potato starch mixture and 40 g of milled sugar to 237 ml of cold (8°C) milk in the bowl of a Mixmaster and blending the resulting mixture for one minute. The pudding was then chilled for one-half hour in a refrigerator at 5° to 8°C after which it was examined. It formed a firm gel having a good flavor and smooth texture. The pudding was examined after 24 hours and found to have undergone no syneresis.

Example 2: The procedure of Example 1 was repeated except that tapioca starch was substituted for the potato starch. The resulting pudding was firm and had a smooth texture and bland taste. The pudding exhibited no syneresis after being stored for 24 hours.

Example 3: The procedure of Example 1 was repeated, except that waxy maize starch was substituted for the potato starch, and sufficient amylose solution was added to the starch to constitute a 2:3 amylose-dispersed starch mixture. A pudding prepared from the product of this example was fast setting and formed a firm gel.

Pregelatinized Amylose Starch as Thickening Agents

N.G. Marotta, H. Bell and P.C. Trubiano; U.S. Patent 3,650,770; March 21, 1972; assigned to National Starch and Chemical Corporation have also produced a gravy-thickening agent wherein the starch product used is not inhibited or crosslinked. In other respects the process is similar to that of U.S. Patent 3,443,964, page 163, particularly with regard to particle size of the starch thickener to be used. This process is illustrated by the following example.

Example: 100 parts of pineapple juice, 10 parts of high-amylose cornstarch and 25 parts of sugar were used in the preparation of the pineapple sauce formulation. The high-amylose cornstarch used contained 70% by weight of amylose, which had been mixed with 2 parts of water and pregelatinized and dried on heated drums for about 30 minutes at a drum temperature of 300°F, and pulverized so as to obtain a product exhibiting 5% by weight of particles retained on a No. 12 mesh screen and 20% by weight of particles passing through a No. 100 mesh screen.

The starch was dry blended with the sugar and then dispersed in the pineapple juice. The resulting slurry was then cooked at 190°F for 5 minutes. The resulting hot sauce was then introduced into No. 2 cans which were sealed and cooled. Since the pH of the resulting pineapple sauce was 3.4, no additional sterilization was required, the initial cooking procedure serving both to produce the desired pulpy texture as well as to sterilize the food product. The resulting pineapple sauce was found to exhibit an excellent, natural pulp-like texture as well as a natural, light-yellow color. There was no evidence of particle deterioration as a result of the sterilization procedure. In addition, neither stirring nor agitation disrupted the conformation of the swollen particles.

AMYLOPECTIN IN CAKE MIXES

D.H. Hughes, W.T. Bedenk and N.B. Howard; U.S. Patent 3,366,487; January 30, 1968; assigned to The Procter & Gamble Company have developed improved dry prepared culinary mixes containing sugar, flour, and shortening from which baking batters can be prepared by the addition of liquid materials. More particularly, the improvement involves the addition of amylopectin and certain starches to such mixes to improve the moisture characteristics of the baked products. These dry mixes will produce layer cakes having increased moisture levels and which at the same time have all of the other desired characteristics commonly associated with delectable cakes.

Surprisingly, even though normal sources of starch such as corn or wheat contain up to about 80% amylopectin and some hybrid sources of starch such as waxy corn can contain up to about 100% amylopectin, starches derived from these sources are not per se satisfactory for use as the amylopectin ingredient in the culinary mixes of this process. Any of the above-named sources of starch can be used but the amylopectin ingredient must be isolated from them. Therefore, as used here, amylopectin refers to the isolated amylopectin fraction of starch.

A wide variety of cakes can be prepared from mixes but, for the purpose of illustration, a specific application of this process to layer cake mixes will be set forth in detail. The composition of the dry mixes suitable for baking layer cakes can vary but representative compositions are within the following ranges.

Ingredient	Dry Mix, Percent by Weight
Amylopectin	0.1-4.0
Starch	1.0-5.0
Starch phosphate	0-3.0
Flour	20-50
Sugar	20-70
Shortening	4-26
Leavening agents	0.5-4
Egg solids	0-5
Hydrolphilic colloids	0-1
Nonfat dried milk solids	0-5
Cocoa	0-10
Flavoring (including spices)	0-2
Coloring	Minor amounts

The exact method of compounding these dry mixes is not critical, although very satisfactory results are obtained by mixing in a ribbon blender. The flour, sugar, and shortening, are blended into a homogeneous premix. This premix can be passed through an impact grinder to eliminate lumps. Additional ingredients can then be added and the whole again mixed. An additional step of impact grinding may be desirable to remove any lumps present in the final dry mix.

Example: Devil's Food Cake — Five dry layer cake mixes (A, B, C, D and E)

were prepared by blending together thoroughly sugar, flour, and shortening in
a conventional heavy-duty mixer, and passing this blend through a standard
roller mill. After the milling, the minor ingredients shown below in addition to
the sugar, flour, and shortening were added. The mixture then was subjected to
an impact grinding to break up any agglomerates or large particles present.

Ingredient	Dry Mix, Percent by Weight
Sugar (industrial fine granulated sucrose and dextrose)	46.8
Flour (soft wheat cake flour)	31.0
Shortening*	11.0
Cocoa	6.0
Nonfat dried milk	1.4
Salt	1.0
Soda	2.1
Monocalcium phosphate	0.2
Sodium acid pyrophosphate	0.2
Carboxymethyl cellulose	0.1
Flavoring	0.2

*Mixture of tallow and directly rearranged lard hydrogenated to an iodine
value of 55 containing 7% by weight rapeseed monoglyceride hardened to
an iodine value of 8 and 2% by weight propylene glycol monostearate.

In addition to the ingredients above, the five cake mixes contained the amylopectin and starch ingredients shown in the table below. These special ingredients
were added to the mix along with the minor ingredients. Batter was then made
by adding water in an amount shown in the table below to 540 g of each mix.
Two eggs were then added and the batter was mixed at medium speed on a home
style electric mixer for two minutes. Two eight-inch round pans were filled with
each batter. One pan of each batter was baked at 350°F for approximately
thirty-five minutes and the other pan of each batter was baked at 400°F for approximately thirty minutes.

After baking, the cakes were allowed to cool at room temperature for thirty
seconds; the center and edge heights were then measured for each cake as this
is an indication of proper cake structure. The crust appearance was also noted
and each cake was then tasted for eating quality; these data are recorded in the
table below.

**Moisture Enhancing Ingredients,
Weight Percent of Dry Mix**

Cake	Starch (from rice)	Amylopectin (from waxy maize)	Amylopectin (from corn-starch)	Water Level, grams
A	–	–	–	320
B	–	–	–	360
C	3.0	0.75	–	360
D	3.0	0.50	1.0	360
E	3.0	–	1.0	360

(continued)

Cake	Center/Edge Height, Inches		Dry	Gummy	Moist	Pasty	Other	Crust Appearance
	400°F	350°F						
A	2.56/1.64	2.40/1.77	Yes	Yes	No	No	–	Smooth
B	2.35/1.56	2.21/1.75	No	No	Yes	Yes	–	Cracks and wrinkles
C	2.46/1.58	2.30/1.74	No	No	Yes	No	Tender; cool	Smooth
D	2.55/1.62	2.38/1.88	No	No	Yes	No	Tender; cool	Smooth
E	2.54/1.61	2.30/1.68	No	No	Yes	No	Tender; cool	Smooth

LOW DE STARCH DEGRADATION PRODUCTS

A variety of starch conversion products are manufactured by the acid, enzyme or acid-enzyme processes. Typically such starch conversion or hydrolysis products are composed of a number of components including: dextrose, a monosaccharide; maltose, a disaccharide, higher sugars, which include the trisaccharides, and tetrasaccharides; and dextrins, which include all polymeric saccharides higher than tetrasaccharides. Dextrose, of course, represents the ultimate degree of conversion or hydrolysis of starch, whereas the higher molecular weight dextrins represent the smallest degree of conversion or hydrolysis. Maltose and the higher sugars, in turn, represent intermediate degrees of conversion of starch.

Chemical and physical properties, such as hygroscopicity and dextrin crystallization at higher concentrations, of any particular starch conversion product, is dependent upon the relative distribution of the saccharide components in the product. Low molecular weight saccharides, particularly dextrose and maltose, for example, tend to render such conversion products hygroscopic. In contrast, an abundance of high molecular weight polysaccharides, such as polymers containing 20 or more saccharide monomeric units, tend to render such conversion products subject to crystallization of the dextrins upon concentration. Although a variety of processes have been employed in an attempt to balance such properties by controlling the distribution of saccharides in the conversion product (or syrup), none has proved entirely satisfactory.

Typically starch conversion products are converted to a degree not to exceed 30% measured by the dextrose equivalent (hereinafter referred to as DE). Very low conversion products will be 20 DE or below. There are three conventional methods for making these low conversion starch products: straight acid conversion, straight enzyme conversion, and a combined acid-enzyme conversion. In all of these processes the starch is dispersed in water and the catalyst added to promote the reaction to the desired degree of conversion or hydrolysis.

Normally, pastes of native starch are so viscous at temperatures which are safe for enzyme action that in the latter two processes a thinning step is employed so that one can work at reasonably high starch concentrations; that is to say,

about 15%. Normally conversions are carried out at 30 to 35% concentrations.
Low DE hydrolysates are used as fillers, gelling agents and the like where sweetness is not a major consideration. The thinning of starch is frequently a pre-
requisite for the preparation of high DE syrup, see U.S. Patents 3,551,293 page
258, 3,783,100 page 260 and 3,592,734 page 262.

Classically dextrins were produced by heating starch at 170° to 195°C for 10 to
20 hours. Modifications were produced by adding alkali or acids. Current usage
of the term dextrin can refer to the higher oligomers (seven glucose units or
more) produced during the degradation of starch.

ACID HYDROLYSIS

Low Ash Hydrolysates Using Cation Exchange Resins

Low DE starch hydrolysates which have low ash content have been prepared by
*R.G.P. Walon; U.S. Patent 3,519,482; July 7, 1970; assigned to CPC International
Inc.* The products are characterized as being clear and stable, bland tasting and
nonhygroscopic at low sweetness levels.

These low DE starch hydrolysates may be formed from aqueous starch dispersions by contacting a solution of gelatinized starch in water, or an aqueous dispersion of granular or ungelatinized starch, with a cation exchange resin for a
time sufficient to hydrolyze the starch to a hydrolysate having a DE ranging
from 10 to 45, and more often from 10 to 40. This starch hydrolysate has low
ash content and is relatively low in inorganic anion content such as chloride. It
is preferred that the cationic exchange resin used be a strong acid cation exchange
resin and particularly used in the hydrogen form.

Optionally, the starch may be first liquefied by heat or enzyme pretreatment or
by both types of treatment in sequence. Thus, for example, starch may be first
liquefied with an enzyme system like α-amylase and converted to a DE of less
than 5. The thinned starch can then be converted by resort to the resin system
while being heated.

Cation resins useful in the process are generally copolymers prepared by reacting styrene with varying amounts of divinylbenzene and then sulfonating the
aromatic nuclei forming the polymer backbone. Examples of suitable monovinyl
aromatic compounds other than styrene are α-methylstyrene, chlorostyrene,
vinyltoluene, vinylnaphthalene and homologues thereof, capable of copolymerizing. Examples of additional suitable polyvinyl aromatic compounds are divinyl-
toluene, divinylxylene, divinylnaphthalene, and divinylethylbenzene. Resins of
this type are manufactured under the name Amberlite or Dowex. Also useful
in place of the above resins are cation exchange resins obtained by condensing
an aldehyde, a phenol and an organic sulfonic acid to give sulfonated polymeric
resins.

Preferred cation exchange resins are prepared by sulfonating polymeric resins
derived by copolymerizing a monomer solution containing 10 to 20% by weight
of divinylbenzene and 80 to 90% by weight of styrene. Most preferred resins
have a divinyl content of 12 to 20% by weight and a styrene content of 80 to

88% by weight. The first step in this process is effected by gelatinizing an appropriate starch source. This may be done independent of the resin treatment or if granular starch is used, occurs as an initial step in the process.

Also, the granular starch may be heated to 80° to 90°C for 10 to 45 minutes along with a cation exchange resin to effect proper gelatinization. The pasted starch is then initially or further contacted with a strong acid cation exchange resin in hydrogen form for a time sufficient to effect the proper degree of conversion. This time may vary considerably depending upon the mode of conversion which may either be effected batchwise or in a continuous manner. Thus, for example, the ion exchange resin in bead or granular form may be mixed with a hot starch paste and maintained in contact for the appropriate amount of time, preferably under agitation.

In still another embodiment, a resin membrane may be used or a series of resin membranes whereby the hot contacted starch slurry is converted. The resin membranes may, for example, be heated by electric heating plates placed between the membranes. In one illustrative technique a hot slurry of resin beads and pasted starch is prepared, and the conversion effected at elevated temperatures, while agitating the slurry. Again, a column technique may be used whereby a heated column of resin is prepared and a pasted or liquefied starch solution flowed through the column in varying rates. Usually the influent batch paste is itself hot.

The low DE products from the resin contact usually have a 10 to 45 DE, more often have a DE of 10 to 40, and most often a DE of 15 to 25. Due to the fact that no mineral acid or enzyme addition is used, the low DE products have relatively low ash contents. Usually ash content is in the range of 0.3% dry basis or lower and the product may be used for its desired end-use without further purification. In many instances a product may be obtained having an ash content of about 0.2% dry basis, that is based on starch solids present, and under optimum conditions has an ash content of about 0.1%. In essence, the salt content of the hydrolysate usually depends upon the original salt content of the starch undergoing conversion.

Example 1: A raw cornstarch slurry is first gelatinized to produce a starch paste by addition of 20% of a strong acid cation exchange resin in hydrogen form based on dry weight of the starch. Gelatinization was effected in an autoclave and the gelatinized starch-resin mixture had a Baumé of 18° and pH of 5.9. This exchanger was prepared by sulfonating a 16% divinylbenzene-84% styrene copolymer. Then, the resin-starch mixture was further autoclaved with constant agitation under pressure. The conversion was effected by raising the pressure in the autoclave to about one-half atmosphere (about 7 psig) for a period of 25 minutes. The pressure was then held at this level for an additional 15 minutes, and thereafter raised to 1 atmosphere over a period of 15 minutes. The desired DE level was then achieved by holding the starch-resin mixture at 30 minutes at the 1 atmosphere level.

The product was then blown off and the resin separated by a sieve. The product had the following characteristics: Baumé, 20.2°; pH, 2.4; DE, 17.9; ash, 0.1%; and dextrose, 3.1%.

Example 2: In another example the starch suspension was first run through a cation exchanger in hydrogen form at a temperature of 45° to 50°C. The decationized starch was dewatered by centrifugation and then washed with substantially iron-free water. The decationized starch was then placed in suspension in demineralized water and then converted essentially according to directions of Example 1. The converted starch product analyzed approximately as that set out above, but in this instance was much lower in ash content after conversion.

Acid Modification with HCl-HF Mixtures

P.J. Ferrara; U.S. Patent 3,692,581; September 19, 1972 has found that starch and flour products are readily acid-modified using as the acidifying agent, mixtures of mineral acids and hydrofluoric acid. The ratio of a mineral acid to hydrofluoric acid may be varied over a wide range. The hydrofluoric acid may be used in combination with various known acidifying mineral acids such as sulfuric, hydrochloric and phosphoric acids. It is not clear why the combination of hydrofluoric acid with other mineral acids achieves the unique acid-modification results which have been found. The following example and data will serve to illustrate the process.

Example: This example shows a comparison of a typical sorghum flour (milo) acid-modified using a commercial 22° Baumé HCl and the same flour subjected to an equivalent amount of acid with a portion of the HCl replaced by HF. The relative acid-modification effectiveness of HCl by itself, and the mixture of HCl and HF are shown in the following table. These data were derived from experiments carried out in a rectangular mixing conveyor 12" wide x 60" long and 14" deep constructed of 316 stainless steel equipped with twin shafts of 1½" stainless pipe, which serve as twin shafts, drilled to accommodate adjustable pitch V-shaped paddles.

A total of 78 paddles were installed in such manner that the faces of the paddles cut through material on the conveyor and simultaneously created a pattern of flow whereby the flour being acidulated travels in a circuit route along one shaft and was cycled back along the opposite shaft. Milo flour charged into the conveyor was limited to 100 lb per run, the amount which permits thorough cycling of the contents. The cover or the top of the conveyor unit consisted of a 2-piece unit made of polyvinyl chloride plastic sheets, each drilled to accommodate the entry of an atomizing nozzle of the 2-fluid type; a nitrogen gas supply to atomize and the acidifying solution. The conveyor had a bottom gate which could be opened to discharge the contents.

The milo flour used was a commercial grade, with the following analysis: 5.9% protein, 12.8% moisture, 0.73% mineral ash, and 2.7% fat.

(A) For the acid modifications 2.61 lb or 22°Bé hydrochloric acid was weighed out and placed in a stoppered polytetrafluoroethylene flask so that it could be fed into the two atomizing nozzles, one at each end of the mixing conveyor. This acid is equal to 0.92 lb of contained HCl or 0.92% of the 100 lb weight of milo flour. This quantity of acid was fed into the milo over a period of 20 minutes. Mixing was then continued for another 20 minutes, and the acid-modified flour was then discharged from the conveyor into a wax-lined, fiberboard drum which was then covered.

(B) A second batch of acid was prepared by adding 0.50 lb of a 98% acid grade of powdered fluorospar to 5.13 lb of 22°Bé HCl and allowing this to react and settle for 24 hours. The same quantity, 2.61 lb of liquid acid was weighed out, care being taken not to disturb the sludge or sedimented solids. This quantity of acid mix is equivalent to 0.61 lb of contained HCl and 0.12 lb of HF. A part of the reduction in the combined acid weight is due to the formation of the calcium salt, and some due to the formation of the lower molecular weight HF. Nevertheless, for the purposes of showing the HF effect, the amount of this mixed acid is equivalent to the 22°Bé HCl in part (A).

When the 2.61 lb mixture of HCl and HF was added, as in Test B, to 100 lb of the same milo as in Test A, it is clear as shown in the table, that the reduction in paste viscosity (used as an index of acid modification) is greatly favored by the substitution of HF for a part of the HCl. Samples of the acid-modified milo of Test A and Test B are removed from the fiberboard drum at 24-hour intervals, and checked for viscosity using a C.W. Brabender Visco/Amylograph. These acid modification tests made with milo flour stored at approximately 80°F indicate that the presence of HF more than doubles the rate of acid modification. The mixed acid HCl-HF is also more easily taken up by the flour without forming wet agglomerates as happens with HCl alone.

The following table shows the peak viscosity of acid-modified milo flour in Brabender units. The test viscosity concentration was 100 grams of flour product per 400 ml of water. In Test A the flours were processed with 0.92% HCl. In Test B the flours were processed with 0.61% HCl and 0.12% HF.

	Viscosity value, B.U.	
	Test A	Test B
Hours from start:		
24	1,860	1,580
48	1,425	1,110
72	1,105	810
96	930	590
120	720	360
144	605	240
168	525	125
192	460	80
216	420	40
240	365	
264	360	
288	345	

A comparison of the data Tests A and B show that the use of HF in combination with HCl makes it possible to lower the paste viscosity of milo flour to a level approximating 400 B.U. in less than half the time interval than when HCl is used alone.

Solvents for Acid Production of Dextrins

Conventional modified starches are produced by immersing starch granules in a diluted solution of an acid or acids where the starch is treated for several days at lower than room temperature, and then deacidifying and drying the resultant. Since the swelling and gelatinizing point of starch granules is around 60°C, the starch is treated at a temperature lower than room temperature for several days to prevent gelatinization. This is especially true in the production of low molecular dextrins, since low molecular dextrins are highly water-soluble, low temperature treatment becomes much more necessary, otherwise the recovery rate decreases.

Dextrins having a low DP have been produced by *K. Sugimoto; U.S. Patent 3,799,805; March 26, 1974; assigned to Hayashibara Company, Japan* using aqueous-organic solvent reaction media. This gives low molecular dextrins in a higher yield and in a shorter period without gelatinizing the starch granules.

With the addition of an organic solvent or organic solvents, such as methanol, ethanol, propanol and acetone, the reaction temperature can be raised to the temperature at which starch granules gelatinize and the reaction time can be short- to a great extent.

To obtain dextrins with a DP around 28 to 29, if starch is reacted at 37°C using 1 N hydrochloric acid or 1 N sulfuric acid the reaction requires 18 days, and if 1 N nitric acid is used a reaction period of 60 days is necessary. On the other hand, with the use of a 1 N hydrochloric acid solution in 70% ethanol the elevation of the reaction temperature to over 65°C and shortening of the reaction period to less than 12 hours is possible. Furthermore, the low molecular dextrins obtained by this process are characterized in that they possess a large amount of linear chained components. Therefore, the product possesses a higher limit of hydrolysis by beta-amylases than those of ordinary dextrins.

Example 1: 20 grams of sweet potato starch (moisture content 18%) and 50 ml of 65% aqueous ethanol solution were mixed, and a sufficient amount of hydrochloric acid to prepare the total mixture to 1 N added to mixture. After allowing the mixture to stand at 65°C for 24 hours, the mixture was deacidified and washed with ethanol and granular dextrins were recovered in a yield of 70%. The mean DP of the product was 24.8. The product was further hydrolyzed, resulting in the finding that the limit of hydrolysis was 70.2% showing an extreme increase from the 58% obtained with starch used as starting material.

Example 2: To 100 grams of cornstarch was added 300 ml of 70% ethanol which was adjusted to 1 N with hydrochloric acid. The mixture was kept in an Erlenmeyer flask equipped with a countercurrent type cooler for 20 hours at 65°C. The reaction mixture was filtered, and the precipitate was immersed in 200 ml of 80% methanol for 2 hours to remove the hydrochloric acid and then dried. A product with a mean DP of 25.5 was recovered in a yield of 67%. The hydrolysis of the product was 75.5% by beta-amylolysis at a concentration of 1%. The product was found warm-water-soluble.

ALPHA-AMYLASE CONVERSION PROCESSES

Alpha-amylase enzymes are among the most useful of industrial enzymes. The greatest industrial consumption of alpha-amylase enzymes occurs in the liquefaction of starch for the preparation of paper and textile sizes, starch coatings, laminating adhesives and as adhesives for the binding of pigments to web products such as paper. Alpha-amylase enzymes are also used to saccharify starch for the manufacture of maltose, glucose and alcohols. In starch liquefaction processes, the alpha-amylase enzymes catalyze the random hydrolyzation of the D-glycosidic linkages in the central portion of the long glucose chains in the starch molecules. In saccharification reactions, alpha-amylase enzymes are used in conjunction with other enzymes, the alpha-amylase acting as a reaction promoter or catalyst. For example, beta-amylase may be utilized in conjunction with alpha-amylase for the purpose of reducing starch to maltose. Beta-amylase splits off single glucose units, attacking at the nonreducing ends of both the long and short chains of the starch molecule. Consequently, a saccharification reaction utilizing only beta-amylase would proceed at a relatively slow rate. Alpha-amylase acts as a reaction promoter by randomly cleaving the long chains of the starch molecules near their centers, thereby substantially increasing the number of active sites at which the beta-amylase may hydrolyze the starch molecules. Alpha-amylase likewise finds use as a reaction promoter with such enzymes as maltase and amyloglucosidase for the degradation of starch into alcohols.

The following processes make use of alpha-amylase for the production of dextrins and for thinning of starch with the production of low DE hydrolysates. Enzyme thinning is also disclosed in U.S. Patents 3,551,293, page 258 and 3,783,100, page 260 where the process is a vital step in the production of high DE syrups.

Fluorosilicate Inactivation of Alpha-Amylase

J. Frankevicz and S.M. Short; U.S. Patent 3,513,072; May 19, 1970; assigned to Oxford Paper Company have described their process whereby alpha-amylase enzymes are inactivated by fluorosilicate compounds, giving a high degree of control over starch liquefaction in a starch slurry.

To overcome the disadvantages inherent in the use of the chemical inactivators and the thermal inactivation methods provided by the prior art, this process provides improved chemical inactivators for alpha-amylase enzymes which consist of fluorosilicate-containing compounds. Fluorosilicate inactivators have the following advantages as enzyme inactivators:

(1)   They are effective at low concentrations.
(2)   They will not discolor the substrate acted on by the alpha-amylase enzyme.
(3)   They are not toxic and may be used in foodstuff applications.
(4)   They have a sharp threshold concentration at which enzyme inactivation occurs. Hence, fluorosilicate inactivators are particularly useful in continuous processes as a means for quickly and accurately controlling enzyme activity. A further advantage is found in batch processes where a residue of previously inactivated enzyme-substrate mixture is mixed with a fresh batch of enzymes and undigested substrate material without any noticeable harmful effects.
(5)   These chemical inactivators are also inexpensive, costing in the neighborhood of 15 to 18 cents a pound.

The effective threshold concentration of fluorosilicate inactivator will depend upon the particular alpha-amylase enzyme used. However, in experiments conducted on pure commercial alpha-amylase enzymes and alpha-amylase containing enzyme mixtures, the effective inactivation concentration is in the range of 0.04 to 0.08% fluorosilicate inactivator by weight based on the dry weight of the starch substrate. In certain applications a degree of inactivation may be accomplished at concentrations of 0.02% based on the dry weight of the starch substrate.

A useful characteristic of fluorosilicate inactivators is the fact that there is a sharp threshold concentration above which the inactivation of alpha-amylase enzymes is substantially complete and below which the inactivation of the alpha-amylase enzymes is substantially negligible. In applications not utilizing high temperature alpha-amylases, the threshold concentration is approximately 0.06% by weight based on the dry weight of the substrate starch. This characteristic may be used with advantage in batch process operations in which a residue of processed and inactivated starch slurry is retained in the kettle and mixed with a new mixture of starch and enzymes to be processed.

Additionally, this sharp cut off characteristic may be used to advantage in continuous processing methods. Since the reaction of the fluorosilicate radical with the alpha-amylase is substantially instantaneous, a very close control over the enzyme reaction may be effected by automatic means. More specifically, in continuous starch liquefaction processes, a high degree of control over the product viscosity may be maintained by automatic metering of fluorosilicate compounds into the reaction slurry, such metering being controlled by a viscosity measuring device. Various fluorosilicate compounds can be used to inactivate the enzymes in this process including alkali metal fluorosilicates such as sodium fluorosilicate (sodium silicofluoride) and ammonium fluorosilicate.

Example 1: Fifty pounds of air-dried Union Pearl starch, 80 lb of deionized water, and 2 lb of Amyliq Jet/enzyme (a high temperature alpha-amylase derived from *Aspergillus niger*) were charged to a conventional cooking vessel. The charge was heated to 174° to 177°F by direct steam injection and held at this temperature for 35 minutes. The steam was turned off, and 20.4 grams (0.1% of the oven-dried starch) of sodium fluorosilicate were added to the reaction mixture.

The mixture was allowed to mix for 10 minutes and was then dumped to another vessel. The solids contained in the mixture were 34.1%, and the overall mixture had a viscosity of 200 cp. The mixture was then covered and allowed to stand in a water bath for 24 hours at 75°C. A sample was taken from the mixture at this time and the solid content remained the same, 34.1%, and the viscosity also remained the same, about 200 cp.

Example 2: Twenty-nine runs were conducted using the same method and quantities used in Example 1, except that the quantity of sodium fluorosilicate inactivator was varied. The starch solids concentration was between 40.7 and 41.3% for eight of the runs, between 18.0 and 18.2% for an additional nine runs and between 31.1 and 32.2% for a final twelve runs. Each of the aqueous starch and enzyme charges was heated from 174° to 177°F for 35 minutes. After heating, a measured quantity of sodium fluorosilicate was added and thoroughly mixed in the slurry for 10 minutes.

After the 10-minute mixing period, a sample of the slurry was analyzed for reducing sugar content expressed in dextrose units. The mixture was transferred to a second vessel and allowed to stand in a water bath at 175°F for 24 hours, after which period a second sample was taken and the reducing sugar content was analyzed. The results of the twenty-nine runs are listed in the following table which indicates for each run the corresponding values of percentage change in reducing sugar content and sodium fluorosilicate concentration based upon the oven-dried weight of the starch.

The data of the table would indicate that there is a sharp cut off concentration around 0.06% sodium fluorosilicate by weight oven-dried starch below which the enzyme inactivation is substantially negligible and above which the enzyme inactivation is substantially complete.

Percent Na_2SiF_6 on Oven Dry Starch	RS_1, %	RS_2, %	ΔRS	Percent Solids
0.032	1.10	22.2	21.1	18.6
0.097	1.10	1.8	0.7	18.6
0.161	1.10	1.9	0.8	18.6
0.325	1.10	0.94	-0.2	18.6
0	1.10	22.4	21.3	18.6
0.032	1.10	20.2	19.1	18.2
0.066	1.10	1.4	0.3	18.2
0.097	1.10	1.2	0.1	18.2
0	1.10	21.4	20.3	18.2
0.048	1.7	22.0	20.3	32.3
0.063	1.7	13.1	11.4	32.3
0.078	1.7	3.5	1.8	32.3
0.093	1.7	2.3	0.6	32.3
0	2.2	23.2	21.0	31.1
0.080	2.2	1.6	-0.6*	31.1**
0.113	2.2	1.8	-0.4*	31.1**
0.162	2.2	1.7	-0.5*	31.1**
0.016	1.5	26.3	24.8	31.1***
0.085	1.5	3.7	2.2	31.1***
0.161	1.5	1.8	0.3	31.1***
0.348	1.5	†	*	31.1***
0.062	3.92	17.9	14.0	40.7
0.086	3.92	3.66	-0.26	40.7
0.123	3.92	3.60	-0.32	40.7
0	3.92	22.3	18.4	40.7
0.012	4.02	23.2	19.2	41.3
0.037	4.02	22.8	18.8	41.3
0.061	4.02	20.7	16.7	41.3
0.121	4.02	5.08	1.06	41.3

*These data indicate that the starch suspension retrograded, indicating that all enzyme activity had ceased.
**Retrogradation.
***These four samples aged 42 hours.
†Too thick to analyze.

RS_1 – Percent reducing sugars (expressed as dextrose) immediately upon completion of cooking.
RS_2 – Percent reducing sugars (as dextrose after aging one day at 75°C).
$\Delta RS = RS_2 - RS_1$.

Alpha-Amylase from *Bacillus coagulans* for Production of Dextrins

α-Amylase enzymes are widely used in the thinning or dextrinization of starch for example in glucose production. Such dextrinization reactions, as is the case with other enzyme-catalyzed reactions, proceed more rapidly at high temperatures provided that the enzyme is not destroyed and it has been found that bacterial α-amylases are especially useful in that their preferred temperature for dextrinization is of the order of 80°C.

H.M. Smalley; U.S. Patent 3,697,378; October 10, 1972; assigned to Glaxo Laboratories Ltd., England has developed a bacterial α-amylase, derived from *B. coagulans,* which is active at even higher temperatures and hence is of particular use in the high temperature dextrinization of starch. The enzyme may be obtained by submerged aerobic culture of an α-amylase producing strain of *B. coagulans* on a nutrient medium therefor followed by isolation of the enzyme therefrom.

Example: Shake Flask Fermentation — 20 ml of saline (0.9% by weight sodium chloride solution) was added to an agar culture of *B. coagulans* strain NCIB 10280 and 1 ml of the cell suspension was used to inoculate a 250 ml baffled conical flask containing 20 ml of a seed medium (P2) having the following composition.

Ingredients	Percent by Weight
Glycerol	0.7
Yeatex granules	1.0
Sodium dihydrogen phosphate	0.16
pH 5.8.	

After 18 hours of growth at 55°C on a rotary shaker at 220 rpm 1 ml of the cell suspension was used as inoculum for 20 ml of the following medium in a 250 ml baffled conical flask:

Ingredient	Percent by Weight
Glycerol	2.0
Yeatex granules	3.5
Na_2HPO_4	0.25
KH_2PO_4	0.1
Chalk	4.0

Amylase was produced in the broth filtrate after a 48 hour fermentation at 55°C on a rotary shaker and was determined by an autoanalyzer method based on the SKB assay (Sandstedt, Kneen and Blish, *Cereal Chemistry*, 1939, 16, 712). A titer of 4.00 SKB unit/milliliter was obtained.

Properties of the enzyme are given in the complete patent. This α-amylase can be used for the dextrinization of starch where starch in an aqueous medium is treated with this α-amylase at a temperature of the order of 80°C or greater, preferably at least 90°C or even 100°C.

Enzyme Liquefaction Using Calcium Salts

R.V. Vance, A.O. Rock and P.W. Carr; U.S. Patent 3,654,081; April 4, 1972; assigned to Miles Laboratories, Inc. have provided an enzyme process for starch liquefaction which liquefies substantially all the starch and produces a product which is substantially free of retrogradation products.

The process comprises the steps of (1) adding to an aqueous slurry of starch-containing material an amount of a water-soluble calcium compound sufficient to provide a calcium content molarity of from 0.003 to 0.03 and rapidly heating the resulting mixture to a temperature of 121°C (250°F) to 177°C (350°F) to gelatinize the starch; (2) cooling the gelatinized starch to a temperature of 93°C (200°F) to 102°C (215°F), adding α-amylase and maintaining the resulting mixture at a temperature of 93° to 102°C for 10 to 60 minutes; (3) cooling the mixture to 85°C (195°F), adding α-amylase and maintaining the resulting mixture at 85°C until the mixture reaches a DE of 9 to 30. Preferably, the process uses an aqueous slurry of starch containing from 35 to 45 weight percent dry solids.

The total amount of α-amylase added is equivalent to 0.05 to 0.25 weight percent, based on dry solids weight in the starch slurry, of an α-amylase having an activity of 3,500 to 5,500 SKB units per gram of enzyme. While the α-amylase need only be added in two portions, the α-amylase is preferably added in three portions. The first portion is added in step (1) prior to rapid heating of the slurry and consists of 0 to 30 weight percent of the total weight of added α-amylase. The second portion is added in step (2) and consists of 10 to 50 weight percent of the total weight of added α-amylase. The third portion is added in step (3) and consists of 30 to 60 weight percent of the total weight of added α-amylase.

Any water-soluble calcium compound, such as calcium acetate, calcium chloride, calcium citrate, calcium hydroxide, calcium hypophosphite, calcium lactate, and the like can be used. Calcium chloride and calcium hydroxide are the preferred calcium compounds to be used in this process. It has also been found suitable to use a mixture of water-soluble calcium and sodium compounds in this process. When the sodium compounds are used they should be present in such amount that the resulting starch slurry has a sodium content molarity of 0.0018 to 0.12 and a molar ratio of sodium to calcium from 0.6:1 to 4:1. The preferred concentrations are a calcium molarity of 0.005 and a sodium molarity of 0.006.

The liquefied starch obtained by this process is especially useful for further enzyme saccharification. In addition liquefaction time has been reduced and yield increased over prior processes.

Example 1: 36 lb of refined cornstarch (10% moisture) were mixed with 6 gal (50 lb) deionized water to form a slurry containing 37.7 weight percent dry solids. To the slurry were then added 19.6 grams of calcium chloride (75 weight percent $CaCl_2$) to form a calcium molarity of 0.005. The pH was adjusted to 7.0 by the addition of 5.84 grams of calcium hydroxide. The calcium molarity based on the calcium hydroxide was 0.003 producing a total calcium molarity of 0.008. An α-amylase having a potency of 3,770 SKB units per gram was added to the slurry in an amount of 0.025 weight percent based on the starch dry solids weight. The starch slurry was then pasted (gelatinized) by passing it through a steam jet mixer at 138°C (280°F) and collecting it in a 30 gal heat-jacketed vessel. The gelatinized starch was flash-cooled to 99° to 100°C (210° to 212°F) and a second portion of the same enzyme added in an amount of 0.075 weight percent based on the original starch dry solids weight. The mixture was then gradually cooled to 85°C (185°F) during a period of 20 to 30 minutes. A third portion of the same enzyme was added in an amount of 0.075 weight percent based on the original starch dry solids weight.

The three portions of enzyme added represented, respectively, 14 weight percent, 43 weight percent and 43 weight percent of the total added enzyme. The starch was then kept at 85°C for 120 minutes at which time the liquefied starch had a DE of 14.3. No starch retrogradation was noted. The insoluble constituents of the liquefied starch coagulated readily and were rapidly removed by means of a small precoat filter. The filtration rate for the liquefied starch on a 2" x 18" drum filter was in excess of 367 ml/min. This is a substantial improvement in filterability of liquefied starch as compared to filtration rates as low as 17 ml/min for prior steam-heated single enzyme addition liquefaction procedures.

Example 2:   100 lb of refined cornstarch (about 10% moisture) were mixed with 17.2 gal (142 lb) of deionized water to form a slurry containing 37% dry solids. To the slurry were then added 0.1 weight percent calcium chloride (based on dry starch) and 0.044 weight percent sodium hydroxide (based on dry starch) to adjust the pH to 6.7. The resulting slurry contained 0.0053 molar calcium and 0.0065 molar sodium. An α-amylase having a potency of 3,770 SKB units per gram was added in an amount of 0.025 weight percent based on the dry starch. The starch slurry was then pasted by passing it through a steam jet mixer at 132° to 138°C (270° to 280°F). The gelatinized starch paste was flash-cooled to 99° to 100°C and collected in a 2 gal insulated tank provided with an efficient agitator.

α-Amylase was continuously metered into the tank in a concentration of 0.1 weight percent based upon the starch dry solids present in the tank. The starch paste was retained in the tank for a nominal time of 4 minutes at 99° to 100°C. Overflow from this tank was pumped continuously into a further agitated reactor tank where it was retained at 99° to 100°C for an additional 1 minute.

The resulting mixture was then cooled to 85°C and 0.1 weight percent of the above α-amylase was added. The resulting mixture was then maintained at 85°C for 90 to 120 minutes at which time the liquefied starch had a DE of 15 to 20. The filtration rate for this liquefied starch on the drum filter described in Example 1 was 367 ml/min. No starch retrogradation was noted.

Dual Temperature Liquefaction of Nonwaxy Starches

A two-stage one-enzyme process has been disclosed by *F.C. Armbruster; U.S. Patent 3,853,706; December 10, 1974; assigned to CPC International Inc.* for producing low DE hydrolysates and syrups.

These hydrolysates, whether in the form of dilute or concentrated syrups, or in the form of dry solids, are characterized by blandness of taste and low sweetness, and they are nonhygroscopic. They are fully and readily soluble in water. When used in food products, they have a minimal effect upon flavor, while providing bulk, stability, and lack of hygroscopicity.

This process comprises subjecting a mixture of starch and water having a solids content less than 50% to the hydrolytic action of bacterial α-amylase to obtain a starch hydrolysate having a DE between 2 and 15, subjecting the starch hydrolysate to heat to substantially inactivate the enzyme, i.e., to a temperature greater than 95°C, cooling the starch hydrolysate to a temperature less than 95°C, and subjecting the hydrolysate to further hydrolytic action of bacterial α-amylase to obtain a starch hydrolysate having a DE between 5 and 20. The resulting product is also characterized by having the sum of the percentages of saccharides therein, dry basis, having a degree of polymerization of 1 to 6 divided by the DE provide a ratio of at least 2.0. This ratio is referred to hereinafter as the characteristic or descriptive ratio.

One preferred method of the process involves the steps of: (1) slurrying cornstarch in water to a solids concentration of between 10 and 50%; (2) solubilizing the starch by gelatinization; (3) subjecting the mixture to treatment with bacterial α-amylase to hydrolyze the starch to a DE between 2 and 15; (4) heating the starch hydrolysate to a temperature greater than 95°C, preferably between 110° and 150°C, to terminate the hydrolytic action of the enzyme; (5) cooling the starch hydrolysate to a temperature less than 95°C; (6) subjecting the hydrolysate to further treatment with bacterial α-amylase to hydrolyze the starch to a DE between 5 and 20; and (7) recovering a starch hydrolysate product characterized by high water-solubility and a descriptive ratio of at least 2.0.

The resulting hydrolysate may be concentrated and/or refined by conventional procedures to yield a stable corn syrup which is substantially haze-free and highly soluble in water. The syrup may be spray dried to yield corn syrup solids with low hygroscopicity and high water-solubility. Suitable starches include cereal starches such as corn, grain sorghum and wheat.

The preferred enzyme used for the conversion of starch to low DE syrups in this process is a starch liquefying, heat resistant hydrolytic α-amylase. Suitable bacterial α-amylases may be produced by certain strains of *Bacillus subtilus, Bacillus mesentericus* and the like by conventional fermentation methods. HT-1000, a proprietary bacterial α-amylase preparation is an example of an enzyme preparation that is suitable for use in the process. Other suitable bacterial α-amylase include Rhozyme H-39 and CPR-8. The process can be illustrated by the following example.

Example: The production of low DE products from cornstarch is as follows. Unmodified cornstarch was slurried in water to provide an aqueous suspension containing 32% by weight of the starch. The pH was at 7.5 to 8.0. To this mixture was added HT-1000 bacterial α-amylase at a concentration of 0.05% based on starch solids. This starch suspension was then transferred over a 30 minute period to an agitated tank containing sufficient water so that the final solids content of the starch slurry was reduced to 28%.

The temperature of the starch slurry was controlled from 90° to 92°C. Liquefaction was then continued for 60 minutes, at which time the hydrolysate was within the DE range of 2 to 5. The liquefied starch was then heated to 150°C and held at that temperature for 8 minutes. The heat treatment destroyed residual enzyme activity. It also resulted in improved filtration rates and decreased yield losses upon filtration.

Further saccharification to the final DE was accomplished by the addition of more HT-1000 bacterial α-amylase after cooling the liquefied starch hydrolysate to a suitable temperature for conversion. Thus, the liquefied starch was cooled to 80° to 85°C and HT-1000 enzyme preparation added in an amount of 0.02% by weight starch solids. After more than 14 hours of conversion, the desired DE of 20 was obtained. The final starch hydrolysate product was analyzed and the following analytical values were obtained.

D.E.	DP_1	DP_2	DP_3	DP_4	DP_5	DP_6	DP_{7+}	Descriptive Ratio
20.7	2.4	7.5	10.8	8.0	6.9	15.1	49.4	2.4

It may be seen from the above that the product resulting from this example has a descriptive ratio of about 2.4. Other examples in the complete specification show the hydrolysis of other starches such as potato, tapioca, wheat, rice, sago, arrowroot, etc.

Dual Temperature Liquefaction of Waxy Starches

F.C. Armbruster and E.R. Kooi; U.S. Patent 3,849,194; November 19, 1974; assigned to CPC International Inc. have also disclosed a process using waxy starches as the starting material. Suitable waxy starches include waxy milo, waxy maize, and waxy rice.

In this process, two temperature stages are used in the hydrolysis, but as the amylase is heat deactivated at the end of the reaction, only one inoculation of the α-amylase is required. The details of the process are shown in the following example.

Example: An aqueous starch slurry was prepared containing 30% solids by weight of waxy milo starch. The temperature of the slurry was raised and held between 85° and 92°C. A bacterial α-amylase preparation was added in an amount just above 0.025% by weight of the starch over a period of slightly more than 30 minutes. The mixture was then held at the same temperature for an additional period of 30 minutes. The waxy starch was liquefied and had a DE of 2.79. The temperature was then reduced to below 80°C and the conversion allowed to continue until the desired DE was reached. The temperature of the mixture was then suddenly raised to about 120°C to inactivate the enzyme and terminate the conversion.

The following table sets forth typical saccharide analyses of low DE hydrolysates obtained in accordance with the above procedure. DP designates the range of polymerization. DP_1 represents the total quantity expressed in percent by weight dry basis of monosaccharides present in the hydrolysate. DP_2 represents the total quantity of disaccharides percent in the hydrolysate and so forth.

Included in the table are analyses of typical acid hydrolysates for comparative purposes.

Typical Saccharide Analyses

Hydrolysate Composition	DE				
	5	10	15	20	25
(A) Enzyme hydrolysis					
DP_1	0.1	0.3	0.7	1.4	2.4
DP_2	1.3	3.4	5.5	7.6	9.7
DP_3	1.8	4.3	6.9	9.4	12.0
DP_4	1.8	3.5	5.2	6.9	8.6
DP_5	1.8	3.6	5.5	7.4	9.3
DP_6	3.3	7.0	10.6	14.3	18.0
DP_7 and higher	89.9	77.9	65.6	53.0	40.0
Total DP_{1-6}	10.1	22.1	34.4	47.0	60.0
Descriptive ratio	2.0	2.2	2.3	2.4	2.4
(B) Acid hydrolysis (prior art process)					
DP_1	–	2.3	3.7	5.5	7.7
DP_2	–	2.8	4.4	5.9	7.5
DP_3	–	2.9	4.4	5.8	7.2
DP_4	–	3.0	4.5	5.8	7.2
DP_5	–	3.0	4.3	5.5	6.5
DP_6	–	2.2	3.3	4.3	5.2
DP_7 and higher	–	83.8	75.4	67.2	60.7
Total DP_{1-6}	–	16.2	24.6	32.8	41.3
Descriptive ratio	–	1.6	1.6	1.6	1.7

It is readily seen from the above table that hydrolysis of a waxy starch with bacterial α-amylase to a DE between 5 and 25 provides a descriptive ratio of at least 2.0 whereas acid hydrolysis fails entirely to produce a hydrolysate having this ratio. It was observed that the hydrolysates prepared by enzyme hydrolysis of waxy starch exhibited extraordinary clarity and substantially complete lack of opaqueness whereas the acid hydrolysates were decidedly opaque and exhibited little clarity except above DE of at least 25.

Hydrolysate Having a Narrow Range of Saccharide Units

By using an inhibited (i.e., crosslinked) starch, *D.P. Langlois; U.S. Patent 3,804,716; April 16, 1974* has enzymatically produced a starch conversion product having at least 80% by weight of its saccharide components containing 5 to 12 saccharide units. Such a saccharide distribution is extremely advantageous because the low molecular weight saccharides (dextrose and maltose) and the high molecular weight polysaccharides (13 to 30 saccharide monomeric units) are virtually eliminated. As a result of the controlled distribution of saccharide components in the conversion product, the product is clear and the dextrins are noncrystallizing under normal conditions of use and concentration and the dried product is nonhygroscopic at normal exposure conditions of 65% relative humidity and 80°F. The product is particularly useful in the production of hard candy and lozenges.

As a first step in the process, starch is treated in a manner sufficient to form granular starch particles which are insoluble in boiling water and can be maintained at a particle size greater than 2μ during enzymolysis. One highly desirable technique involves the preparation of what is known as inhibited starch by treating an uninhibited starch, which is soluble in water upon heating, with a cross-

linking agent, such as epichlorohydrin, until starch granules are rendered insoluble in boiling water. Previously, the use of an inhibited starch incapable of being solubilized even in boiling water, has been considered unsuitable for enzyme
reaction. Quite unexpectedly, it has been discovered that the crosslinked, inhibited
granular starches prepared with controlled crosslinking of this process, while maintaining their insolubility in boiling water are capable of a slight swelling and loss
of birefringence so as to be reactive with enzymes. The swelling which occurs in
the granular starches permits hydrolyzing enzymes to convert the starch to the
desired starch conversion products of this process.

This controlled amount of inhibition has been found to occur when the molar
ratio of glucose units in the starch to crosslinking agent is about 50:1 to 500:1.
Accordingly, for cornstarch as little as 1 mol of epichlorohydrin per 500 mols
of glucose units will give sufficient crosslinking. Similarly, up to 10 mols of
epichlorohydrin can also be used without causing excessive crosslinking to the
extent that the granules are incapable of enzyme hydrolysis. The granular particles which have to be treated to make them insoluble in boiling water, are suspended in a water-containing liquid medium. Typically, the liquid medium is
water, although it has also been found that desirable results can also be obtained
in water-alcohol solutions. Preferably, a pH range of 5.0 to 5.2 is maintained to
insure optimum reaction conditions.

Then the mixture of liquid and insoluble starch granules is heated to a temperature and for a time sufficient to cause a swelling of the starch granules. Typically, a swelling of about 35% by volume is sufficient to allow enzyme conversion of the starch granules. Such swelling, in turn, has been found to take place
at temperatures from 65° to 100°C. At 65°C, swelling can begin to take place,
and at temperatures above 72°C, the starch will tend to lose its crystallinity.

Following the heating step, the insoluble starch granules are reacted with a
starch hydrolyzing enzyme until hydrolysis takes place. The temperature requirements for the enzymolysis reaction are satisfied by maintaining the starch-
liquid mixture at about 65° to 72°C. Enzymes suitable for use in the practice of
this process are the α-amylases, malt α-amylase, and bacterial α-amylase. The
choice of the enzyme will depend upon the desired final product.

After about 25% of the starch is solubilized by reaction with enzyme, the granules begin to lose their shape. After solubilization of 98% of the starch, many
of the granular shells are collapsed; and at no time during the reaction are the
granular shells themselves solubilized. After the enzymolysis reaction is terminated, hydrolyzed starch contained in the liquid medium is separated from insoluble fractions of the liquid-starch mixture. Then the hydrolyzed starch conversion product is refined and concentrated for storage. The starch conversion
products and process are further illustrated in the following examples.

Example 1: A slurry of 32 grams (dry substance material) starch in 50 ml water
was stirred as 5 ml of 1.0 N NaOH was added dropwise. After all the alkali was
added 0.45 grams of epichlorohydrin was added and the slurry stirred as the
temperature was raised to about 40°C and held for 24 hours. The starch was
adjusted to 4.8 pH filtered, washed, dried to equilibrium moisture and stored.
The starch prepared by this procedure is inhibited in the sense that starch granules are not soluble in boiling water.

Example 2: A slurry of 162 grams (dry substance material) of the starch of Example 1 in 250 ml of water is heated to boiling and held at that temperature for about 15 minutes after which it is cooled to 65°C. At this temperature 1,000 units of α-amylase (Novo Units, Novo Industri) are added and the mixture stirred to maintain a suspension. The starch will become solubilized, while the granular shells remain in their insoluble form. The rate of solubilization can be followed by withdrawing a sample, centrifuging to remove the insoluble material and determining the soluble material by any suitable method. The refractive index tables for corn syrup, provides a simple means of doing this. When the desired extent of solubilization has been achieved, the reaction may be terminated by heating the solution to boiling to inactivate the enzyme. One may also wish to control the reaction by measuring the degree of conversion of the solubilized starch.

A sample is withdrawn, centrifuged and a DE determination made on the supernatant solution. When the desired degree of conversion is obtained, the reaction may be terminated as described above. The starch conversion product prepared by this process is nonhygroscopic and not subject to dextrin crystallization at high temperatures. The starch conversion product may be worked up in any suitable fashion. It may be filtered with a filter aid to clarify the solution, decolorized with carbon if a colorless product is desired and concentrated to a heavy syrup. Since the solution is stable at high concentrations, it may be stored at the high concentration. If a powdered product is desired it may be dehydrated by spray drying or by another dehydrating method.

OTHER ENZYME PROCESSES

Beta-Amylase with Crosslinked Starches

A method for the preparation of starch products having improved stability and resistance to syneresis and gelling when exposed to low temperatures and repeated freeze-thaw cycles has been developed by *O.B. Wurzburg and C.D. Szymanski; U.S. Patent 3,525,672; August 25, 1970; assigned to National Starch and Chemical Corporation.*

This method comprises subjecting crosslinked starch to the action of a specific enzyme, namely one which is capable of digesting the outer branches of the amylopectin molecule and whose action will not include or go beyond the 1,6 branching point in the amylopectin molecule as well as any substituent group which may be present in either the amylose or amylopectin molecules of the starch. Inasmuch as β-amylase best meets these requirements, its use is preferred in this process. Although for purposes of brevity and convenience the term β-amylase is used interchangeably with the term enzyme, it is to be noted that other enzymes exhibiting the above characteristics may also be used in the process.

It is further believed that by treating crosslinked starch with the specified type of enzyme, the outermost branches of the starch molecule are shortened or removed. Thus, the possibility of association on the part of these branches is lessened and it is believed that this accounts for the remarkable reduction in the objectionable syneresis and gelling in the starch products of this process as contrasted with conventional crosslinked starches.

The enzyme treatment used in this process is preferably conducted using cross-linked, partially swollen starches which may, in turn, be derived from raw starch bases obtained from corn, potato, wheat, rice, sago, tapioca, sorghum or the like, and preferably those starches which contain, essentially, only amylopectin such as waxy corn, waxy rice and waxy sorghum. It is also possible to use any substituted ether or ester derivative of these starch bases for the preparation of the crosslinked, pregelatinized intermediates.

In order to inhibit, i.e., to crosslink, any of the above raw starch bases, it is ordinarily necessary to react the starch with a crosslinking agent by means of an etherification, esterification, or acetal formation procedure or by a combination of the latter procedures. In order to function effectively in this process the inhibited starches should have a GSP value in the range of from 10 to 31 since within this range they appear to provide the optimum thickening and rheological properties for most food applications.

The degree of starch degradation that is required to improve the low temperature stability of the starch is also subject to variation. Although this value is dependent upon the type of starch used in the reaction as well as upon any substituent groups which may be present upon its molecule, values ranging from 13 to 55%, by weight, of starch degradation will, in most instances, insure improved low temperature stability. The degree of starch degradation is ascertained by determining the amount of free maltose liberated during the enzyme reaction and then employing the following relationship to calculate the percent of starch degradation.

$$\text{Percent starch degradation} = \frac{\text{grams of free maltose} \times 100}{\text{grams of total starch on a dry basis}}$$

In the following example, which further illustrates the process, all parts given are by weight unless otherwise specified.

Example: This example illustrates the preparation of a typical enzyme modified starch product of this process as well as its improved low temperature stability. A sample of waxy maize starch was inhibited by reaction with epichlorohydrin to form an inhibited waxy maize starch having a GSP value of 22. An aqueous slurry of the inhibited waxy maize starch was then subjected to drum drying whereby it was passed over drums heated to a temperature sufficient to gelatinize and simultaneously dry the starch.

A suspension of 30 parts of the above crosslinked, pregelatinized starch in 600 parts of a 0.026 N aqueous acetate buffer solution having a pH of 4.8 was prepared. The temperature of this suspension was raised to 55°C whereupon 94 units of β-amylase were added. The system was kept under continuous agitation for 1 hour and, then, 5 parts of 0.001 N mercuric chloride were added to deactivate the enzyme. The resulting starch suspension was then spray dried to recover the degraded product.

To determine the percent degradables in the resulting starch product, aliquots of the suspension were submitted to a colorimetric analytical procedure which used 3,5-dinitrosalicylic acid as an indicator. It was determined that 16.6%, by weight, of the starch intermediate had been degraded.

To test the low temperature stability of the above prepared starch product, 6.6 parts were slurried in 100 parts of a 1:1 water:cranberry juice mixture. This starch-juice mixture was cooked for 10 minutes in a boiling water bath whereupon 15 parts of sucrose were added. The cooked mixture was transferred to small containers, cooled to room temperature and then stored at 0°C. The samples were removed daily and were on each occasion completely thawed and refrozen. Low temperature instability is indicated by a deterioration of clarity and texture as well as by the tendency towards syneresis. It should be noted that other fruit juices can be used in this procedure, but the use of cranberry juice provides a particularly severe test because of its high degree of acidity.

Upon subjecting the above product to this freeze-thaw procedure, it was found to survive 14 cycles before showing the first indications of an increased opacity. This was in contrast to the control, i.e., the crosslinked, pregelatinized waxy maize starch which had not undergone the enzyme modification, which showed complete deterioration and syneresis after only 3 to 4 cycles. These results clearly indicate the improved low temperature resistance exhibited by the starch products resulting from the process.

The above described procedures were then repeated under identical conditions with the exception, in this instance, that 188 units of β-amylase was used in the reaction and the reaction was allowed to continue for a period of 48 hours at 55°C. Analysis of the resulting starch product indicated that 52%, by weight, of the starch intermediate had been degraded. Upon being submitted to the freeze-thaw procedure, it survived without change for a total of 23 cycles.

Production of Beta-Cyclodextrin

Cyclodextrins, also referred to as Schardinger dextrins, are crystalline, cyclic oligosaccharides consisting of 6, 7 or 8 glucose units linked by α-1,4-bonds and designated by Greek letters α, β, γ according to increasing number of glucose units. The preparation of the three cyclodextrins and their properties have been described by Tilden and Hudson (*JACS* 64, 1432,1942) and D. French (*JACS* 71, 353, 1949).

Cyclodextrins are obtained by hydrolysis of starch by an enzyme found in the cell-free broth of a culture of *Bacillus macerans.* α-Cyclodextrin is formed first, and the β- and γ-forms are formed thereafter. The resulting mixture may be fractionated by precipitation with toluene or trichloroethylene. Repeated precipitation is necessary when approximately pure compounds are to be produced.

It has been found by *S. Okada and N. Tsuyama; U.S. Patent 3,812,011; May 21, 1974; assigned to Hayashibara Biochemical Laboratories, Inc., Japan* that several microorganisms isolated from soil produce a cyclodextrin glycosyltransferase which converts starch to β-cyclodextrin only, at least in the initial stages of fermentation, and that an enzyme produced by *Bacillus megaterium* T 5 (ATCC 21737, FERM-P 935) permits β-cyclodextrin to be produced directly from starch in very high yields before significant amounts of α-cyclodextrin are formed, particularly if the enzymatic hydrolysis of the starch is carried out in the presence of a precipitating agent for the β-cyclodextrin.

Details of the preparation and properties of *B. megaterium* T 5 are given in the complete patent. The following example illustrates use of the enzyme in the production of the cyclodextrin.

Example: A sterilized aqueous medium containing 0.5% corn steep liquor, 1% soluble starch, 0.25% $(NH_4)_2HPO_4$, 0.25% $(NH_4)_2SO_4$ and 1% $CaCO_3$ was inoculated with *B. megaterium* T 5 and incubated at 37°C for 2 days with stirring. The broth contained 18 to 20 u/ml enzyme activity. The microbial cells were removed by centrifuging at 0°C and the supernatant having a volume of 1,450 ml was found to contain 11.2 u/ml of cyclodextrin glycosyltransferase activity, for a total of 16,250 units, and 9,720 mg protein.

After decolorizing with active charcoal and centrifuging, 1,370 ml of the purified liquid had an activity of 9.4 u/ml, a total activity of 12,880 units, and contained 6,040 mg protein. Salting out with 65% saturated ammonium sulfate and dialysis against tap water reduced the volume to 180 ml, and raised enzyme activity to 36.0 u/ml for a total of 6,480, while total protein was reduced to 180 milligrams. After further treatment with SE-Sephadex, the residual 170 ml of purified enzyme solution had a total activity of 2,820 units (16.6 u/ml) and a total protein content of 17.10 mg. In terms of units of specific activity per milligram protein, the purification improved the enzyme concentration from 1.67 to 165 u/mg.

500 ml 10% soluble starch solution, 25 ml enzyme solution, and 100 ml trichloroethylene were mixed and stored at 30°C. The mixture was agitated for 30 minutes every morning. After one day, 10 ml of the enzyme solution, and every day thereafter, 5 ml enzyme solution and 20 ml trichloroethylene were added.

After five days, the mixture was stirred 3 to 4 hours at 5°C and centrifuged to recover the crystals formed which were washed with water and heated on a water bath to evaporate the trichloroethylene. A sample analyzed by paper chromatography (quadruple development with n-butanol-pyridine-water 6:4:3) and thereafter developed for color with an M/50 solution of iodine in KI solution showed 2% impurities.

The product therefore was agitated for 4 hours at 5°C in aqueous solution with bromobenzene. The precipitated complex crystals were decomposed and pure β-cyclodextrin was recovered in an amount corresponding to 67% of the starch initially employed. In a control run with the enzyme of *B. macerans* under the same conditions, it was not possible to obtain a primary product containing more than 85% β-cyclodextrin, and the yield was only 41%.

COMBINED ACID-ENZYME PROCESSES

Low DE Hydrolysates

Previous attempts to produce low DE syrups from starches by acid hydrolysis, have failed by way of extremely poor filtration rates, yield losses and substantial insolubility of syrup solids.

F.C. Armbruster and C.F. Harjes; U.S. Patent 3,560,343; February 2, 1971; assigned to CPC International Inc. have developed a process for preparing a low DE starch hydrolysate. This process comprises subjecting a mixture of starch and water, having a solids content of less than 50%, to the hydrolytic action of acid to obtain a starch hydrolysate having a DE between 5 and 15, subjecting the acid hydrolysate to the hydrolytic action of bacterial α-amylase to a DE between 10 and 25, the increase in DE being at least 5, to produce a starch hydrolysate having a dextrose content of less than 4%. The hydrolysate is also characterized by having a descriptive ratio above 2.

One preferred method of the process involves the steps of slurrying cornstarch in water to a density between 5° and 30°Bé, adjusting the pH of the slurry to between 1 and 3 and raising the temperature of the starch slurry to between 70° and 160°C to solubilize and hydrolyze the starch to a DE between 5 and 15. The pH is adjusted to between 6 and 8 and the hydrolysate is dosed with bacterial α-amylase. The mixture is then hydrolyzed under the proper conditions to a DE between 10 and 25, preferably between 15 and 25 when it is desired that the final product be a haze-free syrup.

The enzyme conversion step is carried out between 50° and 95°C. The mixture is held at the conversion temperature for a time ranging from a few minutes to as long as 1 or 2 hours or perhaps more. The resulting hydrolysate may be concentrated and/or refined by conventional procedures to yield a stable corn syrup, which is substantially haze-free and highly soluble in water. The syrup may be spray dried to yield corn syrup solids with low hygroscopicity and high water solubility. Suitable starches which may be used include cereal starches such as corn, grain sorghum and wheat, waxy starches such as, waxy milo and waxy maize, and root starches, such as potato starch and tapioca starch.

The preferred enzyme for the conversion of the acid hydrolysate to low DE hydrolysates in the process is the bacterial α-amylase produced by certain strains of *Bacillus subtilis, Bacillus mesentericus* and the like. HT-1000, the proprietary name of a bacterial α-amylase is an example of an enzyme preparation that is suitable for this use. Other suitable bacterial α-amylase include Rhozyme H39, and CPR-8.

Example 1: Several samples of cornstarch (A, B and C) were slurried in water providing slurries having Baumés ranging from 14° to 22°. These slurries were partially acid hydrolyzed to a maximum of 15 DE. The particular DE achieved by acid hydrolysis in each of the samples is set forth in Table 1 below. After acid hydrolysis, the slurry was neutralized to a pH between 6 and 7. The neutralized liquor was cooled to 80° to 85°C, and dosed with bacterial α-amylase (HT-1000) in the quantity set forth below.

A final DE of 19 to 21 was obtained in each of the samples in 1 to 3 hours. The final conversion liquors are low in color. These liquors are easily refined and evaporated to 42°Bé to provide syrups. Sample D was a conventionally acid hydrolyzed starch conversion product. Acid hydrolysis was carried out to reach a DE of 20. Tables 1 and 2 below set forth the reaction conditions for conversion and the product analyses respectively.

TABLE 1: ENZYME CONVERSION CONDITIONS

Sample	DE of Acid Hydrolysate	Percent Dry Substance	Temperature, °C	pH	Enzyme Dose	Time, hours	Final DE
A	15.2	38	80	6.5	0.01	1	19.7
B	12.9	37.5	85	6.5	0.05	2	20.2
C	10.3	38.1	85	6.5	0.1	2	21.8
D	20	–	–	–	–	–	–

TABLE 2: PRODUCT ANALYSES

Sample	Final DE	Percent Dry Substance	DP_1	DP_2	DP_3	DP_4	DP_5	DP_6	DP_{7+}	Descriptive Ratio
A	19.7	72	3.9	5.8	8.2	7.2	7.3	10.2	57.4	2.1
B	20.2	72	2.3	5.9	8.5	6.4	6.6	12.6	57.7	2.1
C	21.8	75	2.3	8.3	10.9	8.1	9.2	16.9	44.3	2.5
D	20.0	–	5.5	5.9	5.8	5.8	5.5	4.3	67.2	1.6

The descriptive ratios of each of samples A through C were greater than 2. The descriptive ratio of sample D (conventional acid hydrolysate) was 1.6. After refining, typical analyses of the syrups were as follows: substantially colorless, clarity greater than 90%, sulfated ash 0.42% dry basis, protein 0.03% dry basis. The refined syrups were allowed to stand at room temperature and at the end of a one month period remained haze-free.

Example 2: Waxy milo starch was slurried in water to a Baumé of 20°. The slurry was then partially hydrolyzed by acid to a DE of 6.5 and then neutralized to a pH between 6 and 7. The acid hydrolyzed product was then dosed with 0.0206 bacterial α-amylase (HT-1000). The temperature of the dosed liquor was between 80° and 85°C. Conversion was allowed to proceed for about 1½ hours until a DE of about 21 was obtained. Analysis of the product obtained appears in Table 3 below.

TABLE 3: SAMPLE G

Final DE	21.4
Dry substance percent	75
DP_1	2.3
DP_2	5.9
DP_3	8.4
DP_4	7.3
DP_5	8.0
DP_6	11.3
DP_{7+}	56.8

It may be noted that in Examples 1 and 2 the descriptive ratio exceeds 2.0, i.e., the sum of the percentages of degree of polymerization from 1 to 6 divided by the DE provides a ratio in excess of 2. If the descriptive ratio is at least 2, the product is highly water-soluble and exhibits almost no haze formation. If the descriptive ratio of syrups in the range of 15 to 25 DE is substantially below 2, e.g., 1.6 or less, the products exhibit haze formation and are less water-soluble than products with a ratio of at least 2.

Low DE Bulking Agents

Prior efforts to supply suitable bulking agents for synthetic foods generally have been unsuccessful. To be ideal a filler must be inexpensive, stable, water-soluble, low in caloric content, essentially tasteless, and nonhygroscopic. Few substances previously available have all of these characteristics.

To overcome the disadvantages of starch hydrolysates as fillers and bulking agents, several approaches have been tried. For example, in U.S. Patent 2,610,132, it was disclosed that haze-forming components in low DE starch hydrolysates can be precipitated and removed by cooling the hydrolysates in the presence of fats at temperatures of 60° to 140°F. This approach is not, however, successful in removing the haze-forming materials completely. In U.S. Patent 2,965,520, low DE starch hydrolysate syrups having increased stability toward haze formation are produced by initially acid hydrolyzing a starch paste to a DE content of 18 to 32% and, after the insolubles are removed, treating the acid hydrolysate liquor with α-amylase to slightly increase its DE. These hydrolysates however were relatively sweet and hygroscopic. The best overall process heretofor available for providing low DE hydrolysates of the desired properties uses a combination of the processes described in U.S. Patents 2,610,132 and 2,965,520.

The improved process developed by *R.E. Nisbet and E.A. Allen; U.S. Patent 3,616,220; October 26, 1971; assigned to A.E. Staley Manufacturing Company* produces low DE starch hydrolysates which are essentially completely water-soluble and relatively nonhygroscopic and tasteless.

This method comprises heating an acid-containing aqueous starch slurry at an elevated temperature to convert the slurry to a starch paste and effect the hydrolysis of the resultant paste to an acid hydrolysate having a DE of 8 to 16% then, in the substantial absence of fats, cooling the hydrolysate without freezing to a temperature below 60°F, holding the cooled hydrolysate at that reduced temperature for at least 1 hour to precipitate the polysaccharides from the cooled liquid, and then separating the precipitated material from the cooled liquid.

In preferred embodiments of this process, the cooled liquor is further converted with an enzyme such as α-amylase to further improve hydrolysate properties, e.g., haze stability. In such embodiments, for example, it is typical in a subsequent α-amylase treatment to increase the DE of the precipitate-free liquor to a value above 12%, and more usually in the range of 12 to 22% by weight, dry substance basis. The details of the process are given in the following examples.

Example 1: An 18°Bé aqueous slurry of unmodified cornstarch was acid converted with hydrochloric acid at a pH of 2.0 and a temperature of 270°F in a continuous steam injection heater and then flashed to atmospheric pressure to provide a hot, acid hydrolysate having a DE of 9.4%. The hot, acid hydrolysate, which contained an insoluble yellow fraction of fats, initially was neutralized to pH 4.5 with soda ash and then divided into two portions. One portion remained untreated and served as the control. The other sample was centrifuged at 200°F to remove the insoluble fat fraction. The control and centrifuged hydrolysates then were further divided into 1,600 cc samples and samples of both were held for 16 hours in stainless steel pots placed in separate water baths, one at 32°F and one at 100°F.

At the end of this holding period, the respective cooled samples each were then centrifuged to remove precipitate. The amount of precipitate removed from each sample is shown in the following table. After a pH adjustment to 5.8 with soda ash, samples of each of the liquors were then refrigerated at 35°F for similar periods ranging up to 344 hours. Following the particular refrigeration period used, a sample was centrifuged to determine the amount of precipitate formed.

The amounts observed for the respective samples are listed in the following table.

Sample Type	Cooling Temperature, °F	Acid Hydrolysate DE, % wt, DSB	Precipitate, volume percent						
			0 hr*	10 hr*	39 hr*	56 hr*	128 hr*	152 hr*	344 hr*
A (fat absent)	100	12.3	4.7	20.0	30.0	33.3	37.7	35.3	35.3
B (fat present)	100	12.0	15.4	0.5	19.3	23.3	30.7	30.3	32.7
B (fat present)	32	12.3	20.0	0	4.5	8.0	18.5	20.3	21.0
A (fat absent)	32	11.5	26.0	0	0	0	3.3	4.5	8.0

*Refrigeration time.

As shown by the data in the above table, the run made according to this process by cooling to below 60°F, in the absence of fats, removed significantly more insolubilizable polysaccharides and provided hydrolysate liquors more stable toward subsequent precipitate formation than the other cooling techniques. The data further show that the attainment of the maximum amount of insolubilizable polysaccharide material at temperatures above 60°F requires the presence of fats. This follows what would be expected from the disclosure in U.S. Patent 2,965,520. It further will be noted that although the hydrolysate produced by this method has the greatest stability against precipitate formation, it has the lowest DE, making it suitable for use as a diluent or bulking agent in powdered synthetic food formulations.

Example 2: Additional 300 cc samples of each of the four hydrolysates of Example 1 obtained after cooling at 32° and 100°F were adjusted to a pH of 5.8 with soda ash. About 0.03% of an α-amylase preparation (Aquazyme 120, Novo Industri) dry solids basis, was added to each and hydrolysis effected by placing in separate stainless steel pots held in a 185°F water bath. After conversion periods of 1 to 2 hours, samples of the liquors were recovered, treated with carbon on a steam bath, and then filtered. Samples of the resultant liquors were then refrigerated as in Example 1 and, after refrigeration, centrifuged to determine the amount of precipitate formed. The results of these tests are set forth in the following table.

It will be noted that the data of the table show that the method produced lower DE hydrolysates which were completely stable under the test conditions used with less than 1 hour subsequent enzymatic conversion. The data also show that the alternative procedures required the subsequent use of longer hydrolysis periods with an attendant greater increase in hydrolysate DE to achieve comparable stability.

Sample Type	Cooling Temperature, °F	Enzyme Conversion Time, hr	Enzyme Hydrolysate DE, % wt, DSB	Precipitate, volume percent					
				10 hr*	39 hr*	56 hr*	128 hr*	152 hr*	344 hr*
A (fat absent)	100	1	13.2	0.0	0.0**	0.3	6.0	7.6	10.0
A (fat absent)	100	2	15.0	0.0	0.0	0.0	0.0	0.0**	1.1
B (fat present)	100	1	13.1	0.0	0.0	0.0	0.0	0.0	5.7
B (fat present)	100	2	15.7	0.0	0.0	0.0	0.2	0.7	5.7
B (fat present)	32	1	13.9	0.0	0.0	0.0	0.0	0.0	0.0
B (fat present)	32	2	15.4	0.0	0.0	0.0	0.0	0.0	0.0
A (fat absent)	32	1	13.8	0.0	0.0	0.0	0.0	0.0	0.0
A (fat absent)	32	2	14.9	0.0	0.0	0.0	0.0	0.0	0.0

*Refrigeration time.   **Visual cloudiness, but no measurable precipitate after 1 hr centrifuge.

Two-Stage Dextrinization

Retrogradation of low DE starch hydrolysates is largely avoided in the two-step process developed by *A.L. Morehouse, R.C. Malzahn and J.T. Day; U.S. Patent 3,663,369; May 16, 1972; assigned to Grain Processing Corporation.* In the first step, the starch is liquefied with acid or enzyme to a dispersion substantially free of residual starch but having a DE of 3 or lower. In the second step, a dextrinizing enzyme is used to give a product having a DE not above 18.

In the first step of the process, starch is slurried in water to a solids concentration from 10 to 40, preferably 20 to 30% by weight. All varieties of starch or starch products or amylaceous materials can be used, however pure cornstarch is preferred. To the aqueous starch slurry is added α-amylase, preferably bacterial α-amylase, or an acid such as hydrochloric, oxalic or sulfuric acid and the like. The slurry is then maintained at a relatively high temperature for a short time to achieve substantially complete liquefaction of the starch. In this stage, starch is liquefied, that is, the starch granules are pasted and the starch hydrated and dispersed to such an extent that hydrolytic cleavage of the starch molecules can be readily accomplished. During this liquefaction stage, dextrinization of the starch is not permitted to progress to a stage where the dextrose equivalent value is above 3.

The starch liquefaction can be carried out in any suitable equipment which permits the slurry to be maintained at an elevated temperature and preferably provided with means for agitating the slurry. A high temperature, i.e., one above 90°C and preferably 92° to 95°C, is used in this step.

To accomplish liquefaction with α-amylase, the enzyme is used in amounts from 2,500 to 9,000, preferably 3,000 to 6,000 SKB units per pound of starch. The pH of the slurry is adjusted to 6 to 8, preferably 6.5 to 7.5, and maintained above 90°C for a short period to achieve substantially complete liquefaction of the starch. For example, with a purified α-amylase in amounts from 3,600 to 7,200 SKB units per pound, substantially complete liquefaction of a starch slurry having a solids content of 20 to 30% and having a pH of 6.5 to 7.5 is generally accomplished in 5 to 20 minutes at 90° to 95°C.

When an acid such as hydrochloric acid is used for liquefaction, the acid is used in amounts to provide a hydrogen chloride normality in the slurry of 0.02 to 0.12, preferably 0.02 to 0.04. The slurry is maintained at 100° to 160°C for a short period to achieve substantially complete liquefaction of the starch. For example, with a starch slurry containing 20 to 30% by weight solids, hydrochloric acid is added to adjust the acidity to 0.06 N. The acidified slurry is then pumped into a reactor and heated to 100°C for 3 or 4 minutes. The slurry is then cooled and neutralized by addition of an alkaline material such as sodium hydroxide.

The dextrinization step is accomplished by treating the liquefied starch slurry with a dextrinizing enzyme such as α-amylase. It is preferable to carry out the dextrinization step as a batch reaction. The dextrinizing enzyme is generally used to provide from 300 to 3,000 SKB units. Any excess of the agent employed for the liquefaction is permitted to remain in the slurry during the dextrinization step, which step is conducted at a temperature of 65° to 85°C, preferably 74° to 80°C, and at a pH of from 7.0 to 9.0, preferably 7.5 to 8.5, for a time sufficient to produce a hydrolysate product having the desired DE value, preferably a value between 8 and 18. An important characteristic of starch hydrolysates prepared by the two-stage process of this method is the relatively low amount of high molecular weight fragments present which are believed to cause retrogradation, gel formation and haze in the final product.

DE of Hydrolysate	Maximum Solid Concentration Remaining Fluid and Free of Opacity*
8	10%
10	20
12	35
14	50
16	65
18	80

*By free of opacity is meant that the transmittance of light rays through the hydrolysate maintained at $10°C$ over a period of 48 hours is at least 70%.

The starch hydrolysates of the process are unique in that they have a DE value below 18 and contain not more than 1% by weight of glucose and not more than 6% maltose and yet remain fluid and free of opacity. The oligosaccharide composition of the final hydrolysate product having DE values between 8 and 18 is generally as follows, as determined by chromatography:

DE of Hydrolysate	- - - - - -Degree of Polymerization (DP*) in Percent by Weight - - - - - -										
	1	2	3	4	5	6	7	8	9	10	Over 10
8	1	3.5	5	4.5	4	6.5	6	4	2.5	2	65
10	1	5	6.5	5.5	4.5	9	8	5	3	1	52
12	1	5.5	7.5	6.5	5.0	11	10	5.5	3	1	40
14	1	5.5	8.5	6.5	5.5	13	12	6	3	1	35
16	1	5.5	9.5	7	6	15	13.5	6	2.5	1	30
18	1	6.0	10	7	6.5	17	15.5	6	2.5	1	24

*DP means degree of polymerization of glucose. For example, DP_1 = glucose, DP_2 = maltose, etc.

Example: A starch slurry was prepared using unmodified industrial cornstarch, water, α-amylase and calcium chloride. The slurry had the following composition:

Pearl starch	1,200 grams
Tap water	4,800 grams
$CaCl_2$	0.6 gram
α-Amylase (3,300 SKB u/g)	5 grams*
pH	7.7

*Equals 6,000 SKB u/lb starch.

The starch was pumped at a controlled rate through a two-stage converter which consisted of two round bottom flasks with heating mantles and stirrers. Peristaltic pumps were employed to move material through the system. The following steady state conditions were established for each stage:

Volume	First Stage (Liquefaction) 500 ml	Second Stage (Dextrinization) 1,000 ml)
Rate of addition and withdrawal	25 ml/min	25 ml/min
Residence time	20 min	40 min
Temperature, °C	91	76

A solution of α-amylase was added dropwise to the second stage of the converter at a rate providing 600 SKB units amylase per pound of starch. The converted hydrolysate which was removed continuously from the second stage flask was adjusted to pH 3.5 by the addition of dilute hydrochloric acid to stop the dextrinization. The hydrolysate had a DE of 12.3. A sample of the crude hydrolysate filtered readily using filter aid and standard equipment.

NONHAZING LOW DE HYDROLYSATES

Typically, starch conversion syrups having a DE below 15 are subject to haze development upon standing. In extreme cases, such syrups became completely opaque and gelled into a paste. In less extreme cases, haze particles were found to agglomerate and settle to the bottom of the syrup, resembling a sludge. In other cases, haze particles are too fine and dispersed to agglomerate and merely remain in suspension, lending the syrup a cloudy appearance.

Use of Reverse Osmosis

G.R. Meyer; U.S. Patent 3,756,853; September 4, 1973; assigned to CPC International Inc. has found that starch conversion syrups having a DE of 5 to 15 can be prepared which do not haze or form suspended matter upon standing. More particularly, it has been found that nonhazing starch conversion syrups having a solids content of 60 to 85% by weight and a DE of 5 to 18 can be prepared by a process which comprises first hydrolyzing starch to a DE of 20 to 40 and then subjecting the resulting starch conversion syrup to reverse osmosis until the DE of the syrup has been reduced to 5 to 18. The hydrolyzing of the starch to a DE of 20 to 43 can be carried out by acid, enzyme or a combination of acid and enzyme conversions.

In this process the starch conversion syrup having a DE of from 20 to 40 is subjected to reverse osmosis by contacting the syrup with one side of a semipermeable membrane at a pressure in excess of the osmotic pressure across the membrane, until the DE of the syrup has been reduced to 5 to 18. While the application of any pressure to the concentrate stream in excess of the osmotic pressure across the membrane will produce reverse osmosis, a pressure in excess of 50 psi is usually required. Furthermore, since the rate of mass transfer is directly proportional to pressure, exceedingly high pressures, such as those approaching the breaking point of the membranes used and typically ranging from 500 to 2,500 psi, are preferred.

The rate of permeation in reverse osmosis varies directly with temperature. An increase in the operating temperature of 10°C can increase the rate of permeation by as much as 100%. However, as the operating temperature is increased there is an increase in the tendency of the membrane to soften to the point of rupture. While operating temperatures ranging from 10° to 150°C can generally be utilized depending on the type of membrane used, an operating temperature of 30° to 100°C is preferred.

The rate of permeation in the reverse osmosis step of the process, and the efficiency with which the DE of the starch conversion syrups is reduced to within the required range varies somewhat with the concentration of the conversion syrup employed. While high solids concentrations in the syrups are desirable for high yields of product, the lower solids concentrations favor permeation rates. As a result, intermediate solids concentrations in the syrups, such as those ranging from 10 to 40% by weight, are preferred.

The semipermeable membranes which can be used in this process can be of the flat or uniplanar as well as of the tubular or hollow fiber type. Exemplary material which can be used for the membranes are cellulose esters such as cellulose acetate, triacetate, formate, propionate, nitrate and mixtures of such esters; cellulose esters such as methyl, ethyl, hydroxyalkyl, carboxyalkyl and the like; regenerated cellulose; polyvinyl alcohols; casein and its derivatives; and similar polymeric materials such as acrylonitrile polymers. In the following examples, the reverse osmosis cell used is that described in U.S. Patent 3,133,132.

Example 1: A starch conversion syrup having a DE of 20 and a solids content of 40% by weight is pumped through a reverse osmosis cell equipped with a cellulose acetate membrane, at a pressure of 1,200 psi, and at a temperature of 55°C. The concentrated solution is recycled through the cell until the DE has been reduced to 5 on a dry basis to yield a nonhazing starch conversion syrup.

Example 2: A starch conversion syrup having a DE of 25 and a solids content of 20% by weight is pumped through a reverse osmosis cell equipped with a cellulose acetate membrane, at a pressure of 500 psi, and at a temperature of 30°C. The concentrated solution is recycled through the cell until the DE has been reduced to 15 on a dry basis to yield a nonhazing starch conversion syrup.

Example 3: A starch conversion syrup having a DE of 40 and a solids content of 25% by weight is pumped through a reverse osmosis cell equipped with a cellulose acetate membrane, at a pressure of 2,500 psi, and at a temperature of 75°C. The concentrated solution is recycled through the cell until the DE has been reduced to 18 on a dry basis to yield a nonhazing starch conversion syrup.

Use of Molecular Exclusion

I.F. Deaton; U.S. Patent 3,756,919; September 4, 1973; assigned to CPC International Inc. has also found that starch conversion syrups having a DE of 5 to 28 can be prepared which do not haze or form suspended matter upon standing. More particularly, it has been found that nonhazing aqueous starch conversion syrups having a solids content of 60 to 85% by weight and a DE of 5 to 18 can be prepared by a process which comprises first hydrolyzing starch to a DE of 20 to 43 and then subjecting the resulting starch conversion syrup to molecular exclusion until the DE of the syrup has been reduced to 5 to 18.

Molecular exclusion comprises passing the conversion syrup through a column or bed of a porous and generally granular media. The reduction of the DE of the syrup is effected by the difference in the rate at which molecules of varying size diffuse into the pores of this media. The larger molecules representing the lower DE fraction of the syrup diffuse more slowly and, therefore, pass more quickly through the bed than the smaller molecules representing the higher DE fraction.

The reduction of the DE of the syrup by molecular exclusion can be effected in a fixed bed of a porous media having sufficiently large pores to admit the smaller molecules being separated. The reduction of the DE of the syrup by molecular exclusion can also be effected in a countercurrent moving bed of a porous media having sufficient mechanical strength to withstand handling. Examples of porous media that can be used are ion exchange resins, granular forms of dextran, argarose and polyacrylamide gels, porous glass beads and activated carbon or alumina.

The preferred porous media suitable for the molecular exclusion step of this process are ion exchange resins, preferably the metal salts of strongly acid nuclearly sulfonated resins having a crosslinked vinylaromatic resin matrix. Typically, these resins are nuclearly sulfonated polymers or copolymers of vinylaromatic compounds, such as styrene, vinyltoluene, vinylxylene, and the like that have been crosslinked in molecular structure to an extent rendering the sulfonated polymers and copolymers insoluble in aqueous solutions of acids, bases, or salts. An example of a suitable resin for the molecular exclusion step of this process is the sodium or potassium form of Dowex 50W-X4. This resin is a 50 to 100 mesh sulfonated copolymer of styrene and divinylbenzene having 4% divinylbenzene linkages.

The starch conversion syrup is preferably passed through the bed of porous media at elevated temperatures, to reduce the viscosity of the syrup, thereby improving its flow rate through the column. The temperature of the media itself can also be maintained at a level similar to that of the syrup. Useful syrup and media temperatures which can be used in this process range from 100° to 200°F, and preferably from 110° to 170°F.

When the molecular exclusion step of the process is carried out as a batch process, the aqueous starch hydrolysate having a DE of 18 to 43 after being heated to a temperature as heretofore indicated, is applied to the porous media column. The column is then washed with water and the effluent collected. The early portions of the effluent having a DE of 5 to 18 are collected to yield the nonhazing starch conversion syrups. This product can then be concentrated to a solids content ranging from 60 to 85% by weight for use in commercial applications.

When the molecular exclusion step of this process is carried out in a continuous manner, the supply liquor comprising the starch hydrolysate having a DE of 20 to 43 is fed into a column of the porous media in which the media is moved continuously or incrementally in a countercurrent relation to the supply liquor. This procedure permits the continuous flow of supply liquor and results in a continuous product of starch conversion syrup having a DE of 5 to 15, while requiring a constant supply of fresh media. The spent porous media containing the high DE fraction of the supply liquor can be readily regenerated by washing the resin with water and can then be recycled to the fresh resin supply. The molecular exclusion step of the process is more specifically illustrated in the following examples.

Example 1: Molecular Exclusion Using a Fixed Bed Column — A fixed bed molecular exclusion column having a diameter of 7 cm and a length of 4 ft, was filled to a bed depth of 62.5 cm with a sulfonated polystyrene ion exchange resin in the sodium form (Dowex 50W-X4). The column was first filled with water and was heated to a temperature of 140°F. A starch hydrolysate syrup having a DE between 22 and 24, was heated to 140°F and was applied onto the column. The column was then eluted with distilled water (3,200 ml). The effluent from the column was collected in 19 fractions of 200 ml each. The fractions 5, 6 and 7, were combined and were concentrated by evaporating the water to 40°Bé to yield a nonhazing syrup having a DE of 11.

Example 2: Molecular Exclusion Using a Countercurrent Moving Bed — A molecular exclusion column having an inner diameter of 1⅜" and a length of 4 ft was filled to a bed depth of 100 cm with a sulfonated polystyrene ion exchange resin in the sodium form (Dowex 50W-X4). A starch hydrolysate syrup having a DE of 22.5 was fed into the bottom of the column at a rate of 5 ml/min. After every addition of 50 ml of supply liquor, 100 ml of the resin was removed from the bottom of the column and was replaced with a similar amount of regenerated resin at the top of the column. The used resin withdrawn from the column was vacuum filtered to remove the hydrolysate liquor and this liquor was recycled into the supply liquor. The filtered resin was washed to recover the higher DE product. The lower DE product liquor was continuously withdrawn from the top of the column. The column was operated in this manner for a period of 20 hours to yield a nonhazing starch conversion syrup having the following properties.

Sample Number	After Hours of Operation	Dry Substance	DE
1	5.25	4.5	6.0
2	9.5	9.3	9.0
3	13.25	13.2	11.5
4	16.25	14.6	13.7
5	19.0	9.5	12.2
6	20.0	12.7	12.7

Derivatives of Starch Hydrolysates

Haze formation can also be eliminated from low DE starch syrups in the process disclosed by *G.A. Hull; U.S. Patent 3,639,389; February 1, 1972; assigned to CPC International Inc.* by the formation of derivatives or incorporating such derivatives in the syrup.

Broadly described, these derivatives of low DE starch hydrolysates may either be of the cationic, anionic or nonionic type. That is, the parent low DE starch hydrolysate may be reacted with a wide variety of agents to introduce a positive, negative or neutral substituent group. In essence then, the hydrolysates are amenable to reaction with a host of inorganic and organic reactants.

Generally in making the derivatives solution reactions are employed, although one can envision a reaction of a solid low DE starch hydrolysate, say with a gaseous or liquid reactant in absence of external solvent. It is preferred that the derivatization take place in liquid media, such as water. The derivatives can be marketed as liquids or dried and sold as solids. After syrups are derivatized to prevent haze formation they are concentrated to a solids concentration greater than 40% by weight. The nonionic derivatives have the structural formula:

$$\text{low DE starch hydrolysate} -\text{O}-\text{R}_1-\text{Z}$$

where R_1 when present is alkylene, hydroxyl alkylene, halogeno alkylene, aralkylene, cycloalkylene and phenylene, and Z is

$$-\overset{\overset{\text{O}}{\|}}{\text{C}}-\text{N}\overset{\textstyle R_2}{\underset{\textstyle R_2}{\big\langle}} \qquad -\overset{\overset{\text{O}}{\|}}{\text{C}}-\text{R}_2 \qquad -\text{CN} \qquad -\overset{\overset{\text{O}}{\|}}{\underset{\underset{\text{O}}{\|}}{\text{S}}}-\text{R}_2$$

where R_2 is alkyl, aralkyl, alkenyl, hydrogen, cycloalkyl, hydroxyalkyl, halogeno alkyl, and cycloheteryl. The cationic derivatives have the structural formula:

$$\text{low DE starch hydrolysate} -\text{O}-\text{Y}$$

where Y is

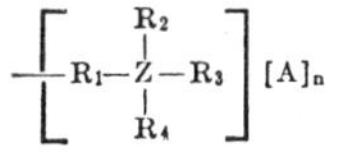

$$-\left[\text{R}_1-\overset{\overset{\textstyle R_2}{|}}{\underset{\underset{\textstyle R_4}{|}}{\text{Z}}}-\text{R}_3\right][\text{A}]_n$$

where R_1 is alkylene, hydroxy alkylene, halogeno alkylene, aralkylene, cycloalkylene, and phenylene; Z is sulfur, phosphorus or nitrogen; R_2, R_3 and R_4 individually are alkyl, aryl, aralkyl, hydrogen, cycloalkyl, hydroxyalkyl, halogeno alkyl and cycloheteryl; A is an anion; and n is 1 or 0; with the further provisos that when Z is sulfur, R_4 is absent, none of R_2 and R_3 is hydrogen, n is 1 and A anion; that when Z is nitrogen and at least one of R_2 and R_3 is hydrogen, then R_4 is absent and A when present is an acid; and that when Z is phosphorus, none of R_2, R_3 or R_4 is hydrogen, n is one and A is an anion. The anionic derivatives have the structural formula:

$$\text{low DE starch hydrolysate} -\text{O}-[\text{X}]\,[\text{M}]^+$$

where X is an organic acid residue and M is a metallic or organic cation; and the syrup has a solids content from 40 to 80%.

Example 1: Acetylated Low DE Starch Hydrolysate — Cornstarch was hydrolyzed in the conventional manner. The resulting low DE starch hydrolysate had about a 17 DE. A 30% solids solution of the above syrup was adjusted to pH 8.0 using 2 N sodium hydroxide. At room temperature 5.1 grams of acetic anhydride was added dropwise while maintaining the pH at about 8.0 by continuous addition of 2 N sodium hydroxide. The resultant product had a DS of 0.05.

Example 2: Sulfopropyl Ether Derivative of Low DE Starch Hydrolysate — One mol (162 grams, dry basis) of a 11 DE corn syrup was dissolved in 300 ml of water. To this was added 0.06 mol of sodium hydroxide and 0.05 mol of propane sultone. The solution was stirred overnight at 50°C. During the reaction the pH dropped from 11.3 to 6.4. The product had a DS of 0.05.

The following work illustrates the efficacy of the process. Low DE starch hydrolysates were derivatized with both acetic anhydride and propane sultone as above.

In the reactions of the starch hydrolysate with acetic anhydride a number of runs were made at different ratios of the mols of acetic anhydride per mol of anhydroglucose unit of the starch hydrolysate. Also, both corn and waxy milo starch hydrolysates of varying low DE's were reacted with the acetic anhydride.

With respect to the sulfopropyl ether derivative of the low DE starch hydrolysate as prepared in Example 2, the derivative was first concentrated to a 50% syrup solids. The syrup remained free of haze for 3 days compared to haze formation in about 2 hours for the control or nontreated syrup. The following table shows that derivatizing various low DE starch hydrolysate syrups with acetic anhydride in varying molar proportions measurably reduced haze formation upon standing compared to the control. Results are given below.

Effect of Acetylation on Haze Formation in Low DE Starch Hydrolysate Syrups

Starch Hydrolysate	Acetic Anhydride Mol/Mol Anhydroglucose Unit	Clear	Slight	Medium	Opaque
		- - - - - Haze Formation (days) - - - - -			
Corn 16–18 DE	None (control)	–	1 hr	5 hr	1
(50% syrup solids)	0.03	–	–	7 hr	3
	0.04	–	–	2	7
	0.05	–	1	6	8
Corn 11–12 DE	None (control)	–	–	–	1
(60% syrup solids)	0.03	–	3	5	7
	0.04	–	4	7	11
	0.05	8	9	12	15
	0.10	120	–	–	–
Waxy milo 11–12 DE	None (control)	1	–	2	5
(60% syrup solids)	0.01	2	3	5	6
	0.02	4	5	9	11
	0.03	7	10	16	18
	0.04	15	16	25	28
	0.05	35	39	46	53

DEFATTING LOW DE SYRUPS

In the preparation of starch conversion syrups, removal of residual fatty and proteinaceous materials is practiced to make subsequent decolorization more efficient or to make the syrups suitable for commercial uses in foods, adhesives, and the like. Commonly used methods of filtration and centrifugation at atmospheric pressures and temperatures below 170°F are difficult and result in a loss of carbohydrates. Furthermore, these conditions require the use of large size equipment, and, particularly in the case of centrifuges, is ineffective. It has been found by *I.F. Deaton; U.S. Patent 3,695,933; October 3, 1972; assigned to CPC International Inc.* that starch conversion products having exceedingly low dextrose equivalents (hereinafter referred to as DE) and which form aqueous syrups of improved clarity can be readily prepared by a process which does not require extensive refining steps.

In its broadest scope, the process comprises heating a starch conversion liquor and defatting the heated liquor. More specifically, the process comprises heating a starch conversion liquor having a solids content of from 10 to 50% by weight and a DE of from 1 to 20, on a dry basis, to a temperature of from 212° to 300°F; and defatting the heated liquor to a fat content below about 0.3% on a dry basis. In a preferred embodiment, the process comprises heating a starch conversion liquor having a solids content of from 10 to 50% by weight and a DE of from 2 to 15, on a dry basis, to a temperature from 212° to 300°F under superatmospheric pressure, defatting the heated liquor to a fat content below 0.2% by weight on a dry basis; and drying the resulting product.

Utilization of the preferred process results in a defatted starch conversion product having a DE of from 2 to 15 and a fat content below 0.2% by weight. The starch conversion liquors, which are used as the starting materials for the process are prepared by hydrolyzing starch to a DE of from 1 to 20 or, preferably, from 2 to 15. The initial starch, which is subjected to hydrolysate treatment, may be derived from a wide variety of starchy materials, such as cereal starches, waxy starches, and/or root starches. Typical of these groups are cornstarch, potato starch, tapioca starch, grain sorghum starch, waxy milo starch, waxy maize starch, and the like.

Example: A starch conversion liquor prepared by acid conversion having a solids content of 11.1% by weight and a DE of 10.9 on a dry basis was heated in an autoclave to a temperature of from 250° to 260°F for a period of 60 minutes. After this time, the fat was decanted by drawing off the conversion liquor from the bottom of the autoclave. This defatted liquor was then sprayed through a high pressure nozzle at 4,000 psig into a heated air chamber having an air inlet temperature of 330°F and an air outlet temperature of 251°F. The resulting product had a solids content of 97.3% by weight and a fat content of 0.03% by weight.

DEXTRINS AND LOW DE HYDROLYSATES IN PROCESSED FOODS

Dehydrated Alcohol Products

It has been previously taken for granted that it is not possible to remove only water from a solution of alcohol and water, because the alcohol is lower in boiling point and higher in volatility in comparison with water.

According to this process developed by *J. Sato and T. Kurusu; U.S. Patent 3,786,159; January 15, 1974; assigned to Sato Shokuhin Kogyo KK, Japan* solid matter such as powder or the like containing alcohol can be mass-produced by simple technical steps and because the product obtained is solid, it can be widely used in foodstuffs, medicines, confectionery or the like.

According to this process, a solution is formed of a mixture of alcohol, water and a water-soluble material, the water-soluble material being present in an amount more than 70% by weight as compared to the water and more than an equal amount by weight as compared to the alcohol, the solution is spray dried at the lowest possible temperature, and there is obtained a solid product in which almost all of the water is removed and the alcohol is covered with and held in the water-soluble material.

The water-soluble material may be gelatin or one or more modified starches which are easily water-soluble such as mizuame, i.e., starch converted syrup, dextrin, gelatinized oxidized starch, ester starch, ether starch or the like or a mixture of the modified starches and gelatin.

The alcohol component of the solution may be an alcoholic beverage such as wine. If, in this case, the alcoholic beverage contains a saccharide, protein, salt or the like, the amount of the water-soluble material to be added can be accordingly decreased. Namely, it is sufficient if the total amount of these materials previously contained in the alcoholic beverage and the water-soluble material which is to be added exceeds 70% by weight of the water component.

It is presumed that if the water-soluble material is added to and dissolved in the water in an amount of about 1.5 or more times as the water, almost all of the existing water is combined with the water-soluble material, whereby a homogeneous system comprising the three components of alcohol, water-soluble material and water becomes a two-component system comprising alcohol and water-containing water-soluble material, that is, one component being a combined body of water and water-soluble material. If this system is mixed and stirred, there is formed a dispersion system such as a kind of colloid solution in which the alcohol is dispersed in the combined body of water and water-soluble material.

If, this solution, which is a thick mixture solution, is sprayed, there are formed fine liquid drops in each of which the alcohol is located at the center surrounded by the combined body of water and water-soluble material. If such a drop is instantaneously dried, the water is evaporated and at the same time the water-soluble material is solidified by drying and forms a coating surrounding the alcohol whereby the alcohol is prevented from being evaporated and is held within the coating.

Example: In this experimental example gelatin and one powdered starch converted syrup (which will be called starch converted powder hereafter) are used as the water-soluble component and ethyl alcohol is used as the alcohol component. Into alcohol–aqueous solutions, each of which has the ratio of alcohol to water as shown in the following table, respective mixtures each comprising powder gelatin and starch converted powder in a mixing ratio of 1:1, are added and dissolved while being agitated in a mixing ratio of 130, 90, 70, 50 and 30% in relation to the amount of water.

Thereafter, each of the solutions is heated to 70°C within a tightly closed container without causing volatilization of the alcohol, and is then spray dried in a drying chamber at a temperature of 75°C. As a result thereof, there can be produced alcohol-containing powders as shown in the following table. The gelatin used is 25 mp (62/3%) in viscosity and has a 5% water content. The starch converted powder comprises 24% direct reducing sugar, 71% dextrin and 5% water.

	Number				
	1	2	3	4	5
Alcohol, kg	105.3	72.9	56.7	40.5	24.3
Water, kg	100	100	100	100	100
Powder gelatin, kg	65	45	35	25	15
Starch converted powder, kg	65	45	35	25	15
Ratio of total amount of powder gelatin and starch converted powder to water, %	130	90	70	50	30
Ratio of total amount of powder gelatin and starch converted powder to alcohol, %	123.5	123.5	123.5	123.5	123.5
Produced powder amount, kg	222.1	153.7	111.1	60.1	30
Alcohol content in produced powder, %	41.5	41.5	37.0	16.8	0
Alcohol remaining ratio, %	87.5	87.3	72.5	25.0	0

Here, the alcohol remaining ratio is obtained from the initial alcohol amount and the remaining alcohol amount (the amount of the powder produced multiplied by the percent of alcohol in the produced powder).

In this experimental example, dextrin obtained by enzymatic amylolysis of starch (hereinafter called enzymatic amylolysis dextrin) is used as the water-soluble material component. Into alcohol-aqueous solutions, each of which has the ratio of alcohol to water as shown in the following table, amounts of the enzymatic amylolysis dextrin (DE = 12%) of 130, 90, 70, 50 and 30% relative to the water amount are added and dissolved while being agitated. Then, the solution is spray dried at a chamber temperature of 75°C, whereby there can be obtained alcohol-containing powders as shown in the following table.

	Number				
	1	2	3	4	5
Alcohol, kg	65	45	35	25	15
Water, kg	100	100	100	100	100
Enzymatic amylolysis dextrin, kg	130	90	70	50	30
Ratio of dextrin to water, %	130	90	70	50	30
Ratio of dextrin to alcohol, %	200	200	200	200	200
Produced powder amount, kg	190.8	131.6	94.4	53.3	30.0
Alcohol content in produced powder, %	31.9	31.6	25.8	6.2	0.0
Alcohol remaining ratio, %	93.5	92.5	69.7	13.1	0.0

Rye, sake, brandy, beer and vodka are also included in other examples of the complete patent. The products can be used as flavoring materials in food products or to be rehydrated to alcoholic beverages. Oxidized starches, starch hydrolysates and starch derivatives (esters and ethers) may also be used in the process.

Dextrin-Sucrose Products

It has been found that a mixture of dextrin, which may include cyclodextrin, and sucrose when subjected to the action of cyclodextrin-glucosyl-transferase (CGT) in an aqueous medium produces predominantly an oligosaccharide (malto-oligosylfructose) in which glucose moieties of the dextrin chain are bound to the glucose moiety in the sucrose, the fructose moiety being terminal.

When an aqueous mixture of starch and sucrose is subjected to the action of cyclodextrin-glucosyl-transferase from *Bacillus maceras,* cyclodextrin is formed as a by-product, and the syrup obtained is turbid. The formation of cyclodextrin can be avoided by replacing starch with an oligosaccharide with a DE of 15 obtained by hydrolyzing the starch before exposing it to CGT. The syrup produced under such conditions, however, contains appreciable quantities of reducing sugars originating from the starch hydrolyzate, and the product is relatively unstable for this reason.

S. Okada, N. Tsuyama, M. Mitsuhashi and J. Ogasawara; U.S. Patent 3,819,484; June 25, 1974; assigned to Ken Hayashibara, Japan found that liquefied (partly hydrolyzed) starch, when subjected to CGT available from *Bacillus macerans* and *Bacillus megaterium* is converted to cyclodextrin and oligosaccharides, whereupon the reaction ends. When a mixture of soluble starch and sucrose is subjected to the same enzyme, cyclodextrin is initially produced, but thereafter is consumed completely, and there is ultimately produced a syrup in which oligosaccharide radicals are attached to the glucose moiety of the sucrose, and no significant amount of reducing sugar is left.

The enzymes used in this process are derived from strains of *B. macerans* and *B. megaterium,* such as *B. macerans* IFO 3490 and IAM 1227, 17-A strains which resemble *Bacillus macerans,* and *Bacillus megaterium* var strain T-5 (FERMP No. 935), T-10, T-26F, etc. Enzymes produced by other strains may also be employed. ATCC numbers of T-5, T-10 and T-26F are 21737, 21738 and 21739, respectively.

The effects of enzymes produced by different strains are illustrated in the following table. The enzyme of 17-A strains was obtained by diluting a culture broth having an activity of 350 u/ml (Tilden-Hudson) to 53 u/ml, and adding 1 ml of the diluted broth to the substrate per gram of starch. The enzyme of *Bacillus megaterium* var strain T-5 was prepared by diluting the broth 1.75 times, and the culture broth of T-26F was used undiluted for the same activity. Each fermentation mixture was adjusted to the optimum pH of 5.5.

All substrates contained 50% solids, the sucrose/starch ratios were as indicated, and the reaction periods were 2 days (approximately 48 hours) at 55°C in each case. The table lists the amount of reducing sugar in the product as a percentage fraction of total sugar present and the transfer ratio in percent. Under the conditions outlined above, cyclodextrin had disappeared completely. The transfer ratio was calculated by subjecting the fermentation mixture to assay by paper chromatography for fructose, and cyclodextrin was similarly isolated by paper chromatography and estimated from the color reaction with iodine.

Sucrose : Starch	17-A		T-26F		T-5	
	Red'g Sugar	Transfer Ratio	Red'g Sugar	Transfer Ratio	Red'g Sugar	Transfer Ratio
2 : 1	1.05	37.7	0.83	43.6	0.88	45.5
3 : 2	0.99	47.0	0.82	50.0	0.85	49.5
1 : 1	1.05	58.1	1.06	60.1	1.01	58.8
1 : 2	1.10	73.3	1.02	74.0	0.89	74.0
1 : 4	1.11	83.2	1.12	85.0	0.99	84.2
1 : 6	1.13	88.0	1.40	90.2	1.24	89.1
1 : 8	1.25	88.0	1.47	91.1	1.37	90.0

An initial mixture of sucrose and starch 1:1 produces a syrup which is intensely sweet. When higher viscosity and high thermal stability are more important than sweetness, as much as 10 parts of starch may be combined with one part of sucrose, all parts being by weight, unless otherwise stated explicitly. With a high starch:sucrose ratio, it is necessary to use more enzyme or to increase the reaction period for complete decomposition of cyclodextrin. It is most advantageous to maintain an initial solids concentration of about 50%.

The reaction temperature and pH are chosen according to the specific enzyme employed. A pH range of 4.5 to 8.0, and a temperature of 45° to 60°C are best for enzymes derived from strains of 17-A or T-5. The reaction period depends on the enzyme activity. It is preferred to select the enzyme concentration so that the reaction is completed within two days with complete disappearance of insoluble cyclodextrin. The solution obtained by the enzyme action is heated to inactivate the enzyme, and is then purified in a conventional manner before it is used in food. The use of active carbon followed by removal of cations by ion exchange is typical of the purification methods employed. A colorless, transparent, and odorless starch syrup is obtained and may be partly evaporated to increase its viscosity, or spray dried to produce a water-soluble powder.

Example: Twenty parts soluble starch and 20 parts powdered sucrose were dissolved in 60 parts water, and the solution was adjusted to pH 5.5 at 60°C. The enzyme salted out from a culture of 17-A strains was added to the hot solution in an amount of 53 Tilden-Hudson u/g of starch, and the mixture was incubated for 2 days, whereupon the cyclodextrin initially present in the solution had disappeared completely, as indicated by a negative iodine reaction.

The broth was decolorized with activated carbon, and then purified by passage over ion exchange resin at low temperature. The purified liquid so obtained was partly evaporated to a syrup which was colorless, transparent and odorless, and had an intensive sweetness characteristically different from that of sucrose syrup. The transfer ratio was found to exceed 60%. When the syrup was further evaporated by direct heating to drive off practically all water present, there were obtained colorless hard candies which were fully transparent, of low hygroscopicity, and a pleasant strong sweetness.

Examples are also included in the complete patent showing the use of such sweetening materials in condensed milk, hard candies and sponge cake.

Low DE Hydrolysates in Gum Confections

Gum confections are generally comprised of three basic constituents (a) a sweetening constituent, (b) starch, and (c) water. Other ingredients include various flavoring materials, preservatives, coloring agents and the like. The term sweetening constituent is used here to define those ingredients which are included to add sweetness to the gum confection. For most conventional gum confections, the sweetening constituent is sucrose and/or corn syrup. In special situations, other sweeteners, including artificial sweeteners, may be used.

Gum confections comprised of the above described three basic ingredients are usually formed by first cooking starch with water and the sweetening constituent until the starch is gelatinized, forming a liquid referred to as a syrup. The starch used may be any well-known starch, cornstarch being preferred. As stated above, the sweetening constituent is usually sucrose and/or corn syrup. The hot syrup so formed by the cooking step is next shaped and dried to set up a gel, whereupon a final product is produced. Even under the most rigorously and efficiently controlled commercial techniques used in the above process for making gum confections, drying times are excessively long. Such inordinately prolonged drying procedures not only reduce production but necessitate the use of excessive space and drying molds which could otherwise be recycled into the production line.

The gum confections produced by the method developed by *H.E. Horn and B.A. Kimball; U.S. Patent 3,582,359; June 1, 1971; assigned to CPC International Inc.* not only require materially shorter drying times but also exhibit acceptable and, in many instances, superior levels of firmness, resiliency, tenderness and strength. Basically, these gum confections comprise starch, water, a sweetening constituent, and a significant portion of a low DE starch hydrolysate. Conventional additives such as flavoring agents, preservatives, coloring agents and the like may also be included.

Generally speaking, these gum confections may be formed by using water, starch and other ingredients with the substitution of up to 70% by weight of the sweetening constituent with low DE starch hydrolysate such that the total weight of the sweetening constituent and the hydrolysate equals the amount of sweetening constituent previously used. Preferably the hydrolysate replaces from 6 to 60% by weight of the sweetening constituent and most preferably 12% by weight.

Example: The first gum confection was formulated from a standard and well accepted confection recipe using by weight composition, 47.4% corn syrup (82% solids), 31.2% sucrose, 10.4% (dry basis) cornstarch (67 fluidity), and 11.0% water. This confection is referred to in the following table as the control.

The second type of gum confection was formulated by replacing different amounts of the sucrose and/or corn syrup in the control composition with a starch hydrolysate produced by a one enzyme method from waxy milo starch and having a DE of 10. This confection is referred to in the following table as confection A.

The third type of gum confection (composition B in the table) was formulated by replacing 30% by weight of either the sucrose or the corn syrup in the control composition with a starch hydrolysate produced from cornstarch according to a two step enzyme hydrolysis and having a DE of 10 and a descriptive ratio greater than 2.0.

All confections were formulated by the same technique of initially dispersing the corn syrup in the water. Next the sucrose, and hydrolysate (if any) were added to the solution and thoroughly dispersed. The starch was then added and the resulting slurry mixed with constant agitation for 45 minutes at 185° to 195°F to insure complete mixing and wetting of the starch. The slurry was then cooked in a continuous cooker at 295°F and at a slurry feed rate of 0.2 gal/min. The cooked compositions at approximately 79% solids, were then placed in predried starch molds and were dried at room temperature until the gum confections were capable of being removed and sanded in a conventional manner. The following comparative data were recorded. The characteristics of resiliency, strength, firmness, and tenderness are rated on a four point system as follows: (1) poor, (2) fair, (3) good, and (4) excellent. The symbol "a" following a rating in the tenderness test signifies excessive skin toughness (suitable for the chewy type gum confection), the symbol "b" excessive core tenderness.

	Control	Confection A				Confection A				Confection B	Confection A		
Percent sucrose replacement (by wt)	0	15	30	45	60	0	0	0	0	30	0	15	0
Percent corn syrup replacement (by wt)	0	0	0	0	0	15	30	45	60	0	30	15	100
Percent deposited solids	78.9	79.0	79.5	80.5	78.7	78.4	79.2	79.5	78.7	79.5	79.2	79.5	77.7
Slurry tank temp, °F	190	190	190	195	185	195	185	195	190	185	190	190	190
Cook temp, °F	295	295	295	295	295	295	295	295	295	295	295	295	295
Drying time, hours	55	30	24	22	21	44	44	44	44	24	44	44	18
Comments:													
Resiliency	2	3	4	3	3	1	1	1	1	3	1	1	3
Strength	3	3	4	3	3	2	3	3	3	4	3	1	3
Firmness	2	3	4	3	3	2	2	2	2	3	2	2	3
Tenderness	3	3	4	1a	1a	1b	1b	1a	1a	3	1b	1b	1a

In all instances in the above data the drying times of the confections of this process were materially less than those of the control which used exactly the same confection formulation but without the partial replacement of its sweetening constituent with hydrolysate. In many instances, the confections of this process had drying times of 24 hours or less while the control had a drying time of 55 hours. Such improved drying characteristics result in substantial savings in time, money, production space, and tray equipment.

In many instances all or some of the characteristics of resiliency, strength, firmness and tenderness of these confections are superior to those of the control. In this respect the replacement of 30% of the sucrose in Confection A resulted in a gum confection product which was not only excellent, but superior in every way to the control. In those confections where the tenderness and resiliency tests are rated poor, it should be understood that though rated poor they are still quite acceptable and in some instances more desirable for some specialty gum confection products, such as the chewy type.

Low DE Hydrolysates plus Dextrose in Gum Confections

Dextrose in combination with a starch hydrolysate having a DE of 5 to 25 are used in the method by *M.M. Godzicki and B.A. Kimball; U.S. Patent 3,589,909; June 29, 1971; assigned to Corn Products Company* to replace sucrose in gum confections. Thus also reduces drying time and improves resiliency, strength, firmness, tenderness and skin formation.

Example: In order to demonstrate the superiority of the confections of this process various sample gum confections were made. The first gum confection was formulated from a standard and well accepted sucrose based confection recipe using by weight composition, 47.4% corn syrup (82% solids), 31.2% sucrose, 10.4% (dry basis) cornstarch (67 fluidity), and 11.0% water. This confection is referred to in the following table as the control.

The second type of gum confection was formulated by using different mixtures of dextrose and a starch hydrolysate produced by the one enzyme technique from waxy milo starch and having a DE of about 10. Different combinations were used as a total replacement for the sucrose of the control composition and all other ingredients remained the same.

All confections were formulated by the same technique of initially forming a solution of the corn syrup in the water. Next, the sucrose, and mixture of hydrolysate and dextrose, were dispersed in the solution and thoroughly mixed. The starch was then added and the resulting slurry was mixed with constant agitation for 45 minutes at a temperature ranging from 185° to 195°F to insure complete mixing and wetting of the starch. The slurry was then cooked in a continuous cooker at 295°F and at a slurry feed rate of 0.2 gal/min. The cooked compositions at approximately 79% solids, were then placed in predried starch molds and were dried at room temperature until the starch gum confections were capable of being removed from the starch molds and sanded in a conventional manner.

The following comparative data were recorded. The characteristics of resiliency, strength, firmness and tenderness were rated on a four point system as follows: (1) poor, (2) fair, (3) good, and (4) excellent. Skin formation was rated on a three point basis as follows: (a) excellent, (b) sticky, and (c) tough.

Sucrose (percent replaced)	0	100	100	100	100
hydrolysate (percent of hydrolysate and dextrose mixture)		80	60	40	20
Dextrose (percent of hydrolysate and dextrose mixture)		20	40	60	80
Slurry tank temp., ° F	190	180	180	180	180
Cook temp., ° F	295	295	295	295	295
Percent deposited solids	78.9	78.0	78.7	78.1	78.5
Drying time, hrs	55	22½	22	21½	20½
Product analysis:					
Resiliency	2	3	4.	4	3
Strength	3	3	4	3	3
Firmness	2	3	4	2	3
Tenderness	3	1	4	3	3
Skin	A	C	A	A	B

In all instances, the drying times of the gum confections of this process were less than half the drying times of the sucrose-based control which employed the same confection formulation except for its sweetening constituent. Such improved drying chracteristics result in substantial savings in time, money, production space, and tray equipment. All of the above confections of this process, furthermore exhibit superiority in one or more of their resiliency, strength, firmness, and tenderness characteristics. This coupled with such substantially improved drying times, makes them more desirable than the control. In this respect it is noted that the confection which used 80% hydrolysate and 20% dextrose exhibited a tough skin formation.

This confection is particularly suitable for the chewy type gum confection. However, of the five confections representative of this process, it may be the least desirable for most purposes. Particular attention is drawn to the gum confection which used as its sweetening constituent including hydrolysate, corn syrup along with a mixture of 40% dextrose and 60% hydrolysate. This product is superior in every way to the control and constitutes a preferred example of the compositions of this process.

Aerated Confections Containing Low DE Hydrolysates

Examples of aerated confections include nougats, marshmallows, macaroons, whipped toppings, icings, chiffons, angel food cakes, sponge cakes, whipped frozen desserts, fondant creams and the like. Examples of conventional whipping agents generally used in these confections include proteins such as vegetable, animal, and dairy proteins. Such proteins include albumin, gelatin, soy protein, casein, whey, other milk products and the like. Preferably the proteins used are colloids which are also water-soluble. Examples of conventional sweeteners used include sugars generally, and more specifically, usually include sucrose, dextrose, maltose, corn syrup, mixtures thereof and the like.

Improved whipped confections are produced more readily and economically by the process developed by *H.E. Horn, E.R. Jensen and B.A. Kimball; U.S. Patent 3,586,513; June 22, 1971; assigned to CPC International Inc.* The basic whipping agent compositions as contemplated by this process comprise a conventional whipping agent in combination with a low DE starch hydrolysate having a DE of 5 to 25. In practice, and when using these compositions in a whipped confection, the low DE starch hydrolysate is used to replace up to 70% by weight

of the conventional whipping agent. Preferably the hydrolysate replaces from
5 to 50% by weight of the conventional whipping agent and most preferably replaces 15% by weight of the conventional whipping agent.

Example: Various whipping agent compositions are formulated and aerated into
nougat frappes. The frappes (i.e., foams) are evaluated for density by the weighed
cup method in which the weight of foam in a container of known volume is measured. Foam stability is evaluated by the separation procedure which involves
placing the foam in a 6" diameter funnel and measuring the milliliters of liquid
separation after 30 minutes. The foams (frappes) are made by first reconstituting dry blends of spray dried egg whites and a low DE starch hydrolysate made
by a one step enzyme technique and having a DE of 10.

These solutions are then whipped as follows. A volume of 250 ml, of a solution
is placed in a Hobart mixer and blended for 1½ minutes at 107 rpm and 1½ minutes at 361 rpm. The foams are then evaluated using the above procedure with
the following results:

Foam Characteristics

Percent Replacement*	0	3	7	10	15	20	25
Foam density (g/ml)	0.0612	0.0617	0.0644	0.0664	0.0610	0.0659	0.0660
Foam height (in)	5.3	5.3	5.0	4.9	5.8	5.2	5.1
Foam separation (ml)	4.9	4.7	4.5	4.0	3.0	2.5	2.1

*Percent by weight, solids basis of hydrolysate replacement of egg whites. The foam
whipping agent compositions used consist essentially of egg whites and hydrolysate.

As can be seen, the addition of increasing amounts of hydrolysate as a replacement for egg whites in a frappe foam materially increases the stability of the
foam. This is evidenced by the decreasing amounts of liquid which separate from
the foam as indicated in the last line of the above table.

All foam densities and foam heights are acceptable and all foams produced may
be used in a nougat composition. However, since the starch hydrolysate is only
about one-tenth of the cost of egg whites, a significant economic advantage arises
through replacement of egg whites with hydrolysates. Further attention is directed to the results indicated for 15% replacement of egg whites with hydrolysate. In addition to providing 40% more stability to a foam than if only egg
whites are used, the foam whips more quickly and to a materially increased peak
volume and lower density (higher fluff). When this whipping agent composition
and foamed frappe made therefrom is used in nougats, substantially improved
nougat candies and centers result.

Low Calorie Sweetener

*W.H. Schmitt and R.A. Lukey; U.S. Patent 3,653,922; April 4, 1972; assigned
to Alberto-Culver Company* have disclosed a process for preparing pulverulent or
granular free-flowing water-soluble low calorie sweetening compositions having
the general appearance of sucrose. An aqueous composition or solution containing 60 to 80% solids in the form of a distinctly major proportion of a water-soluble starch hydrolysate having a DE up to 25 and a small amount of an essentially
noncaloric artificial sweetener are dried under atmospheric conditions on a double

drum dryer under controlled conditions. The essentially noncaloric artificial sweeteners are cyclamates and saccharin or mixtures thereof, generally in the form of salts such as the sodium, potassium, calcium and magnesium salts or other innocuous or pharmacologically acceptable salts. Essential to this process is the drying operation and Figure 7.1 shows a double drum dryer that can be used in this process.

FIGURE 7.1: DOUBLE DRUM DRYER FOR SWEETENER

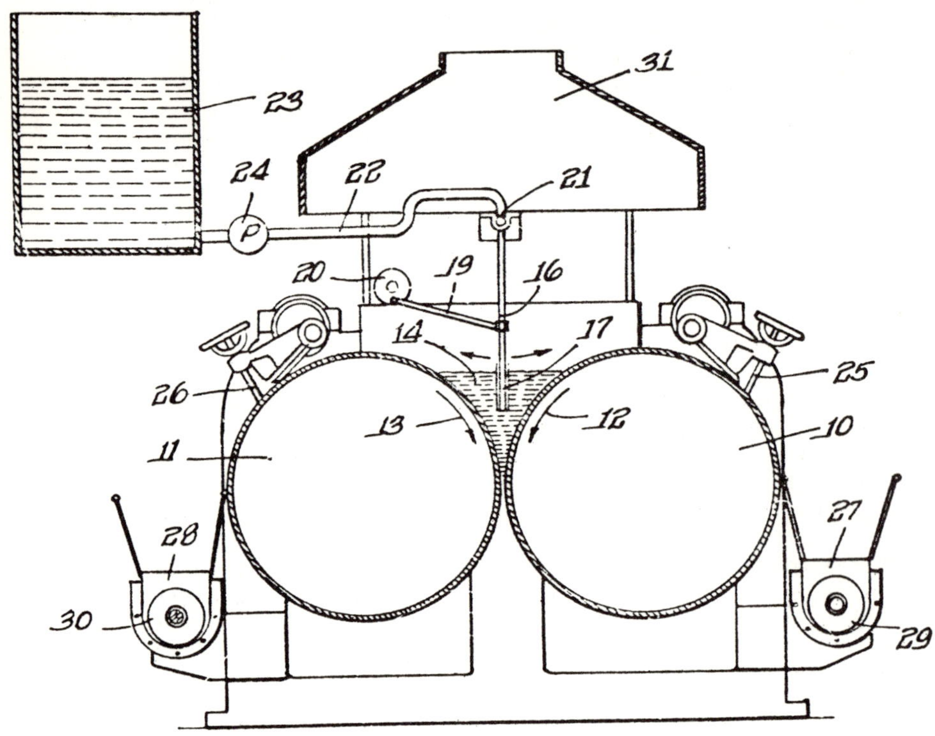

Source: W.H. Schmitt and R.A. Lukey; U.S. Patent 3,653,922; April 4, 1972

As shown illustratively and schematically in Figure 7.1, the dryer comprises a pair of internally steam heated cylindrical drums 10 and 11 suitably supported for rotation towards each other as indicated by arrows 12 and 13. A pool 14 of the aqueous starch hydrolysate-essentially noncaloric artificial sweetener composition is maintained in the space between the drums and the pool is maintained by feeding from the feed pipe 16, or by any other suitable means, preferably continuously, additional quantities of the aqueous composition. The feed pipe preferably dips into the pool and is attached to an agitator vane 17 having a plurality of perforations 18 therein, the feed pipe 17-agitator vane 18 assembly being adapted to be oscillated or moved in the pool by any suitable means, as, for instance, by a link **19** connected to the feed pipe **16** and operated by an eccentric **20** so as to perform the additional function of agitating or mixing the aqueous composition sufficiently to inhibit formation of a skin on the surface of the aqueous composition and/or to break such skin as may form.

The inhibition of skin formation and/or the breaking of skin formation is particularly advantageous since it brings about increased production or rates of production of the finished dried low calorie sweetening composition over what results where such skin inhibition or breaking is not effected. The feed pipe communicates with a pendulum feed structure **21** which, in turn, is fed through a flexible supply tube **22** from a reservoir **23**, containing the aqueous composition or solution to be dried, by means of a pump **24**.

The thin films of the aqueous composition which are picked up as a coating by each of the drums **10** and **11** become confluent in the pinch between the drums and the resulting confluent bead is pulled apart by the motion of the counter-rotating drums in their normal paths. This forms a disrupted hot film on each drum which is carried along and the water evaporated therefrom, producing a dried film which is removed by the scrapers or doctor knives **25** and **26** which ride on the respective drums. The dried sweetener composition is removed from the drums by scrapers or doctor knives in the general form of flakes having a moisture content which can be controlled within desired limits but is preferably in the range of 1 to 4%.

The dried flakes fall into troughs **27** and **28** and are conveyed by suitable screw conveyors or the like **29** and **30** and discharged into a mill or grinder (not shown) where the dried flakes are milled or ground to a desired particle size, preferably granules of a size of the order of ordinary granulated sugar. A conventional vapor hood **31** is provided to remove the vapors from the drying operation.

In order to obtain the full benefits of this process, it is important that there be a mechanical tearing action of the thin films concomitant with the simultaneous application of heat. While this is best achieved by means of a double drum dryer arrangement, it can also be accomplished by other means as, for instance, by a belt drying arrangement modified to bring about a compression of the dried film and the tearing or disrupting thereof in a drying environment in which the starch hydrolysate-substantially noncaloric artificial sweetener aqueous composition is applied to a drying surface so as to cause a substantially uniform temperature throughout the material being dried, such temperature being equal to or above that required to effect drying and providing a means for mechanically disrupting the interior of the drying material.

Th following are examples of aqueous compositions or solutions prior to drying in the manner and under the conditions of this process. The percentages stated are by weight.

Example 1: 72% milo starch hydrolysate (DE 11.5), 3% calcium cyclamate, and 25% water.

Example 2: 69.75% cornstarch hydrolysate (DE 19.5), 4.2% calcium cyclamate, 1.05% saccharin, and 25% water.

Example 3: 62% cornstarch hydrolysate (DE 12), 3% sodium cyclamate and 35% water.

Example 4: 64.4% cornstarch hydrolysate (DE 12), 4.48% sodium cyclamate, 1.12% sodium saccharin and 30% water.

Example 5: 69.05% milo starch hydrolysate (DE 11.5), 5.474% calcium cyclamate, 1.37% calcium saccharin, 0.076% potassium sorbate and 24% water.

Sucrose-Low DE Syrup Blends as Cereal Coatings

A.A. Lyall and C.N. Lundy; U.S. Patent 3,792,183; February 12, 1974; assigned to Nabisco, Inc. have disclosed a candy coating for cereal particles which is not susceptible to moisture pickup even under humid conditions.

The syrup containing, by weight, 60 to 85% sugar solids, is made up, and is applied to a cereal base without heating. The sugar solids comprise by weight, on a dry basis, from 80 to 68% sucrose and 20 to 32% low dextrose equivalent glucose solids. After coating, the product is subjected to substantial heat in a drying apparatus for 20 to 28 minutes with agitation of the coated cereal particles. A hard, glossy, nonhygroscopic transparent cereal coating is obtained which does not crystallize or decrease in glossiness even after extended shelf life.

Low dextrose equivalent (low DE) glucose solids are glucose solids having a dextrose equivalency of 15 to 28%, based on the total weight of glucose solids. Preferably the dextrose equivalency of the glucose solids is in the range of 24 to 28%, based on the total weight of such solids. Such solids are conveniently obtained by dehydrating low conversion corn syrups.

The first step in the coating operation is the preparation of the syrup. A normal batch size is 625 lb, consisting of 400 lb sugar, 100 lb low DE glucose solids and 125 lb water. The ingredients are placed in a steam jacketed stainless steel kettle equipped with agitator and heated to 180°F. When this temperature is reached the solution is then pumped through a heat exchanger, the hot end of which is at 240°F and the cold end of which is at 180°F. The syrup is continuously circulated through the heat exchanger until both the hot and cold ends stabilize at their respective temperatures (240° and 180°F). This normally requires a period of about 30 ± 5 minutes, for the batch size, proportions of ingredients, stated above, and the particular equipment used for this purpose.

However, the length of time during which the syrup solution is cycled through the heat exchanger is variable, since it depends upon a number of factors, such as the particular equipment used, syrup solids content, intake product temperature, steam pressure, etc. At this point the syrup solution is considered to be conditioned, i.e., the sugar and glucose solids are in the most minute state without caramelizing. The syrup solution is then pumped into a stainless steel steam jacketed holding kettle and held at 180°F under agitation.

Of course, the amounts of ingredients, and to some extent the proportions used as well, need not be those specified above; any desired amounts of sucrose solids and low DE glucose solids and water may be used in making up the syrup solution, as long as the prepared solution contains 60 to 85% by weight of sugar solids, and the latter are comprised of 68 to 80% by weight sucrose solids and 20 to 32% by weight low DE glucose solids.

The details of the cereal coating process and apparatus used, e.g., the Manley Coater (Model 2070) are given in the complete patent. After coating, it is essential that the coated product be subjected, at least initially, to a substantial amount of heat in order to maintain the gloss on the product and also to attain complete separation of the cereal particles during the drying stage. By a substantial amount of heat, in this context, is meant a temperature of at least 200°F. Since exposure of candy coated cereal particles to elevated temperatures for prolonged periods of time can result in damage to the product, such as by causing carmelizing of the coating, it is necessary that such exposure to elevated temperatures during drying be as short as possible, consistent with the requirement for drying of the syrup to a hard, transparent, glossy candy coating with a low moisture content (2 to 3%). This is achieved by carrying out the drying in a vertical turbo dryer such as the Wyssmont Model SR-20 dryer.

The circulation of heated air over the revolving trays causes some agitation, but most of the agitation of the cereal particles is effected by the 20 consecutive drops through radial slots from one tray to the next, as the coated cereal passes from the top to the bottom of the dryer. Regardless, however, of the amount of agitation, even to the point of product breakage, the high content of nonsucrose solids in the sugar syrup suppresses any tendency to crystallization, either during drying, or after drying is completed.

The temperature at which the drying is carried out ranges from 200° to 260°F, with the upper portion of the dryer being maintained, desirably, at a temperature of 220° to 260°F and the lower portion at a temperature of 200° to 250°F. The drying time is from 20 to 28 minutes, and preferably 26 minutes, based on a dryer output of 1,300 lb/hr of dry coated cereal particles having a final moisture content of 2%.

STARCH SYRUPS

Corn syrup and dextrose are used primarily in the baking, beverage (beer and malt liquor), canning, confectionery and dairy industries. They are most often used in combination with sucrose as supplementary sweeteners and to prevent the crystallization of the surcrose. Depending on the method and extent of hydrolysis, the physical properties of starch syrups can vary extensively. Syrups which have almost identical DE values can vary considerably in the actual dextrose content and the amounts of maltose (DP = 2), maltotriose (DP = 3) and higher oligosaccharides that are produced. One method of the classification of corn syrup is based on the DE values:

Conversion	DE
Low	28 – 38
Regular	38 – 48
Intermediate	48 – 58
High	59 – 68
Extra-high	Over 68

The syrups can also be classified by their method of preparation such as acid conversion syrups, enzyme conversion syrups, acid-enzyme conversion syrups or the high maltose syrups which are covered in the next chapter.

GLUCOAMYLASE PRODUCTION

Although the presence of starch hydrolyzing enzymes is widespread within the plant and animal kingdom, sources of microbiological origin are used most commonly in industry in the enzymatic saccharification of liquefied starch to form dextrose-containing syrups. The culture filtrates of *Aspergillus phoenicis, Aspergillus diastaticus, Aspergillus usamii* and *Aspergillus niger* produce excellent enzyme systems which hydrolyze liquefied starch to dextrose. Cultures of *Aspergillus niger* are particularly advantageous. These enzymes are referred to as glucoamylase or amyloglucosidase. The broth resulting from the fermentation of the above organisms generally contains several enzymes having different activities, some of which interfere with the production of dextrose when the enzyme preparation is employed to hydrolyze starch. Thus, for example, in the culture broth of *Aspergillus niger* three predominant enzyme systems have been identified, namely, alpha-amylase, glucoamylase (amyloglucosidase) and transglucosidase. Alpha-amylase attacks gelatinized starch by a random splitting of the starch molecule which reduces the molecular size appreciably and thus causes a desirable reduction in viscosity of the dispersion.

In contrast to the multichain action of alpha-amylase, the action of glucoamylase is thought to be a single-chain action where an enzyme molecule attaches to the dextrin and splits off one glucose unit at a time and thus can theoretically convert it quantitatively to dextrose. The action of gluocoamylase on dextrin polymers is much more specific at the alpha-1,4-gluocosidic bonds than the alpha-1,6-gluocosidic bonds in that it will cleave the former type bond approximately 30 times as fast as the latter type bond. One unit of glucoamylase saccharifies soluble starch at a rate equivalent to one gram dextrose per hour at 60°C and pH 4.3 providing not more than 25% of the substrate is saccharified.

The presence of transglucosidase with glucoamylase in enzyme preparations detracts from the potential yield of dextrose in the hydrolysate. Transglucosidase is known to catalyze transglucosylation reaction between dextrose, maltose and other intermediate saccharified products. As a result, upon completion of the saccharification reaction saccharides other than dextrose are still present in substantial amounts.

Recent methods for removing transglucosidase include: addition of clay minerals (U.S. Patent 3,042,584); use of sulfonated higher alcohols (U.S. Patent 3,067,108); use of magnesium oxide (U.S. Patent 3,108,928); lignin or tannic acid mixed with a sulfonated higher alcohol, an alkylated aromatic sulfonic acid or bis(2-ethylhexyl)phosphate (U.S. Patent 3,117,063); treatment with protease (U.S. Patent 3,268,417); inactivation by acid conditions (U.S. Patent 3,303,102); treatment with colloidal alumina monohydrate (U.S. Patent 3,332,851); use of ion exchange resins (U.S. Patent 3,335,066); the use of a combination of pH and carbonates (U.S. Patent 3,345,268); the use of a detergent (U.S. Patent 3,380,891); and the use of cation to precipitate oxalates (U.S. Patent 3,380,892).

Production Using Active Strains of *Aspergillus niger*

In the prior art, a selected or derived culture of Aspergillus has been combined with particular substrate and growth conditions to yield ferments of high glucoamylase enzyme activity. In accordance with the process described by *J.M. Van Lanen and M.B. Smith; U.S. Patent 3,418,211; December 24, 1968; assigned to Hiram Walker & Sons, Inc.* two particular very active Aspergillus cultures are combined with substrates which are preferably of the nature described for the production of glucoamylase and other enzymes essential to near-maximum rates of yeast fermentation and to near-complete conversion of starch to fermentable sugars.

In general, an enzyme system for saccharifying and fermenting grain mashes can be produced by this process in the following manner. In the preferred approach, a medium is prepared from ground cereal grain, other beneficial nutrient supplements identified below, and water, with the pH adjusted to 4.0 to 7.0. The medium is then sterilized and cooled to 80° to 95°F. Finally, the sterilized medium is inoculated with a growing culture of one of the highly active strains identified below under submerged, aerobic conditions (i.e., preferably vigorously aerating and agitating the medium for a period of 3 to 7 days at 85° to 95°F). To obtain the high enzyme potencies that characterize the process, the grain concentration should be relatively high as compared to prior methods and should preferably range between 12 and 20 grams of grain per 100 ml of medium.

The nutrient supplements that have been found to greatly increase the rate of glucoamylase production comprise grain stillage (the dealcoholized, liquid-grain residue from a previous grain alcohol fermentation) and distillers dried solubles (the product obtained by removing alcohol and grain screenings from grain stillage to produce thin grain stillage and then evaporating and drying the fraction). These supplements are used on a dry substance basis at the rate of 1 to 10% based on the total weight of the final medium, with the preferred level being 3 to 6%.

To achieve high yields of glucoamylase, aeration and agitation must be of such an intensity as to provide an excess (i.e., 3 ppm or more) of dissolved oxygen in the ferment throughout the fermentation cycle. This is accomplished by an aeration rate between 0.25 and 1.5 volumes of air per volume of medium per minutes and by agitation corresponding to that provided by a power input of from 0.5 to 2.5 horsepower per 100 gallons of medium. Pure culture conditions should be used at all stages of inoculum development and fermentation, and the air used should be sterilized.

Two strains of *Aspergillus niger* have been found to be most suitable for the saccharification and fermentation, viz, *Aspergillus niger* NRRL 3112 and *Aspergillus niger* NRRL 3122. These two cultures are mutants that were obtained by irradiation. When grown in the media and under the conditions described here ferments of these cultures are more active and efficacious than any previously utilized in the production of grain alcohol, as is more fully indicated by the following examples.

Example 1: This experiment deals with the glucoamylase yield of a high grain medium-*Aspergillus niger* 3112 combination. A stainless steel fermentor equipped with an air sparger and an agitator (430 rpm and 8" diameter) powered by a by a ¾ horsepower motor was used. The total capacity of the fermentor was 75 gal, and the normal medium volume was about 50 gal. A glucoamylase medium consisting of 100 lb of ground corn, 0.5 lb ground barley malt, and 40 gal of tap water was used. This medium was raised to 250°F by introducing steam into the fermentor. After holding at 250°F for 5 minutes, the medium was cooled to 150°F by cooling the outer surface of the fermentor. One lb of ground barley malt was added, and the temperature was held at 150°F for 10 minutes to allow for partial liquefaction of the cornstarch. The temperature was again raised to 250°F, and the medium was held at this temperature for 60 minutes, cooled to 94°F as previously described, and then was inoculated with 6 liters of a growing culture of *Aspergillus niger* 3112.

The indicated inoculum was prepared as follows. Spores from a stock spore culture grown on Czapek solution agar were transferred to two 500 ml Erlenmeyer flasks containing 100 ml of medium consisting of 4.5 grams of ground corn, 0.5 grams of ground barley malt, 0.25 grams of yeast extract, and 95 ml of tap water. This medium was adjusted to pH 5.0 with sulfuric acid before steriliza-

tion and was then sterilized. After 24 hours' incubation of the flask cultures
on a mechanical shaker at 84°F, each 100 ml culture was transferred aseptically
to 3 liters of medium in two 6 liter Florence flasks equipped with an air filter
and air sparger. The medium contained per flask 135 grams of ground corn,
1.5 grams of ground barley malt, 7.5 grams yeast extract, and 2,850 ml of tap
water. The pH was adjusted to 5.0 with sulfuric acid before sterilization and
was then sterilized. The flask cultures were incubated for 24 hours at 84°F
during which time sterile air was continuously sparged into the medium. After
inoculation of the fermentor with the contents of the 2 Florence flasks, the
agitator was turned on, sterile air was introduced at the rate of 2 cfm, and the
temperature was controlled at 94°F. Glucoamylase production and pH are shown
in the table below as a function of time.

Hours of Fermentation	pH	Glucoamylase, units/ml
Set*	5.6	–
18	4.4	0.3
46	3.2	2.4
73	3.0	4.3
90	2.9	5.3
114	2.9	6.7
144	2.8	7.9

*Set solids = 15.3%

Example 2: This experiment determines the effect of addition to the medium
of the unique nutrient supplements referred to above. A fermentation was
carried out in the same fermentor using the same culture (*Apergillus niger* 3112),
medium, inoculum, and fermentation conditions as in Example 1 except that
grain stillage (dealcoholized beer from a previous grain alcohol fermentation)
was used in place of the 40 gal of tap water. The stillage solids supplied was
about 3.5% of the total medium weight. The results of this glucoamylase fermentation are shown in the following table.

Hours of Fermentation	pH	Glucoamylase, units/ml
Set*	4.7	–
17	4.5	0.2
41	3.7	1.9
67	3.4	4.3
91	3.2	6.3
112	3.2	7.7
136	3.0	9.3
160	3.0	12.0

*Set solids = 18.8%

HIGH DE HYDROLYSATES USING ONE ENZYME

Using Alpha-Amylase from *Bacillus subtilis* Strain

Dextrose solutions having (DE's of up to 42 have been prepared by a one-step
procedure involving the treatment of an aqueous starch suspension with alpha-

amylase preparations obtained from various microorganisms including strains of *Bacillus subtilis.* Such a procedure is described in U.S. Patent 3,265,586. The dextrose content (D) of such solutions is typically approximately one half the DE of the solution.

Using the process disclosed by *L. Keay; U.S. Patent 3,622,454; November 23, 1971; assigned to Monsanto Company* solutions having dextrose contents approaching 100% (85 to 99%) may be obtained. In carrying out this process, an aqueous suspension of starch, typically containing up to about 50% starch dry substance, is treated with alpha-amylase prepared from the organism *Bacillus subtilis* var. *amylosacchariticus* for a period of time and under hydrolytic conditions to obtain a dextrose solution having a dextrose content at least 50%.

The alpha-amylase enzyme composition which has been found effective in this process is alpha-amylase (saccharifying) *Bacillus subtilis* var. *amylosacchariticus* FUKOMOTO. The temperature employed in the process may vary over a wide range with a temperature of 30° to 95°C being typically employed. Temperatures below 30°C are not preferred because of the slow rate of hydrolysis while temperatures over 95°C should be avoided to minimize inactivation of the alpha-amylase. A temperature of 60° to 90°C is generally preferred.

The pH of the aqueous starch suspension may also vary and is typically maintained from 5 to 8.0 and preferably from 6.0 to 7.0. The reaction time required in the process is a critical feature and in all cases must be sufficient to permit a dextrose content of at least 50 to be obtained. The time will vary and depend principally on the temperature of the starch suspension, the amount of enzyme used, the starch concentration in the suspension and the particular starch source (corn, wheat, rice, etc.) A reaction time in the range of 12 to 96 hours is generally employed.

The amount of alpha-amylase used will vary depending among other things, upon the concentration of starch in the suspension, the final dextrose content desired and the activity of the particular alpha-amylase sample used. It is generally preferred to include in the aqueous starch suspension small amounts of a material, or materials, for improving the stability of the alpha-amylase, especially at higher temperatures. Examples of such materials include water soluble salts of sodium such as for example, sodium chloride, and the like. In the examples the enzyme used is twice crystallized alpha-amylase prepared by fermentation of the organism *Bacillus subtilis* var. *amylosacchariticus.* The enzyme was obtained from Miles Laboratories Incorporated.

Example 1: A starch gel (aqueous suspension) was prepared by adding 35 grams of powdered pearl starch (corn) to 80 ml of boiling water. Sufficient boiling water was then added to give a total volume of 100 ml. The thus-formed gel was allowed to cool on standing to about 85° to 90°C. Alpha-amylase enzyme (25 mg) was then added with stirring to the gel. The gel liquefied immediately upon addition of the alpha-amylase. After addition of the enzyme, the resulting liquefied suspension was allowed to cool to about 70°C. The cooled suspension was then stirred while maintaining the temperature at approximately 70°C for about 44 hours to obtain a syrup having a DE of 96 and a D of 94.

Example 2: A starch gel was prepared by adding 20 grams powdered pearly starch (corn) to 90 ml of boiling water. Sufficient water was then added to give a volume of 100 ml. The gel was allowed to cool on standing to about 85°C. Alpha-amylase (5 mg) was added with stirring to the gel. The gel liquefied almost immediately upon addition of the alpha-amylase. After addition of enzyme, the resulting liquid suspension was allowed to cool to about 70°C. The cooled material was then stirred at approximately 70°C to obtain a syrup having a dextrose content of about 99.

Using Glucoamylase for Thinning and Saccharifying

L.P. Hayes; U.S. Patent 3,806,415; April 23, 1974; assigned to A.E. Staley Manufacturing Company has found that dextrose conversion syrups can be obtained from starch pastes without requiring an acid and/or alpha-amylase prethinning step. A substrate suitable for saccharification is obtained by initially treating a starch paste at a temperature in excess of 170°F with glucoamylase. The paste containing the glucoamylase is then vigorously agitated and cooled to provide a hydrolysate substrate suitable for direct saccharification to a dextrose conversion syrup. The method comprises the steps of:

(a) providing a starch paste from an aqueous slurry containing at least 20% by weight starch solids at a pH ranging from 3.5 to 9.0, by heating under superatomospheric conditions to a temperature of at least 250°F for a period of time and under conditions sufficient to provide a starch paste characterized as being essentially free from insoluble starch granules and having a DE of less than 2.0,

(b) initially treating the starch paste at a temperature of at least 170° to 210°F with an effective amount of a glucoamylase preparation sufficient to substantially reduce the viscosity of the paste,

(c) hydrolyzing the starch by vigorously agitating and cooling the treated starch paste to a temperature of less than 150°F with the period of time between initial treatment of the starch paste by the glucoamylase preparation and cooling to a temperature less than 150°F ranging from 3 to 40 minutes, and

(d) saccharifying the starch hydrolysate under conditions and for a period of time sufficient to provide a conversion syrup which contains on a hydrolysate solids weight basis dextrose as a principal hydrolysate constituent.

The following examples will illustrate the process in more detail.

Example 1: (A) Starch Paste Preparation — A starch slurry comprised of 74 parts by weight water and 26 parts by weight unmodified cornstarch (pearl starch) was prepared. The pH of the slurry was adjusted to 5.5 with 3.0 molar calcium hydroxide. The aqueous slurry was then pumped into a steam injection heater, maintained at 325°F and 96 psi (absolute) steam pressure. The resultant starch paste was collected in a 3" diameter tailpipe assembly, a retention zone, operatively connected to the steam injecter heater and adapted to maintain the paste therein at substantially the same pressure and temperature employed in the heater. The tailpipe assembly was vertically inclined at about 45° angle and provided with a ball valve in the lower portion of the tailpipe assembly and in close proximity to the entry point of the paste from the steam injection heater. After 5 minutes retention time in the tailpipe assembly, excess steam was intro-

duced into the tailpipe assembly chamber through a ball valve from a steam
source maintained at a pressure higher than that of the steam injection heater.
The excess steam expelled the starch paste at a high velocity through a pressure
regulation valve whereupon the paste was flash cooled to a temperature of about
208°F. The resultant paste had a dextrose equivalent of less than 0.5 and a
Brookfield viscosity of 1,400 cp at 150°F, 20 rpm with a No. 1 spindle.

(B) Starch Paste Hydrolysate Preparation — The flash cooled starch paste was
then adjusted to a pH of 4.0 with dilute hydrochloric acid and transferred to a
water-jacketed mixing vessel maintained at 160°F. While vigorously agitating the
paste with a high speed mixer at 12,000 rpm with a high shear and high lift
mixing blade, the starch paste was allowed to cool. When the starch paste had
been cooled to 175°F, it was initially treated with 2,000 units of glucoamylase
75 preparation for each 100 grams of starch paste solids.

Hydrolysis of the paste was allowed to continue under vigorous agitation. After
initiating the hydrolysis with the glucoamylase preparation, the viscosity of the
paste began to decrease rapidly. Within less than about 5 minutes, the starch
paste had cooled to less than 170°F. By regulating the temperature of the water
coolant, the hydrolysis was allowed to proceed under vigorous agitation conditions such that at about 10 minutes after the initial glucoamylase treatment,
the hydrolysate had been cooled to a temperature of less than about 150°F.

(C) Preparation of the Dextrose Conversion Syrup — The cooled paste hydrolysate
was then saccharified in a conventional dextrose fermenter at 140°F and pH 4.0
for 72 hours. The resultant conversion syrup contained 97.4% by weight dextrose (on a dry weight solids basis) without any evidence of retrograded starch.
The dextrose conversion syrup was then filtered through a conventional filter
medium with the syrup product indicating a rapid rate of filtration. The filtered
conversion syrup has an exceptionally high degree of clarity.

Example 2: Employing the apparatus and method of Example 1, starch pastes
of diverse viscosities were prepared. Each starch paste sample was held in the
retention zone for two minutes. Two runs (identified as Runs A and B) were
prepared employing steam injection heater temperatures of 325°F and 280°F
with pressures respectively of 96 and 49.5 psi. In runs A and B, excess steam
pressure was used to expel the product from the retention zone chamber 1
through the regulating valve orifice. Runs C and D, respectively corresponded to
Runs A and B with the exception that these runs were not expelled from the
retention zone with excess steam. Viscosities of the resultant starch pastes were
determined for Runs A, B, C and D and found respectively to be 1,400 cp,
2,200 cp, 8,000 cp and 76,000 cp.

The starch paste of Runs A through D were then subjected to the starch paste
hydrolysate preparation and dextrose conversion steps of Example 1. Runs A
and B provided conversion syrups having significantly greater dextrose yields
comparative to the conversion syrups prepared from the pastes of Runs C and
D (e.g., Run A contained 97.4% dextrose as opposed to 94.6% dextrose for
Run C). The conversion product derived from Run D exhibited an extremely
poor rate of filterability and significant amount of retrograded starch. The higher
dextrose yields for the conversion syrups prepared from the starch pastes of Runs
A and B was primarily due to the significantly lower viscosity which in turn
permitted more uniform dispersion of enzyme and hydrolysis of the starch paste

at temperatures in excess of 170°F. Apparently, the uses of excess steam coupled with additional high shearing effect of the regulating valve orifice upon the starch paste results in a substantial reduction in paste viscosity without imparting a substantial increase in starch paste DE.

OTHER ENYZME SACCHARIFICATION PROCESSES

Sugars from Hydroxypropylated Starch

According to a process developed by *C.C. Kesler, P.L. Carey and O.G. Wilson; U.S. Patent 3,505,110; April 7, 1970; assigned to Penick & Ford, Limited*, sugar products having desirable properties are prepared from hydroxypropylated starches. The starch is etherified with propylene oxide to a predetermined level of hydroxypropyl substitution. The hydroxypropylated starch is then hydrolyzed to produce a syrup having a specified dextrose equivalent (DE) and fermentables (FE). If desired, the syrup can be converted to a dry product by spray drying.

The resulting product either as a syrup or in dry form can advantageously be used as a substitute for ordinary sugars. There has long been a need for relatively nonfermentable sugars for use in special foods. There is evidence that amylase-resistant saccharides may be of value in controlling dental caries, and such products can also be used to reduce calorie intake. Since these saccharide products are highly resistant to the action of digestive enzymes, including salivary amylases, pancreatic enzymes, and animal diastases, there will be substantially no enzymatic digestion of the saccharides in the mouth or in the gastrointestinal tract. Furthermore, since the products are nontoxic and highly palatable, they should be more readily acceptable as sugar substitutes than artificial sweetening agents.

The hydrolysis of the starch derivative may be accomplished by the regular pressure and acid conversion process used to convert cornstarch to corn syrup or the starch may be liquefied with a liquefying enzyme and then treated with a saccharifying enzyme to produce a syrup containing sugar ethers. A combination of acid-conversion followed by enzyme saccharification may also be used. Suitable starch liquefying enzymes are bacterial amylases, such as, Rohm and Haas' Rhozyme H-39, Miles Chemical's Tenase or HT-1000, and Wallerstein's WC-8. Saccharifying enzymes of different types may be used including fungal amylases from *Aspergillus oryzae,* such as Rohm and Haas' Rhozyme K-2 and Miles Chemical's Dextrinase A; amyloglucosidases, such as, Miles Chemical's Diazyme L-30, or Rohm & Haas' Diastase 73, and Wallerstein's Amygase; and the saccharifying enzymes from malted grains.

While any of the standard hydrolysis procedures can be used, providing the hydrolysis is controlled, as will subsequently be described, it has been found preferable to use an enzyme-enzyme hydrolysis procedure, first treating the starch with a liquefying enzyme (e.g., bacterial amylase), and then with a saccharifying enzyme (e.g., fungal amylase). The hydrolysis is continued until the starch is converted to the form of a syrup. The hydrolysis may conveniently be carried out at a solids concentration at 25 to 40%. The resulting syrup can be concentrated to produce a final syrup containing from 60 to 80% solids, such as 70 to 75% solids. Alternatively, the hydrolysate can be spray or drum dried to produce a dry product. In practice, the hydrolysate product preferably has a DE of 2 to 25% and an FE of 15 to 30%. In some embodiments, the dextrose equivalent may range from 1 to 30% and the fermentables from 5 to 35%.

Example: 34 grams of NaOH were dissolved in 2,040 grams of isopropyl alcohol and 1,490 grams of cornstarch (10.1% moisture) were added with stirring. To the slurry were added 210 grams of water and then 390 grams of propylene oxide. The slurry was stirred at 123° to 125°F for 91 hours. The alkali was then neutralized with glacial acetic acid to a point that an aqueous solution of the starch had a pH of 5.5. The starch was filtered and washed by reslurrying with 3,000 ml of alcohol. The filtered starch was air dried. Analysis of the starch indicated 13.91% hydroxypropyl substitution.

To a 20 gallon stainless steel vessel equipped with heater, agitator and thermometer were charged 42½ lb of water. Over a period of an hour 20¼ lb of starch (made by the above procedure) and 7 grams of Tenase liquefying enzyme were added alternately.

The temperature was slowly increased during this period to 140°F. It was held at 140°F for one hour and then increased to 187°F. The solution was cooled to 140°F over a one hour period. Then 7 grams of Tenase were added and two of the heating and cooling cycles made followed by one hour at 180° to 186°F.

After cooling to 140°F the pH was adjusted to 4.8 with HCl and 80 grams of Diazyme L-30 (amyloglucosidase) added. The solution was placed into a 140°F oven for 86 hours. The syrup was heated with live steam at 210°F for ½ hour. The syrup was filtered through a mixture of Sil-Flo and Celite 110. The syrup was passed through a 2" x 24" column packed with granular decolorizing carbon, filtered to remove the carbon fines and then passed through two series of 2" x 24" cationic and anionic columns. The syrup was concentrated to 76% solids in a vacuum evaporator. The DE was approximately 24%, the fermentables (FE) 27%, and the product analyzed nil maltose. Digestion with salivary amylase for four hours at pH 6.8 and 37.5°C did not increase the glucose, maltose, or DE.

Three-Enzyme Saccharification Process

The disclosure by *T.L. Hurst, R.F. Larson and A.W. Turner; U.S. Patent 3,630,844; December 28, 1971; assigned to A.E. Staley Manufacturing Company* relates to an enzyme composition and to an improved method for producing starch conversion syrups having a high DE (dextrose equivalent), a high FE (fermentable extract) value and a limited D (actual dextrose dry substance basis) content. More particularly, this process is directed to an improved and more reliable method for consistently producing starch conversion syrups having a minimum DE value of 68% and a maximum D content of 47%.

These syrups are important to the brewing and baking industries primarily because they contain a relatively high concentration (above 77% and preferably above 80%) of fermentable saccharides (principally dextrose and maltose), yet remain clear and fluid under normal storage conditions. This process is characterized generally by its efficiency, reliability, reproducibility in either small or large scale operations, its adaptability to either batch type or continuous type conversions and equipment, and the ease with which the enzyme conversion product may be refined and finished off into finished syrup. These special syrups

can be conveniently prepared by the following steps and under the following
conditions: A slurry of starch is prepared by mixing granular starch with water
to a Baumé of between 20° to 25°. The starch slurry is then thinned to a DE
of preferably between 15 to 20, as by the use of an acid or an enzyme or by
combinations thereof. The thinned starch slurry is then adjusted, if necessary,
to a pH of between 4.0 and 6.5, preferably, to a pH of between 4.5 and 5.9,
and a solids content of between 20 and 55%.

The adjusted starch hydrolysate is saccharified by adding an enzyme composition
comprising a diastase or amylase such as alpha- and/or beta-amylase, refined glucoamylase substantially free of transglucosidase and amylo-1,6-glucosidase and maintaining the hydrolysate at a temperature of 128° to 132°F for 24 to 100 hours
or for a time sufficient to obtain a syrup having a DE of between 68 to 75%, a
minimum FE of about 77% and a maximum D content of about 47%. After
the desired syrup is obtained the enzymes are deactivated by heating the mixture to a temperature of between 165° to 200°F. Finally, the syrup is refined
and concentrated to a Baumé of 40° to 45° and a solids content of 78 to 85%.

One of the six steps above described, the most important is the saccharification
step. The saccharification step, that is the step wherein a thinned starch hydrolysate is converted to a syrup having a particular composition, comprises contacting a starch hydrolysate with an enzyme composition comprising (a) diastase
(a diastatic enzyme capable of hydrolyzing starch to maltose and higher saccharides)
such as beta- and/or alpha-amylase, (b) glucoamylase (an enzyme capable of hydrolyzing starch to dextrose) and (c) amylo-1,6-glucosidase (an enzyme capable
of hydrolyzing the amylopectin fraction of starch at its 1,6-glucosidic linkages).

Examples 1 through 10: To a 20° to 22°Bé slurry of cornstarch in water, 0.1%
of HCl based on the dry substance weight of the starch was added with stirring.
This proportion was obtained by mixing 1.16 lb of 20°Bé HCl with 100 gallons
of a 21°Bé starch slurry. The resulting pH of the slurry was 2.2 pH. The slurry
was converted to 15 to 19 DE using steam at 55 psig in a steam injection heater
to raise the temperature to 302°F which was maintained for 3 to 5 minutes.
After discharge to atmospheric pressure, the hot batch was neutralized to
4.5 to 5.0 pH by the addition of a solution of soda ash. This required approximately 0.15% DS of soda ash based on the DS weight of the syrup solids. A
soda ash solution containing 1.25 lb of soda ash per gallon is satisfactory. After
neutralization, the batch was cooled to 128° to 132°F before further processing.

The DS value of the cooled hydrolysate liquor was determined, and predetermined
amounts, as reported in Table 1, of highly active malted barley (Wallerstein's
malt amylase), glucoamylase, and amylo-1,6-glucosidase were added. The glucoamylase preparation (Rohm & Haas' Diastase 73) was diluted with water to about
100 u/ml and refined with 0.2% w/v of lignin at pH 4 to remove fungal spores,
color and transglucosidase.

The amylo-1,6-glucosidase was added as an aqueous solution. The amylo-1,6-
glucosidase was derived from the organism *Aerobacter aerogenes* by the process
reported in *Biochemische Zeitschrift*, Vol 334, pages 79 to 95 (1961). The malt
was added in dry powdered form. With the enzymes added, the starch slurry
was converted at 128° to 132°F and at 5.6 to 5.9 pH for 48 hours. At the end
of 48 hours, the enzyme conversion was arrested by sparging with steam sufficient
to raise the temperature to 175°F in 1 hour, which temperature was maintained

for about 1 hour. The syrup batch was then refined with resins and/or carbon and evaporated to a 43°Bé (82.5%DS). The syrups' D, DE and FE values for each of the enzyme combinations were determined and are reported below.

| | | Glucoamylase, | Amylo-1,6-Glucosidase, | - - - - - - - - - - -48 Hours - - - - - - - - - - | | | |
Example	Malt, %	units/g	units/g	Percent DE	Percent D	Percent FE	Percent D/DE
1	2.0	2.6	–	72.1	46.6	81.8	0.65
2	2.0	2.4	–	68.9	42.1	79.1	0.62
3	1.0	2.8	–	72.5	53.4	–	0.74
4	0.5	3.0	–	74.3	57.2	–	0.77
5	0.25	3.0	–	70.0	54.8	–	0.78
6	0.25	2.5	0.5	70.0	43.0	78.2	0.61
7	0.5	1.5	0.25	68.2	42.1	81.9	0.62
8	0.5	2.0	0.50	69.7	44.6	79.5	0.64
9	0.5	3.0	1.0	69.3	37.7	86.4	0.54
10	1.0	2.8	0.5	71.3	43.2	83.8	0.61

It can be seen from the above that, in the absence of amylo-1,6-glucosidase, about 2.0% malt in combination with glucoamylase is required if a syrup having DE and FE values around 70 and 80 respectively and a D content of 40 to 45 is to be obtained. It can further be seen that as the malt content (beta-amylase) present during saccharification is reduced (Examples 1 through 5) the dextrose content of the syrup increases. At a dextrose content of 45 to 47% syrup crystallization or setting up generally occurs.

However, as is demonstrated by Examples 6 through 10, the presence of amylo-1,6-glucosidase controls the production of dextrose even at lower malt concentrations. This is further demonstrated by the D/DE ratio also reported in the above table. A specialty type syrup having D/DE ratios of between 0.5 and 0.65 and FE values of about 80% is the type of syrup best suited for use in the brewing and baking industry.

Chlorine Dioxide Inactivation of Glucoamylase

A process for preparing dextrose containing syrups having a predictable dextrose content is disclosed by *R.G. Dworschack and C.A. Nelson; U.S. Patent 3,630,845; December 28, 1971; assigned to Standard Brands Incorporated.* During enzymatic hydrolysis of starch to dextrose, CIO_2 is incorporated into the hydrolysate to inactivate the enzyme when the dextrose content of the hydrolysate reaches a predetermined level.

The chlorine dioxide may be provided in the hydrolysate by introducing gaseous CIO_2, a water solution of CIO_2, a salt of chlorous acid or a combination of sodium chlorite and a peroxygen compound such as hydrogen peroxide, sodium peroxide and sodium percarbonate. In a preferred embodiment liquefied starch is enzymatically converted to a high fermentable hydrolysate, for instance having a dextrose content of from 40 to 45% and a maltose content of from 40 to 45%.

This enzymatic conversion may be accomplished by first treating a starch slurry with a starch liquefying enzyme and then with enzymes which produce maltose, for instance by the addition of distillers' barley malt, to obtain a high content of maltose, generally in the range of from 50 to 65%. Then a glucoamylase preparation, such as one derived from *Aspergillus niger,* is added and conversion allowed to proceed until a predetermined amount of dextrose is formed in the hydrolysate. ClO_2 is then provided in the hydrolysate to inactivate the enzymes.

Typically, the ClO_2 is provided in the hydrolysate when a level of from 40 to 45% dextrose and from 40 to 45% maltose is reached. Preferably, the glucoamylase should be inactivated when the dextrose content of the hydrolysate reaches a level of 43 to 45%. These hydrolysates may be filtered, refined and concentrated to obtain high fermentable syrups. These high fermentable syrups find particular application in the baking industry. The most desirable syrups for the baking industry from the standpoint of handling are those having a maximum dextrose content at or below that at which dextrose will crystallize out of solution.

Example 1: This example illustrates the use of various amounts of ClO_2 to substantially inactivate glucoamylase in cornstarch hydrolysates having various pH's. A slurry of cornstarch (about 29.6% dry substance) was liquefied with *Bacillus subtilis* alpha-amylase at 88°C and at a pH 6.8 to 7.0. The liquefied starch slurry was autoclaved, the temperature of the slurry lowered to 55°C and the pH adjusted to 5.7. 1% ground distillers' malt was added to the slurry, and hydrolysis carried out for 21 hours. A glucoamylase preparation produced by a culture of *Aspergillus awamorii* was then added at a level of 3 glucoamylase units per 100 g of substrate dry basis. The temperature was maintained at 55°C for an additional 48 hours at which time the DE was 66.3 and pH was 4.4.

The hydrolysate was divided into three 400 ml portions. The pH of two of them was adjusted with dilute H_2SO_4 to 3.0 and 3.5, respectively, the third was not adjusted. Each of the three portions was next divided into 4 portions of 100 ml. One of the 100 ml portions at each pH level served as a control. Into each of the others were introduced various amounts of ClO_2 shown in the following table. Then the hydrolysates were placed in a 55°C water bath for 24 hr and the DE determined.

pH	DE of hydrol-ysate [1]	ClO_2			
		None (DE)	60 p.p.m. (DE)	150 p.p.m. (DE)	300 p.p.m. (DE)
4.4	66.3				
4.4	66.3	75.8	74.8	73.5	69.7
3.5	66.3		74.1	70.4	67.8
3.0	66.3	74.4	73.7	68.2	67.5

[1] Prior to 25 hours in water bath at 55° C.

From the above table, it is apparent that as the amount of ClO_2 introduced was increased the glucoamylase was inactivated to a greater degree.

Example 2: This example illustrates the inactivation of glucoamylase in a cornstarch hydrolysate by in situ generation of ClO_2 from $NaClO_2$. An enzyme liquefied cornstarch slurry was prepared in the manner described in Example 1. 1% ground distillers' malt and 3 glucoamylase units per 100 grams dry substance were added. The liquor was maintained at 55°C and at a pH of 5.4 for 22 hours. The resulting hydrolysate had a DE of 50.2 and was divided into 4 portions.

The portions were treated in the manner shown in the following table and maintained at 55°C for 47.5 hours. The DE of the portions was determined after 23.5 hours and after 47.5 hours. The results of the experiment are shown in the table on the following page.

		23.5 Hours After Treatment	47.5 Hours After Treatment
Treatment	DE	DE	DE
Control, pH 4.5	50.2	59.9	68.2
Control, pH 3.0 adjusted with HCl	50.2	58.4	64.3
0.1% $NaClO_2$, pH 3.0	50.2	50.8	50.8
0.05% $NaClO_2$, pH 3.0	50.2	51.9	52.0

THINNING STEPS FOR HIGH DE SYRUPS

High-Low Temperature Steps Using Thermostable Enzymes

In many enzymatic processes for converting starch to a saccharide, a preliminary thinning step is carried out with either acid or alpha-amylase. The initial thinning step in such two-step methods is designed to provide a substrate which is more readily attacked by the subsequently employed product-specific enzymes than the starting starch.

M. Seidman and C.L. Royal; U.S. Patent 3,551,293; December 29, 1970; assigned to A.E. Staley Manufacturing Company have developed an improved enzyme thinning process done at two temperature levels. Briefly the process comprises initially pasting a starch slurry containing alpha-amylase by uniformly heating the starch mass for a short time period, e.g., about 15 seconds, at an elevated temperature at which alpha-amylase is unstable, e.g., about 215°F, and subsequently completing thinning of the resultant partially hydrolyzed starch mass at a temperature at which alpha-amylase is stable, e.g., 185°F.

Broadly, the process consists of enzymatically hydrolyzing starch where a thermostable alpha-amylase in an amount effective for thinning is introduced into an aqueous starch slurry in a mixing zone, the resultant slurry is passed in a continuous stream through a pasting zone in which the starch simultaneously is gelatinized by heat and partially hydrolyzed by the alpha-amylase, and the partial hydrolysate then is passed to a thinning zone where it is kept at a temperature at which alpha-amylase is essentially thermostable and above the gelatinization range of the starch and further hydrolyzed by (1) alpha-amylase introduced into the starting slurry or (2) a combination of alpha-amylase and thermostable alpha-amylase introduced into the starch subsequent to heating of the starch in the pasting zone to provide a resultant liquid digest of thinned starch; the improvements comprise:

(a)   Rapidly and uniformly heating the entering slurry to an elevated temperature up to 235°F and in the range at which the alpha-amylase is unstable in the pasting zone;

(b)   Maintaining the resultant hot starch at the elevated temperature for a time needed for the solubilizable polysaccharide content of the starch to be substantially solubilized, whereby a portion of the alpha-amylase is deactivated, and to provide a hot partial hydrolysate stream, which when cooled to the temperatures used in the thinning zone, undergoes a rapid reduction in viscosity;

(c)   Thereafter uniformly cooling the hot partial hydrolysate stream to a temperature in the range used in the thinning zone to provide a cooled starch hydrolysate which undergoes a rapid reduction in viscosity and which is essentially free of polysaccharides derived from the solubilizable polysaccharides in the starting starch which are incapable of conversion by alpha-amylase; and

(d)   Maintaining the cooled hydrolysate in the thinning zone at the thinning zone temperature to continue thinning and to provide the thinned-starch digest.

The above method is applicable in any process where thinning and solubilization of starch can be carried out or completed with alpha-amylase. The present basic technique, for example, is useful in the preparation of 6 to 18% DE complex carbohydrate solutions from unmodified starch and in the preparation of 26 to 30% DE syrups from partially acid-solubilized starch of less than about 16% DE. A further application of the present basic technique is in so-called dual-enzyme processes for producing simple and complex sugars employing an initial thinning of starch with alpha-amylase and a subsequent additional hydrolysis of the intermediate thinned starch with a saccharifying enzyme.

Example 1:   This example illustrates a typical application of the present method in dual-enzyme conversion of starch to dextrose. An aqueous 18°Bé (60°F) slurry of cornstarch was placed in a continuously stirred vessel, and based on the starch weight content, about 0.01% sodium bisulfite, about 0.06% dicalcium orthophosphate dihydrate, and about 0.36 unit per gram starch (about 0.144%) of a bacterial alpha-amylase enzyme preparation Aquazyme 120 were added to the agitated slurry. The pH of the slurry was adjusted to 6.6 by the addition of soda ash. The slurry was then pumped under 20 psig at a rate of 3.8-gallons per minute through a steam injection heater (jet heater) and heated to 215°F with 80 psig steam.

The resultant paste stream was held in a tail pipe connected to the jet heater at this temperature and a pressure of 5 to 7 psig for 15 seconds and then discharged into a vacuum flash cooler wherein it was flashed and instantly cooled to 195°F. The resultant solubilized liquid starch obtained in the flash cooler then was withdrawn and six 1,500-cc samples of it were placed separately in 2,000 cc stainless steel beakers which were located in a water bath at 195°F and adapted with stirrers driven by a common shaft. The samples were held in the water bath at 195°F for 45 minutes. The viscosities of the samples rapidly decreased during the first fifteen minutes to below 100 centipoises.

At the end of the 45-minute period the resultant thinned liquid samples were essentially completely solubilized, as evidenced by a negative iodine test for starch, and an autogenic separation of proteinaceous and fatty impurities. The thinned liquids had an average Brookfield viscosity of about 30 centipoises at 195°F and an average DE of about 5.0%. The thinned liquid samples were combined, the pH of the combined liquid was adjusted to about 4.0 with hydrochloric acid, and the liquid was heated in a stainless steel autoclave at 300°F for 10 minutes. Steam was injected into the autoclave through its cover during the heating period. At the end of the heating period the liquid was blown out with steam through a dip leg. The first 1,500 cc of liquid which was not diluted appreciably was collected

and cooled to 145°F in a water bath. About 9.5 units of refined glucoamylase
per gram dry solids in the heat-treated liquid then were added to the liquid, and
the mass maintained at 140°F and a pH of 4.0 for 72 hours. The resultant glucoamylase digest was characterized by about 5% by volume insolubles (corresponded
to a weight percent insolubles of 1.5%) and a layer filtration rate for 100 cc of
digest of 3 minutes 35 seconds. The filtrate obtained from the layer filtration
of the digest had a DE of about 100.0% and a dextrose content of about 97.5%.

Example 2: To demonstrate the effect of varying the pasting zone heater temperature in the present method, the dual-enzyme conversion procedure of Example 1 was repeated for several runs with the exception of, in the pasting zone,
heating the starting slurry in the steam jet to the various temperatures listed below in the table and determining the characteristics of the thinned digests and
glucoamylase digests obtained as in Example 1.

| Steam jet temp., °F. | Thinned digest (45 min. at 195° F.) | | Glucoamylase Digest | | | |
	Brookfield viscosity, cps.	D.E., percent	D.E., percent	D., percent	Volume percent insolubles	Filtration rate min./ 100 cc.
195	126	6. 5	100. 0	98. 6	22	ª 10:00
200	102	5. 9	99. 8	98. 6	7	5:53
210	41	5. 3	99. 9	98. 0	4	2:54
215	30	5. 0	100. 0	97. 5	5	3:35
220	34	4. 3	100. 0	99. 4	4	3:35
230	56	4. 0	97. 8	96. 9	11	10:19

ªFiltration rate shown is for 75 cc. of filtrate collected.

The data in the above table indicate that although the thinning technique of the
present method results in some loss of thinning enzyme activity (exemplified by
decreasing DE of thinned digest with increasing jet temperature), the method
surprisingly produces more uniformly converted thinned digests as indicated by
the glucoamylase digests prepared therefrom which contain lower amounts of
insolubles and are filterable at considerably faster rates than corresponding digests
obtained by the prior art processes.

*R.F. Larson and A.W. Turner; U.S. Patent 3,783,100; January 1, 1974; assigned
to A.E. Staley Manufacturing Company* have also prepared thinned hydrolysates
by initially hydrolyzing a partial hydrolysate of a DE less than 2.0 with alpha-
amylase at 200° to 212°F and continuing the hydrolysis at 185° to 200°F under
nonretrograding process conditions. The alpha-amylase requirements necessary
to provide the thinned hydrolysates are significantly reduced by the process. The
resultant thinned hydrolysates exhibit improved saccharification properties and
rapid filtration characteristics.

The method comprises the steps of partially hydrolyzing an aqueous starch slurry
maintained at a pH of 3.6 to 6.5 by heating under superatmospheric conditions
to a temperature of at least 250°F for a period of time and under conditions sufficient to provide a partial starch hydrolysate characterized as essentially free from
insoluble starch granules and a DE of less than 2.0, and further hydrolyzing the
partial starch hydrolysate by initially treating with an effective amount of alpha-
amylase and allowing the hydrolysis to proceed while maintaining a temperature
of at least 200°F. A thinned hydrolysate is then prepared by cooling the hydrolysate to a temperature within the range of at least 185°F to less than 200°F and

allowing the alpha-amylase to continue to hydrolyze the hydrolysate to a DE value of greater than about 5. Saccharified or conversion sugars prepared from the thinned hydrolysate herein exhibit significantly improved filtration rates over those sugars prepared from conventional enzyme-thinned hydrolysates. The significantly improved filtration rates are obtained because this method provides a thinned hydrolysate wherein the amount of retrograded starch during the thinning and subsequent saccharification is maintained at a nominal and/or essentially free, nonretrograded level. As a result, conversion syrups prepared from the thinned hydrolysates are essentially free from retrograded starch. The method optimizes both the effectiveness of the alpha-amylase during the thinning of the starch containing materials as well as the enzyme preparation used in saccharifying the thinned hydrolysates.

Example: (A) Preparation of the Partial Hydrolysate — An 18°Bé aqueous slurry of unmodified cornstarch at a pH of 4.2 was prepared. The slurry was pumped at a rate of 2.5 gal/min through a steam injection heater operated at 317°F under 75 psi steam (absolute). The resultant paste stream was retained, in a tail pipe connected to the steam heater, at about 317°F and 75 psi for 8 minutes. Immediately prior to the discharging the paste into a vacuum flash cooler, the paste was neutralized to a pH 6.5 with 3.1 M calcium hydroxide solution. The neutralized paste was then flash cooled to a temperature of 210°F. The resultant partial hydrolysate was essentially free from starch granules and retrograded starch. The viscosity of the hydrolysate was 600 cp as determined by Brookfield viscometer at 205°F with a No. 1 spindle operated at 20 rpm and found to have a DE of 0.4.

(B) Preparation of the Thinned Hydrolysate — The resultant partial hydrolysate was initially treated with 25 units of alpha-amylase per 100 grams of dry hydrolysate solids at a temperature of 206°F. By gradually cooling the hydrolysate under controlled conditions such that the hydrolysate had cooled to a temperature of 202°F, 198°F, 191°F, and 185°F, respectively at 5, 12, 23, and 180 minutes after initial treatment with alpha-amylase (Ban 120), a thinned hydrolysate characterized as having a DE of 9.0 was obtained. The thinned hydrolysate was substantially free from starch granules and retrograded starch as evidenced by a yellow color when a sample of the thinned hydrolysate was tested via the iodine starch test.

(C) Preparation of a Dextrose Conversion Syrup — In preparing the conversion syrup, there was used conventional dextrose conversion equipment with the syrup medium being maintained at 4.0 pH and 60°C. After 72 hours of saccharification with 900 units of glucoamylase per 100 grams of dry hydrolysate solids, the saccharification was terminated by carbon refining and filtration. The filterability of the conversion syrup was determined by adding a 400 cc sample of the syrup at 170°F to a Buchner funnel coated with a carbon layer (layer having been applied using 2.32 grams Darco S-51 carbon in 100 ml cc water on 9.0 cm Whatman No. 2 filter paper) and connected to a water aspirator and measruing the time required for the indicated volume to collect in a calibrated receiving flask. Upon the basis of recovered filtrate for a given period of time, the filtration rate of conversion syrup was ascertained as being equivalent 25 gal/hr/ft^2 of filter medium.

The resultant conversion product was analyzed for weight percent of insolubles by precoating a Buchner funnel with 5.0 grams of Hyflo filter aid, filtering 250 cc of the conversion syrup, washing the resultant cake with 50.0 cc water, drying the washed cake in an oven at 140° to 160°F, subtracting the weight of Hyflo from dried filter cake weight obtained, and dividing this value by the dry weight of solids in another 250 cc of the conversion syrup. The conversion syrup was also analyzed for volume percent insolubles by centrifuging at maximum speed on a Model 65828 H International laboratory centrifuge, a 15 cc sample of the conversion syrup in a 15 cc graduated centrifuge tube for 20 minutes.

The weight percent of insolubles test indicated 0.5% weight insoluble residue with the weight percent of insoluble starch being essentially zero. No detectable insolubles were observed by the volume percent insolubles test. The DE value of the conversion syrup was 98 and the percent dextrose solids was 96%.

Acid-Enzyme Thinning

An improved process for converting starch and starch-containing material into dextrose and maltose containing products having a high DE value is disclosed by *K.K.K. Kroyer; U.S. Patent 3,592,734; July 13, 1971*. This process gives an end product having a high DE-value without a simultaneous formation of products tending to cause difficulties in the filtration of the end product.

According to the process starch milk is subjected to a mild acid treatment under pressure and is at the same time heated to 100°C whereafter the acid treated product is neutralized and cooled to a pH-value and temperature suitable for liquefaction, and the product thus formed is subjected to an enzymatic lique-faction and finally to an enzymatic saccharification. As used here mild acid treatment defines a hydrolysis of the starch or polysaccharide such that the DE-value does not exceed 10 when the treatment has been completed. Such a mild acid treatment may be carried out either as a short treatment with a strong acid or as a longer treatment with a weaker acid. The acid treatment is preferably carried out for 5 minutes at a pH of 1.8 to 2.5 while heating to 140°C.

During the mild acid treatment an initial hydrolysis of the starch grains takes place thereby decreasing the viscosity of the heated starch milk so that it may be pumped through a tubular heat exchanger. During the heating of the starch grains to above 100°C the so-called crystallites, are broken down. The molecule chains or threads of these crystallites are not broken down during a normal en-zymatic liquefaction and it is believed that these nonconverted or only partially converted molecules cause the problems encountered during the filtration of the end products of the prior processes.

It is important that the cooling of the product subjected to the mild acid treatment is carried out quickly and that the liquefaction enzyme is added at once. This is due to the fact that starch tends to retrograde at a low temperature to form products which cannot be broken down during the enzymatic liquefaction and the subsequent enzymatic saccharification. Therefore, the cooling of the neu-tralized product is preferably carried out in a flash cyclone and liquefaction enzyme is added at the same time at the bottom of the cyclone. Thereby the liquefaction enzymes are thoroughly mixed with the cooled product and the enzymes begin to function immediately after the cooling to the liquefaction tem-perature. The enzymatic liquefaction is preferably carried out at a temperature of about 88°C using bacterial alpha-amylase while the product is passed through a converter pipe. When the liquefaction has been completed, the product is cooled to saccharification temperature, e.g., a temperature of about 60°C and is introduced into a saccharification tank to which the saccharification enzyme is added. In a preferred embodiment of the process, e.g. amyloglucosidase the acid treated product is subjected to a treatment at a temperature of above 100°C when the neutralization has taken place, but before the cooling and the addition of liquefaction enzyme. This results in a further decomposition of the crystallites without a further formation of reversion products.

The process can be described in further detail with reference to the accompanying Figure 8.1 which shows a flow diagram of the apparatus used. In Figure 8.1, 1 is a vessel containing starch milk and acid, which is supplied thereto through pipes 2 and 3, respectively. The vessel 1 is also provided with a stirrer 4 and is connected through a pipe 5, in which a diaphragm pump is inserted to a heat exchanger 7 of the type disclosed in British Patent 715,425. The heat exchanger 7 is provided with a pipe 8 for supplying heating steam and another pipe 9 for discharging noncondensed steam and condensate. The outlet end of the heat exchanger 7 is connected to a converter pipe 10, the other end of which is connected to a flash cyclone 11.

FIGURE 8.1: APPARATUS FOR STARCH CONVERSION

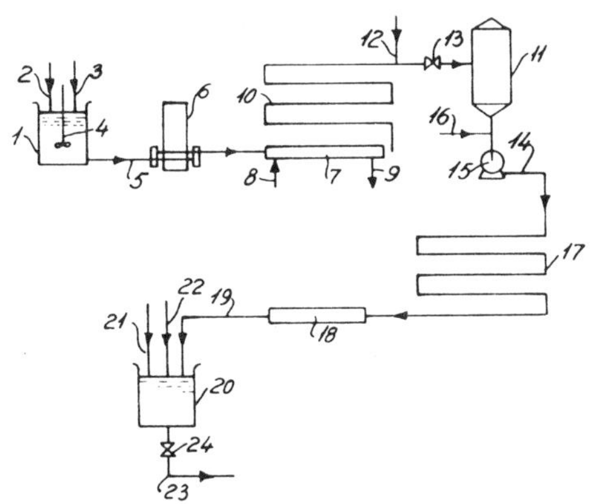

Source: K.K.K. Kroyer; U.S. Patent 3,592,734; July 13, 1971

In front of the flash cyclone 11 there is provided a pipe 12 for introducing a base into the converter pipe, and a back pressure valve 13. The lower end of the flash cyclone is connected to a pipe 14 in which a pump 15 is inserted. The pipe 14 is also provided with a branch pipe 16 for introducing liquefaction enzyme. The pipe 14 connects the flash cyclone with one end of another converter pipe 17 the other end of which is connected to a cooler 18 which in turn is connected through a pipe 19 to saccharification tank 20. The tank 20 is provided with pipes 20 and 21 for introducing saccharification enzyme and acid, respectively. A pipe 23 in which a valve 24 is inserted is connected to the bottom of the tank 20.

IN SITU HYDROLYSIS IN PROCESSED FOODS

Increased Fat-Absorptivity in Pet Foods Using Dual Enzymes

There are several factors which affect the shaping by extrusion of a fat containing amylaceous system. Two of the most important factors are the fat absorptiveness of the system and the amount of water in the system. If the particular system involved contains material which is highly fat absorptive and/or adsorptive, relatively high levels of fat may be introduced into the system and the material as a whole will retain its cohesive nature and therefore be capable of being shaped by extrusion. However, there is a point, regardless of the level of fat absorptiveness and/or adsorptiveness of the system, where all of the fat will be absorbed. The material then lacks cohesiveness and becomes hard to shape and form.

Sometimes it is possible to add water to the system to enable the forming and extrusion of the system when fat is present at levels high enough to begin to degrade the cohesiveness of the system as a whole. Water, however, may have undesirable effects on the starch in the material being extruded and the extrudate may be tough and hard to shape and form and may lack desirable eating qualities in other ways.

The process disclosed by *H.N. Dunning, E.H. Borochoff and H. Olevksy; U.S. Patent 3,595,666; July 27, 1971; assigned to General Mills, Inc.* is the conversion of starch in an amylaceous food material containing fat to its various degradation products by the simultaneous contacting of the amylaceous material with an alpha-amylase and amyloglucosidase. This conversion allows the fat containing amylaceous material to be more readily shaped by extrusion in situations where the fat level of the specific material in question is too high to allow for its shaping by conventional extrusion techniques.

The enzymes used in the process are alpha-amylase and amyloglucosidase. The alpha-amylase randomly attacks the alpha 1→4 bonds in starch or dextrin resulting in the fragmentation of both linear and branched fragments of starch, i.e., dextrinization. If sufficient time is allotted, under certain conditions, pure alpha-amylase can be made to convert linear glucose chains entirely to a mixture of maltose and dextrose, and branched chains to a mixture of maltose, dextrose, and pannose. (The latter being a trisaccharide containing an alpha 1→6 linkage.) Alpha-amylase is known as a liquefying enzyme because, as commonly used for dextrose conversion, it converts a starch slurry to dextrins thereby thinning it.

The second enzyme used in this process is an amyloglucosidase. Amyloglucosidase works directly from the nonreducing ends of starch chains, but splits off single glucose units primarily. It acts on alpha 1→4 linkages and alpha 1→6 linkages, the rate of reaction being about 20 times faster on the former than on the latter. The details and use of the products are shown in the following examples.

Example 1: This example illustrates the use of the modified starch in the production of a typical grain based pet food. The pet food contained the following ingredients (percent rounded off to the nearest 0.01). The ingredients are given in percent by weight of the final product.

Ground soybean oil meal	19.73
Wheat flour (70% starch)	41.28
Bone meal (steamed)	3.24
Whey	1.85
Propylene glycol	3.57
Soybean oil	1.21
NaCl	0.65
Vitamin and mineral mix	0.72
Color	0.01
Tallow (beef)	6.36
H_2O	21.28
Flavor (meat)	0.03

The procedure for preparing the pet food for extrusion comprises mixing all of
the ingredients with the exception of water and then mixing water with the rest
of the ingredients. Two lots were prepared in this manner with one lot having
0.008% of alpha-amylase and 0.008% of amyloglucosidase as measured by weight
of the final product. After mixing the samples were held at 80° to 85°F for 15
minutes. Each of the samples was placed in a 6 inch diameter water jacketed
extruder having two ½" x ½" orifices. Water was fed into the jacket at temperatures ranging from 150° to 160°F during the extrusion. A 2-bladed rotary
cutter was used to cut the extrudate into ¾" lengths. The extrusion was carried
out with the auger feeding mechanism set at 24 rpm and pressure developed was
300 psig.

While both samples came through the extruder orifice, the nonenzymatically
treated sample was crumbly, lacked cohesiveness and broke apart when handled
while the enzymatically treated product had excellent cohesiveness, plasticity
and resilience and was handled easily without any evidence of crumbling. The
temperature of extrusion was elevated, and this temperature was maintained
after extrusion to promote the conversion to dextrose in the final product because sweetness is desired in certain types of pet food.

Example 2: This example illustrates the application of this process to the snack
industry. A snack containing fat levels typical of certain snacks, is made without a frying step. Duplicate samples were made by mixing 200 grams of ground
puffed rice with 47 grams of coconut oil for 2 minutes. 53 grams of water was
added and mixing was continued for 2 minutes. 0.25 gram of alpha-amylase and
0.25 gram of amyloglucosidase was added to one of the samples with the water.
The dough was allowed to stand at about 70°F for 5 minutes and extrusion was
attempted. The sample without the enzymes was not satisfactory because it
lacked the cohesiveness necessary for forming when pushed through the orifice,
the enzymatically treated sample produced an excellent cohesive, plastic, resilient
extrudate. Extrusion was carried out under 250 psig using a fan shaped macaroni
die. The sample which has been contacted with the enzymes produced a fan
shaped configuration upon cutting.

Conversion of Starch in Pet Food Using Two Enzymes

Many pet foods on the market contain a high percentage of amylaceous material
derived from a grain or grain product such as flour or starch, some sugar and
added proteinaceous material. Proteinaceous material includes such things as
vegetable protein, animal protein, fish and meat by-products. The cereal based pet

food products often have some sweetener present to enhance palatability for the animal. In this regard, dextrose is often used. While dextrose is a lower level sweetener than sucrose, it is relatively inexpensive and there is some evidence that pets may actually prefer dextrose from a palatability standpoint. The addition of any sugar in a standard pet food operation is a difficult and costly one. The handling of the sugar along with the other ingredients is messy and difficult and is therefore desirable to avoid.

The process disclosed by *E.H. Borochoff, T.W. Craig and H.N. Dunning; U.S. Patent 3,617,300; November 2, 1971; assigned to General Mills, Inc.* enables the manufacturer to avoid the direct handling of sugar in combination with the other materials normally used in pet food manufacture. The preferred embodiment of the process comprises contacting a solid amylaceous system simultaneously with an alpha-amylase and an amyloglucosidase. However, the amyloglucosidase may be added before there is substantial dextrinization by the alpha-amylase but this makes an additional step in the process.

To accomplish the desired enzyme conversion the temperature after the enzyme has been included should not exceed the inactivation temperature of the enzyme, between 80° to 90°C. However, elevating the temperature beyond that point after the desired amount of dextrose has been produced is an excellent method for controlling the amount of conversion.

Other factors which influence the effect, the process and more specifically the rate of enzymatic conversion are the concentration of the enzyme, the moisture level of the solid amylaceous system, and the time as related to the temperature during conversion. In general, the higher the temperature up to the inactivation temperature, the faster the conversion and, of course, the longer the time that the enzyme is allowed to act, the higher the level of conversion in the final product. The particular choice of time and temperature may be based on the compatibility of these conditions with the processing conditions used in the normal manufacture of the particular product. The concentration of enzymes will also effect the amount and rate of conversion. Levels of each enzyme approximately equal to each other are preferred. In general, an enzyme concentration greater than 0.01% by weight of starch, and preferably an enzyme concentration level above 0.1% is used.

A solid amylaceous system converted by this process needs water at a level of 25% by weight of the product as a whole to obtain a reaction. A maximum conversion figure needs about 40% moisture.

Example 1: This example is designed to show the effect of time and enzyme concentration on the level of conversion. The corn used was corn grits containing 75% starch. The wheat was partially dehulled and contained 66.7% starch. A solid cohesive mass consisting of the grain containing 40% moisture, and an amylase and an amyloglucosidase was formed by adding enzymes and water to the finely ground grain to make doughs. Then temperatures sufficient to inactivate the enzymes were applied to the doughs. (The time referred to below is the time between enzyme addition and toasting). Five lots were made in this manner, three of corn, two of wheat. The results are shown in the following table. Reducing sugars as listed provide a fairly accurate measure of the level of dextrose. The percent of dextrose present in the reducing sugar varies between 70 to 90% as determined by paper chromatographic studies.

Lot No.	Grain	Enzyme Level*, %	Temperature, °C	Time, Hr	Total Sugar, %	Reducing Sugar**
1	Corn	0.5	60	4	49.0	33.0
2	Corn	0.25	60	18	51.4	50.0
3	Corn	0	60	16	3.1	2.7
4	Wheat	0.5	60	4	56.3	25.0
5	Wheat	0	60	4	8.8	1.1

*The level of each enzyme is expressed by weight of the grain.
**Percent by weight of final product.

Example 2: A typical grain based pet food was made using the ingredients listed below which are given in percent by weight. These numbers are rounded off to the nearest 0.01.

Soybean meal (solvent extracted)	19.34
Flour (about 70% starch)	40.87
Bone meal (steamed)	3.18
Whey	1.82
Propylene glycol	3.18
Soybean oil (crude, degummed)	1.18
NaCl	0.63
Vitamin and mineral mix	0.69
Red color	0.02
Beef tallow mixed with formula	2.86
Beef tallow added to surface	3.36
Meat by-products (dehydrated)	2.07
Water	20.84
Alpha-amylase	0.1
Amyloglucosidase	0.1

With the exception of the meat by-products, surface tallow, water and enzyme, the ingredients were mixed together in a ribbon blender. Water and enzymes were blended with the mixture and the mixture was then formed. The formed mass was cut and enrobed with tallow and dehydrated meat by-products. The product was split into two lots and held at room temperature. The first lot was held for 3 days and the second for 6. After holding total sugar was measured with the following results.

Fat	Total Sugar by Wt of Final Product
3 days hold	7.95
6 days hold	12.0

Self-Hydrolysis of Pureed Sweet Potatoes

Sweet potatoes when first harvested contain a relatively high proportion of starch to soluble solids in the form of sugars and the like. The freshly harvested potatoes are comprised mainly of cells which are rather dry, starchy and quite firm in texture and the potatoes lack sweetness and a good flavor. These adverse characteristics create problems in subsequent processing unless the roots are conditioned by curing and storing for a period of three weeks or longer. Moreover, some varieties of sweet potatoes never attain a degree of conditioning that will eliminate these adverse characteristics. *M.W. Hoover; U.S. Patent 3,404,074; October 22, 1968; assigned to North Carolina State University at Raleigh* has developed a process for producing from freshly harvested and starchy roots a high quality sweet potato puree having a low starch to soluble solids ratio. This process effects the hydrolysis of the starch in freshly harvested roots by activating and utilizing the naturally occurring saccharifying enzymes.

In carrying out this process, either freshly harvested, starchy sweet potatoes or cured sweet potatoes are washed, peeled and trimmed. This is a step common in the prior art. The trimmed roots are then conveyed to a comminutor which reduces the particle size to approximately 0.03 inch in diameter. It was found that a particle size of this diameter gives a fine texture which insures an intimate contact between the enzymes and the starch thereby hastening the process of reducing the starch to soluble solids. The particle size may be comminuted to a lesser diameter without affecting the process and may have a diameter as large as 0.06 inch; however, a diameter larger than 0.06 inch causes the end product to be too coarse.

After being comminuted to a puree, the total solids in the mixture ranges from 20 to 30% with the standard being 26%. The comminuting as described above takes place at room or ambient temperature after which it is force-fed by means of a sanitary pump to the first cooker. The first cooker comprises an inline constant flow apparatus in which the puree is force-fed through a Venturi-type restriction where steam of 20 to 50 lb/in^2 is injected. The steam pressure is adjusted to raise the puree to the desired temperature at a given flow rate. The rate of flow through this restriction is about 30 ft/min with the range being from 7 to 70 ft/min.

The steam instantly and thoroughly mixes with the puree, i.e., the potato particles, which reaches the optimum cooking temperature of 176°F in approximately 4 seconds; however, depending upon the design of the restriciton, it may take less than 4 or as much as 8 to 10 seconds to reach that temperature. As the puree travels from the first steam injection cooker, the temperature may range from 160° to 185°F. At temperatures below 160°F, the puree is not warm enough to activate the enzymes which is necessary to hydrolyze the starch. At temperatures above 186°F, the enzymes are killed and saccharification of the starch in the puree stops.

After being raised to a temperature of approximately 176°F, the puree is force-fed into a conversion tank whereupon the puree is held as a bulk mass and the starch continues to be converted into dextrins and soluble solids, the soluble solids containing trace amounts of dextrose and other similar sugars. By saccharification, approximately 46% of the total potato solids is converted into sugar, and since the sucrose and hexose comprise approximately 12% of the total solids, the total sugar content after saccharification is about 58%.

The remainder of the potato has been either converted into dextrins or remains as an insoluble solid; in other words, it has been hydrolyzed. The temperature in the conversion tank ranges from 140° to 190°F. Beyond these ranges, the temperature is either not high enough for proper saccharification or is so high that it kills the enzyme action in the puree. (Note here that the temperature which activates the enzyme activity is obtained substantially instantaneously, e.g., in 4 seconds. However, once the enzyme activity has started and is self-sustaining the temperature of the puree mass in the conversion tank may gradually drop to 140°F or gradually rise to 190°F and the enzyme activity will still sustain itself.) It is to be noted that at the end of the holding or converting step, the puree consists of from 14 to 20% soluble solids, soluble solids meaning the various sugars. Where the resulting puree does not contain 14% soluble solids, the resulting product is too light and fluffy and will not stick to the drum dryer whereas if the puree has a soluble solids content of above 20%, the mass is too sticky and cannot be adequately dried on the drum dryer nor can it be easily separated therefrom. Therefore, it is important to gauge the conversion of starches to soluble solids so that the puree leaving the conversion tank will be comprised of from 14 to 20% of soluble solids.

After the proper saccharification has taken place, the puree is then force-fed through a line where it enters a second steam inline injection cooker. The cooker is similar to the first with the exception that the temperature of the puree is raised to a minimum of 190°F and preferably to around 210°F. This temperature is sufficient to kill the enzymes and finish any remaining cooking thereby keeping the puree from having a raw or starchy flavor. The upper limits of this cooker is controlled by the degree of caramelization which may take place. Caramelization is the actual burning of the sugar which discolors the final product as well as making the same untasty. It is obvious that in this step the puree may be cooked under pressure or the like in order to complete the same; however, care must be taken not to overcook the puree or to localize the heat supplied thereto so that caramelization will be avoided.

Example: Sweet potatoes which were of the Nugget variety and had an average processing solids content were peeled, trimmed and washed whereupon they were comminuted through a screen having 0.03 inch opening thereby forming a puree. The comminuted material was then continuously pumped through a first steam inline injection cooker at a flow rate of approximately 60 ft/min where the temperature of the product was raised to 176°F in substantially 4 seconds. This represents the first cooking stage. After the first cooking stage, the comminuted potato was continuously fed into a conditioning tank in which it remained for 20 minutes to allow further saccharification of the starch and the occurring of other beneficial effects.

After the 20-minute hydrolyzation period, the puree was then pumped through a second steam inline injection cooker where the temperature of the product was rapidly raised to 212°F. The puree thus processed was in condition to be canned, frozen or dried by a conventional process. It was found that the total solids content consisted of 58% sugar and that the total soluble solids in the puree was 17% based on the weight thereof.

FRUCTOSE END GROUPS ON STARCH SYRUPS BY TRANSFER REACTIONS

Using Enzymes from *Bacillus macerans*

S. Okada, N. Tsuyama and K. Sugimoto; U.S. Patent 3,703,440; November 21, 1972; assigned to Hayashibara Company, Japan have prepared starch syrups which have oligo-glucosyl-fructose as their main constituents by subjecting the enzyme of *Bacillus macerans* to a solution of liquefied starch and sucrose. The oligo-glucosyl-fructose are sugars of oligosaccharides which are bonded with sucrose on their reducing ends in α-1,4-linkages. The formation of these oligosaccharides by subjecting mixture solutions of cyclodextrins and sucrose to the enzyme of *Bacillus macerans* is known (D. French et al, *JBC,* 76, 2387, 1953). However this is an equilibrium reaction and it is impossible to transfer the cyclodextrins completely into sucrose by the reaction. The unreacted cyclodextrins, which are difficult to dissolve in water and apt to cause turbidity of the starch syrup products, can be removed by the employment of organic solvents, such as trichloroethylene, which becomes a problem in the food processing industry.

In the production of oligo-glucosyl-fructose by adding the enzyme of *Bacillus macerans* to a mixture of liquefied starch and sucrose, the transferring of liquefied starch to sucrose could be accomplished without formation of cyclodextrins. Further, the sweetness and physical properties of the produced oligosaccharides are superior and had lower coloration, even by heating with protein or peptide, owing to their lower concentration degrees of reducing sugar than common starch syrups.

The enzyme of *Bacillus macerans* has been known many years as an enzyme to form cyclodextrins. The subjection of the enzyme on linear chained oligosaccharides, DP of over 10 caused intramolecular transfer, disintegrated the oligosaccharides to oligosaccharides with 6 glucose residues and then effected formation of cyclomolecules, i.e., cyclodextrins. However on the subjection to oligosaccharides with lower DP, less than 8, it was found that intramolecular transfer was not effected because of their lower DP, and with no formation of cyclodextrins. In the presence of desirable sugar acceptors, the formed oligo-saccharides transferred to the acceptors and a characterized transfer product was produced.

When the DE of starch obtained by liquefaction of starch employing enzyme is more than 10%, the enzyme of *Bacillus macerans* hardly effects the formation of cyclodextrins. However it was found that some amount of cyclodextrins are obtained as by-products when DE of the liquefied starch is 7%. Also in the case of acid liquefied starch with DE of over 10%, the action with *Bacillus macerans* hardly forms cyclodextrins, thus the liquefied starch is usable.

Since the reaction is a transfer reaction, reactions may be performed with starch concentration DS 5 to 45%. (DS represents dry substance.) In case the starch concentration is over DS 45%, the reaction of the enzyme of *Bacillus macerans* tends to be slightly inhibited.

Though the sucrose concentration depends on the employed starch concentration, the production of oligo-glucosyl-fructose may be performed with sucrose concentration in the range of DS 1 to 60%. However in case the liquefied starch concentration is high and the sucrose concentration is relatively low, the liquefied starch cannot completely transfer to sucrose, and has tendencies to hydrolysis and to slightly increase the amount of reducing sugars in the starch syrup products.

Using Alpha-Amylase from *Bacillus subtilis*

S. Okada and N. Tsuyama; U.S. Patent 3,701,714; October 31, 1972; assigned to Hayashibara Company, Japan have also found cheaper enzymes for the production of oligosyl fructose molecules using starch as the starting material. Employable enzymes in the process include saccharogenic alpha-amylase derived from *Bacillus subtilis* (BSA), alpha-amylase secreted generally from fungi, such as *Aspergillus niger, Rhizopus niveus,* and Taka-Diastase, alpha-amylase of *Endomycopsis* and pancreatic or salivary alpha-amylase. These enzymes are of the type that split or hydrolyze the second, third and fourth glucosidic bonds from the nonreducing ends upon subjecting on oligosaccharides. On the other hand, when any of the enzymes of bacterial liquefying alpha-amylase (BLA), malt or alpha-amylase of *Bacillus stearothermophilus* is subjected on oligosaccharides, the enzymes hydrolyze the fifth and sixth linkages from the nonreducing ends, but hardly effect sugar transfer. Thus the latter enzymes are not applicable in the process.

By subjecting mixtures of starch, sucrose or fructose to the actions of the specific alpha-amylase, one can obtain simultaneous hydrolysis of starch and transfer of the formed oligosaccharides into sucrose or fructose. Employable starches include potato, sweet potato, tapioca, corn, waxy-maize, wheat, and rice starches, or short chain amylase obtained by hydrolyzing amylopectin with isoamylases. One or more than one variety of the starches are admixed with sucrose or fructose, and then to the mixture is added the above described BSA, or fungal alpha-amylase or pancreatic alpha-amylase. The mixture is then liquefied at pH 5.0 to 7.0, at 60° to 90°C continuously or by batch method, then allowed to stand at 40° to 60°C. If necessary, enzyme may be further added subsequent to the liquefaction.

In case BSA or pancreatic amylase is used as the enzyme preparation, there is a jeopardy of causing a redecomposition of the once formed fructose-containing oligosaccharides into sucrose or fructose and malto-oligosaccharides. However, in case funal enzyme is employed G2F (2 glucose residues and a fructose molecule) accumulates without formation into sucrose or fructose caused by redecomposition. The saccharified and transferred sugar solution is decolorized with 1% of pulverized active carbon, filtered, and then deionized and decolored by strong acidic and weak basic ion exchangers. The liquid sugar like starch syrup may be prepared by condensing the product to moisture content of 15 to 25%. The product is colorless, transparent and possesses a relatively low viscosity. If preferable, the product can be pulverized by spray drying or by drying in vacuo.

Example 1: Strain of *Bacillus subtilis* var. *amylosacchariticus* (K2) was cultivated on a medium comprising 5% of soybean cake, 1% of ammonium phosphate, and 3% of starch for three days in a shaking incubator. The filtered culture broth (enzyme activity 40 units/ml) was used as an enzyme preparation (BSA).

To a mixture of 1 kg of starch and 500 grams of sucrose was added 4 liters of water and 1.25 ml of the above mentioned enzyme solution (total activities, 5,000 units), and heated in a boiling water bath with stirring to 80°C. At the completion of liquefaction the solution was cooled at 55°C and after being allowed to stand for 3 days the hydrolysis ratio became approximately 30%. To the reaction solution was added 1% of active carbon and boiled. Subsequent to filtration the solution was passed through ion exchanger layers, and then decolorized, deionized and concentrated in vacuo. Thus a starch syrup product was obtained. 1 gram of syrup was diluted with 20 ml of water. 0.01 ml of the dilution was spotted on the end of a sheet of filter paper, developed by using n-butanol, pyridine and H_2O (6:4:3, in volume) as a solvent, and then visualized with the employment of phloroglucinol as a colorating reagent. Thus spots that correspond to G2F, G3F, G4F, etc, besides sucrose were obtained.

Example 2: As described in Example 1, to a mixture comprising 1 kg of starch, 500 grams of sucrose and 4 liters of water was added 50 ml of culture broth (2,000 units) and liquefied by heating. When the temperature of the solution declined to 60°C, the solution was further incubated with a further addition of 30 ml of enzyme. Thus an excellent starch syrup as the one described in Example 1 was obtained.

MEMBRANE SEPARATION OF STARCH HYDROLYSATES

Generally, in commercial practice regular corn syrups are marketed at a DE of 43 ± 1.5 and are obtained by acid hydrolysis. Likewise, high corn syrups are marketed at a DE of 63.5 ± 1.5 and are generally obtained by a combination of acid and enzyme hydrolysis. It is a particular object to the process, disclosed by *C.T. Egger, G.B. Pfundstein and D.L. Gillenwater; U.S. Patent 3,668,007; June 6, 1972; assigned to Grain Processing Corporation* whereby regular corn syrups and high corn syrups are obtained simultaneously without the use of enzymes.

These two starch hydrolysates are obtained by acid hydrolysis alone without the products exhibiting undesirable bitterness or undesirable color. In accordance with this process starch is subjected to hydrolysis with an acid such as as hydrochloric or sulfuric. The syrup is then fractionated to obtain a low conversion portion and a high conversion portion. The fractionation is achieved by means of membrane separation. The known principle of this procedure is to apply pressure to transfer liquid or liquid containing relatively small molecules of solute through a semipermeable membrane while retaining the relatively larger molecules.

Various types of membranes are available which will permit certain solute materials to pass through the membrane into the diffusate while other solute materials are retained thereon as retentate. In this process a membrane is chosen which is capable of permitting a high conversion material (high DE material) to pass through the membrane into the diffusate while low conversion materials (low DE materials) are retained. The diffusate and retentate are then separately concentrated to yield, respectively, high and low DE syrups or the diffusate and retentate can be further processed to yield, respectively, high and low DE syrup solids.

In carrying out the process cornstarch, for example, is subjected to acid hydrolysis according to conventional techniques. Thus, for example, cornstarch is slurried in water, the solids concentration ranging from about 15° to 22°Bé, preferably 20° to 22°Bé and an acid such as sulfuric or hydrochloric acid is added. Hydrolysis is effected at 240° to 330°F, preferably 300° to 310°F, for a period sufficient to produce a medium conversion hydrolysate or syrup, i.e., one having a DE of 45 to 60. The hydrolytic action is terminated by neutralization.

After acid hydrolysis, the medium conversion syrup is subjected to membrane separation. If desired, the syrup prior to the membrane separation operation can be clarified by filtration or centrifugation to remove insolubles so as to prevent membrane fouling. Conventional membrane separation equipment which is designed for either low or high pressure operation can be employed and a semipermeable membrane is employed which permits the high conversion (high DE) material in the syrup to pass into the diffusate while retaining the low conversion (low DE) material in the syrup. Accordingly, the choice of membrane is routine based on these considerations. A suitable representative membrane is Permasep 3598-136-1.

After membrane separation the diffusate and the retentate can be concentrated (evaporated) by conventional procedures to obtain a high conversion syrup and a low conversion syrup, respectively, having desired viscosities. Alternatively, the diffusate and retentate can be processed to obtain syrup solids.

Example: A 30% solids corn syrup which was hydrolyzed with acid to a DE of 56 was subjected to membrane separation. The membrane used was Permasep 3598-136-1 available from the du Pont Chemical Company and a pressure of 500 psi was applied to the corn syrup. The retentate represented 45.6% of the initial syrup solids and assayed 43.1 DE. This syrup meets the usual specifications (43.5 ± 1.5) of a commercial regular corn syrup. The diffusate representing 54.4% of the initial syrup solids assayed 67.5 DE. Minimum blending of the diffusate with a lower DE syrup produced a corn syrup meeting the usual specifications (63.5 ± 1.5) of a high commercial corn syrup. The above procedure was repeated utilizing a 10% solids 56 DE acid hydrolyzed syrup. The diffusate in this case assayed at 63.9 DE while the retentate assayed at 40.3 DE which, with a minimum of blending, meets the usual specifications for a regular corn syrup.

PURIFICATION OF STARCH HYDROLYSATES

A method of purifying and decolorizing sugar liquors, such as starch hydrolysates has been disclosed by *F.L. Corson and D.R. Johnson; U.S. Patent 3,551,203; December 29, 1970; assigned to CPC International Inc.* This process includes the steps of passing these liquors through a plurality of filters whereby the downstream filter contains the freshest carbon filter cake. The filters are periodically substituted by cutting in downstream a fresh filter while simultaneously cutting out the filter that has been in service for the longest time. The filters contain cakes of finely divided powdered activated carbon and are of a critical depth, $\frac{1}{16}$ to 1.5 inches.

In order to make this new technique work to best advantage, the filters must contain cakes of powdered activated carbon. The particulate carbon must be of a subdivision such that at least 90%, and preferably 95%, of the particles are less than 100 mesh in diameter and at least 70% less than 325 mesh. In addition, each filter cake should have a thickness ranging from ¼ inch to 1 inch. If there is a substantial deviation from the particle size of carbon or from the cake thickness suggested here, considerable over-all process efficiency is lost even though the techniques of plural filters, periodic replacement of upstream filters, and repositioning of remaining filters are practiced, in conjunction with insertion of fresh carbon filter cake downstream in the series.

A typical useful carbon is Nuchar CEE. For example, this carbon is particularly useful when a filter cake is used having a thickness ranging from

0.5 to 1.0 inch of cake with each square foot of a 1 inch cake containing approximately 1.2 lb of carbon. Another useful carbon is supplied by Atlas Chemical Industries Incorporated as DARCO S-51. A material of this type is most preferably used when a cake thickness of 0.25 to 0.5 inch of cake is prepared. Here, one square foot of 0.5 inch cake is approximately equal to 1.2 lb of carbon.

Figure 8.2 is a simplified schematic drawing showing one arrangement of the equipment for use in the practice of one preferred embodiment of the process. The path of flow of the sugar liquor, that is to be treated, is generally from left to right, and is shown by the solid lines, with arrows indicating the direction of flow. Although the filter cakes do not move from one filter to another, the relative positions of the filters in the process changes, and the progression of the carbon filter cake, from use in one operation to use in a succeeding operation upstream, is shown by the broken lines, with arrows indicating the direction of the progression. For simplicity, the arrangement of piping and valves, that permits shifting the relative positions of the filters is not shown; it can readily be understood.

FIGURE 8.2: METHOD OF PURIFYING SUGAR LIQUORS

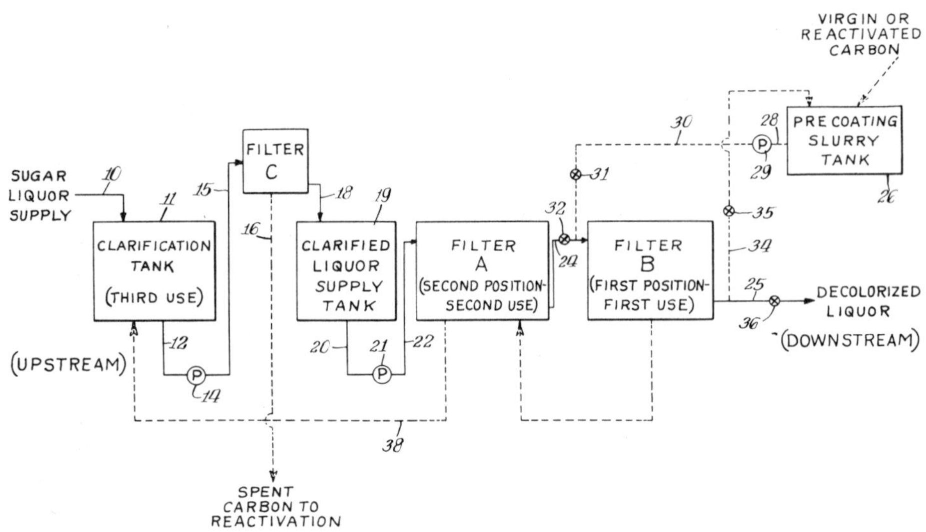

Source: F.L. Corson and D.R. Johnson; U.S. Patent 3,551,203; December 29, 1970

Referring now in detail to Figure 8.2, sugar liquor, such as corn syrup, is delivered through a sugar liquor supply line 10 to a clarification tank 11. A syrup-used carbon slurry is formed in the clarification tank using the used carbon filter A. The tank 11 is connected by a drain line 12 to a pump 14, which delivers slurry from the tank 11 through a line 15 to the filter C. The filter C may be of any desired type.

preferably a rotary drum filter with an endless cloth filter element. The filter **C** separates the carbon from the clarified liquor. The separated, spent carbon is sweetened off and is then discharged from the filter **C**, as indicated by the dashed line **16**, and may be reactivated or discarded. The clarified liquor is transferred through a discharge line **18** into a supply tank **19**.

The clarified liquor supply tank **19** is connected through a line **20** to a pump **21**, that delivers the liquor through a line **22** to the upstream decolorizing filter **A**. The filtrate from filter **A** flows through a line **24** to the downstream decolorizing filter **B**. The filtrate from filter **B** is discharged through a line **25** as decolorized liquor. To place the decolorizing filters in operation, initially, the precoating slurry tank **26** is used. Either virgin or reactivated carbon is slurried in this tank with water or decolorized liquor. To permit precoating of filter **B** with carbon, the slurry tank is connected through a line **28** with a pump **29** that discharges through a line **30** that connects to the inlet of filter **B**. A valve **31** is disposed in the line **30**, to close it off when it is not in use.

Similarly, a valve **32** is disposed in the line **24**, to permit the filter **B** to be isolated from filter **A**. A line **34** is connected to the outlet of filter **B**, to permit recirculation to the slurry tank **26** during precoating. A valve **35** is disposed in the line **34**, to permit closing of that line during normal operations. Similarly, a valve **36** is disposed in the line **25**, to permit closing that line during precoating.

It is preferred to use at least three decolorizing filters. Only two are shown in the drawing, for simplicity in presentation. The third filter is out of use while the other two are in use. The decolorizing operation follows the illustration in the drawing. When a fresh filter must be cut in, filter **A** is cut out. The necessary piping is conventional and is not shown, to simplify this explanation. The carbon filter cake from filter **A** is transferred, as indicated by the broken line **38**, to the clarification tank, for slurrying with the incoming sugar liquor in the tank **11**. The valve settings (not shown) are adjusted to place filter **B** in the upstream operating position formerly filled by filter **A**. A fresh filter (not shown) is cut in to place the fresh filter in the operating position previously filled by filter **B**.

To demonstrate the process, the system was started up by preparing a carbon slurry with supply sugar liquor, precoating the filters with carbon, filling the system with sugar liquor and heating the system to operating temperature. The tanks which held carbon slurries or supply liquors were agitated, and those tanks holding supply liquor were steam jacketed to bring and maintain the liquor at the processing temperature of 160°F.

Once the system was in operation, it was run long enough for the carbon in each of the decolorizing filters and slurries to reach the degree of color adsorption (exhaustion) that it would have in a conventional countercurrent adsorption system. This involved two filter changes, so that the third-use carbon, as used in clarification after use in filter **B** in the downstream position, then in filter **B** in the upstream position, was mixed in with the supply liquor after initial use as first-use and second-use carbon in decolorizing. Three cycle changes were run, before sampling, to insure this proper degree of carbon usage. Several additional cycles were run to obtain samples and data that would be representative of normal operation of the system. During the operation, the supply liquor was fed into a 50-gallon, steam-jacketed, agitated, round-bottom clarification tank.

Spent carbon was added, and the resulting slurry was pumped through the clarification filter **C**. The clarified filtrate was then fed into the supply tank **19**. The clarified filtrate was then pumped from the clarified liquor supply tank **19** through the decolorizing filter **A**. The filtrate from filter **A** was then passed through filter **B**, which contained a relatively fresh carbon filter cake. The filtrate from the final decolorization step was then evaluated.

Example: An acid converted corn syrup was used in the process. A 20°Bé supply liquor of a corn syrup, having a DE of 40 and a color of 5, was reduced to a final color of 0.35. A flow rate of 12 gal/hr/ft^2 filter area was employed through filter cakes containing 0.6 pound of carbon per square foot of filter cake. A total carbon dosage of 0.31 pound of carbon per 100 pounds of solids was used. The process is equally applicable to other starch syrups such as corn syrup having a DE ranging from 32 to about 65.

CORN SYRUPS IN PROCESSED FOODS

With Vegetable Meal in Animal Feeds

It is known that vegetable protein meals, e.g., corn (maize) gluten meal, in combination with carbohydrates, such as molasses, are useful in animal feeding. For example, it is known to mix 25% of molasses with certain feed products, but when corn gluten meal is mixed with such an amount of molasses, the resultant mixture is sticky and may set up into a hard cake which is difficult to handle. Moreover, there is a need for certain feedstuffs which contain a much higher amount of carbohydrate than 25%. For example, it has been found that as much as 50% of soluble carbohydrates combined with 50% of gluten meal improves animal feeds in certain respects. Crystalline sugars such as sucrose and dextrose are eminently satisfactory but too expensive for this purpose.

Starch hydrolysate liquors, e.g., mother liquors from the manufacture of crystalline dextrose, are particularly useful as soluble carbohydrates in feedstuffs, but heretofore no satisfactory method of mixing such carbohydrates with gluten meal in the desired amount has been found.

M.L.E. van Tittelboom; U.S. Patent 3,372,032; March 5, 1968; assigned to Corn Products Company has disclosed a process whereby 50 to 60%, dry basis, of starch hydrolysate liquors, may be mixed with vegetable protein meal to form a dry, powdered product. In carrying out this process, it is necessary to intimately mix the starch hydrolysate liquor, e.g., mother liquor from the manufacture of crystalline dextrose, throughout the protein meal, e.g., corn (maize) gluten meal. To accomplish this, for example, the liquor, in finely divided form, may be sprayed onto the gluten meal while it is being mixed or agitated slowly.

After all of the liquor has been added to the gluten meal, the resultant sticky mass is agitated slowly until crystallization of the dextrose has been completed. Inasmuch as heat is generated by crystallization of the dextrose, cooling means should be provided to keep the product at a temperature favorable to crystallization. One method of accomplishing this is to cool the top layer of the mass in movement in the mixer by means of air. The temperature of the starch hydrolysate liquor used should be high enough to prevent crystallization of dextrose and to permit it to be sprayed onto or mixed easily with the gluten meal. The temperature of the mixture of starch hydrolysate liquor and gluten meal should be within the range of 30° to 50°C to permit crystallization of the dextrose. A practical operating and preferred range is 38° to 40°C. When the temperature of the starch hydrolysate liquor is 70° to 80°C and that of the gluten meal is room temperature, the temperature of the resultant mixture is generally within the preferred range.

The moisture content of the mixture of gluten meal and starch hydrolysate liquor must be adjusted so that not more than 13% is present in order that the dry product may be formed. A practical operating moisture content of gluten meal is 9% and of the starch hydrolysate liquor 17% (about 43.5°Bé). When the greater amount of starch hydrolysate liquor specified is used, it may be necessary to dry the gluten meal to a lower moisture content than 9%. In addition to corn (maize) gluten meal, one may use sorghum meal or other vegetable protein meals such as commercial deoiled soya meal or deoiled peanut meal containing not more than about 10% oil.

Any starch hydrolysate liquor having a DE value of at least 75%, for example, 79 to 84%, and containing at least 50% of dextrose may be used. Particularly useful are the mother liquors, with or without reconversion, obtained from various stages in the manufacture of crystalline dextrose. The ratio of vegetable protein meal to starch hydrolysate liquor may vary from 40 to 50% (meal) to 60 to 50% (liquid), all on a dry substance basis, the preferred ratio being 45 to 50% (meal) to about 55 to 50% (liquor).

Example: 4,500 kg of corn gluten meal having a moisture content of 9% was placed in a vertical conveyor mixer (10,000 liters volume) operating at 60 rpm. A pipe with a spray nozzle was attached to the bottom part of the mobile arm (operating at 1.33 rpm) in order to introduce the starch hydrolysate liquor under pressure and spray the same onto the gluten meal. Five thousand kilograms of mother liquor from the manufacture of crystallization of dextrose which had been reconverted to a DE value of 83 to 84% and having a density of 43.5°Bé (83% dry substance) was sprayed onto the gluten meal while the same was slowly agitated. The temperature of the liquor was 80°C.

The spraying was done in two stages, although this is not necessary. In the first stage 3,250 kg of the liquor was sprayed onto the gluten meal while it was agitated for 1.5 hours. Then, the resultant mixture was agitated for 0.5 hours Next, 1,250 kg of liquor was sprayed onto the mixture during a period of 1 hour. At the end of this period, the mixture was a sticky mass. Agitation of the mixture was continued for 19 hours, the temperature of the mixture being maintained at 38° to 40°C, cool air being introduced as required at the top of the mixer so that the top layer of the mixture was exposed to the cool air. At the end of this period, the sticky mass had changed into a dry powdery mass which was cooled, screened, and packed. The moisture content of the product was about 13%, hence no drying was required. The tailings from the screening operation can be ground and recycled, e.g., mixed with the crystallized product at the outlet of the mixer.

Fondant from Sucrose and Corn Syrups

Mixtures of sugar and sugar syrups, generally known as soft sugars, are important compositions in commerce. Two principal examples are the brown sugars and the dry sugar mixtures used in the preparation of fondants. Brown sugar is a mixture of sucrose crystals and molasses, the latter being a crude syrup containing sugars (a portion of which are of the noncrystallizing type), impurities and water. Fondants may be broadly described as mixtures of extremely fine crystalline sucrose particles and syrups which contain noncrystallizing sugars. Fondants are useful as icings, chocolate cream centers, fudges and the like. They are characterized by being extremely smooth in texture as a result of the fine particle size of the sugar cystals. An important function of the syrups which are contained in the fondants is to prevent crystal growth and the accompanying loss of the characteristic smooth texture.

Both the brown sugars and the dry fondant mixes are extremely hygroscopic and and as a result tend to cake severely when stored for long periods of time. Attempts to solve this problem by incorporating anticaking agents such as starch, tricalcium phosphate and the like. The disadvantage of this approach are obvious. *R.G. Gidlow, J.A. Stein, W.L. Ganske and A.M. Zenzes; U.S. Patent 3,506,457; April 14, 1970; assigned to The Pillsbury Company* have now developed a process for preparing soft sugars, particularly brown sugars and fondant mixtures, which remain noncaking, free-flowing and readily dispersible over relatively long periods of time in storage. This is obtained by:

(a)   forming a mixture of fractured sugar crystals and syrup, the syrup coating substantially the entire surface area of the fractured sugar crystals, and the weight ratio of syrup to fractured sugar crystals being greater than the sugar crystals could accommodate prior to fracturing and yet be free-flowing, but no greater than the fractured sugar crystals can accommodate and yet be free-flowing; and
(b)   forming porous agglomerates from the resultant mixture by adding moisture to increase the tackiness of the surfaces of the individual coated particles and agitating the tacky particles to form agglomerates.

The term sugar crystals as used here means dextrose, sucrose, and similar sugars or their mixtures in the crystalline form of the type resulting from conventional sugar refining processes. The term fractured sugar crystals as used here means the product resulting from comminution of sugar crystals. The term syrup as used here means aqueous solutions of sugars, at least a portion of which do not crystallize or are very difficult to crystallize. The most commercially important examples of syrups containing noncrystallizing sugars are invert sugar syrups, molasses, and glucose syrups, the latter term being used to describe corn syrup rather than aqueous solutions of pure dextrose. The term agglomerate as used here refers to a stable aggregate of small individual particles held together at points of contact between contiguous particles, and defining a multiplicity of voids and interstices which extend throughout the aggregate. The process is illustrated using corn syrup as the syrup.

Example: Instant Agglomerated Fondant Mix (Sucrose and Corn Syrup) —
The following is a typical example of the preparation of a stable dry free-flowing agglomerated fondant mix made by this process. A mixture consisting of 85% sucrose and 15% liquid corn syrup (43°Bé) was prepared and subjected to a fine grinding operation reducing the sucrose to a size of less than 40 microns. The fine pulverulent sucrose particles coated with a thin film of the corn syrup were then agglomerated in apparatus similar to that described in U.S. Patent 2,995,773.

The agglomerating section was 18 inches long and the agglomerator was vibrated with a ½ inch stroke at a frequency of 750 cycles per minute. The pulverulent coated particles were fed to the agglomerating section at a rate of 500 lb/hr. The fluidizing and moisturizing humid air was passed upwardly through the bed of pulverulent material in the agglomeration section at a rate of 98 cfm, the incoming humid air having a wet bulb temperature of 160°F, and a dry bulb temperature of 300°F. The drying section was 81 inches long, and drying air at 270°F was fed through the agglomerates in the drying section at 196 cfm.

The major portion of the agglomerates was smaller than 10 mesh and larger than 40 mesh in size. The agglomerates were of good quality and were free-flowing noncaking and readily dispersible and soluble in water, and they exhibited the typical characteristics of agglomerated products of this process. The agglomerates also formed a fondant comparable in quality to a boiled fondant when reconstituted with the required amount of water.

Corn Syrup-Iron Compositions

Among the best sources of iron for food uses are salts such as ferrous sulfate, ferric ammonium citrate, ferrous fumarate, ferric choline citrate, and ferrous gluconate. Commercial preparations of these compounds include the relatively insoluble $FeSO_4 \cdot 1.2H_2O$ (32.1% Fe). When the ferrous sulfate compound is combined with wheat flour, off flavors develop during storage. Therefore, it is usually added during the baking process at the dough mixing stage in tablet form. The tablets, added at the rate of one per 100 pounds of flour, are broken up by hand which causes some doubt as to the completeness of distribution. This method is not easily adaptable to automation. For baking and other food processing, a liquid iron composition would have great advantages. However, present commercial liquid preparations are not particularly suitable for food uses.

A method for preparing highly stable, liquid, iron-fortifying compositions containing up to 1% by weight of ferrous or ferric ions is disclosed by *G.N. Bookwalter; U.S. Patent 3,809,773; May 7, 1974; assigned to U.S. Secretary of Agriculture*. An aseptic and deaerated aqueous solution of an iron salt is prepared and blended with a high DE corn syrup. Food products such as baby formulas, bread and fruit drinks may be iron-supplemented with the iron-fortifying compositions.

Iron salts useful in preparing liquid iron-fortifying compositions must be water-soluble and nontoxic at addition levels described herein and include ferric ammonium citrate, ferric choline citrate, ferrous gluconate, and ferrous sulfate. Any iron salt within the above limits will be considered equivalent for the purpose of the process. Ferrous sulfate heptahydrate ($FeSO_4 \cdot 7H_2O$) is preferred primarily because it provides compositions which range from colorless (compositions containing up to 0.1% ferrous ions) to very slightly blue (compositions containing 0.1 to 1.0% ferrous ions) while other ferrous or ferric salts produce compositions which range from medium to dark amber. These amber colors are easily detected in food products which are not themselves highly colored.

Ferrous sulfate is preferred also because it is the least expensive source of iron for physiological purposes. High conversion corn syrups used in this process have a DE of 60 to 68 and a solids content of 80% by weight. An extra high conversion corn syrup having a DE of 70 may also be used to prepare these compositions.

The first step of the process is the preparation of an aqueous solution of ferrous or ferric salts. It is preferred that the water be boiled before dissolving the iron salts to provide an aseptic solution. When ferrous salts are used, boiled de-aerated water cooled to 70° to 90°F eliminates clouding of the resulting solution, probably caused by oxidation of the ferrous ion. Clear solutions containing up to 26% by weight $FeSO_4 \cdot 7H_2O$ can be prepared in this manner. The concentration of iron in these solutions is varied in accordance with the amounts of iron and water desired in the final composition. Final compositions containing more than 30% water by weight tend to be biologically unstable, and at this water content any dissolved ferrous or ferric ions over 1% by weight of the final composition will tend to precipitate.

A small amount of acid can be added, preferably to the aqueous iron solution, to bring the acidity of the final composition down to pH 2. In some cases where the composition is used or stored under drastic conditions, such as high temperatures (e.g., 120°F), stability is decreased. Adjusting the pH downward to about 2 will maintain stability. Without addition of acid a composition containing 1% ferrous ions will have a pH of about 3.5. After the aqueous iron has been prepared, it should immediately be mixed with the syrup. Immediately is defined herein as the length of time before the occurrence of clouding of the iron solution. This can be as long as 1 hour. The preferred temperature range for the syrup during the mixing step is from 65° to 300°F.

Examples 1 through 4: Aqueous iron solutions were prepared by mechanically mixing for 10 minutes a ferrous or ferric salt in distilled water which had previously boiled (to deaerate and make aseptic) and cooled to 70° to 90°F. At this point a mineral acid was added to some of the solutions to lower the pH of the final composition.

The iron solutions were then added by means of a funnel having its exit and protruding beneath the surface of the liquid to a syrup which had been previously heated to between 180° and 200°F. During the addition and for about 10 to 30 seconds after all the iron solution had been added, the syrups were mechanically stirred to insure complete mixing. The compositions, still at a temperature of 180° to 200°F, were transferred to clear glass pint bottles which were then screw-capped and inverted to eliminate fermentation during storage. Amounts and kinds of starting materials and final products for several composition preparations are described in the following table.

	Examples			
	1	2	3	4
Starting material, g.:				
High conversion corn syrup DE 65, 82% solids by weight	3,619	3,570	3,938	3,938
Sucrose	524	523	0	0
$FeSO_4 \cdot 7H_2O$ solution:				
Salt	3.5	164.5	230	207
Deaerated distilled water	100	400	380	360
Water	412	115	0	0
Acid (1 N HCl)	0	0	0	20
Final composition:				
Total solids content, wt. percent	75.5	74.9	75.4	75.1
Ferrous ion, wt. percent [1]	0.015	0.69	1.01	0.92
Water, wt. percent		Remainder		
pH	5.0	3.1	3.6	2.3

[1] Calculated from total amount of iron salt added.

All compositions were colorless and were stable for at least 1 year at room temperature (70°F). At storage temperature of 120°F, Examples 1 through 3 showed a tendency to discolor indicating some unstability. Example 4, on the other hand, showed complete stability at that temperature.

Citrus Product

Various attempts have been made in the past to combine milk protein and citrus juices, however, these attempts have not generally been successful due principally to the fact that the acidity of the citrus juice causes coagulation of the milk protein and separation on standing. However, it has been found by *M.D. Maraulja, J.A. Attaway and C.D. Atkins; U.S. Patent 3,862,342; January 21, 1975; assigned to State of Florida, Department of Citrus* that a superior food product combining both milk protein and citrus juices, such as, for example, orange, grapefruit, lemon or lime juice, can be prepared having improved storage life and texture.

A major portion of citrus juice is combined with suitable protein material and starch along with a soluble food stabilizer and the mixture then treated to effect mixing and dissolution of the starch. The resulting product is a creamy liquid which is stabilized by the interaction of the soluble protein material and the starch dissolution products and which can be further processed as desired into a number of delicious food preparations.

The dissolution or partial hydrolyses of the starch into the soluble amylopectin and relatively insoluble amylose fractions and the complexing of these fractions with the milk protein, according to the process is effected by boiling the starch in sufficient water for about 10 to 15 minutes at a temperature of about 210° to 260°F. The necessary water for the complete dissolution of the starch can either be added to the starch-protein mixture and boiling carried out prior to the addition of the citrus juice, or the citrus juice, which will generally contain sufficient water for reaction with the starch, is added to the starch-protein first and the entire mixture boiled.

Preferably, however, the water is added to the starch-protein mixture and boiling carried out prior to the addition of the citrus juice since the risk of impairing the flavor of the juice is not thereby encountered. Although the amount of water required for the starch is not critical, it is advisable that about 4 to 8 liters of water be present per 100 grams of starch to ensure complete dissolution as well as complete mixing and solution of the other ingredients.

When relatively high protein and starch levels are used, hydrolyses of the starch can be facilitated by heating of the mixture at a temperature below 140°F in the presence of about 0.10 to 0.15% by weight, based on the starch, of suitable enzymes to effect additional solubilization of the relatively insoluble amylose fraction to soluble maltose. While this partial hydrolysis of the amylose fraction takes place rapidly below 140°F, above this temperature inactivation of the enzyme occurs and the hydrolysis stops.

Example: 12 quarts of distilled water were added to a steam jacketed kettle and while stirring, the following ingredients slowly added: (a) 250 grams of milk protein material containing 89 weight percent glycoprotein and soluble in the pH range 2 to 3.2; (b) 210 grams of starch; and (c) 105 grams of low viscosity sodium carboxymethylcellulose having a degree of carboxymethyl substitution of 0.65 to 0.85 DS. The mixture was slowly heated using steam pressures of 25 to 28 lb/in^2 until the mixture had been brought to a boil. Boiling was then maintained for 15 minutes to insure the complete hydrolysis of the starch and complete solution of the sodium carboxymethylcellulose polymer. During this period of time a temperature of 210° to 225°F was maintained. The mixture was then transferred to a larger tank and while stirring, a mixture of 2 quarts of distilled water and 1,281 grams of corn syrup added, followed by 3 quarts of 68°Brix concentrate and 1.45 quarts of water.

Finally, 2.5 ml of Valencia orange essence oil, 2.5 ml of Valencia orange cold-pressed oil and 3.0 ml of vanilla extract were added for flavor adjustment. The product temperature was below 120°F at this point. The beverage was then passed through a heat exchange unit and the orange juice constituent heat stabilized at 185° to 210°F, filled into 12 oz cans, cooled under a water spray and stored at approximately 40°F.

Ingredients by Weight Percent

	Wt. in Grams	Percent by Weight
18.45 Qts. of 14.0° Brix juice (3 qts. 68° Brix conc. + 15.45 qts. water) of oranges	18,472.0	90.89
Milk protein material:Milk protein	228	1.12
Ash and moisture	22	
Starch	210	1.03
Sodium carboxymethylcellulose	105	.52
Corn syrup	1,281.0	6.30
Orange essence oil (.009%)	2	
Cold-pressed orange oil (.009%)	2	0.03
Vanilla extract (.013%)	3	
	20,325	

CORN SYRUP SOLIDS AND DEXTROSE

The dried corn syrups, particularly of limited hydrolysis, are referred to in the industry as corn syrup solids. They may be composed of dextrose, maltose and and higher oligosaccharides depending on the method of hydrolysis used in preparing the corn syrup.

The major use of corn syrup solids and dextrose is in food processing, mainly in the baking, canning, beverage, confectionary and dairy industries where they supply sweetness, body and regulate texture or crystallization of other sugars present.

SPRAY DRYING CORN SYRUPS

Prior methods for producing dry dextrose-containing products include spray drying with air at 140° to 150°F and quickly cooling the product (U.S. Patent 2,324,113); hot air drying, in a rotary dryer for 30 to 45 minutes and with inlet air at 300° to 350°F, a bed of dextrose crystals wetted with a dextrose solution (U.S. Patent 2,369,231); a like process of rotary drying with inlet air at 180° to 200°F, a special blend of dextrose and polysaccharides (U.S. Patent 2,854,359); spraying a hot dextrose (105° to 150°C) liquor onto a cold (10° to 40°C), agitated dextrose seed bed, mixing to complete crystallization and finally drying with cold air (10° to 40°C) (U.S. Patent 3,239,378); spontaneous cooling and nucleation of a dextrose solution at 50° to 70°C by rapidly and instantaneously mixing with a cold bed of dextrose crystals (U.S. Patent 3,265,533); and drying intimate blends of dextrose solids and dextrose solutions at 50° to 100°C using cold air or hot air at dryer inlet temperatures up to 95°C (U.S. Patent 3,271,194).

Using Recycled Glucose with High DE Syrups

The spray drying of a glucose syrup, i.e., a glucose product having a DE of 30 to 40 is well known and can be performed without any difficulties and entirely without recycling. Moreover, it is well known to recycle the fines fraction of the spray dried product in order to agglomerate these fine particles with the coarser particles so as to obtain a less dusty product. However, both of these known processes have failed if an attempt is made at applying them to a glucose solution of high DE value because the product formed in the spray drying apparatus will be so sticky as to clog the apparatus in a very short time.

A continuous spray drying process has now been developed by *I. Repsdorph, K.K.K. Kroyer and J. Damgard-Iversen; U.S. Patent 3,477,874; November 11, 1969; assigned to Karl Kristian Kobs Kroyer, Denmark and Aktieselskabet Niro Atomizer, Denmark* to form a free-flowing product. This process comprises the steps of supplying the syrup to a special spray drying apparatus, withdrawing from the spray drying apparatus the solidified particles formed, subjecting the solidifed particles to a flow of drying and cooling air, and then recycling a portion of the solidified particles to the spray drying apparatus at a rate higher than the rate of supply of the syrup calculated as dry matter. The following Figure 9.1 and explanation which shows the flow diagram will further illustrate the process.

FIGURE 9.1: USING RECYCLED GLUCOSE WITH HIGH DE SYRUPS

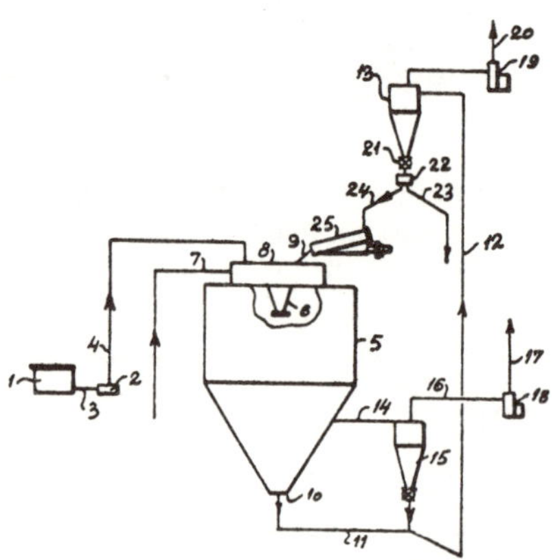

Source: I. Repsdorph, K.K.K. Kroyer and J. Damgard-Iversen; U.S. Patent
3,477,874; November 11, 1969

In Figure 9.1, **1** is a vessel containing an aqueous solution of glucose of high DE value having a solids content of about 75%. A pump **2** supplies the glucose solution through pipes **3** and **4** to an atomizer **6** mounted in a spray drying chamber **5**. The atomizer may e.g. in known manner consist of an atomizing wheel rotating at high speed. By means of the atomizer, the solution is split up into fine particles forming a mist in the spray drying chamber. Hot air is at the same time supplied through a pipe **7** to an air distributor **8** in a quantity and at a temperature such that an evaporation of the aqueous solvent will take place before the mist particles strike the conical bottom of the spray drying chamber.

At the same time, a solid, nonsticky recycled glucose product is supplied to the spray drying chamber through a separate pipe **9** and is atomized in the chamber.

The recycled glucose product is supplied in an amount exceeding the amount of dry matter in the fresh solution supplied to the spray drying apparatus, and it has been found that by proceeding in this manner, the glucose particles formed by dehydration of the fresh liquid will be powdered by the recycled product in such a manner as to loose the stickiness which they would otherwise possess and which would result in clogging of the spray drying apparatus in a minimum of time.

The recycled product also promotes crystallization of the fresh particles during drying in the spray drying apparatus which further contributes towards preventing stickiness. As a result of these combined influences, a solid free-flowing glucose product may be taken out through an outlet **10** at the bottom of the spray drying chamber.

This product is supplied to a conveying member **11** and is then pneumatically conveyed through a pipe **12** to a cyclone **13**, while at the same time being subjected to the dry air by means of which the pneumatic conveying is performed. Thereby the product is further dried, and at the same time is conditioned by being maintained in a dry and cool atmosphere. Preferably, the air used for the pneumatic conveying, the drying and the conditioning of the product is supplied at a relatively low temperature such as 30°C so as also to cool the product.

The drying air used in the spray drying chamber **5** is conducted through a pipe **14** to a cyclone **15** in which the powder carried along by the air is separated and discharged from the bottom of the cyclone to the conveying member, while the air is removed from the cyclone through a passage **16** and a blow off pipe **17** by means of a blower **18**. In the cyclone **13**, the glucose is separated from the conveying air and collects in the lower portion of the cyclone while the air is removed by a blower **19** and discharged through a pipe **20**.

At the bottom of the cyclone **13** there is arranged a sluice **21** and below the latter a fractioning device **22** is provided, which may e.g., separate the product into a minor fraction of relatively coarse particles, which are passed through a pipe **23** through a storing or packing station, and a major fraction of relatively finer particles which are passed through a pipe **24** to a vibrator **25** from which the pulverulent product is passed to the spray drying chamber **5** through the pipe **9** as previously mentioned. The proportion of the product passed through the pipe **24** should at any rate be greater than the proportion passed through the pipe **23**, and preferably the relative proportions should be selected at a predetermined value and while this distribution may be accompanied by a certain fractioning according to size as above mentioned this is not essential. If desired, the product recycled to the spray drying apparatus may be comminuted before entering the latter.

If the conveying member **11** is in the form of a pipe having an air intake at one end, and connected to the bottom outlets of the spray drying chamber **5** and the cyclone **15** through branch pipes, preferably through a sluice, as far as the cyclone is concerned, the pipe being connected to the pipe **12** at the other end, the whole conveying of the spray dried product may be performed pneumatically in a closed system, either by means of the blower **19** alone, or additionally by means of a blower provided at the junction between the pipes **11** and **12** and serving at the same time to subject the product to a certain disintegrating effect so as to break up any lumps. In that case, the vacuum in the spray drying chamber and in the pipe **11** should be relatively adjusted in such a manner as to permit the transfer of spray dried particles from the bottom ends of the spray drying chamber **5** and the cyclone **15** to the pipe **11**.

The process of U.S. Patent 3,477,874 has been modified by *R.H. Gray, Jr.; U.S. Patent 3,674,556; July 4, 1972; assigned to W.R. Grace & Co.,* to improve the efficiency. In the present process droplets of glucose-containing solutions are dried in the presence of recycled product solids, using hot air having an inlet temperature of at least 400°F (205°C) and up to as high as 500°F (260°C) while maintaining the average air residence time in the drying zone from 2 to 60 seconds (calculated by dividing the drying zone volume by the volume flow rate of the hot drying air therethrough). While the relationship between air inlet temperature and average residence time is not necessarily linear, it is generally observed when other variables are substantially constant that shorter residence times should be used as the drier air inlet temperature increases.

A preferred embodiment of the present process includes, the use of recycled product solids having an average particle size of not greater than 200 microns and preferably not greater than 150 microns. The most surprising aspect of the present process is that the unusually high drying air inlet temperatures do not result in any noticeable burning and/or darkening of the dry product. This is a truly unexpected result when considered in light of the closest known prior art, in which the air temperatures are at least 30° to 40°F and mostly 100° to 200°F lower than those used in the present process.

Example 1: A starch hydrolyzate liquor having a DE 97 is prepared in the usual manner, refined, concentrated to 67° Brix and then dried in a spray dryer having a 7.5 foot diameter and a 60° product collection cone, providing an overall dryer volume of 240 cubic feet. The liquor is atomized with a high vane centrifugal wheel type atomizer rotating at 21,600 revolutions per minute. Recycle solids are fed to the drying chamber through four separate one-inch outside diameter tubes equally spaced about the periphery of the atomizer.

In one exemplary run in this dryer the glucose liquor is fed to the atomizer at 160°F and at a rate of 20 gallons per hour (equivalent to a solids feed rate of 3.0 lb/min). The ratio of recycled solids to liquor solids is approximately 3.2. For start-up, ground product from a previous run is used. Then product from the run in progress is used. Air is fed to the dryer at an inlet temperature of 440°F and at a rate of 1,000 cubic feet per minute, giving an average dryer residence time of 14 to 15 seconds. The outlet air temperature was 160°F.

Product recovered from the above run is a white powder and has a moisture content of 2.5%. The particle size of the product was mostly in the range of 50 to 100 microns with some agglomerates in the range of 150 to 300 microns. The product was not noticeably burned and had no noticeable burned taste.

Example 2: Another run is conducted in the same equipment and under conditions similar to those in Example 1. In this run all recycled solids are ground in a Cumberland mill (a commercially available attrition mill) to an average particle size of 100 to 150 microns prior to being fed to the dryer. The inlet air temperature was 440°F, outlet air temperature 161°F and product temperature 151°F. Average air residence time was 15 seconds. The recovered product has an average particle size of 200 to 300 microns and a moisture content of 2.5 to 2.8 wt %. Again there is no noticeable darkening or burned taste.

During the course of this run the dryer walls remained relatively clean, with no heavy wall accumulations. No difficulty is experienced with large solids masses plugging the outlet. The general condition of the dryer is satisfactory for prolonged continuous operation. Operation in accordance with the present process provides quality product at higher production rates (for any given dryer capacity) than previously thought possible. This results in substantial technical and economic advantages over any prior processes.

Spray Crystallization Drying

A free-flowing total sugar product has been produced by *M. Niimi, T. Furukawa and H. Masada; U.S. Patent 3,540,927; November 17, 1970; assigned to Nippon Shiryo Kogyo Co., Ltd., Japan* using spray drying. The total sugar product, is a mixture of sugar crystals with uncrystallized solid solution of sugar and impurities.

This process involves the step of spray-crystallization, which is distinguished from ordinary spray drying. In the spray-crystallization step a sugar massecuite containing preformed sugar crystals is subjected to spray drying to form aggregates of sugar crystals. The complete process consists of a sequence of steps in which the conversion of the crystallizable sugar to crystalline form occurs in sequential stages, including crystallization stages both before and after the spray drying step. Because of the short time involved in the spray drying step itself, most of the crystallization can occur before and after the spray drying. The spray-crystallization essentially forms the aggregates from preformed microcrystals.

In the preferred process, a water solution of the crystallizable sugar is first subjected to partial crystallization to form a flowable or pumpable massecuite composed essentially of microcrystals of the sugar dispersed in a water solution of the sugar, which will usually be a substantially saturated solution. This massecuite is then subjected to spray drying. In the spray drying operation, atomized droplets of the massecuite are formed and part of the water is removed from the droplets while the droplets are air-borne, the water passing into the drying air stream, and the droplets being converted to soft granules, comprising aggregates of the sugar microcrystals while containing residual crystallizable sugar solution.

The soft granules are then aged and dried to form additional sugar crystals therein and to further reduce their water content. The final product may be substantially dry, that is, it may contain little or no free water, but it may contain water of crystallization, such as when the microcrystals are in the form of dextrose monohydrate. Any remaining trace of free water is distributed within the granular aggregates, and therefore has little tendency to make the aggregates sticky or subject to caking.

The granular aggregates produced by this process are of a distinctive physical configuration. They are generally spherical and are formed of cohered sugar microcrystals, providing a hard, firm granular structure. As water evaporates from the sprayed droplets, the solution phase causes the droplets to assume spherical shapes due to the effect of liquid surface tension. With further contruction of the liquid phase during the spray drying, the residual solution is withdrawn within the aggregates, being distributed between and over the surfaces of the sugar crystals.

After aging and further drying, additional sugar crystals are formed within the interstices of the aggregates. In the final product, substantially all of the crystallizable sugar may be converted to crystalline form, leaving only the oligosaccharides in noncrystalline form. Since the oligosaccharides are distributed in thin films on the interior surfaces of the sugar crystals of the aggregates, there is little or no tendency for the noncrystalline oligosaccharides to cause the product to become hygroscopic.

Example: A starch hydrolyzate liquor having a DE of 97 prepared by any usual method is refined and evaporated in an ordinary manner to a Brix of 67°, and allowed to crystallize by agitating slowly in a crystallizer at 20°C. It is usually desirable to mix 0.5% of dextrose monohydrate crystals or substances containing dextrose monohydrate as the main component as the seed of crystallization. The time required for completion of the crystallization (the establishment of an equilibrium between the crystalline dextrose and the saturated solution at this temperature) is not fixed and varies depending on the grade or purity of the liquor, but is usually attained in eight hours. The crystallization process may be either the usual batch type process or if desired may be a continuous procedure.

FIGURE 9.2: SPRAY CRYSTALLIZATION DRYING

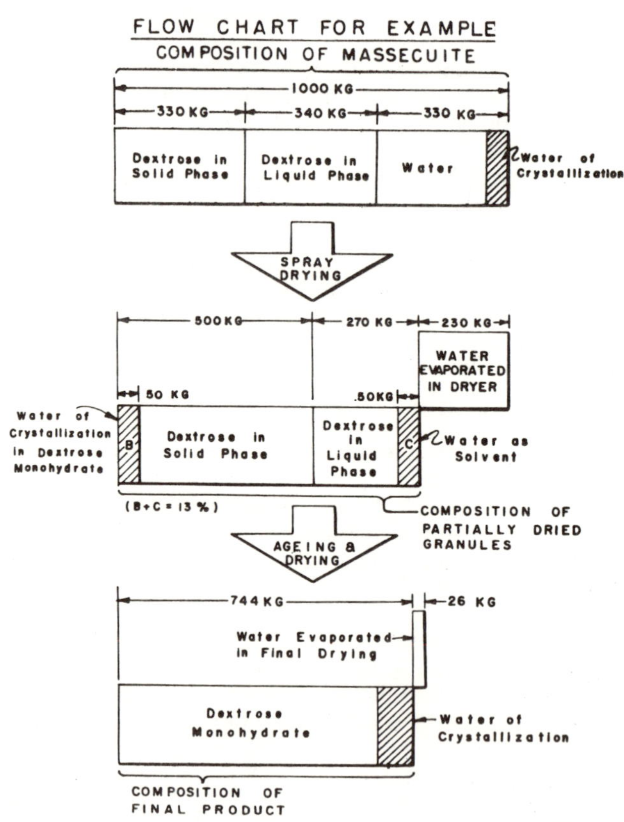

Source: M. Niimi, T. Furukawa and H. Masada; U.S. Patent 3,540,927; Nov 17, 1970

The massecuite obtained is composed of the developed crystals suspended in a saturated solution and is of such fluidity and stability that it may be readily pumped through pipes to the spraying apparatus. The massecuite containing the microcrystals is spray dried in small droplets which are allowed to drop through an air stream in the drying chamber. The temperature of the air is preferably controlled to below 50°C to obtain a product containing dextrose monohydrate, which does not adversely affect the subsequent additional crystallization of the granules formed from the massecuite droplets or cause caking of the final product.

The spray drying chamber can consist of a vertical cylindrical upper section 7,700 mm in diameter and 7,700 mm high and a lower conical collection section 7,700 mm high. At the top center of the upper section is an atomizing device 350 mm in diameter rotated at 6,500 rpm, to spray the massecuite at a rate of 1,000 kg/hr. A stream of drying air is admitted to the chamber at 55°C, but drops to 45°C due to the heat of evaporation of the water removed. The particles deposited on the bottom of the spray dryer section contain 13% moisture on a dry basis, and as the result of evaporation of 230 kg of free water per 100 kg of charged massecuite there is obtained 770 kg of partially dried material.

This material which resembles damp snow is composed of 65 to 75% of crystalline dextrose monohydrate and 35 to 30% of a liquid dextrose solution. When aged for 3 to 5 hours and with slight additional drying, additional crystallization occurs in the solution portion and a free-flowing product results. The foregoing process example is diagrammatically summarized in Figure 9.2, which is simplified to the extent that that the small percent of oligosaccharide is not graphically represented.

Drying Seeded Glucose Solutions

O. Hansen; U.S. Patent 3,567,513; March 2, 1971; assigned to Aktieselskabet Niro Atomizer, Demark has developed a spray drying process which eliminates the recirculation of the spray dried product as disclosed in U.S. Patent 3,477,874. In the present process, part of the dried product is mixed with a saturated glucose solution and left standing until a transformation from a gritty to a smooth homogeneous consistency has taken place, then the mixture is recirculated to the spray dryer.

The explanation of the favorable effect which the recirculated product has on the course of the spray drying is not known with any certainty, but it is presumed that in the suspension formed of recirculated material in the saturated solution, the crystals of the solid matter dissolve, while at the same time new crystals of another shape and size are formed. It has been observed through a microscope, that a marked change of the crystals occurs during the period in which the mixture is being left standing in that larger and more uniform crystals are formed. The transformation can also be observed macroscopically in that a freshly prepared suspension of the recirculated product and a saturated solution have a gritty consistency, while the suspension, after having been left standing, gradually acquires a smooth and homogeneous consistency.

Moreover, the ratio between alpha and beta glucoses shifts with a reduction of beta glucose and an increase of the alpha in the solid phase. The ratio between alpha and beta glucoses in a solid glucose product, which originally had a ratio of 59:41, after a period of 24 hours, during which the product was in contact with a saturated glucose syrup at 22°C, shifted to 99:1. The fact, that the transformation which takes place during the dynamic equilibrium, results in the formation of a greater quantity of alpha glucose at the expense of the beta glucose, is a natural consequence of the fact that the alpha glucose (or alpha glucose monohydrate) is the stable, solid form at temperatures of up to approximately 113°C.

It is, furthermore, presumed that the transformed product contains crystals of a shape and size particularly suited as a base for crystallization of the dry matter in the over-saturated drops formed by the spray drying and that this is the reason for a faster and more complete crystallization being obtained, which results in the formation of a free-flowing product.

During the spray drying there also occurs a shift in the ratio between alpha and beta glucoses, which also is presumed to be a result that the temperature of the product during treatment is below 113°C, which is the lowest temperature at which beta glucose is stable. If the added, recirculated product is used in a quantity equal to the glucose content in the syrup, and it is presumed that the transformation into alpha glucose has been complete, this implies that the ratio between alpha and beta glucoses in the bulk end product is at least 3:1.

If the recirculated quantity is not sufficiently large in relation to the quantity of dry matter in the solution, a sticky product is obtained after the drying. It has been found in practice, that to obtain a satisfactory drying, the recirculated product must be mixed with a quantity being at least equal to one fifth of the dry matter content in the solution. According to this process, the recirculated quantity of glucose product should not be larger than the dry matter content in the solution, for by using larger quantities a viscous mixture is formed, which is difficult to atomize.

In order to obtain the largest possible production, the recirculated quantity should, however, come as closely to the upper limit as possible. It has been established, that a satisfactory production is obtained when the recirculated quantity is equal to 0.8 to 1.0 times the quantity of dry matter in the solution. To promote the crystallization during the spray drying, it has proved expedient for the mixture, after having been standing, but prior to the spray drying, to be ground. By this step, the crystal size of the product is kept very small, so that the product is particularly suited for recirculation. The grinding can, for instance, be carried out in a ball mill or a colloidal mill. The following Figure 9.3 shows a schematic diagram of the process.

The figure shows a mixing tank **1** for glucose syrup which is fed through line **2**, and recirculated dry product coming through line **3**. The suspension formed in the tank is kept in motion by the stirrer **4**. From tank **1** the suspension is pumped by pump **5** through line **6** to storage tank **7**, which is also equipped with a stirrer **8**. From the storage tank the suspension is pumped by pump **9** through line **10** to a ball mill **11**. The ground suspension leaves the mill through line **12** and is fed by pump **13** into the spray dryer **14** in which it is centrifugally atomized into hot air coming from the blower **15** and heater **16** through the air disperser **17**. The dried product leaves the spray dryer through powder valve **18** and the drying air together with the fine particles of dry product is taken out through line **19** to cyclone **20** which separates the fine particles from the air.

FIGURE 9.3: DRYING SEEDED GLUCOSE SOLUTIONS

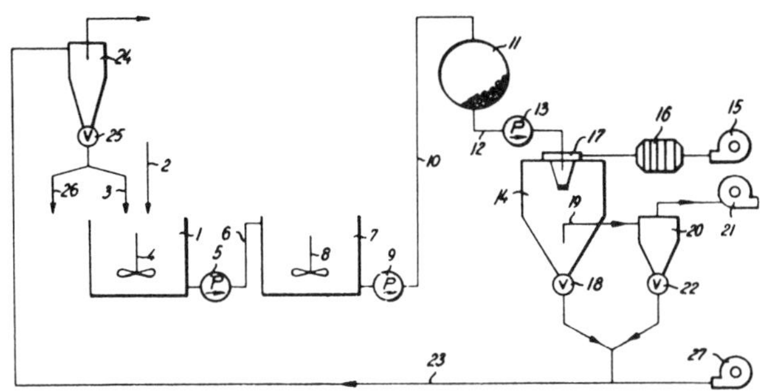

Source: O. Hansen; U.S. Patent 3,567,513; March 2, 1971

The air is sucked out by the ventilator **21** and the fine particles fraction leaves the cyclone through the powder valve **22**. The combined chamber powder fraction and cyclone powder fraction are by means of a blower **27** transported pneumatically through the line **23** to the cyclone **24**. The dry product which leaves the cyclone through the powder valve **25** is divided into two fractions, one being recirculated to the process through the line **3** and the other one being the final product produced leaving through line **26**.

Example: In a mixing apparatus glucose syrup having a dry matter content of 50% is mixed with so-called total sugar in a proportion of 300 kg syrup per 150 kg total sugar dry matter. The mixture is left standing (while being stirred) in a tank for 15 hours at a temperature of 15° to 20°C.

Thereafter the mixture is spray dried, while hot air having a temperature of 130° to 180°C is introduced at a rate of 1,500 kg/hr. The outlet temperature of the air was around 50° to 55°C. A powder was formed, which initially adhered to the drying chamber, but which after some time fell out evenly. The powder had a discharge temperature of 40°C. The ratio between the alpha and beta glucoses in the end product was 93:7. The product was a free-flowing powder having a water content of approximately 12 to 15%.

Dried Hydrolysates of Low Bulk Density

Starch hydrolysates having a low bulk density and rapid dispersibility in water are produced by a spray drying process developed by *G.R. Meyer and R.J. Everett; U.S. Patent 3,674,555; July 4, 1972; assigned to CPC International Inc.* The particles formed are substantially hollow spheres.

More specifically, the process provides a continuous method for producing spray dried starch hydrolysates having a bulk density below 20 lb/ft^3 which comprises the steps shown in the listing on the following page.

(a) Heating an aqueous slurry containing from 20 to 45% by weight of a starch hydrolysate having a DE of 2 to 30 in a first, confined zone under pressure to a temperature of 210° to 300°F

(b) Injecting the slurry into a second, large volume zone into contact with heated air having a temperature of from 320° to 450°F with a sudden release of the pressure on the slurry upon injection

(c) Maintaining conditions in the second zone such that drying to a moisture content below 3% is achieved before the resultant dry solids engage against a collection surface

(d) Recovering the dried solids in particulate form.

A preferred embodiment is a continuous process for producing spray dried starch hydrolysates having a bulk density below 15 lb/ft³ which comprises:

(a) Heating an aqueous slurry containing from 20 to 45% by weight of a starch hydrolysate having a DE of 5 to 25 in a first, confined zone under pressure to a temperature of 212° to 250°F

(b) Injecting the slurry into a second, large volume zone into contact with heated air having a temperature of 350° to 400°F with a sudden release of pressure on the slurry upon injection

(c) Maintaining conditions in the second zone such that drying to a moisture content between 0.5 and 3.0% is achieved before the resultant dry solids engage against a collection surface

(d) Recovering the dried solids in particulate form.

The injection of the superheated slurry, which is referred to as a syrup in the examples that follow, is done with the use of a nozzle to yield a particulate product. Injection through means of a centrifugal atomizer provides a product similar to cotton candy, wherein the syrup remains in substantially continuous form, resulting in a stringy type product.

Example 1: This example illustrates the feasibility of obtaining a low bulk density from a syrup of 35 to 40% dry substance and having a DE of 12. A cornstarch hydrolysate having a DE of 12 and a dry substance content of 37.2% was heated to 232°F in a tubular heat exchanger that was directly connected to the nozzle of a spray drying atomizer. The pressure at the time that the syrup reached 220°F was 3,300 psig. The temperature of the surrounding drying zone was maintained at 390°F. The resulting product consisted primarily of fragments of hollow spheres, and had a bulk density of 11.7 lb/ft³.

The same syrup product, when processed through the same atomizing equipment at an injection temperature of 140°F yielded a product of 23 lb/ft³ bulk density. The product solids were instantly soluble in cold water. They had a bland flavor and a clean, white appearance. They are particularly useful as a carrier material for synthetic sweeteners, to form a low calorie substitute for sucrose that can be used in place of sucrose on an approximately equal volume of replacement basis.

Examples 2 through 6: Other syrups of differing dry substance concentrations were processed at different temperatures. The results are summarized in the table on the following page.

Ex.	Dry substance concentration, percent	Syrup temperature, °F.	Dextrose equivalent	Pressure, p.s.i.g.	Air temperature for spray drier °F.		Product bulk density, lbs./ cu. ft.
					Inlet	Outlet	
2	35.3	233	22	5,000	390	280	10.6
3	37.2	242	12	2,500	390	308	12.3
4	37.0	140	20	4,000	333	255	23.3
5	49.0	140	12	4,000	312	250	28.3
6	74.0	223	22	5,000	365	330	23.3

OTHER DRYING PROCESSES

Stable Powders from Acid Modified Syrups

Hydrolyzed starch solids are normally very hygroscopic materials which, after being dried, will take on water and become sticky and damp when exposed to even normal room conditions. However, *W.A. Mitchell and W.C. Siedel; U.S. Patent 3,657,010; April 18, 1972; assigned to General Foods Corporation* have found that when hydrolyzed starch materials, such as concentrated syrups, are heated in the presence of acid up to temperatures of 300° to 370°F and the molten mass is allowed to slowly cool and solidify, the resulting solid may be ground into a powder which is substantially nonhygroscopic and which has a moisture content of less than 1%.

It is believed that the chemical mechanism that effects the above is basically a combination reaction, probably catalyzed by the acid, in which relatively stable anhydrides are formed. The reaction involved in this process is dependent upon the acid concentration and also upon the heating rate and maximum temperatures. A series of closely related but differing products can be produced by varying the above conditions. The hydrolyzed starch materials used in this process may be either in syrup or solid form. The syrup may be a commercially available syrup (e.g., corn syrup) or may be prepared by dissolving hydrolyzed starch solids in a liquid medium. The hydrolyzed starch material preferably has a DE of 5 to 70.

The concentrated syrups are the preferred hydrolyzed starch materials for use in this process since the acid material may be homogeneously blended with liquid syrups much easier than with dry solids. The concentrated syrups also possess the advantage of having a smaller amount of water to vaporize off before the desired reaction will proceed. The water component of these syrups, which normally boil below 250°F, must be eliminated to permit the temperature of the hydrolyzed starch material to reach the elevated levels which are required for the acid treatment.

Special hydrolyzed starch syrups or solids (e.g., modified dextrins) such as those produced by enzymes can be used to obtain bland-tasting, moisture-stable powders. Extensive acid and/or heat treatment can be used to obtain products which may be less susceptible to amylase enzyme attack. These products could have possible applications in low calorie foods.

Surprisingly the acid-heat treatment of hydrolyzed starch products such as corn syrup solids results in products which are not substantially changed in terms of color, taste, solubility, or viscosity.

The dry powders formed by this process have been kept as a free-flowing powder at normal room conditions for periods of at least several months. These dry powders have also been stored in paper envelopes under such drastic conditions as 90°F and 85% relative humidity (RH) and been found to remain as dry powders for several weeks. This is never possible when corn syrup solids not produced by this process are so exposed.

Example: 810 g of Corn Products Globe Syrup No. 1,121, corn syrup (78% solids, 43 DE), were placed with 250 g of finely ground (400 mesh) fumaric acid in a 5-quart bowl. By means of an electric jacket the mixture was heated until it reached a temperature of 318°F. The mass was spread on an aluminum foil belt and allowed to solidify by air-cooling to room conditions. The material was then ground into a powder and sieved.

Using the powder of the above example the fraction through a No. 70 and on a No. 140 U.S. standard screen was taken and packed in paper envelopes so that each pack contained 3.2 g of fumaric acid. These packs were stored at 90°F/85% RH and at 100°F/30% RH. The material stored at 90°F/85% RH was free flowing and possessed acceptable solubility after three weeks of storage. The material stored at 100°F/30% RH was not clumped and possessed good solubility after 5 weeks of storage. The dry powder produced according to this example is adapted to be incorporated into a powdered mix (beverage mix) and packaged in and distributed in paper envelopes.

Using Plate Heat Exchanger

Solid, glass-like starch hydrolysates are produced using plate heat exchangers in this process developed by *P.K. Carrell; U.S. Patent 3,706,598; December 19, 1972; assigned to CPC International Inc.* These solid, glass-like starch hydrolysates have a DE of 10 to 25, a water content less than 15% by weight, a bulk density of at least 40 lb/ft^3 and have excellent wetability and solubility in water.

The products are prepared by first hydrolyzing starch to a DE of from 10 to 25; concentrating the resulting product to a water content of less than 15% by weight in the liquid phase; and then cooling the product to solidification temperatures. The solidified product can then be ground or milled to a desired particle size for application in the food industry. The low DE starch hydrolysates or conversion liquors used in this process may be produced by the known one-step or two-step enzyme techniques or by the two-step acid enzyme technique.

The starch hydrolysate products are then prepared by concentrating the aqueous enzyme conversion liquors to a water content less than 15% by weight. This concentration can be carried out by using a vapor swept heat exchanger to evaporate the water from the conversion liquor. In the vapor swept exchanger, the starch conversion liquor centers at a concentration of 25 to 50% solids into a plate heat exchanger close to its boiling point. The liquor than passes through a series of plates reaching its boiling point.

The liquor is then transferred into minute droplets in a vapor phase. The stream is subjected to progressively decreasing pressures or progressively increasing temperatures along succeeding heating plates thereby causing the vapor to be superheated. This superheated vapor transfers heat to the liquor droplets by direct contact, effecting the concentration of the liquor. By this method, starch conversion syrups having low DE values, can be evaporated to a water content less than 15% by weight.

The preferred water content of a particular starch hydrolysate product of this process, while it must be below 15%, varies inversely with the DE of the product. Thus, the products having the lower DE values have higher water contents. A product having a DE of from 10 to 15, for example, can have a water content of from 10 to 15% by weight while a product having a DE of from 20 to 25 preferably has a water content of from 5 to 10% by weight or less.

Example 1: A refined corn starch hydrolysate which was prepared by the one enzyme technique, and having a solids content of 30% by weight and a DE of 12 is concentrated in a vapor swept heat exchanger until the water content in the product was 13%. The product was then cooled and milled to pass a 20 mesh screen to yield a glass-like starch hydrolysate product having the following properties. Dextrose equivalent—10, water content—13, bulk density—54.

Example 2: A refined corn starch hydrolysate which is prepared by the two enzyme technique, and having a solids content of 35% by weight and a DE of 17 is concentrated in a vapor swept heat exchanger until the water content in the product is 10%. The product is then cooled and milled to pass a 20 mesh screen to yield a glass-like starch hydrolysate product having the following properties. Dextrose equivalent—17, water content—10, bulk density—54.

Example 3: A refined corn starch hydrolysate which was prepared by the acid-enzyme technique, and having a solids content of 40% by weight and a DE of 22 is concentrated in a vapor swept heat exchanger until the water content in the product was 5%. The product was then cooled and milled to pass a 20 mesh screen to yield a glass-like starch hydrolysate product having the following properties. Dextrose equivalent—22, water content—5, bulk density—54.

CRYSTALLIZATION PROCESSES

Dextrose is the common commercial name for the pure crystalline solid recovered from almost completely hydrolyzed starch. The terms dextrose and glucose are used interchangeably in sugar technology and have the same meaning in reference to starch hydroylsates. Dextrose (or glucose) is a white crystalline monosaccharide which exists in three crystalline forms, named α-D-glucose, β-D-glucose or α-D-glucose hydrate.

Dextrose Hydrate Crystals from Starch Syrups

A process for forming dextrose hydrate crystals without the formation of a mother liquor has been developed by *M.L.E. van Tittelboom; U.S. Patent 3,406,046; October 15, 1968; assigned to Corn Products Company.* This method eliminates the crystallizers and also eliminates centrifuging the crystals to separate them from the mother liquor.

This process for obtaining dextrose crystals from a dextrose-containing liquor comprises first forming a thin layer of the liquor on agglomerates of dextrose crystals. A convenient way to accomplish this is to spray the liquor onto the agglomerates. Then, the wetted agglomerates are contacted with dextrose particles which have a predetermined temperature, e.g., the agglomerates are coated with a thin layer of dextrose crystals. The contact of the crystals (which act as seed) with the wetted agglomerates results in substantially instantaneous crystallization of a substantial amount of the dextrose in the film and the agglomerates become free-flowing and substantially free of stickiness.

The above operation may be repeated several times to crystallize a large amount of dextrose liquor. Because of the high rate of crystallization of the dextrose at the temperature used, a substantial proportion of the crystals is believed to be in the anhydrous form. The agglomerates now coated with a layer of a mixture of anhydrous and hydrate crystals are then subjected to a conditioning to complete crystallization of the dextrose in the film or liquor and to convert substantially all of the anhydrous crystals present in the layer or film to the hydrate form.

In carrying out the process, any dextrose containing liquor which is supersaturated with dextrose and which contains 75 to 88%, dry substance, may be used. The liquor to which the process is particularly applicable is the total hydrolysate obtained by the enzymic hydrolysis of starch. Such liquors are well known in the industry and have concentrations of dextrose of 90 to 100%, dry basis. The process is illustrated by the drawing of Figure 9.4.

In Figure 9.4 the numeral **1** represents a supply of the dextrose liquor. In a typical embodiment, this liquor had a density of 43.5°Bé, a dextrose equivalent of 94.5%, dextrose content of 90% and temperature of 80°C. The liquor is fed to a dosing pump **2**, and then, optionally but preferably, into agitator **3** where the liquor is agitated or whipped to incorporate air into the liquor which increases the total surface within the mass resulting in an accelerated rate of crystallization. The liquor is next sprayed, by means not shown, onto agglomerates of dextrose crystals which may be obtained from a previous run.

These agglomerates, having a temperature of 23° to 25°C are fed by shaker gutter **4** into a slanting ribbon conveyor unit **5**. At the outlet of **4**, the agglomerates are sprayed with the liquor from **3** by small jets, not shown. The agglomerates in the run referred to consisted of dextrose hydrate crystals and had a diameter of 2.5 to 3.5 mm, and a moisture content of 9%. The temperature of the wetted agglomerates was 40°C at this stage. The amount of dextrose liquor sprayed onto the agglomerates was about 20%, by weight, dry basis. The wetted agglomerates are conveyed forward in conveyor **5** and, as they are moved forward, are coated with dextrose seed by seed distributor **6** which consists of a shaker gutter.

The amount of seed used in the run was about 20%, by weight, of agglomerates, dry basis. The coated agglomerates are transferred from the conveyor into another slanting ribbon conveyor **7** where they are mixed, for 15 to 20 minutes, to permit substantially all of the dextrose in the liquor to crystallize, resulting in agglomerates which are dry and substantially free of stickiness. The temperature of the agglomerates was maintained at 30°C during the mixing operation. Cooling air from unit **8** is used in both units **6** and **7** to maintain the temperature at the desired level.

The dry agglomerates are then fed into a conditioning tube **9** where, in this run, they were held with slow agitation for 2.5 hours to permit completion of the crystallization and to allow any anhydrous crystals present to be converted to the hydrate form. Cooling air is introduced at **10** to maintain the temperature at the desired level which in this run was 26°C.

FIGURE 9.4: DEXTROSE HYDRATE CRYSTALS FROM STARCH SYRUPS

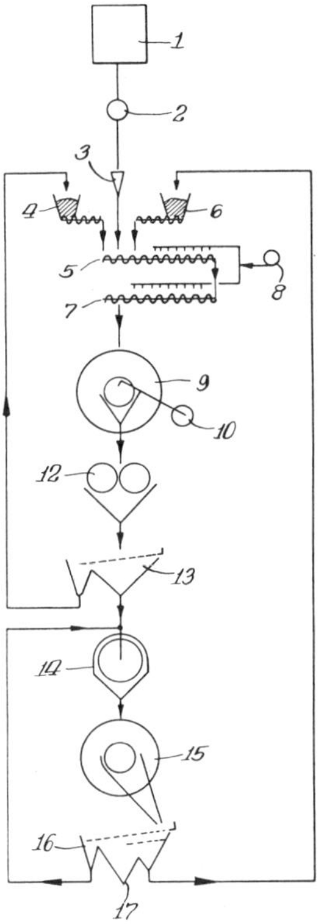

Source: M.L.E. van Tittelboom; U.S. Patent 3,406,046; October 15, 1968

The agglomerates are passed through a calibrating roll **12** and sifter **13**. About 60% of the agglomerates is returned to the process at **4**. The remainder is ground in mill **14** and dried in dryer **15** to a moisture content of 9%. The dried material is screened at **16**. A portion of the screened material is returned to the process for seed, at **6**. The tailings are reground at **14**. The third portion is packed in bags at **17**. The final product consisted of dextrose hydrate crystals.

In another embodiment, the steps of spraying the liquor onto the agglomerates and then coating with seed were repeated three times before the agglomerates were fed into conditioning tube **9**. The following table shows the amount of liquor and seed used in a run where triple coating was used.

	Kg. Recycled Agglomerates	Kg. Seed	Kg. Liquor	Percent DS	Kg.
First coating	34.4			91	31.304
			7.35	83	6.100
		7.35		91.5	6.725
After 15 minutes: Second coating			10.55	83	8.756
		10.55		91.5	9.653
After 15 minutes: Third coating			14.9	83	12.367
		14.9		91.5	13.633
	34.4	32.8	32.8		88.528
	67.2		32.8		or 11.46% H$_2$O
Ratio		2.04/1			

Crystalline Dextrose Monohydrate

A process for the production of crystalline dextrose monohydrate is disclosed by *W.G. Kingma; U.S. Patent 3,709,731; January 9, 1973; assigned to Continental Engineering, Netherlands*. A dextrose mass which initially comprises mainly a concentrated dextrose solution is continuously passed in prop flow through an elongated horizontal crystallization zone to effect crystal formation and then through an aging zone to effect crystal growth, whereupon the dextrose mass is separated into crystals and mother liquor and the mother liquor is concentrated and subjected again to crystallization and aging. A vacuum is created above the dextrose mass flowing through the crystallization zone to cause a temperature fall by self-evaporation and effect crystal formation.

At the same time, the mass in the crystallization zone is stirred in planes perpendicular to its direction of flow and substantially no heat is supplied from the outside. Crystal growth within the aging zone is effected at constant temperature and atmospheric pressure. The concentrated mother liquor may be worked up by blending part of it with initial dextrose solution and recycling it to the beginning of the process or by subjecting at least part of it to a second cycle of operations comprising passing the concentrated mother liquor through a second crystallization zone and a second aging zone and separating the resulting mass into a second portion of crystals and a second mother liquor. The second portion of crystals may be combined with crystals resulting from the first cycle of operations or recovered separately or blended with an initial dextrose solution and recycled to the beginning of the process.

The appropriate temperature for crystal formation and crystal growth of dextrose monohydrate is in general a function of the concentration and temperature of the starting mass and will be between 34° and 28°C at an initial concentration of 70 to 72% and an initial temperature of 45°C. It is preferred, therefore, to create a vacuum within the crystallization zone of sufficient height to reach a temperature between 34° and 28°C when the initial concentration and temperature are as just specified.

In most cases, the crystallization zone is an entity and a vacuum of predetermined height is created above the surface of the mass flowing therein. In a special embodiment, however, the crystallization zone comprises two or more parts situated one behind the other in longitudinal direction of the crystallization zone. Vacuums of different height are created and maintained in these parts, i.e., the vacuum in a part later to be passed by the flowing mass is of greater height than the vacuum in a preceding part.

When the crystallization zone has two parts, it is preferred to create in the first part a vacuum of sufficient height to reach an intermediate temperature of the mass between 40° and 34°C, when the initial concentration was 70 to 72% and the initial temperature was 45°C, and in the second part to create a vacuum of sufficient height to reach a temperature between 34° and 28°C, starting with the intermediate temperature.

In the aging zone, a relatively high amount of heat is produced during crystal growth (25 kcal/kg of dextrose monohydrate crystals). Therefore, it is preferred to absorb heat from the aging zone in order to keep the temperature of the mass constant. After leaving the aging zone, the mass is separated into crystals and mother liquor, the mother liquor is concentrated and the above operations are repeated one or more times with the concentrated mother liquor. This may be effected in several ways. Thus, at least a portion of the concentrated mother liquor may be recycled to the beginning of the process and blended with a fresh dextrose solution. It is subjected then together with this fresh dextrose solution to the above crystallization, aging and separation steps. The process can be illustrated by the following Figure 9.5 which is explained in the example.

FIGURE 9.5: CRYSTALLINE DEXTROSE MONOHYDRATE

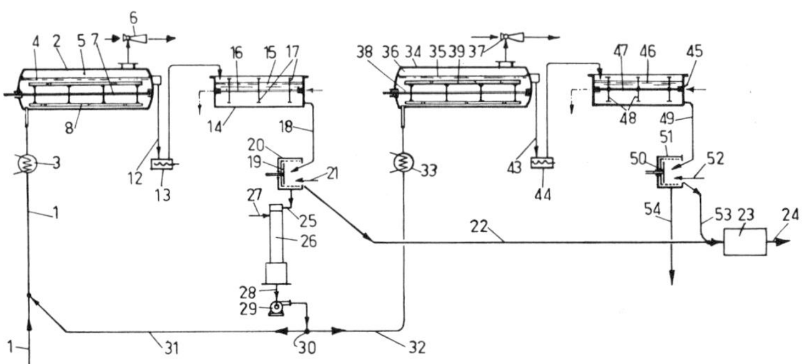

Source: W.G. Kingma; U.S. Patent 3,709,731; January 9, 1973

Example: About 1,550 kg/h of fresh dextrose solution, prepared from starch by a combined acid-enzymatic conversion treatment and having a DS of 72% and a DE of 97%, are supplied through conduit **1** and mixed in crystalline **2** with 720 kg/h of concentrated mother liquor from conduit **31**, this mother liquor having a DS of 73 to 74% and a DE of 89%. The resulting mixture (about 2,270 kg/h of 74% DS and 94.4% DE) is passed to cooler **3** and adjusted to a temperature between 45° and 50°C. Then, the mixture is passed through vacuum crystallizer **2**.

This crystallizer has a capacity of 32 m³, being 8 meters in length and 2.25 meters in diameter, and 24 m³ thereof is occupied by the flowing mass. Using vacuum source **6**, a vacuum of 19 millimeters Hg is created in vapor space **5** and the temperature of the flowing mass falls to 32°C by the self-evaporation, vapor bubbles being formed only at the surface of the mass. Meanwhile, the mass is stirred by rotors **7**. The time of passing through the crystallizer is 15 hours.

A crystal-bearing slurry leaving crystallizer **2** is passed to aging trough **14** which has a length of **6** meters, a diameter of 1.750 meters and a capacity of 14.5 m³. Part of this capacity, viz 13 m³ is occupied by slurry. The temperature of the flowing mass is kept below 32°C by cooling and the time of flow of the mass through the trough is 8 hours. The product leaving trough **14** is passed to centrifuge **20** where crystals of dextrose monohydrate are separated from mother liquor. The crystals are produced in an amount of 870 kg/h and passed to drier **23**.

The mother liquor obtained in centrifuge **20** in an amount of 1,330 kg/h has a DS of 60% and is passed to evaporator **26**, where it is concentrated to 1,140 kg/h having a DS of 73 to 74%. A portion of 420 kg/h of this concentrated mother liquor (89% DE) is passed via a cooler **33** to a second crystallizer **34**. This crystallizer has a length of 5.5 meters, a diameter of 1.3 meters and a capacity of 7.2 m³, the greater part thereof, viz 5.4 m³, being occupied by flowing mass. A vacuum of 15 millimeters Hg is created in vapor space **36** by means of vacuum source **37** and the temperature of the flowing mass is reduced by self-evaporation (thanks to this vacuum) from a value between 45° and 50°C to a temperature of 28°C. The time of passing through this crystallizer **34** is 18 hours under continuous stirring by means of stirring means **38**.

A crystal-bearing slurry is leaving crystallizer **34** and is passed to aging trough **45** which has a length of 3.5 meters, a diameter of 1.1 meters and a capacity of 3.3 m³, a portion of 2.5 m³ being occupied by flowing slurry. The temperature of the mass or slurry is kept at 28°C by cooling through shaft **47** and the time of flowing through the trough is 8 hours. The mass leaving trough **45** is passed to centrifuge **51** where it is separated into a second portion of dextrose monohydrate crystals and a second mother liquor. The resulting crystals (120 kg/h calculated as dry solids) are passed to drier **23**. The mother liquor comprises 175 kg/h of dry solids and has a DE of 81.2%.

The crystals resulting from centrifuge **20** and centrifuge **51** are dried in drier **23** and withdrawn through conduit **24**. The result is a total amount of 1,000 kg/h of dextrose monohydrate crystals (corresponding to 910 kg/h of dry solids). This means a yield of 84%, based on the solids content of the supplied fresh dextrose solution.

Granular Beta Dextrose

R.G.P. Walon; U.S. Patent 3,650,829; March 21, 1972; assigned to CPC International Inc. has disclosed a process for producing a granular dextrose product which comprises seeding a starch hydrolyzate solution containing at least 90% by weight dry substance and having a DE of at least 90, to obtain a crystallized mass, breaking the mass up into particles, and recovering a granular dextrose product. The granular dextrose product is granular β-dextrose.

It is essential that the seeding of the starch hydrolyzate solution is performed at a temperature from 75° to 120°C, preferably above 93°C. The preferred seeding material is dextrose. Other seeding materials, in particular other sugars (e.g., levulose, sucrose, maltose, etc.), may be used but this leads to a less pure product and longer crystallization times. Most preferably, the seed dextrose is β-dextrose that has been prepared by this process, but it can be secured from any desired source. The amount of seed sugar added to the starch hydrolyzate solution should be at least 1% by weight based on the total solution weight. Preferably, the amount of seed sugar added is from 2% to 10% of the total solution weight. Most preferably, 5% by weight is used.

After seeding has been completed, the solution is mixed and allowed to stand for a time of from 5 minutes to 1 hour to obtain a semisolid mass. The semisolid mass is preferably poured onto an essentially horizontal surface to a depth of 1 to 8 centimeters to obtain a cast dextrose sheet. When seeding temperatures above 100°C are used the seeded solution may be immediately poured onto the horizontal surface without being allowed to first stand. The cast sugar is allowed to stand for at least 5 minutes and preferably up to 45 minutes. Preferably, the depth of the semisolid mass after pouring is from 3 to 5 centimeters, and the cast sugar is allowed to stand for a time of 10 to 20 minutes.

The cast sugar may be granulated to form relatively large dextrose granules. The granulation can be carried out using a Sharples granulator, a Tornado mill, or other equivalent apparatus. Relatively coarse granules of a size which will pass through a 10 mm by 10 mm opening in a sieve may be made by this procedure. If finer granules are desired, the large granules must be allowed to stand for at least 15 minutes; they may then be reground to form a granular crystalline dextrose product having any desired degree of fineness. A uniform size granular product may be obtained by sifting the ground or reground dextrose.

Example 1: Effect of Grinding Followed by Granulation — A starch hydrolyzate solution with a DE of 97 and concentrated to 93% solids was seeded at 98°C with 10% by weight of seed crystals of dextrose. The resulting seeded solution was held in the seeding tank for 15 minutes and then was poured at 94°C to form a layer approximately 4 centimeters thick. The cast layer was allowed to stand for 10 minutes. The solidified sugar was then broken into large blocks and ground to pass through a screen containing 10 mm by 10 mm holes. The granules obtained were cooled and then further ground, to pass through a screen containing 1 mm by 1 mm holes. The powdered product obtained was characterized by fine granules of good appearance, especially well suited for tabletting.

Example 2: Effect of Using Different Amounts of Seed Dextrose — Four samples of a starch hydrolyzate liquor with a DE of 97 and containing 93% solids were each seeded with dextrose as follows, the parts being expressed by weight based on the total hydrolyzate weight: (a) 10% dextrose as seed, (b) 5% dextrose as seed, (c) 10% ground cast sugar from Example 1 (crude dextrose) as seed, (d) 5% ground cast sugar from Example 1 as seed.

Samples (a) and (b) were each mixed for 15 minutes before casting. For each, the pouring temperature was 91°C and the standing time, after casting, was 10 minutes with 10% seed [Sample (a)] and 13 minutes with 5% [Sample (b)]. Very good coarse granular dextrose was obtained when both products were ground to pass through a screen containing 10 mm by 10 mm holes.

The mixing time before casting with each of Samples (c) and (d) was 8 minutes. For each, the pouring temperature was 93°C. The casting time was 6 minutes for Sample (c) and 7 minutes for Sample (d). Excellent coarse granular dextrose was obtained when both products were ground to pass through a screen with 10 mm by 10 mm holes. As these demonstrations show, mixing time and casting time can be sharply reduced, if ground sugar from previous runs is utilized as seed in preference to seed crystals from other sources.

Crystalline Anhydrous Beta-Dextrose

It is known that the transition point of the crystalline, anhydrous alpha-dextrose to anhydrous beta-dextrose is 108°C; the transition point of the anhydrous alpha-dextrose to the hydrated alpha-dextrose is 55°C; when there are a liquid phase and a solid phase in a system in a temperature range from 55° to 108°C, the solid phase consists of anhydrous alpha-dextrose; and the anhydrous beta-dextrose starts to deposit only at a temperature of 108°C or higher. Thus, in order to obtain the crystalline, anhydrous beta-dextrose from an aqueous solution of dextrose, it has been thought necessary to carry out the crystallization at a temperature of at least 110°C. However, the crystallizing operation at such high temperatures is very difficult due to the poor stability of the dextrose to heat, and thus it has been deemed impossible to obtain crystalline, anhydrous beta-dextrose on a commercial scale.

T. Kawamura, K. Yamashita, K. Hattori and Y. Ito, U.S. Patent 3,748,175; July 24, 1973; assigned to Tokai Togyo Co., Ltd., Japan have developed a process for preparing a high purity crystalline, anhydrous beta-dextrose on a commercial scale, from an aqueous solution of dextrose such as an aqueous solution of redissolved dextrose or starch hydrolyzate.

These workers have found a specific temperature range for obtaining the beta-dextrose crystals or a mixture of the beta- and alpha-crystals in a better purgeable state from the massecuite containing the same without any thermal deterioration. It is found that the specific temperature range can be divided into three ranges: stable seeding and boiling range (85° to 110°C), metastable boiling range (80° to 85°C), and unstable crystallization range (70° to 80°C). Explanation is made with reference to the accompanying Figure 9.6.

Figure 9.6 is a graph showing saturation curves of anhydrous alpha-dextrose crystals and anhydrous beta-dextrose crystals and imaginary saturation curves of anhydrous beta-dextrose crystals below the transition point and alpha-dextrose crystals above the transition point. That is, the curve **A—B** is the saturation curve of anhydrous beta-dextrose crystals; curve **B—C** that of anhydrous alpha-dextrose crystals; dotted curve **B—D** the imaginary saturation curve of anhydrous beta-dextrose crystals below the transition point; and dotted curve **B—E** that of anhydrous alpha-dextrose crystals above the transition point. The axis of abscissa represents the concentration of the aqueous solution of dextrose in Brix degrees and the axis of ordinate the temperature of the solution in degrees centigrade. The process is summarized as follows.

    (1)   Production of high purity crystalline, anhydrous dextrose:
- (a) Concentration of aqueous solution of dextrose at the seeding must be Brix 88° to 91°.
- (b) Seeding temperature must be 80° to 110°C according to the full seeding method or 85° to 110°C according to the shock seeding temperature.
- (c) Seed crystals must be comprised of anhydrous beta-dextrose crystals.

(d)　Boiling must be conducted at 70° to 110°C, preferably 70°C to the seeding temperature, more preferably a temperature near 70°C.

(2)　Production of a mixture of crystalline, anhydrous beta- and alpha-dextrose at a desired mixing ratio:

(a)　Concentration of aqueous solution of dextrose at the seeding must be Brix 88° to 91°.
(b)　Seeding temperature must be 85° to 110°C.
(c)　Seed crystals must be comprised of a mixture of anhydrous beta- and alpha-dextrose crystals at the desired mixing ratio.
(d)　Boiling must be conducted at 85° to 110°C, preferably 85°C to the seeding temperature, more preferably a temperature near 85°C.

FIGURE 9.6: CRYSTALLINE ANHYDROUS BETA-DEXTROSE

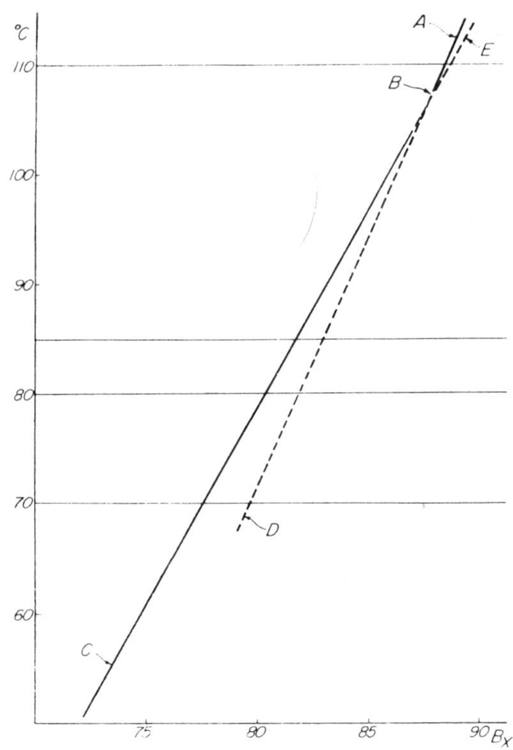

Source:　T. Kawamura, K. Yamashita, K. Hattori and Y. Ito, U.S. Patent 3,748,175; July 24, 1973

Example: An aqueous solution of a starch hydrolyzate (DE: 95.0 to 97%) having a concentration of Brix 50° was concentrated in a vacuum evaporator having a capacity of 25 m³ and when the concentration of the solution reached approximately Brix 80°, the absolute inside pressure of the evaporator was adjusted to 180 mm Hg. Concentrating operation was further continued, and when the concentration of the solution reached Brix 89° and the temperature 90°C, 2 kg of seed crystals consisting of pulverized crystalline, anhydrous beta-dextrose were seeded to the concentrated solution having a concentration of Brix 91° at a temperature of 90°C. The liquid volume in the evaporator at the seeding was 9 m³. After 30 minutes from the seeding, the boiling temperature was reduced to 80°C by reducing the absolute pressure of the evaporator.

The boiling was further continued, while keeping the degree of supersaturation constant. After 8 hours of boiling at 80°C, the crystals grew considerably in the massecuite and the massecuite was transferred to a crystallizer preheated at 80°C, where the crystallization was conducted by gradually lowering the temperature to 78°C, while keeping the degree of supersaturation constant (1.05 to 1.20). When the massecuite temperature reached 78°C, the massecuite was discharged into a centrifuge, where the grown crystals were separated from mother liquor while keeping the temperature at 78°C.

The separated crystals were washed with a small amount of hot water, and the washed crystals were dried in a drier, where crystalline, anhydrous beta-dextrose was obtained as a product. The required total boiling and crystallizing time was 12 hours, and the product yield was 52% on the basis of the weight of total solid matter. The product had a DE of 99.7% and 0.10% of water content, and was of uniform grain size.

Free-Flowing Granular Glucose

K.K.K. Kroyer and L.O. Thomsen; U.S. Patent 3,743,539; July 3, 1973 have developed apparatus and process for producing a free-flowing granular glucose from a glucose solution having a DE of at least 88 and a dry matter content of at least 80 which has been prepared by starch hydrolysis.

The process comprises the steps of continuously supplying the glucose solution to a solid granular glucose mass of a temperature of 50° to 100°C, the glucose solution being supplied at an hourly rate of less than twice the amount of the glucose mass present, mechanically stirring the mixture of glucose solution and glucose mass and continuously removing the free-flowing granular glucose product formed. The reason it is possible to obtain a free-flowing granular glucose product in one step by this process is not known with certainty.

However, it appears that when mechanically stirring the mixture of glucose solution and glucose mass, the glucose solution is distributed over the surfaces of the glucose particles either as thin films or as droplets. The solid glucose particles contacting the film or droplets act as seeds which initiate crystallization in the solution. The mixture of glucose solution and glucose mass which initially is tacky tends to cement the glucose particles together and during the progressing crystallization crystal bridges are formed between adjacent particles to form compact and hard but also brittle agglomerates.

When the agglomerates are stirred, the brittle crystal bridges will be broken down to form a glucose product consisting of particles of different sizes. However, when such a mixture is stirred a fractionation takes place due to the fact that the finer particles descend and the greater particles rise relative to the total mixture. During the upward movement of the larger particles they are continuously stirred and under this treatment and presumably to a still higher degree under the influence of the surrounding particles the surfaces of the agglomerates are abraded. The small crystals or fragments thus formed will also move downwardly.

These fine particles are crystallized to such an extent that they are free-flowing and have a large surface area on which the glucose solution can be distributed. Consequently, the fines moving downward through the glucose mass increase the seeding efficiency. If the amount of fines produced by the stirring is too small, a further mechanical treatment of the agglomerates to disintegrate them should be effected, e.g., by means of a disintegrator. The apparatus used is shown in Figure 9.7.

FIGURE 9.7: FREE-FLOWING GRANULAR GLUCOSE

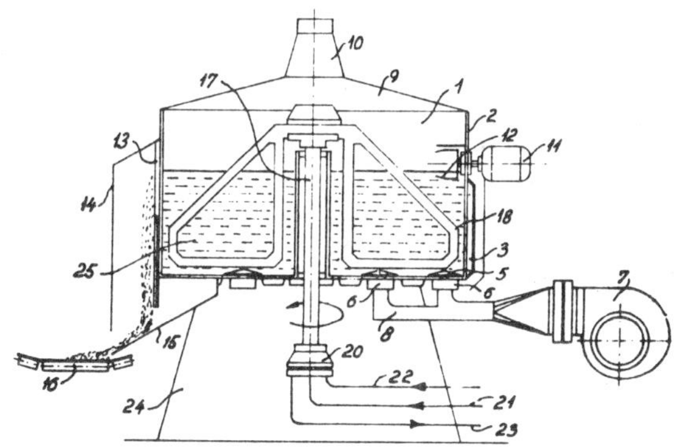

Source: K.K.K. Droyer and L.O. Thomsen; U.S. Patent 3,743,539; July 3, 1973

Referring to Figure 9.7 **1** designates a crystallizer comprising a cylindrical vertical vessel **2** which at the lower portion of the side wall is surrounded by an isolation **3**. The bottom of the vessel is provided with a number of air inlet openings **4**, (not shown in Figure 9.7) each covered by a cover **5**. The air inlet openings are connected with two circular ducts **6** connected to fan **7** by a branch pipe **8**. The vessel is closed at its upper end by cover **9** having air outlet **10**. At one side of the vessel there is mounted a highspeed motor **11** having a motor shaft on which an impeller **12** is mounted. At the opposite side of the vessel there is opening **13** covered by a screen. The opening communicates with a vertical conduit **14**. Below conduit **14** there is an inclined plate **15** leading to conveyor belt **16**.

The crystallizer shown also comprises an agitator means comprising a central hollow shaft **17**, the top of which supports hollow agitator arms **18**. Syrup tubes surrounded by steam jackets and condensate return tubes are provided in these hollow agitator arms **18**. At the lower edge of the agitator arms there are provided syrup discharge openings which are not shown in the drawing.

At the lower end of the central shaft **17** there is a driving means (not shown) for rotating the shaft. Furthermore, a rotating distributor box **20** is located at the lower end of the hollow shaft **17**. The distributor box is connected with conduit **21** for supplying syrup and conduit **22** for supplying steam to the crystallizer. Furthermore, the distributor box is connected to conduit **23** for discharging condensate from the crystallizer. The crystallizer shown in the drawing is mounted on a support **24**.

The operation of the crystallizer is as follows. Syrup is supplied to the crystallizer through conduit **21** and flows through distributor box **20**, hollow shaft **17** and hollow agitator arms **18** to the syrup discharge openings at the lower edge of the agitator arms. During the flow the syrup is maintained at a high temperature by the steam supplied through the conduit **22** and also flowing through the distributor box **20**, the hollow shaft and the hollow agitator arms. The condensate formed in the steam tubes is discharged through distributor box **20** and conduit **23**.

When the agitator arms are rotating through the glucose mass **25** contained in the vessel **2** the syrup is intimately mixed with the glucose mass which initiates a rapid crystallization of the syrup. At the same time air is blown into the bottom of container **2** through air inlets **4**. This air causes the water in the syrup to be evaporated and removed from the crystallizer through the air outlet **10**. The air outlet is preferably connected to a cyclone (not shown) in which entrained glucose particles are separated.

Agglomerates will have a great tendency to concentrate at the surface where they are subjected to impacts from the rotating impeller **12** and are consequently broken down. The fines have a tendency to move towards the bottom and combine with the syrup droplets. Glucose particles having such a size that they pass through the screen in the opening **13** will be removed from the container **2** when the agitator arms are rotating. Such particles fall down through the conduit **14** onto the plate **15** and pass from the plate onto conveyer belt **16**. When being transported away from the crystallizer the glucose product is preferably conditioned in contact only with dry air before it is put into sacks for storage.

Example: 25 kg milled glucose sugar having the following particle sizes: 45% less than 500 μm, 34% between 500 and 1,000 μm, 17% between 1,000 and 2,000 μm and 1% above 2,000 μm was filled into a crystallizer of the type shown in Figure 9.7. The agitator was rotating at 25 rpm and the disintegrator at 4,000 rpm. Before supplying syrup to the crystallizer the glucose sugar was heated to 80°C by blowing hot air into the mass through the air inlets. Then syrup was pumped into the crystallizer at a rate of 10 l/hr (12 kg of dry substance per hour). The syrup used had been prepared by an enzymatic hydrolysis of potato starch and subsequently treated with carbon and with an ion exchanger.

The syrup had been concentrated to obtain a Brix value of 88° to 90° and a temperature of about 100°C. The temperature within the steam jackets surrounding the syrup tubes was 110° to 115°C. The opening in the side wall of the vessel was covered by a screen having 5 mm openings. The removal of glucose product from the vessel was adjusted so that the contents of solid glucose sugar was constantly 25 kg. These conditions were maintained during the whole process which lasted for 10 hours during which about 12 kg of the final product was discharged per hour.

The temperature of the air was maintained constant and the temperature within the crystallizer varied between 80° and 85°C. The product discharged from the apparatus after 1 hour had the following particle sizes: 2% less than 500 μm, 60% between 500 and 1,000 μm, 32% between 1,000 and 2,000 μm, and 5% above 2,000 μm. After 10 hours of operation the particle sizes were as follows: 2% <500 μm, 38% 500 to 1,000 μm, 59% 1,000 to 2,000 μm and 1% >2,000 μm Equilibrium conditions were obtained or almost obtained after ½ to 1 hour and after this time very small variations of the product as far as particle size and shape are concerned were observed.

Analysis of the product showed that the dry matter content was 99.3% and only small deviations were found during the operation. The DE value of the final product was 98 and the ratio of alpha to beta dextrose which initially was 30:60 was changed during the process and equilibrium conditions were established at a ratio of 25:75.

Crystallization of Sucrose-Corn Syrup Mixture

It is known that sugar can be made when any of the liquors and syrups from a sugar refinery as well as other material such as corn syrup solids, dextrin and dextrose are evaporated to remove most of the water to form a viscous syrup containing 93% or more of solids. After the evaporation, the concentrated syrup is then transferred to a beating or creaming apparatus which induces rapid crystallization by agitation. During agitation, sufficient heat of crystallization is released to further evaporate the moisture so that a solid, fondant-like product containing less than 1% water is obtained.

The most commonly used beating apparatus is a trough equipped with a shaft to which arms have been attached. Stationary fingers attached to the trough set between the arms make the equipment self-cleaning and improve the agitation. This produces a sugar product which contains less than 1% water which is essentially a mass of fondant-like agglomerates. This has a rather broad particle size distribution and is unsuited for sale.

The process developed by *F.W. Schwer and C.E. Kean; U.S. Patent 3,655,442; April 11, 1972; assigned to California and Hawaiian Sugar Company* provides an effective way of utilizing substantially 100% of the sugar and producing a product of uniform particle size which is essentially nonhygroscopic.

This is accomplished by providing a screen where the product from the beaters or creamers is divided into three fractions. One of these is an intermediate fraction of a desired particle size, the second fraction is too fine for use and a third fraction is too coarse for use. The fine fraction is passed through a pelletizer and cooler and then combined with a coarse or scalped fraction. The combined fractions are then passed through a comminutor and the comminuted product is now returned to the screening operation.

Thus the particles which are too small are pelletized and then ground to a proper size while those which are too large are ground to the desired size. Naturally such a grinding operation will again produce some fines but these will be picked up in the process and repelletized so that there is no net loss of sugar materials.

In addition to sugar products normally found in a sugar refinery, particularly a cane sugar refinery, other additives such as corn syrup solids and dextrins may be incorporated in the feed stock for conversion to solid granular products.

Example: 1,000 gallons of a liquor having the composition set forth below was cooked at atmospheric pressure to a temperature of 280° to 290°F (138° to 143°C) by passing through a continuous candy cooker.

	Case 1	Case 2
Sugar	49.0	49.0
Corn Syrup Solids	21.0	21.0
(DE of corn syrup)	(24)	(42)
Water	30.0	30.0
Temp. of Cook, °F	285	285

The liquor was concentrated by this operation to a solids content of 93 to 96%. It was introduced into a precreaming device consisting of a horizontal trough equipped with rotating arms disposed within it on a shaft which served to induce crystallization because of the agitation and cooling. Once started either by seeding or mechanical agitation crystallization occurred spontaneously and continuously. Heat given off by the act of crystallization resulted in considerable water vapor being released which was vented through a duct under suction.

The still soft fondant-like mass was transferred to a second creaming device of the same general design and function as the one just described. Here the beating continued, resulting in particles being developed and further liberation and dissipation of water. The product approached 1% in water content at this point and consisted of large agglomerates of sugar particles ranging up to three-fourths of an inch in diameter.

This material was then run between two smooth rollers whose surfaces are continuously cleaned by a third saw tooth roller mounted beneath it. The large agglomerates are thus broken down into smaller particles depending on the spacing between the rollers. However, the product was not sufficiently uniform at this point to be considered finished. The range of particle sizes was too large varying from 100 mesh to 4 mesh in size. Further sizing was required. An optional step following the saw rollers was a drying operation to remove additional moisture from the particles.

The material was then passed through a conventional screen, in this case a Rotex, having a U.S. 12 mesh screen on its upper deck and a U.S. 40 mesh screen in the lower deck. Other mesh sizes may also be used if desired. Three fractions were obtained (1) the plus 12 fraction was considered too coarse and had to be further broken down by passing through a comminuting mill, (2) the –12 +40 fraction which was the product and (3) the –40 fraction which had to be built up into larger particles or pellets which were then in turn comminuted as in (1).

After screening, the coarse fraction (scalpings) was comminuted in a mill. The –40 fraction was a mixture of fine granular material and powder which was then fed into a pelletizing mill of the same type as that used for pelletizing animal feed. Pellets roughly cylindrical in shape (¼" diam x ¼" long) were produced which initially were plastic due to the high temperature developed by the compression. After passing through a cooler (pellets are cooled by a stream of ambient air), the pellets became brittle and were then comminuted in the same mill as used on the scalpings. Conversion of the scalpings and fines into usable material represented by the mid fraction from the screen was accomplished by these two functions of pelletizing and comminuting.

DEXTROSE SOLUBILIZATION OF ACIDS

Citric acid has been employed as an acidulant in dry beverage mixes capable of being dissolved in cold water. In addition to citric acid, such mixes usually contain other hygroscopic materials, such as sugar. These compositions are relatively unstable when stored for extended periods of time since they readily absorb moisture and cake upon standing. Storage problems encountered with such dry beverage mixes have always been of great concern, particularly in climates which are warm and humid. Adipic acid and fumaric acid have many properties which make them desirable for commercial use in such products. However, such uses are limited since both these acids have a very low rate of solubility in cold water.

While the dry beverage mixes of commerce must dissolve in cold water having a temperature of 35° to 45°F in one minute or less, the use of adipic or fumaric acid in such mixes has been impossible due to the fact that these acids do not dissolve rapidly in such water, periods as long as several hours at times being insufficient to put all of the acid into solution.

Since dextrose is stable to moisture absorption upon long storage, it was suggested to dry blend fumaric or adipic acid with dextrose to improve the solubility rate of the acid. However, this method was not successful. It was also attempted to form a heated mixture of acid and dextrose which when cooled to a hard mass was ground to a suitable particle size. This material while having an improved rate of solubility initially was found to have a tendency to absorb moisture during storage thereby rendering the product useless upon reconstitution due to clumping and loss of solubility rate.

W.A. Mitchell and W.C. Seidel; U.S. Patent 3,480,444; November 25, 1969; assigned to General Foods Corporation have developed a cold water soluble adipic acid or fumaric acid composition prepared by heating dextrose to form a melt, uniformly dispersing the acid into the melt and cooling to effect crystallization of the mixture. Crystallization may be promoted by seeding the acid-dextrose melt with a sugar. The resulting mixture can then be comminuted to produce a powdered composition which dissolves rapidly in cold water.

According to a preferred method of this process the crystallization of the acid-dextrose melt is promoted by seeding the melt with a sugar. This can be done by cooling the acid-dextrose melt to between 150° to 250°F and dispersing the sugar into the melt. The seed sugar is preferably dextrose or lactose.

In producing the acid compositions, the acid is first ground to effect a size reduction, typically a particle size of less than 200 U.S. standard mesh is necessary and preferably a particle size smaller than 400 U.S. standard mesh is desirable. The subdivided acid is then dispersed into the melted dextrose to achieve a uniform distribution of small acid particles throughout the melted dextrose.

The acid-dextrose melt is then cooled, preferably to below 230°F, and most preferably to between 200° to 210°F before incorporating powdered lactose or dextrose. Enough dextrose or lactose should be added to achieve a level of between 10 and 40% acid by weight of the total product. Usually 10 to 15% by weight of seed is added to crystallize the dextrose in proper form. However, any amount of seed may be added as long as the final acid content is between 10 to 40%.

As the dextrose forms into crystals, the fine particles of acid will be forced between the surfaces of the crystals as they form. During crystallization some dextrose, due to impurities being present in the melt, will not crystallize and will remain in an amorphous state. A small percentage of amorphous sugar is desirable as long as this is kept to below 20% and preferably 3 to 10% by weight of the total composition.

After crystallization of the dextrose the hardened mass is subdivided or ground to a particle size suitable for quick/solubility in cold water. This size will be between 40 and 140 U.S. standard mesh and preferably 40 to 100 mesh. Particles greater than 40 mesh will not dissolve in less than 3 minutes and particles smaller than 140 mesh will not retain the acid particles in an embedded state and will tend to clump and float upon rehydration.

Example: Powdered dextrose (95% purity) was dissolved in warm water (80°F) until a 70% concentration of dextrose was achieved. The dextrose solution was further concentrated in a low-temperature scraped-surface vacuum evaporator at a pressure of 20" mercury vacuum and a heating jacket temperature of 340°F. The jackets were heated by steam at 100 psig. The dextrose solution entered the evaporator at room temperature and was removed at a temperature of 300°F as a free-flowing liquid. Residence time in the evaporator was about 1 minute and the liquid was concentrated to 98% dextrose solids.

Fumaric acid, finely divided to a mesh size of less than 400 U.S. standard mesh (37 microns), was then dispersed uniformly into the dextrose melt at a level of 3 parts dextrose to 1 part fumaric acid. Uniform dispersion was accomplished by means of a colloid mill having a clearance of between 0.005" to 0.060". The liquid solution entered the colloid mill at a temperature of 260°F and was extruded at a temperature of 270°F. Residence time in the colloid mill was 30 seconds.

The product issuing from the colloid mill was then crystallized by dispersing powdered dextrose as seed material into the fumaric acid-dextrose melt. The melt was first cooled to 200°F, before adding the powdered dextrose in order to prevent melting of the seed. Enough dextrose was dispersed throughout the melt to achieve a concentration of 18% fumaric acid by weight in the final product. The product hardened to a crystallized mass in less than 1 minute. The crystallized mass was then ground in a Fitzpatrick mill to a mesh size of less than 40 U.S. standard mesh and above 100 U.S. standard mesh.

Untreated fumaric acid having a particle size of less than 400 U.S. standard mesh has a solubility rate of 3.2 g/2 qt of water in 15 to 20 minutes when dissolved in water at 45°F whereas the fumaric acid composition produced in accordance with this process had a solubility rate of 3.2 g/2 qt of water in less than 1 minute at the same temperature.

Storage of this composition at 90°F and 85% RH for 6 months revealed no caking or degradation; upon reconstitution it had the same taste, solubility rate, and clean appearance of the freshly prepared sample.

GLUCONATE-CONTAINING PRODUCTS FROM STARCH HYDROLYSATES

The transportation and storage of starch hydrolysates is relatively expensive since they generally are not spray dryable, due to their high glucose content, and must be handled in the liquid state. This expense has to some extent limited the direct use of these materials in foods. It would be a distinct advantage if simple and economical treatment of starch hydrolysates could lead to spray dryable products without the above-mentioned drawback. It would also be advantageous if the products contained high calcium or iron contents since calcium is a dietetic necessity for the formation of bones, teeth, shells, milk etc., in animals and iron is a dietetic necessary for the prevention of anemia.

R.G.P. Walon; U.S. Patent 3,625,701; December 7, 1971; assigned to CPC International Inc. has disclosed a process for preparing carbohydrate-gluconate products which comprises selectively oxidizing the glucose in a starch hydrolysate solution to form a carbohydrate-gluconic acid product. A carbohydrate-gluconate product may be obtained by addition to the starch hydrolysate solution of an amount of base stoichiometrically calculated to give the desired amount of metal gluconate in the final product through neutralization of the gluconic acid produced.

The oxidation may be carried out enzymatically, by electrolysis, or by other chemical methods well known to the art. It is essential that the method of oxidation be chosen so that glucose is selectively oxidized to gluconate, without the concurrent oxidation of other saccharides.

In the preferred procedure, the glucose in starch hydrolysates is selectively oxidized enzymatically to a controlled extent by adding quantities of basic substances to hydrolysate-enzyme mixtures and incubating the solutions. Potentially edible products are thereby obtained. The amount of base added controls the degree of reaction in the enzymatic procedure because once this base has been used up in neutralizing any gluconic acid formed, the pH of the solution begins to drop and the enzyme is inactivated. This generally occurs at a pH of below 4.5. The glucose oxidase enzyme will not operate to convert glucose to gluconic acid above a pH of 8.

Hence, during oxidation, the pH must be controlled to fall within the range from above 4.5 to 8. Most preferably, the pH is maintained to fall within the range from 5 to 6.5 since the enzyme action is most efficient in this range. One preferred procedure consists of using calcium carbonate as the base to control the extent of reaction of enzymatically oxidizing the glucose in greens. The greens are obtained as the mother liquor after a dextrose crystallization process, and have a solids content adjusted by dilution if necessary to 10 to 70% by weight. A product is recovered that is potentially usable as a poultry or cattle feed. In the most preferred procedure, components for feeds usable for poultry, cattle, etc., are produced from greens having a DE falling within the range from about 70 to 80.

In another preferred embodiment, calcium carbonate is used as the base and the glucose in a starch hydrolysate solution having a DE of 10 to 70 is enzymatically oxidized to obtain a product which may be used as a baby food, a dietetic food, a clinical food, for poultry feeding, for cattle feeding, etc. A starch hydrolysate with a DE falling within the range of from 25 to 40 is the preferred starting material for preparing these carbohydrate-gluconate products.

Example 1: Preparation of Gluconic Acid Product from High DS Starch Hydrolysate — A starch hydrolyzate was coverted to a dextrose-fructose solution which exhibited a DE of 96.8, the solids therein consisting of 66.4% dextrose, and 27.5% fructose and the remainder higher saccharides by standard techniques. The solution was placed in a closed reaction vessel with glucose oxidase enzyme. Sufficient sodium hydroxide was added dropwise to the reaction during the reaction to transform 15% of the dextrose into sodium gluconate, based on total sodium hydroxide added. The pH was maintained at 5 to 6.5 during NaOH addition. Oxygen was added to the reaction vessel and a positive pressure of 1 psig was maintained.

When all of the sodium hydroxide had been consumed, the reaction ceased due to inactivation of the enzyme as the pH dropped. The resulting solution was treated by electrodialysis to convert the sodium gluconate into gluconic acid. An acid sweet syrup was obtained having a composition of the solids therein of 51% dextrose, 27.5% fructose, 15% gluconic acid, and the remainder higher saccharides. The sweetening power of this syrup was 90% of that of sucrose.

This example demonstrates the preparation of an acid sweet syrup, the acidity of which is due to the presence of gluconic acid. This product is useful in the preparation of lemon and orange flavored drinks as illustrated in additional examples in the complete patent.

Example 2: Preparation of Calcium Gluconate Product from Starch Hydrolysate — An alpha amylase treated hydrolysate consisting of 32% dextrose and 68% saccharides which exhibited a DE of 45, was put in a closed reaction vessel with glucose oxidase enzyme and with sufficient calcium carbonate to convert 30% of the 32% of the dextrose into calcium gluconate. The reaction was allowed to continue for seven hours. The product obtained showed an analysis of 2% dextrose, 30% calcium gluconate and 68% higher saccharides. It exhibited a DE of 13.

The product of Example 2, made from a starch hydrolysate with a DE of 45, due to its relatively low glucose content, is easy to handle in a spray drying tower. This product should be useful as an additive in poultry feeds, cattle feeds, in milk replacement substitutes, as a constituent of dietetic foods, baby foods, desserts, etc.

MALTOSE AND LOWER OLIGOSACCHARIDE SYRUPS

Maltose, also known as malt sugar and maltobiose is the disaccharide 4-(α-D-glucosido)-D-glucose. Maltose, like dextrose, is a reducing sugar. Its syrups, like the syrups of dextrose, inevitably contain other saccharides. High maltose syrups are valuable for many applications because they exhibit a decreased tendency to crystallize, as compared to high dextrose corn syrups, and tend to be nonhygroscopic. Syrups having high maltose contents have previously been produced industrially by the saccharification of starch or starch hydrolysates with malt enzymes. The conversion action of malt enzymes results in the production of a conversion product in which maltose is the most abundant saccharide present. A typical procedure, for making a high maltose syrup, involved solubilization of starch, followed by treatment with malt enzymes to saccharify the solubilized starch so as to obtain a starch hydrolysate having a high maltose content.

Suitable enzymes for this saccharification process have been limited to malt enzymes. The saccharification or conversion of the starch hydrolysate has usually been carried out with malt enzymes to a DE of 35 to 55, depending upon the extent of conversion desired. The upper limit for economical malt saccharification is usually considered to be 50 to 55 DE. Such a conversion liquor will usually contain 60 to 65% maltose. Saccharification to this extent is dependent upon the starch solubilization step used.

If high maltose syrups of greater than 55 DE are desired, further saccharification may be accomplished with other saccharifying enzymes such as fungal amylase. Such saccharifying enzymes will form predominantly dextrose from the saccharides present. Such a second saccharification step will usually result in a slight decrease in the total amount of maltose present but nevertheless yields syrups of high maltose content, which may be sweeter and usually contain a higher amount of fermentables than the conventional malt conversion syrup.

High maltose syrups have become increasingly important in commercial applications. Syrup compositions having a high maltose content furnish desirable nonhygroscopic properties to hard candies. They are also useful in controlling crystal formation in frozen dessert formulations. Similarly, the high fermentables content of high maltose syrups is of value in the baking and brewing industries.

Reference should also be made to U.S. Patent 3,525,672, page 217, where beta-amylase is used to thin crosslinked starch with maltose production.

PRODUCTION OF MALTOSE USING AMYLASE

β-Amylase from Wheat Bran

M. Mitsuhashi; U.S. Patent 3,492,203; January 27, 1970; assigned to Hayashibara Company, Ltd., Japan has developed the process for extracting pure β-amylase from wheat bran, saccharifying starch and obtaining a saccharified product which contains the large amount of transparent, odorless maltose. In this process wheat bran is treated with water at 20°C to 40°C, and an extract containing β-amylase is obtained. The wheat bran was treated by water without previous pasteurization. 500 grams of various kinds of wheat brans were steeped in 4 times their weight of warm water at 40°C for 1.5 hours and centrifuged. The results of the activity tests of β-amylase on these extracted solutions are indicated in the table below.

Wheat Bran as Raw Material	Starch Value (%)	Total Nitrogen (%)	Activity of β-Amylase, units/g
Canadian wheat* and American wheat**	41	2.30	274
Canadian wheat:			
A	41.6	2.55	243
B	41.3	2.55	319
C	41.7	2.85	448
E	41.6	3.4	441
Nisshin Seifun K.K. product (Flower brand)	58.2	3.47	350
Marusho Seifun K.K. product (AA brand)	48.4	2.47	471

*Manitoba wheat
**Hard winter; western white

The β-amylase activity is determined as follows: A mixture of 5 ml of a 1% soluble starch solution; 4 ml of 0.1 M acetic acid buffer solution, pH 5.0; and 1 ml enzymatic solution was reacted for 30 minutes at 40°C, the reduced sugar produced determined quantitatively as maltose and enzyme value at the time of producing 10 mg glucose was made 1 saccharification unit. Wheat bran materials are as follows: Canadian Manitoba wheat; American hard winter; Nisshin Seifun (Flower brand) a mixture of Canadian product, American product and Japanese product; and Marusho Seifun (AA brand) a mixture of Canadian product, American product, and Japanese product.

As shown in the table above, activity values of β-amylase varies considerably by kinds of wheat bran used, but are approximately one-half of that of malt amylase. The amount of water used is appropriately equal or four times as large as amount of bran. If it is too large, an enzymatic solution becomes dilute. Furthermore, properties of enzyme of those extract solutions were investigated and show that the power to decompose starch into maltose, is strong but the power to decompose starch into dextrin is entirely absent. This differs from the properties of the malt enzyme.

Example: Four times as much water heated to 40°C was added to 500 grams of wheat bran obtained from Canadian wheat and the resultant solution was lixiviated and then centrifuged giving 890 grams of extracted residue (moisture content: 62.5% and dried substance: 383.8 grams) and 1,550 ml of extract (dried substance: 89.7 grams; enzyme activity SA 105 units/ml). Then 0.4% of oxalic acid based on the amount of starch used was added to an emulsion of sweet potato refined starch of 1.15 of SG. The mixture was converted to a liquefied solution having a DE of 25 under one atmospheric pressure and for 15 minutes.

The liquefied solution was neutralized by calcium carbonate to 5.2 pH, and the enzyme solution (2 units/g starch) was added for saccharification at a pH of 5 and for 24 hours at 55°C. A saccharified solution having a DE of 51 was obtained. The saccharified solution was decolored by Carbonatine, a product of Takeda Kagaku K.K., and 0.2% of the Special-Flow, filter aid, a product of Showa Kagaku K.K. was added to the solution. This solution was decolorized by active carbon and then was purified through ion exchange resin (Amberlite IR-128, IR-68 and IR-120, IR-411 in mixture) and there was produced a clarified saccharified solution in high purity, containing a large amount of colorless maltose. Composition of sugar components is as follows.

	Glucose	Maltose	Maltotriose	Tetraose	Dextrin
	- - - - - - - - - - - - - - - Percent - - - - - - - - - - - - - - - -				
High maltose syrup produced by this process DE 50	7	47	17	5	24
Corn syrup produced by acid saccharification DE 50	26	17	13	10	34

Use of Amylase from *Bacillus polymyxa*

In the process developed by *F.C. Armbruster and W.A. Jacaway, Jr.; U.S. Patent 3,549,496; December 22, 1970; assigned to CPC International Inc.*, a starch hydrolysate having a DE less than 20 is subjected to the hydrolytic action of *Bacillus polymyxa* amylase to produce a starch conversion product having a higher maltose content than DE value. *Bacillus polymyxa* amylase is the maltogenic enzyme produced by members of the bacterial species *Bacillus polymyxa* when suitably incubated under conditions of aerobic culture. The characteristics by which members of the species *Bacillus polymyxa* may be distinguished are described in *Bergey's Manual of Determinative Bacteriology*, 7th edition, p. 625. However, it is well known that mutant strains may be isolated from time to time, that do not conform completely to this description.

The partially hydrolyzed starch, that is used as the starting material in this process, is obtained by acid hydrolysis or enzyme hydrolysis of any conventional starch. Prior to hydrolysis, the starch is solubilized by gelatinization by heating to a temperature exceeding 60°C in the presence of moisture. Acid hydrolysis of starch is carried out in a conventional manner to a DE not exceeding 20, preferably to a DE between 10 and 20. Acid hydrolysis of starch is carried out in a conventional manner to a DE not exceeding 20, preferably to a DE between 10 and 20. Enzyme hydrolysis of starch is carried out using suitable liquefying enzymes to attain a DE not exceeding about 20, perferably between 5 and 20.

Example: This example illustrates a preferred procedure for purifying and concentrating *Bacillus polymyxa* amylase, and for the application of the enzyme preparations obtained to convert a partially hydrolyzed starch to obtain high maltose starch conversion products. It demonstrates the effects of temperature and of enzyme concentration upon the conversion.

Bacillus polymyxa amylase preparation is as follows. A shake flask fermentation employing *Bacillus polymyxa* ATCC 8523 was conducted, following the previously described procedure. The culture filtrate obtained was adjusted to pH 6.8, agitated, and to it was added 1,500 ml acetone and 15 grams diatomaceous earth filter aid per 1,000 ml of culture filtrate. After agitation, the suspension obtained was then filtered, and the filtrate was discarded. The filter cake was suspended in a volume of water equivalent to one-tenth the volume of the original culture filtrate, and the resulting suspension was filtered, yielding a purified enzyme solution possessing approximately 10 times the activity per unit volume of the original culture filtrate.

Application of the *Bacillus polymyxa* amylase preparation to produce high maltose starch conversion products is as follows. The above purified enzyme concentrate was used to convert several separate portions of a 35% suspension of an acid hydrolyzed, 16 DE cornstarch hydrolysate. This hydrolysate, nominally 16 DE, analyzed as follows: DE, 16.3; maltose content, 4.2% DB; and yeast fermentables, 10.7% DB. The conversions were conducted at pH 6.0 to 6.5 and respectively at 40°, 50°, 55°, or 60°C, at enzyme dosages ranging from 1.1 to 33.6 units of maltogenic enzyme activity per 100 grams of substrate, dry substance basis. The conversions were terminated after 72 hours and the conversion products were then analyzed for DE, maltose by paper chromatography, and yeast fermentables. The results obtained were as follows.

		Conversion results		
	Enzyme dosage (units/100 g.d.s.)	D.E.	Maltose content (percent d.b.)	Yeast ferment ables (percent d.b.)
Conversion temp. (° C.):				
40	4.2	40.9		
	8.4	48.0	56.7	74.8
	16.8	53.9	59.5	78.6
50	1.1	38.2	43.7	60.4
	2.1	44.7	51.1	72.0
	4.2	48.0	55.0	75.7
	8.4	52.5	59.7	77.6
	16.8	56.3	61.4	79.0
	33.6	58.2	61.2	79.8
55	1.1	37.7	47.2	63.6
	2.1	40.1	50.4	70.4
	4.2	44.8	54.4	72.3
	8.4	47.4	53.3	75.5
	16.8	53.3	60.3	78.0
60	4.2	44.6	52.8	
	8.4	46.9	52.7	
	16.8	52.4	58.3	

As may be seen in the above table, high maltose conversion products were obtained in every instance where the *Bacillus polymyxa* amylase was applied. It may also be seen that the enzyme dosage may be varied over a wide range.

Improved Amylase Enzymes from Streptomyces

Seven specific species of Streptomyces have been used by *Y. Koaze, Y. Nakajima,*

H. Hidaka, T. Niwa, T. Adachi, K. Yoshida, J. Ito, T. Niida, T. Shomura, and M. Ueda; U.S. Patent 3,804,717; April 16, 1974; assigned to Meiji Seika Kaisha, Ltd., Japan to produce enzymes useful for the production of maltose from starch. The starch hydrolysate which is obtained by reacting these amylases with starch in aqueous dispersion contains a predominant proportion of maltose and is useful as a natural sweetening material. Further, this starch hydrolysate mainly composed of maltose may be converted into a mixture of sugar alcohols mainly composed of maltitol by hydrogenating in the presence of a nickel catalyst.

For the production of these amylases it is preferable to use and cultivate any one of the particular strains of Streptomyces which are listed in the table below.

Strains	F.R.I. deposit number [1]	ATCC deposit number [2]
Streptomyces tosaensis SF-1085 (this strain was first isolated from soil and discovered by the present inventors	601	21723
Streptomyces hygroscopicus SF-1084 (this strain is coincident with *Streptomyces hygroscopicus* as described in Waksman's "The Actinomycetes," vol. 2 (1961) and in the "Applied Microbiology," vol. 10, pp. 258-263 (1962))	602	21722
Streptomyces viridochromogenes SF-1087 (this strain is coincident with *Streptomyces viridochromogenes* as described in Waksman's "The Actinomycetes," vol. 2 (1961) and in the "Journal of Bacteriology," vol. 85, pp. 676-690 (1963))	603	21724
Streptomyces albus SF-1089 (this strain is coincident with *Streptomyces albus* as described in Waksman's "The Actinomycetes," vol. 2 (1961))	604	21725
Streptomyces flavus	605	
Streptomyces aureofaciens	606	
Streptomyces hygroscopicus var. *angustomycetes*	607	

[1] F.R.I. is an abbreviation of "Fermentation Research Institute," Agency of Industrial Science & Technology of Ministry of International Trade & Industry of Japan, residing in Inage, Chiba City, Japan.
[2] ATCC is an abbreviation of American Type Culture Collection, Washington, D.C., U.S.A.

The above specified strains can efficiently produce and accumulate the improved amylases when they are cultivated in a known manner under aerobic conditions.

The particular enzymatic activities and characteristics of the improved amylases of streptomyces as produced by this process are summarized in the table below. For comparison, properties of some known amylases of bacteria, fungi, plant and animals are shown together in the table.

Sources of amylase	Optimum pH	Limit in hydrolysis of starch by action of amylase	Ratio of glucose to maltose produced, by weight	Optimum temperature (° C.)	Actibation by— Calcium cation	Chloride anion
Streptomyces albus	4.5-5.0	78	0.058:1	50-60	Negative	Negative.
Streptomyces aureofaciens	4.5-5.0	76	0.058:1	50-60	do	Do.
Streptomyces hygroscopicus	4.5-5.0	82	0.055:1	50-60	do	Do.
Streptomyces hygroscopicus var. *angustomycetes*	4-5.5.0	80	0.053:1	50-60	do	Do.
Streptomyces viridochromogenes	4.5-5.0	79	0.051:1	50-60	do	Do.
Streptomyces flavus	4.5-5.0	76	0.058:1	50-60	do	Do.
Streptomyces tosaensis nov. sp	4.5-5.0	79	0.053:1	50-60	do	Do.
Bacillus substilis (bacteria)	5.3-6.8	[1] 70	0.71:1		do	Positive.
Aspergillus oryzae (fungi)	5.5-5.9	[1] 48	0.64:1		do	Negative.
Germinated barley (plant)	4.7-5.8	80			Positive	Do.
Hog pancreas (animal)		6.9	80	0.024:1	Negative	Positive.
Human saliva		6.9	80		do	Do.

[1] Calculated as glucose.

By comparing the data shown in the above table, it is evident that the improved amylases of these seven Streptomyces species are not coincident with any of the known amylases particularly in view of their optimum pH range, the limit in the hydrolysis of starch, activation by calcium cation and chloride anion. That these amylases of the above seven Streptomyces species exhibit a relatively lower optimum pH and higher optimum temperature, a relatively higher value for the limit in the hydrolysis of starch in combination with their lower ratio of glucose to maltose, is very advantageous for the commercial production of maltose from starch. This allows the hydrolysis of starch to take place at a pH and temperature that would prevent the contamination by growth of undesired microorganisms, enables the hydrolysis of starch to proceed more completely and more efficiently, and leads to a more efficient and preferential formation of maltose.

Example 1: A strain of *Streptomyces hygroscopicus* identified as FRI 602 or ATCC 21722 was inoculated to 9 liters of a seed medium comprising 2% corn meal, 1% wheat embryo and 0.5% ferma media, adjusted to pH 7.0, and the incubation was made at 28°C for 24 hours in a jar-fermenter with agitating and aeration to prepare a seed culture. On the other hand, 300 liters of a production medium of such composition comprising 3% cornstarch, 1% skimmed milk, 0.2% potassium dihydrogen phosphate, 0.05% magnesium sulfate, 0.01% manganese sulfate and a small proportion of defoaming agent were charged in a 600 liter capacity fermenter, sterilized at 121°C for 30 minutes and cooled.

The seed culture prepared in the above was then inoculated to the sterilized production medium, and production medium was incubated at 28°C for 85 hours with aeration and agitation. The resulting culture broth was filtered to give the culture filtrate which was subsequently concentrated below 40°C under reduced pressure to a liquid volume of $^1/_5$ times the original volume. To the concentrate was then added a Z-fold larger volume of cold ethanol, precipitating the amylase which had been produced and accumulated in the culture broth. Drying of the precipitate gave 341 grams of a crude enzyme which showed a potency of 43,800 units per gram.

Example 2: 6 kg of cornstarch was dispersed in 8 liters of tap water, adjusted to pH 6.0 and then added 0.2% of the known bacterial liquefying amylase. The aqueous mixture was adjusted to pH 6.0 by addition of calcium hydroxide and then added dropwise into 6 liters of hot water. Primary liquefaction was carried out at 92° to 93°C. After the completed primary dextrinization, the liquefied starch solution was heat-treated at 120°C. Secondary liquefaction was then carried out adding 0.05% of the same bacterial liquefying amylase.

When the liquefaction of starch was finished, the enzyme was inactivated by heating at 100°C for 5 minutes. When the liquefied starch solution had cooled to a liquid temperature of 55°C, the solution was added with the dry powder amylase preparation of *Streptomyces hygroscopicus* which was prepared in Example 1, in an amount of 400 units per gram of starch. The saccharification reaction was carried out at an initial pH of 6.0 and at a temperature of 55°C for 40 hours.

After the saccharification was completed in the above way, the reaction mixture was filtered to give 24 liters of a filtrate which was subsequently decolorized and desalted in a routine way. By concentrating the decolorized and desalted solution to a solid content of 50%, there was obtained a syrup of the starch hydrolysate comprising 4.1% glucose, 71.3% maltose, 11.7% oligo-sugars and 12.9% dextrin.

Example 3: 250 grams of Raney nickel catalyst was added to the syrup of Example 2 and the initial pH of the syrup was adjusted to pH 9.0 and the syrup was placed in an autoclave of a capacity of 30 liters and fitted with a stirrer. Hydrogen was fed into the autoclave so that the syrup was subjected to the catalytic hydrogenation reaction at 130°C for 3 hours and under a hydrogen pressure of 100 kg/cm^2. After the completed hydrogenation, the reaction mixture was freed from the catalyst. The separated or recovered catalyst was added to a second batch of the syrup of the starch hydrolysate which was prepared in the next operation of the saccharification of starch, and the catalytic hydrogenation was repeated with the second batch of the syrup and under the same hydrogenation conditions as mentioned above.

The filtrate of the reaction mixture from the hydrogenation process was determined for the residual amount of the reducing sugars by the Somogyi's method. It was found that the residual amount of the reducing sugars as defined above was 0.21% in the filtrate, namely the catalyst-free reaction mixture which was obtained from the first run of the hydrogenation process. Further batches of the syrup of the starch hydrolysate were hydrogenated in successive runs under the same hydrogenation conditions, and the recovery and reuse of the catalyst was repeated in these successive runs of the hydrogenation. The residual amount of the reducing sugars increased gradually as the runs succeeded, and it reached 1.03% in the catalyst-free reaction mixture which was obtained from the sixth run of the hydrogenation.

After this sixth run of the hydrogenation, 50 grams of a fresh Raney nickel catalyst was supplemented to the recovered catalyst, and the recovered catalyst so reactivated was reused in the seventh run of the hydrogenation process. The recovery and reuse of the catalyst was again performed in the further runs of the hydrogenation process.

The reaction mixtures which were obtained from the successive runs of the hydrogenation process were each filtered, desalted and concentrated to yield an aqueous, clear and colorless composition of the sugar alcohols which had a 75% solid content and comprised no less than 70% of maltitol and minor amounts of sorbitol, maltotriitol and others. This aqueous composition mainly composed of maltitol was obtained at a yield of about 6.6 kg in each run of the above-mentioned hydrogenation procedure.

Dialysis of β-Amylase Produced Maltose

It has been found by *M. Kurimoto; U.S. Patent 3,832,285; August 27, 1974; assigned to Hayashibara Biochemical Laboratories Incorporated, Japan* that purer maltose than previously available in a practical manner can be prepared under conditions suitable for industrial application by subjecting a starch consisting predominantly of amylopectin to the action of pure β-amylase, and by dialyzing the reaction mixture against water.

The pure β-amylase required for this process is obtained by known methods from amylases rich in β-amylase recovered from vegetable matter such as wheat bran, wheat malt, soy beans, potatoes and the like. The β-amylase is purified by adsorption, salting out, solvent precipitation, or separation by means of Sephadex Gel, or the α-amylase present may be inactivated by means of acid.

The starch used as a starting material should predominantly consist of amylopectin. Substances forming mono- or trisaccharides under reaction conditions, such as maltooligosaccharides, maltodextrins, and amylose should be avoided as far as possible. The preferred starting materials thus are waxy cornstarch, waxy rice starch, and amylopectin separated from other types of starches, but at least some of the advantages of this process are available as long as amylopectin predominates in the starch used.

The starch is dispersed in water to a concentration of 2 to 30% (by weight), and liquefied or solubilized by heating the slurry to a temperature not exceeding 130°C. The process is more easily controlled when the concentration does not exceed 20% and the temperature does not exceed 120°C in order to attain a uniform dispersion and a DE of 5 or less. Acids and enzymes are to be avoided at this stage, but a slight alkalinity of the slurry can be helpful.

The solution so obtained is brought into contact with the β-amylase at the optimum reaction conditions for the latter, that is, a pH of 5.0 to 6.0, and a temperature of 50° to 60°C. If the starting material consists entirely or almost entirely of amylopectin, the β-amylolysis may be permitted to proceed until the reaction mixture has a DE of 50 to 60. If much amylose is initially present, the reaction should be interrupted at DE 30 to 40.

The time required to reach the desired reaction stage depends on the amount of β-amylase employed. When 5 to 100 units of enzyme are used per gram of starch, the reaction period may be between 15 and 5 hours. When the desired DE value is reached, the reaction may be interrupted by inactivating the enzyme at high temperature.

The mixture is then dialyzed against water. The amount of water employed is not critical, but it is desirable to use the smallest amount of water consistent with the desired result to reduce the amount of liquid which has to be evaporated. A compromise between the contradictory requirements which is usually practical involves the use of an outer solution in a volume five to ten times that of the aqueous reaction mixture. The period required for dialysis can vary greatly depending on the temperature and concentration of the liquid, and is usually between 5 and 15 hours. The dialysis temperature may vary between ambient temperature (approximately 20°C) and 50°C, and agitation or circulation of the outer liquid is beneficial in maintaining a desirable concentration gradient across the dialysis membrane.

Suitable membranes include tubes of regenerated cellulose (Visking), ultrafiltration membranes (Amicon Diaflo), and other materials employed for molecular sieving such as gel filtration or reverse osmosis. It is preferred to employ a membrane having an effective pore diameter of 5 A to 30 A.

The dialysate, which is an aqueous maltose solution, is concentrated by partial evaporation of the water, and purified by treatment with active charcoal which removes coloring matter, and by ion exchange to remove ions. The solute is maltose having a purity greater than 95% and as high as 98%, if the starting material is essentially amylopectin, without any need for recrystallization. A purity greater than 99% is readily achieved by a single recrystallization step. If a common starch containing some amylose is employed as a starting material, very pure maltose can still be obtained if the reaction is interrupted before oligosaccharides of low molecular weight are produced. Percentage values in the example are based on weight unless specifically stated otherwise.

Example: Waxy cornstarch which was practically pure amylopectin was suspended in enough water to produce 500 grams of a suspension containing 6.15% starch, and gelatinized by careful heating. The solution was then adjusted to a pH between 5 and 6, and incubated at 50° to 60°C with pure β-amylase free of α-amylase, maltase, glucoamylase, and isoamylase. When the mixture reached a DE value of 50, it was heated to terminate the reaction, and the solution of maltose and oligosaccharides was transferred to a seamless cellulose tube (Visking) which was immersed in a much greater volume of pure water which was stirred at room temperature for 15 hours.

The liquid outside the cellulose tube was analyzed and found to contain 16.10 grams solids (52.4% of the starting material) which consisted of 98.02% disaccharide, 1.87% trisaccharide, and 0.11% monosaccharide. No tetrasaccharide or higher saccharide (dextrin) could be detected. When the same procedure was repeated with a starting suspension of 9.22 grams waxy starch (1.84%), 61.3% (5.65 grams) of the solids was found in the dialysate and consisted of 98.25% disaccharide, 1.65% trisaccharide, and 0.10% monosaccharide.

Potato starch in an initial concentration 6.13% yielded 61.2% of the initial solids in the dialysate which consisted of 93.64% disaccharide, 2.54% dextrin (tetrasaccharide or higher), 2.03% trisaccharide, and 1.78% monosaccharide. The yield was increased to 68.7% when the initial potato starch suspension had a concentration of only 1.80%, and consisted of 96.15% disaccharide, 1.31% each of dextrin and monosaccharide, and 1.23% trisaccharide. The amounts of oligosaccharides and dextrins found in the dialysate are due to the relatively large pore diameter (24 A) of the cellulose employed.

DUAL ENZYME PROCESSES FOR PRODUCING MALTOSE

It is a known process to liquefy starch by the use of β-amylase and then add malt to accomplish saccharification by the action of the malt amylase (i.e., a mixture of α- and β-amylases). By such process, the saccharified solution obtained would contain at most 70% of pure maltose even if the amount of β-amylase was increased by the use of malt containing a fairly large amount of β-amylase. However, the addition of α-1,6-glucosidase which is an enzyme capable of specifically acting in the cleavage of α-1,6-glucosidic linkage in the amylopectin of starch produces a high purity maltose solution.

The term α-1,6-glucosidase is generic and would include both pullulanase produced by the bacteria of the genus Aerobacter, and also isoamylase yielded by the bacteria of the genus Pseudomonas, for example the variety obtained by culturing *Pseudomonas amyloderamosa* (ATCC No. 21262). Experiments have proven that the latter enzyme combined with β-amylase can give a maltose solution as pure as that produced by the combination of pullulanase and β-amylase.

High Maltose Hydrolysates

A process is disclosed by *R.E. Heady and F.C. Armbruster; U.S. Patent 3,565,765; February 23, 1971; assigned to CPC International Inc.* for the preparation of a high maltose-containing starch conversion product which comprises subjecting partially hydrolyzed starch, having a DE not exceeding 20, to conversion with a maltogenic enzyme and a pullulanase enzyme to obtain a product having a DE of at least 45, a maltose content of at least 50%, and yeast fermentables content of at least 80%.

The partially hydrolyzed starch that is used as the starting material is obtained by acid or enzyme hydrolysis of any conventional starch. Prior to hydrolysis, the starch is solubilized by gelatinization by heating above 60°C in the presence of moisture. Acid hydrolysis of starch is carried out in a conventional manner to a DE not exceeding 20, preferably to a DE between 10 and 20. The debranching enzyme, pullulanase, used in the process is produced by members of the bacterial species *Aerobacter aerogenes* when suitably incubated under conditions of aerobic culture.

The method of producing the enzyme pullulanase is described by Hans Bender and Kurt Wallenfels in an article entitled "Specific Decomposition by a Bacterial Enzyme." The article appeared in *Biochemische Zeitschrift,* 334, 79-95 (1961). The source of the *Aerobacter aerogenes* culture used in the present process was the ATCC. The designation of the culture was *Enterobacter aerogenes* ATCC 8724.

In the commercial production of high maltose syrups, it is common for the conversions to be performed at relatively high dry substance levels, usually within the range of 15 to 40% to reduce tank size requirements and evaporation costs, and at relatively high temperatures such as 50° to 60°C to retard or prevent microbial spoilage of the conversion liquors. Many enzymes are inhibited at high substrate concentrations and at high temperatures such as those mentioned above. Surprisingly, however, pullulanase prepared in accordance with the above, is quite amenable for use under the conditions required for successful and economical industrial operations for the production of high maltose conversion products from starch.

The conversions may be performed within the pH range of 4.5 to 8.0, the preferred range being 5.0 to 6.5. The time required for conversion of partially hydrolyzed starch to a high maltose starch conversion product will depend upon the enzyme dosage used and the extent of conversion desired. However, desirable high maltose starch conversion products can be produced conveniently at reasonable enzyme dosages in 24 to 48 hours. The conversion of the partially hydrolyzed starch may be effected with the simultaneous use of the maltogenic enzyme and the pullulanase, or the pullulanase may be applied first or after application of the maltogenic enzyme. The preferred process is simultaneous application of the maltogenic enzyme and the pullulanase.

The starch conversion products may be concentrated and/or refined to produce high maltose syrups. These syrups are substantially noncrystallizing, and exhibit nonhygroscopic properties. To obtain these syrups, the starch conversion products are concentrated to a solids content in excess of 50%. The products may be refined by conventional methods such as carbon refining, ion exchange treatment, and the like, to obtain syrups that are substantially noncrystallizing and have maltose contents between 50 and 90%, dry basis. In the following example all percentages are by weight, dry basis, and all temperatures are in degrees centigrade.

Example: This example illustrates the simultaneous use of a maltogenic enzyme and pullulanase in the preparation of high maltose starch conversion products from partially enzyme-hydrolyzed starch. This example also illustrates varied dosages of pullulanase and malt diastase in combinations to produce the high maltose, high fermentables conversion products.

A 30% by weight suspension of cornstarch was enzymatically liquefied at 91°C, pH 5.5, using HT-1000 (a bacterial alpha-amylase preparation) at a dosage of 0.05% on a dry basis. The enzyme liquefying action was stopped by heating the liquefied starch to 121°C for 15 minutes when partially hydrolyzed starch reached a DE between 2 and 5. The partially hydrolyzed starch was cooled to 55°C and the pH adjusted to 5.8. The partially hydrolyzed starch was then dosed with the respective levels of pullulanase and maltogenic enzymes shown in the table below. The table also sets forth the DE values obtained, and the carbohydrate compositions after both 24 and 48 hours of conversion.

Pullulanase dosage,[a] u./100 g.D.S.	Malt diastase dosage, u./100 g.D.S.	D.E.	Dextrose (percent D.B.)	Maltose (percent D.B.)	Malto triose (percent D.B.)	DP₄ and higher (percent D.B.)	Fermentables[b] (percent D.B.)
			24-HOURS				
0	25	42.3	0.6	60.0	11.8	27.6	72.4
0	50	45.5	2.7	61.5	11.9	23.9	76.1
0	100	47.3	3.2	63.5	12.3	21.0	79.0
0	200	49.4	3.6	66.5	12.0	17.9	82.1
50	25	48.6	0.9	70.0	13.7	15.4	84.6
50	50	49.9	2.9	70.4	13.6	13.1	86.9
50	100	51.7	3.5	71.9	14.6	10.0	90.0
50	200	52.5	2.7	73.4	14.2	9.7	90.3
100	25	49.7	1.7	71.5	15.5	11.3	88.7
100	50	51.4	1.4	72.8	16.3	9.5	90.5
100	100	52.2	2.5	72.2	16.6	8.7	91.3
100	200	53.4	4.2	73.4	15.0	7.4	92.6
200	25	50.5	0.5	74.0	15.9	9.6	90.4
200	50	51.5	2.3	72.4	16.4	8.9	91.1
200	100	52.5	1.6	76.0	15.3	7.1	92.9
200	200	53.6	1.6	76.4	16.4	5.6	94.4
			48 HOURS				
0	25	43.5	1.9	61.6	9.8	26.7	73.3
0	50	47.6	3.3	64.5	11.6	20.6	79.4
0	100	51.1	4.0	67.0	12.5	16.5	83.5
0	200	51.9	3.8	71.1	11.7	13.5	86.5
50	25	51.5	4.1	72.0	14.4	9.5	90.5
50	50	52.8	2.4	75.0	14.8	7.8	92.2
50	100	54.7	3.4	75.0	15.2	6.4	93.6
50	200	56.2	4.5	75.5	14.0	6.0	94.0
100	25	52.9	1.5	75.4	15.9	7.2	92.8
100	50	54.3	1.9	76.5	16.0	5.6	94.4
100	100	54.2	2.8	77.2	14.6	5.4	94.6
100	200	55.7	3.5	77.6	13.8	5.1	94.9
200	25	52.7	1.2	76.0	15.9	6.9	93.1
200	50	52.4	1.5	76.5	16.0	6.0	94.0
200	100	54.3	2.4	77.0	15.3	5.3	94.7
200	200	54.9	3.7	77.6	13.8	4.9	95.1

[a] Forty units of malt diastase is equivalent to 1% distillers barley malt having a Lintner Value of 200-220° L.
[b] Dextrose, maltose, and maltotriose.

The maltogenic enzyme produced by *Bacillus polymyxa* may be substituted for the malt diastase used in the example with good results.

D.A. Bodnar, C.W. Hinman and W.J. Nelson; U.S. Patent 3,644,126; February 22, 1972; assigned to Standard Brands Incorporated has also now developed a method of producing a syrup having a high total concentration of fermentable sugars with a high maltose and low dextrose content. This is obtained by treating an aqueous slurry of starch with a starch-liquefying enzyme to obtain a DE of less than 35 and digesting the liquefied starch slurry with glucoamylase and malt enzymes to obtain a syrup having less than 45% dextrose and sufficient maltose to provide at least 85% total fermentable sugars.

The first step in the process includes treating an aqueous slurry of starch with a starch-liquefying enzyme. It is preferred that the starch slurry have a concentration of from 10° to 25° Brix and most preferably 18° Brix. The pH of the slurry should be adjusted to 5.5 to 7 and preferably to 6.5. This pH adjustment may be accomplished by the addition of any suitable alkali, for example, hydrated lime. A starch-liquefying enzyme is then incorporated and the slurry subjected to starch-liquefying conditions. The preferred starch-liquefying enzyme is α-amylase derived from *Bacillus subtilis* in an amount to provide from 10,000 to 30,000 liquefons per lb of dry substance. The term liquefon is defined as the amount of enzyme which will dextrinize 0.351 mg of starch per minute at 30°C and a pH of 6.2 in a 30 ml test solution containing 400 mg of starch.

The preferred process for liquefying the starch is a two-step procedure where α-amylase is incorporated into the pH-adjusted slurry and heated as quickly as possible to 170° to 190°F preferably about 188°F and held there for a sufficient time (e.g., one-half hour) to paste and thin the starch. This step may be carried out in a number of ways but typically it will entail heating the slurry by means of a steam jet. This slurry is then passed through a steam jet autoclave where it is heated at 250° to 320°F to disintegrate any heat-resistant starch granules which may be present.

The holding time of the starch slurry may be from 10 to 20 seconds. After the slurry is autoclaved it is cooled to 170° to 190°F, preferably 188°F and α-amylase added. Typically, sufficient α-amylase will be added to provide from 1,000 to 10,000 liquefons per lb of dry substance and preferably 5,000 liquefons. The starch slurry will be maintained under these conditions until liquefaction of the starch is substantially completed. This may take about 2 hours. The liquefied starch after this step should be a DE of not over 35 and preferably a DE of 10 to 20. The dry substance of the slurry at this point in the process is not critical but typically it will be 30 to 35%.

The temperature of the liquefied starch slurry is reduced to between 122° and 140°F and preferably to 131°F prior to being digested with malt enzymes and glucoamylase. If required, the pH of the starch slurry is adjusted to 5 to 6.5, preferably to 5.5 to 6. Ground barley malt (a source of β-amylase) and glucoamylase are added to the slurry.

Typically, the ground barley malt will be added at a level of 0.5 to 2% based on the substrate solids, preferably at a level of 1%. Typically, sufficient glucoamylase is added to attain a level of 0.5 to 10 glucoamylase units per 100 grams of dry substance, preferably 3 units. The concentration of the glucoamylase used is to some extent dependent upon the amount of malt used and upon the ratio of dextrose to fermentables desired in the syrup. Malt enzymes and glucoamylase may be added simultaneously or sequentially with the malt enzymes being added first.

Conditions are maintained until the dextrose concentration reaches a level from 33 to 45% and the total fermentables reach a level of above 85%. The faccharification action of the enzymes may be terminated by adjusting the pH of the conversion syrup downward and/or heating the syrup to high temperatures, e.g., above 165°F. Other means of terminating saccharification may be used, for instance, passing the syrup through an adsorbent which removes the enzymes or by introducing into the syrup chemical agents which will inactivate the enzymes.

After the starch is liquefied and prior to its being digested with malt enzymes and glucoamylase, it is preferred that a microbial inhibitor be added. A suitable microbial inhibitor is sodium bisulfite. The microbial inhibitor will prevent the growth of acid-producing microorganisms which may be present and thus prevent the pH of the starch slurry from drifting downward during saccharification to a level where inactivation of malt enzymes may occur. pH maintenance with an alkali during saccharification is essentially impractical because of the large amounts of alkali that are required. These large amounts of alkali will increase the ash content of the syrup to an undesirable level.

Example: A 22° Brix slurry of cornstarch in water was diluted to 18° Brix and adjusted to a pH of 6.8 to 7.0 by addition of hydrated lime. 15,000 liquefons of α-amylase (produced from a *B. subtilis* fermentation) was then added. The mixture was heated to 188°F and held for one-half hour after which it was autoclaved at 300°F for 10 seconds. The autoclaved mixture was cooled to 188°F and additional 5,000 liquefons of α-amylase added and the mixture held at 188°F for 2 hours. At this point the DE of the mixture was 16.2, the pH was 6.3 and contained 30.5% solids.

The temperature of this partial hydrolysate was then reduced to 131°F and sodium bisulfite equivalent to 0.05% of the solids was added to inhibit growth of acid-producing microorganisms. Ground barley malt equivalent to 1% of the substrate solids was added and saccharification allowed to continue for 24 hours. Three units of glucoamylase derived from *A. niger* per 100 grams of substrate solids were then added and, periodically, samples of the conversion syrup were withdrawn and the enzymes inactivated by heating to at least 190°F for at least 15 minutes. The samples were then analyzed for monosaccharides (dextrose) and disaccharides (maltose) content by paper chromatography. The temperature of the digest was maintained at 131°F throughout the saccharification and no pH adjustments were made. The results of the analysis are shown below in Table 1.

TABLE 1: HYDROLYSIS OF 16.2 DE ENZYME LIQUEFIED STARCH USING BARLEY MALT AND GLUCOAMYLASE

Sample	DE	Percent Dextrose	Percent Maltose	Percent Total Fermentables (Dextrose+Maltose)
1	45.8	4.3	55.5	59.8
2	48.4	6.1	57.1	63.2
3	53.8	13.1	58.7	71.8
4	58.9	19.6	59.0	78.6
5	66.1	33.5	51.6	85.1
6	68.0	37.4	48.9	86.3
7	70.2	42.5	44.9	87.4

The products of this process as given in Table 1 show higher maltose and total fermentables when compared with products obtained using acid liquefied starch as starting materials as shown in Table 2.

TABLE 2: HYDROLYSIS OF 15.0 DE ACID LIQUEFIED STARCH USING BARLEY MALT AND GLUCOAMYLASE

Sample	DE	Percent Dextrose	Percent Maltose	Percent Total Fermentables (Dextrose+Maltose)
1	55.6	20.1	55.2	75.3
2	62.6	30.3	50.2	80.5
3	66.0	37.1	45.0	82.1
4	70.1	47.7	36.8	84.5
5	77.8	61.2	25.0	86.2

Using α-1,6-Glucosidase from *Pseudomonas amyloderamosa*

Two modifications on the use of two enzymes in preparing maltose syrups have been developed by *M. Mitsuhashi, K. Masuda, M. Shiosaka, M. Hirao, K. Sugimoto, Y. Yokobayashi, S. Yuen and M. Yoshida; U.S. Patent 3,795,584; March 5, 1974; assigned to Hayashibara Co., Japan.*

When a saccharified solution obtained by the saccharification of liquefied starch with β-amylase and α-1,6-glucosidase is purified in the usual manner, extremely small amounts of high-molecular dextrins remaining render the purification with an ion exchange resin highly difficult and produce the possibility of the product becoming turbid. Here the addition of α-amylase greatly facilitates the purification of the saccharified solution with an ion exchange resin.

By way of exemplification, a solution (C) saccharified by the combined use of β-amylase and pullulanase (saccharified solution concentration 39.4%) and a solution (D) saccharified by the combined use of three converting agents, i.e., β-amylase, pullulanase and α-amylase (saccharified solution concentration 39.4%) were compared with respect to their abilities to be treated with an ion exchange resin. The results were as shown in the following table. As will be seen, (C) is almost incapable of being purified with the ion exchange resin, while (D) can be readily purified.

	Strong acid-medium-base two-bed pretreatment	Strong acid-strong base mixed-bed treatment
C..	8 times by volume of soln. on basis of acid resin volume passed, pH 5.2 Sp. resis. 5×10^4 cm.	2 times by volume of soln. on basis of resin volume passed, pH 3.7 Sp. resis. 8×10^3 cm.
D..	20 times by volume of soln. on basis of acid resin volume passed, pH 8.0 Sp. resis. 1×10^5 cm.	40 times by volume of soln. on basis of resin volume passed, pH 5.0 Sp. resis. 1×10^3 cm.

Further study has also led to a finding that the use of a mixture of different varieties of α-1,6-glucosidase instead of only one in combination with β-amylase for the saccharification of liquefied starch permits an increased output of maltose as shown in the following table.

Enzyme added		Sugar composition (percent)			
		Glucose	Maltose	Malt-triose	Dextrin
E	{Pullulanase Beta-amylase	0.6	93.0	5.1	1.9
F	{Pullulanase Isoamylase of Pseudomonas beta-amylase	0.8	95.1	4.0	0.9

NOTE.—In each experiment a gelatinized solution containing 2% sweet potato starch with the same amount of enzyme (the amount of pullulanase used in E being equal to the combined amount of pullulanase and isoamylase produced by the bacteria of the genus Pseudomonas) at 45° C. for 64 hours. The pH during conversion was 6.0 for E and 5.5 for F.

The increased production of maltose by the combined use of different varieties of α-1,6-glucosidase as can be seen from the above table is presumably attributed to different cleavage abilities of pullulanase and the isoamylase produced by the Pseudomonas bacteria against different α-1,6 linkages.

Example 1: 100 grams of cornstarch was added to 4,500 ml of boiling water and gelatinized and dispersed therein. The aqueous dispersion was then subjected to a pressure at 130°C for 5 minutes, cooled down to 45°C, and the pH was adjusted to 6.0. After the addition of 20 units of pullulanase-salting-out enzyme and 100 units of β-amylase/g of the starch, the mixture was saccharified at 45°C for 48 hours. The saccharified solution was then heated, filtered, concentrated and was purified by decoloration in the usual manner. Upon concentration to a water content of 15%, it gave colorless crystals. Analysis on dry basis showed that the product was high purity maltose consisting of 93.0% maltose, 1.5% glucose, 4.0% malttiose, and 1.5% others.

Example 2: After 100 grams of waxy cornstarch was gelatinized in the same manner as described in Example 1, 20 units of pullulanase and 100 units of β-amylase were added per gram of the starch. The mixture was saccharified at 45°C for 10 hours, and then 5 units of α-amylase was added per gram of the starch. Saccharification was repeated for a further period of 45 hours. The resultant solution was boiled, decolored, and concentrated. Upon further purification with Amberlite IR-120, IRA-68, and IRA-411 followed by concentration, it became completely solid with a water content of 13%. Analysis showed that the product contained 93.5% of maltose on a dry basis. In the form of a 70% solution, it was crystallized and microcrystals were formed. Fractionation of the crystals yielded maltose with a purity of 96 to 97%.

Example 3: 100 grams of waxy cornstarch was gelatinized in the same manner as in Example 1. Next, 20 units of pullulanase, 40 units of α-1,6-glucosidase obtained from a certain variety of bacteria of the genus Pseudomonas and 25 units of β-amylase were added per gram of the starch. The pH of the mixture was adjusted to 5.5, and the mixture was saccharified at 45°C for 46 hours, boiled, and then purified with active carbon and concentrated. The syrup thus obtained which had a water content of 13% immediately crystallized. Upon analysis it proved to have a maltose content of 95% (on dry basis), the anhydride yield being 95%.

Use of Heat Resistant α-1,6-Glucosidase

A series of heat resistant α-1,6-glucosidase enzymes have been found by *K. Sugimoto and M. Hirao; U.S. Patent 3,804,715; April 16, 1974; assigned to Hayashibara Co., Japan* which give high yields of maltose when used in combination with β-amylase. These new strains are produced from *Pseudomonas amylo-*

deramosa ATCC 21262, *Escherichia intermedia* ATCC 21073, *Agrobacterium tumefaciens* IFO 3058, *Azotobacter indicus* IFO 3744, *Bacillus cereus* IFO 3001, *Erwinia aroideae* IFO 3057, *Leuconostoc mesenteroides* IFO 3426, *Mycobacterium phlei* IFO 3158, *Micrococcus lysodeikticus* IFO 3333, *Pedicoccus acidilactici* IFO 3884, *Sarcina lutea* IFO 3232, *Serratia indica* IFO 3759, *Staphylococcus aureus* IFO 3061 or *Streptococcus feacalis* IFO 3128.

Each α-1,6-glucosidase has some specific action on starches. In particular, an enzyme produced from Pseudomonas is somewhat different from enzymes produced from other strains. Therefore, if the former enzyme is used in combination with one of the latter enzymes, a great effect on β-amylolysis is obtained. Also the similar or more preferable effects may be obtained by using β-amylose in combination with two or more α-1,6-glucosidases. In this case, any α-1,6-glucosidase selected from more than ten enzymes can be used. The overall process can be illustrated by the following example.

Example: For obtaining the highest maltose content of about 90%, DE in the liquefaction must be as low as possible. Therefore, the reaction is carried out at DE 0.5 to 5, preferably 1 to 12. Consequently, the viscosity becomes very high, and the difficulty in mixing of enzyme under cooling and residence of non-reacted one due to retrogradation are feared. Accordingly, the cooling and mixing of enzyme are performed in 2 to 3 steps. Liquefied solution is sprayed in vacuum to cool it instantaneously. Then cooling and mixing of enzyme are performed instantaneously in a first mixing tank. For caution's sake, the cooling is stopped at a temperature of 50° to 60°C, and first preliminary saccharification is performed with an enzyme obtained from Lactobacillus or Streptomyces, or β-amylase having high heat resistance.

During several hours of residence time, the viscosity is lowered and fear of retrogradation is reduced. Thereafter, the mixture is introduced into a second preliminary saccharification tank (second mixing tank), in which is then added α-1,6-glucosidase or β-amylase. In both cases, it is introduced into a large quantity of highly saccharified sugar solution, and therefore, the low-sugar solution is diluted and contacted with the fresh enzyme to prevent retrogradation again.

Temperature in the first step is kept as high as possible, and in the second step is lowered to working temperature. In the last saccharification tank, the saccharification is performed at an optimum temperature for two days in batch system. The reaction tank and the stirring and mixing tank are kept free from other germs, since the saccharification is performed at a relatively low temperature of 45° to 50°C. Starches used as raw materials are cornstarch, waxy cornstarch and sweet potato starch. As with α-1,6-glucosidase, all of the above-mentioned enzymes are tested. The enzyme from Pseudomonas, having different activity from that of other enzymes, is also tested on the effect of mixing with another enzyme.

Run No.	I	II	III	IV	V	VI	VII	VIII
Kind of Starch	Corn Starch	Waxy Corn Starch	Corn Starch	Sweet Potato Starch	Sweet Potato Starch	Sweet Potato Starch	Potato Starch	Corn Starch
Temperature of liquefaction (°C)	150 – 160	150 – 160	150 – 160	150 – 160	88 – 95	88 – 95	88 – 95	88 – 95

(continued)

Run No.	I	II	III	IV	V	VI	VII	VIII
Kind of Starch	Corn Starch	Waxy Corn Starch	Corn Starch	Sweet Potato Starch	Sweet Potato Starch	Sweet Potato Starch	Potato Starch	Corn Starch
Enzyme for Liquefaction u	—	—	—	—	α15	α15	α15	α15
Starch (g) pH	5.5	5.5	4.0	5.0	6.0	6.0	6.0	6.0
Liquefied Solution D.E.	1.0	2.1	3.0	2.0	2.5	0.5	2.1	2.5
Concentration of Liquefied Solution %	10	15	20	20	12	15	20	15
Enzyme of First Preliminary Saccharification u	E20 β20	β20	L25	β25	E15 P20	β25	S25	β30
Starch (g) pH	6.0	6.0	6.0	6.0	6.0	6.0	6.0	6.0
Temperature °C	45	60	60	60	45	60	60	60
Enzyme for Second Preliminary Saccharification u	—	Ps30	β20	Ps20 E10	—	E20	β20	Ps15 L15
Starch (g) pH	—	5.5	6.0	5.5	—	6.0	6.0	5.5
Temperature °C	—	45	50	45	—	45	50	45
pH in the Saccharification	6.0	5.5	6.0	5.5	6.0	6.0	6.0	5.5
Temperature °C	45	45	45 50	45	45	45	50	45
Time	48	48	50	45	45	50	45	45
Maltose %	93	92	90	92.5	83	92	92	93
			Finally β5 is added at 80°C			After completion, β5 is added at 80°C		After standing for 48 hrs. β5 is added at 80°C

α: α-amylase
β: β-amylase from wheat bran
Ps: Enzyme from Pseudomonas
E: Enzyme from Escherichia
L: Enzyme from Lactobacillus
S: Enzyme from ray fungus
P: β-amylase from polymixa

As shown in the above table, α-1,6-glucosidase is effective even if it is used alone. In a liquefied solution having a high concentration and a low DE, it is very effective to accelerate the decomposition by adding a highly heat resistant enzyme in two steps. Difficulty is observed in run No. VI, but subsequent purification is facilitated by adding a small quantity of α-amylase at the final stage of the reaction. Use of polymixa enzyme as in the above run No. V limits maltose production and is sometimes unsuitable.

Crystalline Maltose Preparation

A process for preparing a high-purity maltose solution by the use of β-amylase and α-1,6-glucosidase has been disclosed by *M. Kurimoto, K. Sugimoto and M. Hirao; U.S. Patent 3,677,896; July 18, 1972; assigned to Hayashibara Company, Japan*. By concentrating the resulting solution to a massecuite containing maltose crystals, and then spraying the massecuite in a drying column a crystalline maltose is produced. In this process, the α-1,6-glucosidase to be added to the liquefied starch is selected from the group consisting of varieties produced by strains of the genera Escherichia (ATCC 21073); Pseudomonas (ATCC 21262); Lactobacillus (ATCC 8008); Micrococcus (IFO 3333); Nocardia (IFO 3384); and Aerobacter (ATCC 8724).

In the overall process the step of gelatinizing the starch should be carried out with care taken to limit the decomposition of 10 to 40% starch slurry to a dextrose equivalent value which is low enough to avoid further retrogradation of the starch. To meet the requirements, the starch in a concentration between 10 and 40% may be dispersed at 160°C and preferably decomposed to a DE between 1 and 5. Where a liquefying enzyme is employed, the starch slurry containing such enzyme is continuously liquefied at a high temperature between 85° and 95°C and the DE is limited to below 5. The same applies where an acid is used instead.

In the step of saccharification, either β-amylase or α-1,6-glucosidase may be added first. However, it should be noted that the gelatinized starch solution of a low DE value is highly viscous, easy to retrograde on cooling, and once retrograded it becomes inert to the enzymatic reaction. Thus, it is advisable to cool the gelatinized solution rapidly to a desired temperature as by a flash cooler and add an enzyme having a great stability against heat, as promptly and at as high a temperature as possible to avoid retrogradation, thereby promoting the amylolysis and decreasing the viscosity of the sugar solution, and then add the second enzyme and effect the reaction.

In view of this, the enzymes to be added first at elevated temperature are preferably the heat-resistant β-amylase and α-1,6-glucosidase produced by Lactobacillus bacteria. The temperature for this purpose ranges from about 60° to 70°C. After the drop of the viscosity, the second enzyme is added to promote the saccharification. The temperature is about 45°C, and the pH is adjusted to a value suitable for the enzymes. The reaction is concluded in one or two days. When there is any possibility of the saccharified solution containing some retrograded dextrin, it is desirable to add a small amount of α-amylase to decompose the residual dextrin in the course of saccharification, because this facilitates the subsequent purification.

The saccharified solution thus obtained is heated for deactivation of the enzymes, concentrated, decolored with active charcoal, and purified with an ion exchange resin to a pure sugar solution. Fractional estimation by paper chromatography indicates that the resulting maltose, comprising less than 1% glucose, 90 to 95% maltose, and 3 to 4% maltotriose, is well comparable in purity to the existing reagents. When this sugar solution is concentrated to 70 to 80% and is agitated with the addition of seed crystals at about 35°C, hydrated crystals of maltose readily deposit. Upon gradual cooling with stirring over a period of one to three days, an increasing number of microcrystals in the form of triangular flakes grow to a creamy state. Such a cream with a crystal deposition rate of over 35% was sprayed from the upper part of a spray drying column through a high pressure nozzle. The dry air, in parallel flow, was kept at 80° to 90°C. The lower part of the column was provided with a moving bed of wire netting, and hot air was forced upward through the netting and the material is taken out of the column at a low speed. After a period of 30 to 40 minutes of holding, maltose in the form of hydrated crystalline powder was taken out. It was a fine, nonsticking powdery product of small globular shape.

The product was found on analysis to consist of complete hydrated crystals having a water content of 5 to 6%. The maltose content of the anhydrous material was 95%. It was a high purity maltose which did not absorb moisture and remained stable in air.

Example 1: Preparation of Maltose — 100 grams of cornstarch was incorporated into 300 ml of boiling water for gelatinization and dispersion. The solution was subjected to a pressure at 130°C for 5 minutes, cooled to 80°C while the pH was adjusted to 6.0, and β-amylase was added at a rate of 50 units/g of the starch. The solution, following the reduction of the viscosity, was cooled to 45°C and a pullulanase salted-out enzyme (ATCC 8724) was added at a rate of 20 units/g of the starch.

After saccharification at 45°C for 48 hours, the resultant solution was heated, filtered, concentrated and purified by decoloration with active charcoal in the usual manner. The solution upon concentration to a water content of 15% gave colorless hydrated crystals. Analysis of the sugar composition showed that the product was a high purity maltose consisting, on the dry basis, of maltose, 92.5%; glucose, 2.0%; maltotriose, 4.0%; and other substances, 1.5%. Other examples were included in the complete patent showing the production of maltose solutions from various corn and potato starches.

Example 2: Preparation of Crystalline Powder of Maltose — The purified solution of maltose obtained by the procedure of Example 1 was concentrated to 70%, poured into a crystallizer, and, with the addition of 2% of seed crystals of crystalline maltose, it was cooled from 35°C downward with stirring. Over three days the solution was cooled to 20°C and attained a crystallization rate of 47%. The resultant solution was sprayed from the upper part of a drying column through a 1.5 mm diameter nozzle, by means of a high pressure pump at 150 kg/cm².

Hot air at 85°C was supplied from the upper part of the drying column, so that the charge was collected on a moving conveyor of wire netting in the lower part of the column. From below the conveyor warm air at 40°C was blown up, while the charge was gradually conveyed out of the drying column. Thus, over a period of 40 minutes, crystalline powder was taken out. It was then packed in an aging column, where it was aerated with warm air for 10 hours to complete the crystallization and drying. The product was powdery crystals of maltose having a water content of 6%.

Powdery Starch Sugars by Spray Drying

Spray drying is used by *M. Kurimoto and M. Hirao; U.S. Patent 3,713,978; January 30, 1973; assigned to Harashibara Company, Japan* to prepare solid starch sugars. Here again the two enzymes are used in the saccharifying step to produce syrups of primarily malt dextrin or maltose. These syrups were purified, concentrated, and then subjected to spray drying. Into a parallel flow of hot air, each sugar solution was sprayed by means of triple high pressure pumps at pressures ranging from 100 to 150 kg/cm^2. Further particulars will be given in the following description of examples.

Pulverization of the sugar solutions may be done by a vacuum drying technique as well. Experimentally, each sugar solution in a concentration of 80% heated with steam at 5 kg/cm^2 was extruded into a vacuum of –700 mm for drying, and the dried product was pulverized to a powdery state. The apparent specific gravity of the product obtained by this technique was quite low. The following examples use only the spray drying technique.

Examples: Various raw material starches were used. Each starch was first purified and converted to a starch slurry having a concentration of 30 to 40%. For the liquefaction and gelatinization, continuous liquefying equipment was used. Since the saccharification needs a lengthy period of time, a batch-type saccharification tank was provided for that purpose.

To begin with, the starch slurry was adjusted to pH 5.0, pumped into the continuous liquefier equipped with an agitator and, while being vigorously agitated, the charge was suitably liquefied or gelatinized. For example, it was either rapidly gelatinized by heating to 160°C with the supply of live steam or was liquefied by heating to about 90°C, following the addition of α-amylase. Next, the starch, liquefied to a suitable DE value, was rapidly cooled to a temperature and adjusted to a pH both suitable for the enzyme to be subsequently introduced.

The liquefied starch was either subjected to the action of an α-1,6-glucosidase for decomposition of the branched structure of the starch-like amylopectin to straight-chain malt dextrins which in turn were decomposed with the addition of a suitable saccharogenic amylase; or the liquefied starch was subjected to the action of a suitable saccharogenic amylase for decomposition and then, with the addition of the α-1,6-glucosidase, the branched dextrins were decomposed to malt dextrins. By either method, a starch syrup having a desired composition, viscosity, and sweetness was prepared. The temperature, pH, and time for the amylolysis varied according to the type of enzyme used.

The saccharified solution was then heated to deactivate the enzymes, decolored with active charcoal, and further decolored and desalted with ion exchange resins. Following the purification, the solution was concentrated. The resulting solution, which was less viscous than ordinary acid-converted starch syrups and starch sugars of the same DE values, was further concentrated to a concentration of 70 to 80%, and then spray dried.

The spray dryer was a cylindrical column having an effective height of more than 10 meters. From a high pressure nozzle mounted on the top of the column, the concentrated sugar solution was sprayed. The hot drying air was supplied as a downward flow from the top of the drying column. The air temperature used ranged from 80° to 150°C. The powdery starch sugars dried in this way were collected on a moving bed of wire screen stretched in the lower part of the column and the mass was cooled by cold air introduced from below the moving bed. The powdery product on the moving bed was continuously taken out of the column, collected in a hopper, sieved, and packed in a bag without being exposed to the atmospheric air and humidity. The end products obtained in the manner described were powdery products of small globular shape having water contents of 1 to 3% and were easy to handle, though the moisture absorption varied according to the degree of saccharification or DE attained. The conditions for saccharification and spray drying were as tabled below.

Example Number	1	2	3	4
Starch	Sweet potato	Corn	High amylase	Sago
Concentration, percent	40	35	30	35
Liquefaction:				
Enzyme or acid, units/g starch	None	None	Oxalic acid, pH 4	C15
Temperature, °C	163	165	160	90
DE	2.5	2.0	3.0	15
1st saccharification:				
Enzyme, units/g	L30	L40–B10	B10	C5
Temperature, °C	50	50	60	60
pH	6.0	6.0	5.5	6.0
Hold time, hr	25	20	8	20
2nd saccharification:				
Enzyme, units/g	C20	C20	E20	P20
Temperature, °C	45	50	45	45
pH	6.0	6.0	5.5	6.0
DE	40	60	57	27
Hold time, hr	10	10	10	10
Spray drying:				
Original liquid conc., percent	75	80	77	75
Air temperature, °C	160	150	150	155
Water content, percent	1.0	0.8	0.7	0.5
Main components	Malt dextrin	Maltose	Maltose, 90%	Maltose, malt dextrin

    α-Amylase: (C) commercially available liquefying enzyme
    β-Amylase: (B) extracted from wheat bran
    Aerobacter pullulanase: (P)
    Pseudomonas enzyme: (E)
    Lactobacillus enzyme: (L)

PRODUCTION OF LOWER OLIGOSACCHARIDES

The starch industry is continually striving to produce various starch conversion products, such as syrups which can be tailor-made to meet a vast number of needs. One particular area of endeavor involves production of a relatively high DE syrup which also has a minimal dextrose content and contains a high proportion of lower oligosaccharides, those materials having a degree of polymerization of 2 to 6. Such syrup, if it could be prepared, would have a low level of sweetness, and a sufficiently high viscosity so as to be useful in formulation of a number of products. For example, this syrup would have particular utility as an ingredient in ice cream, hard candies, fondants and baked goods. If the syrup just described could be made available via a simplified and economical technique, the product and its mode of preparation would find ready acceptance in the art.

Hydrolysates Containing Lower Oligosaccharides

A process is disclosed by *R.E. Heady; U.S. Patent 3,535,123; October 20, 1970; assigned to CPC International Inc.* which provides starch hydrolysates having 60 to 85% of relatively low oligosaccharides. More specifically these hydrolysates have a DE of 25 to 55, a high proportion (60 to 85%) of low molecular weight oligosaccharides based on total carbohydrate hydrolysate solids, and a corresponding low proportion (below 35%) of high molecular weight oligosaccharides. The hydrolysate is further characterized by the relationship:

$$\frac{DE}{DP_{2-6}/DP_{7+}} = n$$

where DP_{2-6} is the percent of low molecular weight oligosaccharides having a degree of polymerization of 2 to 6, DP_{7+} is the percent of high molecular weight oligosaccharides having a degree of polymerization of 7 and higher, DE is the dextrose equivalent of the hydrolysate and n is 4 to 12. This hydrolysate may be in the form of a solid or liquid syrup.

The above hydrolysates are best made by enzymatically hydrolyzing a starch substrate by an enzyme system having both α-amylase and pullulanase enzyme activity. These starch syrups have exceptional use in that they have sufficient body in terms of a relatively high viscosity to enable them to be used as food carriers, while they concomitantly exhibit a low degree of sweetness. On the other hand, the syrups have a sufficiently high DE, particularly as derived from their high proportion of DP_{2-6} oligosaccharides to render the syrups sufficiently sweet for their intended use. Thus, these syrups are particularly useful in ice cream formulations and like edible foods requiring syrup characteristics as just described.

The amount of α-amylase used may vary depending upon the amount of pullulanase enzyme used, the source of starch being worked, etc. However, excellent results are achieved when one uses 100 to 20,000 units of α-amylase/kg of dry basis starch. Usually, greater than 100 units of pullulanase/kg of dry basis starch are necessary to effect the desired conversion. Most often, the desirable level of pullulanase activity varies from 250 units to 2,000 units. The process is particularly adaptable for starches which have a relatively high amylopectin content of at least 50%. An exemplary starch of this type is waxy milo.

Prior to enzymatic attack, the starch should be solubilized by gelatinization at a temperature exceeding 60°C in the presence of moisture. It is also preferred that the gelatinized starch be pretreated by conventional thinning techniques such as acid or α-amylase hydrolysis to a DE, not exceeding 15, and preferably less than 5. In production of these syrups, the conversions can be performed at high dry substance levels usually within the range of 15 to 40% to reduce tank size requirements and evaporation costs. The temperature of conversion is preferably 30° to 70°C. Most preferably, in order to retard or prevent microbial spoilage of the conversion liquors, the hydrolysis is run at relatively high temperatures, such as 50° to 60°C.

Again, the conversions may be performed over a relatively wide pH range of 4.0 to 8.0, preferably 5.0 to 7.0. The time required for conversion will depend upon the enzyme dosage employed, and the extent of conversion desired, however, desirable syrups can be produced at reasonable enzyme dosages within 24 to 144 hours, preferably within 48 to 120 hours.

Example: In this example a number of runs were made involving enzymatic conversion of a waxy milo starch utilizing pullulanase and α-amylase enzymes, both separately and in combination. The waxy milo starch was first gelatinized and thinned with α-amylase to 2 to 4 DE. In the several runs the degree of conversion was varied by either adjustment of the period of conversion or amount of enzyme activity used.

After the conversion period was terminated the hydrolysates were analyzed for DE, dextrose, lower oligosaccharides, that is for those oligosaccharides having a DP of 2 to 6, and higher oligosaccharides containing 7 or more units. The measurement of the various components making up the hydrolysate was done by known quantitative paper chromotography methods. Miles HT was the source of α-amylase. The data is presented below in the table.

DE range	Alpha-amylase units/100 g. d.s.	Pullulanase units/100 g. d.s.	DE	DP_1*	DP_{2-6}*	DP_{7+}*	DP_{2-6}/DP_{7+}	$\dfrac{DE}{DP_{2-6}/DP_{7+}}$	Hours of conversion
20–25	100	0	21.7	3.0	49.6	47.4	1.05	20.6	68
25–30	200	0	27.9	5.2	56.8	38.0	1.49	18.7	88
30–35	400	0	34.3	8.2	60.6	31.2	1.94	17.7	95
35–40	800	0	40	11.2	60.0	29.2	20.6	19.4	95
40–45	1,600	0	43.8	14.2	60.7	25.3	2.39	18.3	95
20–25	100	100	24.3	2.7	61.7	35.6	1.73	14.0	68
25–30	100	100	25.7	2.9	68.1	29.0	2.35	10.9	88
30–35	100	100	31.1	5.2	71.2	24.6	2.89	10.8	95
30–35	200	100	34.2	7.1	74.1	17.8	4.16	8.2	88
35–40	200	100	35.5	7.3	71.5	21.2	3.37	10.5	95
35–40	200	200	36.8	7.1	77.1	15.8	4.87	7.6	95
40–45	400	100	40.2	10.0	74.3	15.7	4.75	8.5	95
40–45	400	200	41.5	9.7	78.1	12.2	6.42	6.5	95
Pullulanase		100	2.4	0.0	0.0	100			

* Percent of total carbohydrates.

As is evident from the experimental results, only through use of the combination of enzymes does one realize a hydrolysate having the desired DE range, and lower and higher oligosaccharide contents falling within the defined limits of this process. In particular, it was found that all hydrolysates as produced through the combination use of pullulanase and α-amylase in appropriate amounts had a value of n of 4 to 12. Starch hydrolysates formed by use of α-amylase alone even in gross amounts were well outside the maximum figure of 12. In particular, use of low amounts of α-amylase resulted in a relatively low proportion of lower oligosaccharides and a relatively high proportion of higher oligosaccharides. On the other hand, use of exceptionally high levels of α-amylase resulted in an undesirable high level of dextrose, leading to undue sweetness.

Maltotriose and Maltotetrose

Maltotriose is an amylosaccharide having 3 glucose units (glucose-1,4-glucose-1,4-glucose) and maltotetrose is an amylosaccharide having 4 glucose units (glucose-1,4-glucose-1,4-glucose-glucose). Both these sugars are present in relatively small amounts in aqueous extracts derived from the conversion of starchy materials such as cereal grains into fermentable sugars. For instance, a normal gravity brewers' wort will contain approximately 1.2% by weight maltotriose and approximately 2.0% by weight maltotetrose. An ideal yeast for fermentation purposes is one that is able to ferment maltotriose and maltotetrose at a good rate and it is, therefore, of interest to brewers and distillers to be able to investigate this property in a large number of yeast strains. However, supplies of pure malto-triose and maltotetrose have been extremely limited.

In the process developed by *G.G. Stewart and A.E. Earl; U.S. Patent 3,788,910; January 29, 1974; assigned to Labatt Breweries of Canada Limited, Canada* malto-triose and maltotetrose are extracted from a fermentable source material, e.g., brewers' wort, by first fermenting the material with *Saccharomyces uvarum* to remove lower molecular weight sugars, separating the yeast from the remaining liquid and subjecting the liquid containing maltotriose and maltotetrose to gel filtration chromatography and collecting samples containing maltotriose and/or maltotetrose. The collected fractions containing the desired maltotriose and/or maltotetrose can be subjected to further purification by means of charcoal column chromatography.

The fermentation is normally carried out in a stirred fermenter and is usually complete within 4 to 5 days. The fermentation usually establishes a constant gravity of 4° to 5°P (i.e., degree Plato) after a period of 2 days. When the fermentation has been completed, all of the mono- and disaccharides have been metabolized leaving maltotriose, maltotetrose and higher sugars. At this stage, the yeast is removed, e.g., by skimming, filtration or centrifuging, and the remaining wort is stored for further purification. It may conveniently be freeze-dried for storage.

The gel filtration is a well known technique for separation purposes. It is based on the principle that, by means of gel grains which may, for instance, contain a copolymer of a hydroxyl group-containing substance such, for example, as a poly-saccharide like dextran, starch, dextrin and polyglucose with a bifunctional sub-stance such, for example, as epichlorohydrin, dichlorohydrin, etc. or a copolymer of an alkylidene-bis-acrylamide with an ethylenically unsaturated substance, a separation is carried out between the different ingredients in solution into at least two fractions according to molecular sizes, owing to the varying ability of the substances to penetrate into the interior of the gel grains.

A gel filtration process is carried out in two working steps. In the first of these steps the solution of the substances which are to be separated from each other or to be divided into fractions is introduced into a bed of the gel grains. In the second working step liquid is supplied to the bed for the purpose of causing elution to take place. In the case of the modified dextrans, these are in the form of macromolecules which are crosslinked to give a three-dimensional network of polysaccharide chains. The degree of crosslinking can be varied, thereby varying the swelling properties. The gel grains can swell by imbibing water or other sol-vents, e.g., dimethyl sulfoxide, formamide, glycol and mixtures of water with lower alcohols.

Sephadex G-15 (a dextran gel filter bed material), which has an approximate limit for complete exclusion of about 1,500 molecular weight, is particularly useful since maltotriose and maltotetrose have molecular weights of around 500 and 700 respectively. The substantially pure fractions of maltotriose and malto-tetrose can be still further purified by carbon column chromatography to yield a highly pure final product. The following example gives further details of the process.

Example: (A) A high gravity (15°P) hopless ale wort in a 12 liter fermenter was fermented with *Saccharomyces uvarum.* The fermenter was stirred at about 300 rpm for eight days with continuous aeration by which time all of the mono- and disaccharide had been metabolized leaving maltotriose, maltotetrose and higher sugars. A constant gravity of about 4° to 5°P was established after 2 days of fermentation. At the conclusion of the fermentation the yeast was removed by centrifuging and the wort remaining was freeze-dried and stored at 4°C awaiting purification. The recovered yeast could be used again in subsequent fermentations.

(B) 25 grams of the freeze-dried wort were reconstituted to 150 ml with deionized water and then applied to a chromatography column containing Sephadex G-15. The column had a diameter of 10 cm and a height of 70 cm. The material was then eluted, using upward flow, with deionized water containing 20 ppm sodium metabisulfite as an antioxidant. The column was jacketed and cooled at 5°C at all times and the effluent was collected in 25 ml fractions.

Approximately 200 fractions were collected in tubes and these fractions were monitored by measuring the absorbence at 278 mμ to determine whether successive runs were reproducible. Preliminary chromatograms were run on every tenth tube on silica gel thin layer plates using n-butanol:ethanol:water (5:1:4) as solvent and spraying with alkaline silver nitrate.

It was found that maltotetrose predominated in tubes 1–20, while maltotriose predominated in tubes 20 to 100 together with some contaminating maltotetrose. Individual tubes from runs 1–20 were combined together (Batch A) as were the tubes from runs 20–110 (Batch B). Both batches were freeze-dried. Anthrone analysis of the freeze-dried material of both Batch A and B showed that they contained about 80% carbohydrate by weight. Subsequent chromatographic analysis showed maltotriose to be around 80% of the carbohydrate present in Batch B, and maltotetrose to be around 80% of the carbohydrate present in Batch A.

(C) To further purify the maltotriose, 20 grams of the freeze-dried solids obtained from the Sephadex chromatography of the wort (Batch B) were reconstituted with 50 ml of 15% aqueous ethanol and then applied to a carbon-Celite column (50:50 mixture). The carbon used was activated charcoal (Darco-G60). The material was eluted, using downward flow, with 15% aqueous ethanol and the effluent collected in 25 ml fractions. The actual rotation was read on every fifth fraction and the same fractions were chromatographed and visualized, against standards, on Bakerflex silica gel 1B sheets (20 x 20 cm) using ethyl acetate:dimethylformamide:water (60:30:5) as solvent and diphenylamine as the spray reagent.

The first 40 or so fractions, varying slightly from run to run, showed only one spot, that of pure maltotriose, while the next 80 to 100 fractions contained both maltotriose and maltotetrose in a ratio of about 80:20. The two large portions namely the pure maltotriose and the mixture were bulked separately, the ethanol removed from each, freeze-dried and stored in sealed flasks in a desiccator. The mixture was subjected to further charcoal column chromatography to separate the maltotriose from the maltotetrose. The maltotriose so-obtained was in the form of a white, fluffy powder. It was highly hygroscopic and, therefore, was stored in a desiccator.

(D) Batch A containing predominantly maltotetrose was purified in essentially the same manner as Batch B, to yield substantially pure maltotetrose, also as a white, fluffy, hygroscopic powder.

Maltotetrose Syrups Using *Pseudomonas stutzeri* Amylase

The use of high maltotetrose syrup in hard candy formulation is especially promising due to its low hygroscopicity. The use of high maltotetrose syrup in ice creams should permit the use of quite high corn syrup-to-sucrose ratios.

M. Abudullah; U.S. Patent 3,654,082; April 4, 1972; assigned to CPC International Inc. has produced a high maltotetrose syrup by incubating a water solution of an acid-modified starch, a partial hydrolysate of starch obtained by acid and/or enzyme hydrolysis of starch, or their mixtures with an amylase of the type produced by the bacterium *Pseudomonas stutzeri.* An improved yield of maltotetrose is obtained if a starch debranching enzyme is also used in the incubation process either prior to or simultaneously with the *Pseudomonas stutzeri.* The starch used as starting materials may be obtained by acid-modifying or acid and/or enzyme-hydrolyzing any conventional starch. Preferred starches include acid-modified waxy milo starch, partially enzyme hydrolyzed waxy milo starch, partially acid hydrolyzed cornstarch, and partially enzyme hydrolyzed cornstarch.

Before the starch can be converted to a high maltotetrose syrup, it must first be solubilized in water, preferably to form a solution containing over 5% by weight of starch. Preferably, the water solution should contain from 10 to 55% by weight of starch. This solution may be formulated by any of the methods common in starch technology.

The acid-modified starch, partial acid and/or enzyme hydrolysate of starch, or mixture thereof, is then converted using the *Pseudomonas stutzeri* amylase to a high maltotetrose hydrolysate at a temperature from 40° to 65°C and at a pH from 6 to 9. It is preferred that the incubation be carried out at 50° to 60°C and at a pH of 6.5 to 7.5. A starch conversion product produced in this manner will have a maltotetrose content of the solids of at least 20% by weight and a DE of at least 13.

The high maltotetrose hydrolysate may then be converted to a high maltotetrose syrup by removal of water by evaporation or other equivalent means. Since the maltotetrose concentration in the hydrolysate is higher than that attainable from unmodified starch considerably less water need be removed to arrive at a high maltotetrose syrup product. Improved yields of maltotetrose may be obtained if the starch is also incubated with a starch debranching enzyme. This can be done either prior to or simultaneously with the incubation with the amylase of the type produced by *Pseudomonas stutzeri.*

As little as 100 units of *Pseudomonas stutzeri* amylase per 100 grams of starch is sufficient to produce a maximum maltotetrose yield within 2 days. Larger quantities of the enzyme may, of course, be used but this will only result in additional cost. Smaller quantities of the enzyme may also be used but this may result in longer incubation times.

Pseudomonas stutzeri amylase is produced as follows. Cells from a stock slant were aseptically transferred to 500 ml Erlenmeyer flasks containing 100 ml each of a seed medium composed of 1% Tryptone, 0.5% yeast extract, 0.28% K_2HPO_4, 0.1% KH_2PO_4, 0.2% soluble starch and the remainder water. These flasks were incubated at 30°C for 18 hours on a reciprocal shaker. 10 ml portions of seed medium were used to inoculate 1,000 ml Erlenmeyer flasks containing 200 ml of a production medium composed of 1% Tryptone, 0.5% yeast extract, 0.28% K_2HPO_4, 0.1% KH_2PO_4, 1% soluble starch, and the remainder water. The fermentation flasks were incubated for 24 hours at 30°C on a rotary shaker operated at 220 rpm. After fermentation, the culture broth, which possessed an activity of 1 to 5 units per ml, was harvested by centrifugation. The centrifugate was then used as the source of *Pseudomonas stutzeri* amylase.

To further concentrate the *Pseudomonas stutzeri* amylase, the centrifugate was mixed with a volume of 2-propanol equal to twice the volume of the centrifugate and which contained suspended therein 10 grams of diatomaceous earth per liter of 2-propanol. The enzyme was insoluble in this solution and was physically adsorbed onto the diatomaceous earth. The enzyme-diatomaceous earth precipitate was collected by filtration and dried to give a dried enzyme product which was 5 to 20 times more active per unit weight than was the original culture broth.

Pseudomonas stutzeri amylase activity was determined as follows: (1) a solution of 0.55 weight percent soluble starch (Merck, reagent grade) was dissolved in a 0.02 molar glycerophosphate hydrogen chloride buffer (pH 6.9) which contained 0.02 molar calcium chloride; (2) reagents were prepared for the determination of reducing sugars as described by N. Nelson, *J. Biol. Chem. 153*, 375, (1944); (3) 9 ml of starch solution was pipetted into a test tube, and the tube was placed in a 25°C water bath for 15 minutes to equilibrate to temperature; (4) 1 ml of the *Pseudomonas stutzeri* amylase solution to be assayed, was added to the test tube; (5) the resultant 10 ml solution was mixed and incubated at 25°C for 10 minutes; (6) a 2 ml sample of the solution was withdrawn and the amount of reducing sugars produced was determined by the above-referenced method of Nelson. One unit of enzyme activity is defined as being equal to one micromol of reducing sugar produced, calculated as dextrose, per milliliter of enzyme solution per minute.

Example 1: This example illustrates the use of *Pseudomonas stutzeri* amylase in the conversion of acid-modified waxy milo starch of 80 fluidity to conversion products having high maltotetrose contents. 10% and 20% solutions of acid-modified starch were prepared by dissolving the acid-modified starch in boiling water. The resulting solutions were incubated with *Pseudomonas stutzeri* amylase at 50°C at a pH of 6.5 for 20 hours and for 48 hours, respectively. At the end of this time, the starch hydrolysates had the compositions shown in the table below.

| | | | Saccharide distribution, weight percent, D.B. | | | | |
Pseudomonas stutzeri amylase (u/100 g.)	Weight percent, starch	D.E.	Glucose	Maltose	Malto-triose	Malto-tet-raose	Higher saccharides
1,000	10	26	1.1	2.3	5	55	36.6
200	20	21.1	1.1	1.9	4.4	56.2	36.4

Example 2: The starches listed below were converted simultaneously with *Pseudomonas stutzeri* amylase and pullulanase debranching enzyme from *Aerobacter aerogenes* at 50°C and pH 6.5. The resulting high maltotetrose hydrolysates had the following compositions listed on the next page.

Starch substrate	Weight percent, starch	Pullulanase (u/100 g.)	D.E.	Saccharide distribution, weight percent, D.B.				
				Glucose	Maltose	Malto-triose	Malto-tetraose	Higher Saccharides
Waxy Milo, 80 fluidity	10	300	30	1.6	4.0	7.2	67.0	20.2
Do	20	200	26.8	1.0	5.7	7.7	61.0	24.6
Waxy milo, enzyme-thinned to 5 D.E	30	200	29.4	2.5	7.8	11.5	55.7	24.5
Corn, acid-thinned to 16 D.E	35	200	31.6	4.5	8.1	11.6	50.0	25.8

Maltose-Maltotriose Syrups

A storage stable high maltose syrup is produced by the process developed by *T.L. Hurst and A.W. Turner; U.S. Patent 3,791,865; February 12, 1974; assigned to A.E. Staley Manufacturing Company.* More particularly the syrups produced by this process have a minimum FE of 85%, a minimum DE value of about 50%, a maltose content of 60 to 80% and a maltotriose content of 15 to 35%. Due to the relatively high level of maltotriose, these high FE syrups do not crystallize when concentrated to high solids, and accordingly, are not subject to bacteriological spoilage during shipping or storage.

These products are prepared by forming an aqueous starch slurry of 5 to 55% solids. The starch slurry is then pasted and/or thinned to a DE up to 35 by the use of an acid or enzyme or a combination thereof. The resultant starch paste is then adjusted, if necessary, to a pH between 4.0 and 6.5, preferably between 4.5 and 5.9, and a solids content of 5 to 55% by weight, and saccharified by adding a maltogenic enzyme and amylo-1,6-glucosidase. The composition is maintained at a temperature of 85°F to 170°F for 24 to 100 hours or for a time sufficient to obtain the desired syrup. The syrup is then preferably refined and concentrated.

The amount of maltogenic enzyme required in combination with amylo-1,6-glucosidase to saccharify starch to produce a high FE syrup will vary depending on the source, purity, etc., of the enzyme. For the most part, when an enzymatically highly active malted barley is used as the source of maltogenic enzyme, concentrations of malt can range from 0.1 to 2% by weight of the starch on a dry solids basis. The amount of amyl-1,6-glucosidase is not particularly critical and depends to a large degree on the concentration of starch, the activity of the enzyme, reaction conditions, etc. Generally, from 0.05 to 2 units of amylo-1,6-glucosidase per gram of starch, and preferably from 0.25 to 1 unit per gram of starch, are used.

Example 1: 900 grams of unmodified cornstarch (dry solids basis) was slurried in 6 liters of water, adjusted to pH 5.0 and cooked in a jet cooker at 56 psig for 15 minutes. The cooked starch was cooled to 50°C and 1.2 grams of Wallerstein's malt amylase PF (free of α-amylase) and 2 units of amylo-1,6-glucosidase per gram of starch was added. After saccharification for 48 hours at 50°C, the syrup had a DE of 57.6. After 112 hours, the starch syrup had a DE of 57.7, a D of 1.3, 70% maltose and about 29% maltotriose.

Example 2: A 35% by weight aqueous cornstarch slurry was adjusted to pH 8.1 and cooked in a jet cooker at 66 psig for 5 minutes. The cooked paste was ejected into a solution of α-amylase (*Bacillus subtilis*) containing 2.5 SKB units of amylase per gram of starch and maintained at 190°F for 2 hours. The α-amylase was inactivated by heating at 205°F for 20 minutes forming a 15.7 DE starch paste. After the starch paste was cooled to room temperature and adjusted to pH 5.7, an aqueous extract of distiller's malt equivalent to 1.5% malt based on the starch dry solids and 2 units of amylo-1,6-glucosidase per gram of starch was added. The starch paste was diluted into aliquots of 5.67% solids, 10.94% solids, 16% solids, 20.65% solids, 25.06% solids and 31.44% solids. After dilution, the hydrolysates were incubated at 130°F for 68 hours, steamed and filtered. The properties of the saccharified starch syrups are set forth in the table below.

Solids	D.E.	% Dextrose	% Maltose	% Maltotriose	% F.E.
5.67%	57.7%	Trace	75.83%	21.74%	97.57%
10.94%	57.5%	Trace	76.92%	21.25%	98.17%
16.00%	56.5%	Trace	79.16%	19.52%	98.68%
20.65%	56.9%	1.14%	77.53%	19.22%	97.89%
25.06%	56.8%	1.75%	73.60%	21.80%	97.15%
31.44%	55.1%	1.95%	73.16%	20.70%	95.81%

The above syrups did not crystallize on concentration to 80% solids.

USE OF MALTOSE SYRUPS

High Maltose Syrups in Icings

Icings generally contain, as basic ingredients, water and sugar. Cream or butter-cream icings also contain substantial amounts of fat or shortening which permit the aeration. Aerated icings are generally applied to baked goods, confections such as cakes, sweet rolls, pastries, etc. A considerable amount of sucrose is conventionally used in cream icing to provide a stable aerated structure. Because of the high sweetening power of sucrose, the sucrose-containing cream icing has an objectionable high level of sweetness. From a sweetness viewpoint, it would be desirable to use corn syrup as a cream icing component.

Unfortunately, corn syrup cannot be incorporated into a cream icing without adversely affecting the icing stability. Such corn syrup icings readily absorb water under humid conditions even from bakery products to which they are applied. Eventually, the firm, aerated structure becomes liquefied. Bakery products to which such an icing has been applied tend to become soggy in the proximity of the applied icing.

M.H. Katz; U.S. Patent 3,532,513; October 6, 1970; assigned to The Pillsbury Company has now disclosed an improved icing containing high maltose syrups. This icing comprises: (1) A sweetening portion having less than one part by weight monosaccharide for each 20 parts by weight disaccharide and the sweetening portion comprising: (a) 15 parts by weight of a corn syrup having a disaccharide to monosaccharide ratio of at least 3:1 and (b) from 40 to 80 parts by weight of a disaccharide sugar portion substantially free from sugar crystals having a crystal size greater than 50 microns; (2) From 7 to 17 parts by weight water; and, (3) From about 5 to 30 parts by weight emulsified fat uniformly dispersed throughout the icing in the form of minute fat globules.

It has been found that the above icing possesses superior stability against degradation yet provides an acceptable texture, mouthfeel, sweetness level, appearance and application ease. Under hot, humid conditions over a prolonged period of time this icing retains its firm aerated structure. When applied upon high moisture containing bakery products (e.g., a cake having a moisture level of more than about 40% by weight), the icing does not become syrupy when stored over prolonged periods of time under hot and humid conditions.

In order to provide a stable product, it is essential that the monosaccharide content (e.g., dextrose, galactose, mannose, fructose, etc.) to disaccharide weight ratio in the sweetening portion be less than one part by weight monosaccharide for each 20 parts by weight disaccharide. A superior stable icing suitable for application upon a high moisture containing bakery product is provided by employing a corn syrup icing having monosaccharide in less than one part by weight for each 40 parts by weight disaccharide sugar. The apparatus used in preparing the icing is shown in Figure 10.1.

FIGURE 10.1: HIGH MALTOSE SYRUPS IN ICINGS

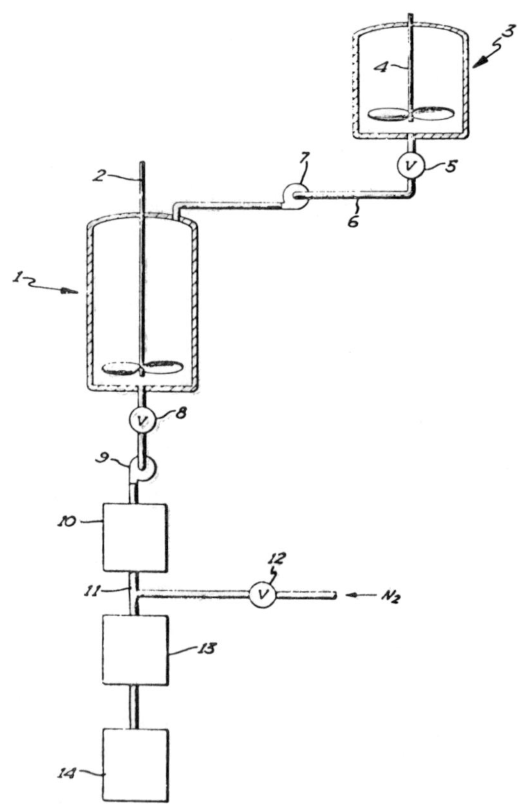

Source: M.H. Katz; U.S. Patent 3,532,513; October 6, 1970

With reference to the drawing, a suitable method for preparing a preferred icing comprises adding 21 lb of water into mixing vessel 1. The mixing vessel is provided with a mixing agitator 2, a surface scraper to prevent scorching of the sugar (not shown) and a thermally controlled heating and cooling means (not shown). The water in the mixing vessel is heated to 144°F whereupon 45 lb of corn syrup is added while maintaining moderate agitation. After the corn syrup has dissolved in the 144°F water, 185 lb of comminuted (10XXX) sugar is added. The aqueous mixture containing the corn syrup and sugar is stirred moderately for 15 minutes at 144°F.

The contents of the mixing vessel are then purged with nitrogen gas by conventional means and the contents are cooled from 144°F to 110°F while maintaining moderate agitation and continual purging. After cooling 9 lb of microbiologically acceptable water at 110°F containing 1 lb of artificial vanilla flavor and 0.5 lb of polyoxyethylene (20) sorbitan monostearate is added to the mixing vessel. The added water and contents of the mixing vessel are then moderately agitated with the mixing vessel agitator for 15 minutes while purging the contents with nitrogen (e.g., at a rate of 8 cu ft/hr).

During preparation and processing of the aqueous mixture, a shortening product substantially free from molecular oxygen is prepared in fat vessel 3. The fat vessel is provided with a nitrogen purging means similar to that employed in the mixing vessel and a fat vessel stirrer 4. In the fat vessel, 37.5 lb of hydrogenated vegetable shortening containing 3.0 weight percent glyceryl monostearate is added. The fat product is heated to 140°F and moderately agitated and purged with nitrogen. Fat valve 5 is then positioned to permit the contents thereof to be pumped through fat conduit 6 by means of pump 7 into the mixing vessel.

The aqueous medium (at 110°F) containing sugar, vanilla flavor, corn syrup and shortening is homogeneously mixed in the mixing vessel while purging the contents thereof with nitrogen. Mixing vessel valve 8 is opened and pump 9 is then engaged whereupon the contents pass through deaerator 10 through pipe 11 which is suitably provided with a nitrogen inlet valve 12.

A suitable gas such as nitrogen, carbon dioxide, etc., is fed into the pipe at a rate sufficient to reduce the specific gravity of the resultant aerated icing to about 1.00. The nitrogen containing product is then passed through a scraped surface heat exchanger 13 which simultaneously whips and cools the product to 85°F. The resultant icing product issuing from the heat exchanger is then conveyed to a conventional filler 14 and packaged such that the gas in the headspace contains less than 3% oxygen by volume.

Syrups for Soft Candy Production

R.G.P. Walon; U.S. Patent 3,692,542; September 19, 1972; assigned to CPC International Inc. has disclosed a confectionery syrup which has an acceptable sweetness, and when used in making confectioneries like candies maintains proper moisture balance; that is, it either hardens due to loss of water or becomes exceedingly soft due to excessive moisture pick-up.

Broadly speaking, this syrup has a DE less than 60, a fructose content of 3 to 16%, a dextrose content of 10 to 35%, a maltose content of 10 to 25%, and 40 to 60% of oligosaccharides having a degree of polymerization of 6 or higher

(DP_{6+}), the percentages being based on total carbohydrate solids present. Preferred syrup compositions have a DE of less than 55, a fructose content of 4 to 12%, a dextrose content of 10 to 30%, a maltose content of 10 to 20%, and 40 to 60% of oligosaccharides having a degree of polymerization of 6 or above.

One excellent way of forming these syrups is to blend a combination of two or three syrups which have been produced by varying techniques of starch conversion. The first syrup used in the blend may be described as a high maltose syrup. One way of producing a syrup of this type is to first acid thin the starch slurry, and follow this step by malt saccharification, and glucamylase saccharification steps. In one run, a high maltose syrup was produced having the following make-up.

Syrup 1 — High Maltose Syrup

Sugar Component	Percent by Weight
DE	37.6
Fructose	–
DP_1	4.8
DP_2	25.0
DP_3	9.8
DP_4	7.0
DP_5	6.5
DP_{6+}	32.5

Another syrup used as a portion of the blend to make the syrups defined here may be described as a low DE, high oligosaccharide syrup, that is, one having a high proportion of DP_{6+} oligosaccharides. One method of making a syrup of this type, also known as a dextrin syrup, is carried out by converting starch with malt diastase, a mixture of α- and β-amylases. Another suitable method of preparing a syrup of this type is effected by either acid thinning a starch slurry, or by thinning a starch slurry with a bacterial α-amylase. In one run, a syrup of this type was produced having the following components.

Syrup 2 — Dextrin Syrup

Sugar Component	Percent by Weight
DE	19.8
Fructose	–
DP_1	4.8
DP_2	8.8
DP_3	6.2
DP_4	4.4
DP_5	3.8
DP_{6+}	72.0

A further syrup used here as a portion of a blend to make the finally sought-after syrup was one high in fructose. Here, a high DE, high dextrose material is first made in the conventional way, by acid thinning a starch slurry followed by enzymatic conversion with bacterial glucamylase. A portion of the dextrose present was then interconverted by known means, such as by ion exchange methods or by base conversion. One typical high fructose syrup is constituted as follows on the next page.

Syrup 3 — High Fructose Syrup

Sugar Component	Percent by Weight
DE	89.7
Fructose	21.4
DP_1	68.0
DP_2	7.2
DP_3	1.0
DP_4	1.1
DP_5	1.2
DP_6	1.1
DP_{7+}	–

Various blends of the foregoing were made by manipulating the ratios of one to another. In same instances, only two of the three syrups were combined, while in others all three were in a number of varying proportions. The following blends of the above syrups yielded a number of compositions having ingredients falling within the ranges set out above. These syrups were made as follows: Syrup A (25% 1, 50% 2 and 25% 3); Syrup B (20% 1, 50% 2 and 30% 3); and Syrup C (10% 1, 75% 2 and 15% 3).

These syrup blends may be used in preparing a wide number of confections, such as coatings for candies, cookies, biscuits, and cereals, as confections for direct consumption, as fragmented chips for baked goods, desserts, and other food items, as spreads, butters, and icings, etc. Further, they may be used in making gum drops, jelly beans, autumn leaves, turkish paste, orange-, lemon- and pineapple-flavored slices, etc. Still further they may be combined with a fat and used for example in chocolate, as chocolate-like coatings for candy, in ice cream, and cookies, and in other confections, such as toffees, caramels, etc. and in icings, fillings, fudges, butter creams, and other compositions of this class. The syrups defined herein are particularly useful in preparing soft candies such as toffees and the like. The following formulation for making a toffee candy uses a syrup comprising 5.3% fructose, 20.6% dextrose, 16.2% maltose, and 44.8% DP_{6+} oligosaccharides.

Example: Candy Preparation —

Ingredients	Amount
Milk, kg	2.5
Fat, kg	0.5
Glyceryl monostearate, g	45.0
Salt, g	37.0
Vanilla, g	4.0
Syrup, kg	7.25

The above candy is prepared in the following manner. The syrup is gently heated and the condensed sweetened milk added thereto. Thereafter, the fat and salt are added, followed by addition of the glyceryl monostearate. When the temperature reaches 110°C, vanilla is added, and then the temperature is raised to 130°C. The candy has good taste and particularly has a sweetness level which is more pleasant than that possessed by toffees currently on the market. The candies do not collapse, and maintain proper moisture content, neither hardening through loss of moisture or becoming excessively soft due to excessive moisture absorption.

CONVERSION PRODUCTS FROM MALTOSE SYRUPS

This section covers ketose sugars and sugar alcohols produced from syrups containing mainly maltose and lower oligosaccharides. Reference should also be made to U.S. Patent 3,804,717, page   which also discloses the reduction of maltose to maltitol.

Production of Maltulose and Maltotriulose

A process for producing ketose sugars composed essentially of maltulose and maltotriulose from a sugar containing maltose and maltotriose has been developed by *S. Sakai, T. Miyake and Y. Sato; U.S. Patent 3,691,013; September 12, 1972; assigned to Hayashibara Company, Japan.*

The isomerization from maltose to maltulose generally occurs in the presence of a basic catalyst, as has long been known as the Lobry de Bruyn reaction. It is an easy way of isomerizing aldose to ketose and, for that purpose, strong alkali, strong basic ion exchange resin or the like is used. In the present process, calcium hydroxide, ammonium hydroxide, sodium hydroxide, or ion exchange resin was used. Regarding the reaction conditions, whether a combination of low temperature and long time, or the opposite combination of elevated temperature and short time is determined by the equipment used.

At a temperature below 40°C, the decomposition by oxidation of sugar is relatively small, but, with the rise of temperature, the loss due to the oxidation and decomposition of the sugar is increased. With these results in view, tests were conducted with equipment capable of providing continuous rapid heating and rapid cooling and neutralizing. The results, as shown in the following table, indicate that the sugar loss increased with the rise of the temperature and the extension of the reaction time. A sugar concentration of 40% and a Ca(OH)$_2$ concentration of 1.5% based on the amount of sugar was used.

	Isomerization Rate, %, and			
	- - - - - - Loss*, %, After Heating Time of - - - - - -			
	5 min	10 min	15 min	3 days
Temperature				
35°C	–	–	–	45(1)
80°C	–	37(1)	–	–
100°C	30(1)	–	45(5)	–

*In parentheses

The reaction suitably proceeded with a sugar concentration of 20 to 40%, and where calcium hydroxide was used, a 0.13% solution was allowed to stand for 3 to 4 days, and the reaction was carried out until no more change was observed in optical rotation. Substantially the same results were obtained with sodium hydroxide and ammonium hydroxide.

When an elevated temperature was used, the reaction was done with a 1 to 2% alkali solution and at pH of 9 to 10. In this case, a short-time reaction caused a sugar loss of only 1 to 2% and a good rate of isomerization although coloring was marked. The rate of isomerization was determined by the estimation of ketose with use of the cysteine-H$_2$SO$_4$-carbazole method. The isomerized sugar

was quickly cooled and neutralized by injecting into a flash chamber with an equivalent amount of hydrochloric acid or dilute sulfuric acid, the flash tank being kept
at a reduced pressure. The calcium sulfate was removed by filtration, and the sodium chloride and ammonium sulfate were directly used for subsequent culture or
enzymatic reaction. The colored matter was obtained in a fairly high concentration,
and the decoloring was effected by passing the matter through a layer of granulated
charcoal.

The paper chromatogram of the purified solution showed that maltose, maltotriose, and a small amount of tetrose were left in a combined residue of at least
50%. As ketose, there were observed maltulose, maltotriulose, and tetrulose.
The solution upon purification with active charcoal and ion exchange resin became a very sweet syrup and was usable as a sweetening agent.

The method of employing a strong basic ion exchange resin for the epimerization
of aldose was such that Amberlite IRA–400 was used for the ion exchanger and
the sugar solution was boiled to remove oxygen and then cycled at 70°C until
the isomerization proceeded. Isomerization to a degree of 30% was possible in
5 hours. The solution was then purified by the same treatment as above described.

Separating the aldose and ketose from the mixed solution could be done satisfactorily by the enzymatic oxidation of the aldose. The mixed sugar solution
is enzymatically oxidized by either culturing a strain of Pseudomonas on the saccharides obtained above as the carbon source or using the bacterial cells obtained
by the cultivation as an enzymatic source. When a Pseudomonas strain was cultured with aeration on a 5 to 10% sugar solution containing a nitrogen source
such as corn steep liquor and some inorganic salts, ketoses are little oxidized
while aldoses such as disaccharide and trisaccharide are oxidized to aldonic acids,
i.e., maltobionic acid and maltotrionic acid, respectively.

The enzyme-producing strains which may be used for the above purpose, *Pseudomonas graveolens* NRRL 14 and *Pseudomonas fragi* NRRL 25, both reported by
F.H. Stodola et al. in *J. Biological Chemistry,* Vol. 171 (1947), pp. 213–221, are
most preferable. Other Pseudomonas strains which do not involve the hydrolysis
of disaccharides may also be used although they extend the reaction time.

Alternatively, the bacterial cells obtained by the above cultivation are added to
the 20 to 30% mixed sugar solution, and the mixture is oxidized by aeration and
agitation while the drop of the pH value is being prevented by the presence of
calcium carbonate and other additives. In this manner the residual disaccharide
and trisaccharide can be oxidized to aldonic acids without any loss of the sugar.
The reaction time ranges from 10 to 20 hours, and the degree of oxidation was
determined by measuring the amount of calcium salt of the aldonic acid produced.

The ketose-aldonic acid mixture, when purified with active carbon and a strong
acidic ion exchange resin, becomes a sweetening agent having some sourness,
which may be used as food additive. The reaction product or aldonic acid-ketose
mixture can be divided into the components by adsorption on a weak basic ion
exchange resin taking advantage of the acidity of aldonic acid. Or, the aldonic
acid can be crystallized and separated out as a calcium salt or basic calcium salt.
In sweetness properties, the maltulose product is ranked between ordinary sugar
and grape sugar and is practically on the same level as maltitol.

Example: Starch slurry of 20% sweet potato at pH 6.0 was liquefied at 90°C by addition of 0.2% of a liquefying enzyme per gram of starch. A homogeneous liquefied starch with a decomposition rate of 3.5% was obtained. This liquefied starch was rapidly cooled to 50°C, β-amylase and α-1,6-glucosidase obtained from an Escherichia strain (*Escherichia intermedia* ATCC 21073) were added in amounts of 20 and 10 units, respectively, per gram of the starch, and the liquefied starch was saccharified at 45°C and pH 6.0 for 35 hours. After the deactivation of the enzymes, the resultant was decolorized and purified with active carbon, and a sugar solution with a maltose purity of 93% on dry basis resulted. Purification of the maltose was done by crystallization.

Two kilograms of this maltose was dissolved in 10 liters of water and, after addition of 13 grams of calcium hydroxide, was reacted at 35°C for 4 days. The optical rotation of the reacted solution was determined and when the value was stabilized the reaction was terminated and concentrated under vacuum. It had a maltose content of about 52%. One half of this mixed solution was put aside, and 1 liter of a solution containing 10% of the reaction product as an anhydride basis was neutralized with CO_2.

Into this solution were dissolved 1% corn steep liquor, 0.2% urea, 0.06% potassium dihydrogen phosphate, 0.025% epsom salt, and 2.5 grams calcium carbonate. The mixture was sterilized with heat and inoculated with *Pseudomonas graveolens* IFO 3460, which was cultured with stirring at 400 rpm and aeration with an equivalent amount of air at 30°C for 50 hours. From the culture fluid the bacterial cells were centrifugally collected and used as an enzyme source.

The enzymatic activity was determined in the following way. 200 μmol of a phosphate buffer (pH 5.6), 20 μmol of maltose, 0.5 μmol of 2,6-dichlorophenolindophenol, and the sonicate of bacterial cells in a total volume of 6.0 ml were mixed and reacted at 30°C for 10 minutes. Activity was measured following the decrease in absorbance at 590 mμ. And one unit of enzyme was defined as the amount which gives a decrease of absorbance of 0.001 in a minute.

The culture fluid freed from the bacteria following the cultivation was purified by decoloration and filtration through active charcoal, concentrated, added to a solution saturated with calcium hydroxide, and while the solution was hot the aldonic acid was precipitated and removed in the form of basic calcium aldonate. The filtrate was freed from the calcium by means of a strong acidic ion exchange resin and, using a slightly basic anion exchange resin (Amberlite IRA 64), the residual bionic acid was removed.

Then it was purified with different ion exchange resins, i.e., strong basic IRA 411 and strong acidic IR 120, to a colorless liquid and concentrated syrup. Analysis by paper chromatography showed that the syrup contains 95% maltulose, 1% maltotriulose, 1% maltose, 2% maltotriose, and the balance some other oligosaccharides. The yield was 45% on the basis of the amount of the maltose.

The bionate separated as an insoluble calcium salt was dissolved in five times by volume of water and, by the gradual addition of dilute sulfuric acid, the calcium was precipitated as calcium sulfate and removed. By means of a cation exchange resin (IR 120) the residual calcium was eradicated, and using a moderately basic ion exchange resin the sulfuric radical was absorbed. The resulting solution was concentrated in vacuum, and maltobionic acid was obtained in a syrupy state.

The yield was 41% on the basis of the maltose amount used. Paper chromatography indicated that this bionic acid contained small amounts of maltulose, maltose, maltotriose, and also maltotrionic acid, while maltobionic acid accounted for more than 95% of the total amount.

Numerous examples are also included in the complete patent showing the use of this maltulose product in production of carbonated drinks, ice cream, bakery products, jellies, candies and the like.

Sugar Alcohols

H. Hijiya and M. Hirao; U.S. Patent 3,838,006; September 24, 1974; assigned to Hayashibara Company, Japan provides a process for producing special starch syrups which are stable against heat, very sweet, and hard to crystallize. Starch is subjected to the action of an α-1,6-glucosidase such as isoamylase or pullulanase, in the manufacture of starch syrups through the use of saccharogenic amylases such as α- or β-amylase, glucoamylase and/or the like, thereby extending the scope of activity of the α- or β-amylase and transforming the structures of the oligosaccharides and dextrins produced to straight-chain structures so as to obtain a starch syrup of a freely chosen composition and properties with the viscosity, flavor, and other attributes modified as desired. The resulting syrup is then reduced to sugar alcohols.

Because starch syrups prepared by the dual enzyme process contain some glucose or maltose, they have limited resistance to heat and, in applications which involve heating or where impurities or mixtures active against carbonyl groups are to be added, the syrups may sometimes be colored to a disadvantage. For this reason, these starch syrups were hydrogenated to convert the same to sugar alcohols.
To be more specific, when starch syrups at a concentration of 50% were reacted with hydrogen gas at a pressure of about 100 kg/cm^2 using a Raney nickel catalyst, sugar alcohol starch syrups resulted which showed no sign of reducibility and contain sorbitol, maltitol and other oligosaccharide alcohols.

In addition to increased sweetness, these syrups were less viscous than ordinary ones of approximately the same DE values, and tended to show higher degrees of threadability. Last but not least important, the heat resistance of the products was such that, when they were concentrated by heating on direct fire by the usual candy test, ordinary starch syrups resisted the heat up to 130°C, whereas, on the same DE basis, the syrups prepared by using isoamylase resisted the heat up to 155°C and the sugar alcohol syrups underwent no decomposition or coloration up to 200°C or nearly to the anhydrous state.

Examples 1 through 5: Starches of corn, white potato, sweet potato, wheat, sago, etc. may be used as raw material. For the purpose of the present process, amylomaize starch or amylose separated from ordinary starch is preferred, but such a material is expensive. It is for this reason that cornstarch or white potato starch was used in this experiment. Because of the nature of the enzymes used, the starch slurry was prepared to a concentration of 30 to 40%, or somewhat lower than the usual concentrations for ordinary starch slurries. The procedure for liquefaction was continuous. Since the saccharification took much time, a batch method was employed for saccharification.

Concerning the conditions for liquefaction, cornstarch, was liquefied at a high temperature of 160°C as shown in the table below to ensure complete liquefaction. White potato starch was continuously liquefied with a liquefying amylase at 90°C and the reaction was stopped at a suitable DE. When the pH was slightly decreased by the addition of a small amount of acid in place of the enzyme, the liquefaction was facilitated.

Example number	I	II	III	IV	V
Kind	Corn starch	Corn starch	Corn starch	High amylose starch.	White potato starch.
Starch conc. (percent)	35	35	35	30	40
Liquefying enzyme or acid			Oxalic acid	Oxalic acid	Liqfg. enzyme, 10 μ/g.
Liqfg. temp. (° C.)	163	165	120	120	90
1st conv.:					
Enzyme (unit/gram. starch)	L 40	L 40 / B 10	L 30 / R 2	B 10	P 30 / B 5
Temp. (° C.)	50	50	45	60	45
pH	6.0	6.0	5.5	5.5	6.0
2nd conv.:					
Enzyme (unit/gram starch)	C 20	C 20	C 20	I 30	
Temp. (° C.)	45	50	45	45	45
pH	6.0	6.0	6.0	5.5	6.0
D.E.	40	48	70	51	55
Main comp.	Oligo. sacc	Maltose olig	Glucose olig	Maltose olig	Maltose olig
Reduction conc. (percent)	50	50	45	45	45
Raney nickel catalyst (percent)	8	8	10	10	8
Temp. (° C.)	50–90	50–90	50–90	50–90	50–90
H pressure (kg./cm.³)	30–100	30–100	30–100	40–100	40–90
CaCO₃ added, percent	0.1	0.1	0.2	0.1	0.2
Residual reducing sugar, percent	0.1	0.1	0.2	0.1	0.1
Viscosity (cp.) after reduc. (65% conc. at 37.5° C.)	85	70	45	72	65
Sweetness against glucose (percent)	40	55	80	54	60

The enzymes used for the conversion purpose were as follows. As the liquefying amylase, a bacterially produced α-amylase (C) which was commercially available was used, and as β-amylase (B) the enzyme extracted from wheat bran was used. The glucoamylase (R) was in the form of a cultured fluid of *Rhizopus delemar*. As α-1,6-glucosidases three varieties were used, i.e., bacterially produced pullulanase (P), isoamylase (I) yielded by the bacteria of *Pseudomonas amyloderamosa* (ATCC 21262), and the α-1,6-glucosidase (L) afforded by *Lactobacillus plantarum* (ATCC 8008).

The experimental results are given in the above table. In Examples I and II the DE of liquefied starch slurries were low (about 1 to 5 DE) and the viscosity very high, and therefore the liquefied starches were quickly cooled to the temperature for the first conversion. Then simultaneously with the addition of amylase, each starch solution was vigorously stirred in the preconverter to progress saccharification before retrogradation takes place. The pH was adjusted to 6.0 and the temperature was kept below 50°C, preferably at about 50°C. In Examples III to V, where the retention time for the first conversion was fully extended to more than 20 hours, the amount of β-amylase or glucoamylase was adjusted to obtain a suitable DE value.

Secondary conversion is the stage which dictates the ultimate composition of the particular starch syrup being made, and hence the amount of enzyme added and the reaction time must be suitably adjusted. In Example I the amylase obtained by breaking down the α-1,6-glucoside bonds of the starch was fully decomposed by the α-amylase at a temperature of 45°C and a pH 6.0 to yield a starch syrup of low saccharinity composed chiefly of oligosaccharides, with little glucose and maltose contents. In Example II a starch syrup comprised mainly of maltose was subjected to the action of α-amylase and residual dextrins were decomposed to lower the viscosity. A starch syrup mostly composed of maltose resulted which was highly resistant to heat, adequately boiled dry, and suitable for confectionery use. In Example III the starch is hydrolyzed by an α-1,6-glucosidase from *Lactobacillus plantarum* and sweetened with glucoamylase at a temperature of about 45°C and a pH of about 5.5 and then decomposed by α-amylase at a temperature of about 45°C and a pH of about 6.0 to give a starch syrup which was very sweet because it was largely composed of glucose. Examples IV and V afforded starch syrups containing maltose as the main constituent.

In the series of experiments, each starch syrup thus prepared and adjusted to a concentration between 40 and 60% was placed in an autoclave. With the addition of 4 to 10% Raney nickel catalyst, hydrogen was pumped into the vessel and the charge was agitated for reaction with the hydrogen. At a hydrogen pressure of 20 to 100 kg/cm^2, the temperature was gradually increased and the charge was reduced until the remaining reducing sugar was less than 0.5%. The reaction with heat was concluded at a temperature not high enough to cause the breakdown of the glucoside linkage, or not higher than 100°C. After the reaction, the catalyst was filtered off and the product was purified by ion exchange and then the nickel was removed to obtain a colorless, clear liquid.

When the pH tended to drop during the reaction, it was maintained at the suitable level by the addition of a base, for example, calcium carbonate. The addition of base is preferably within the range of 0.05 to 0.3%. The compositions of end products so obtained were analyzed by paper chromatography to consist essentially of sorbitol, maltitol and hydrogenated oligosaccharides. The products were very sweet and yet refined and flavory in taste and limited in viscosity. They were remarkedly resistant to heat, remaining unaffected by temperatures up to 200°C. Even when heated in mixture with a nitrogen compound, they underwent no coloration to a great advantage.

Maltitol from Maltose

Maltose produced by the combined β-amylase and α-1,6-glucosidase method is converted to maltitol in the process developed by *M. Mitsuhashi, M. Hirao and K. Sugimoto; U.S. Patent 3,708,396; January 2, 1973; assigned to Hayashibara Company, Japan*. Maltose produced with β-amylase alone must be purified by alcohol fractionation to remove dextrins and other higher oligosaccharides.

In hydrogenating maltose, the maltose was dissolved in water to prepare a 40 to 60% aqueous solution and Raney nickel was used as a catalyst in an amount of 8 to 10% of the amount of the aqueous solution. Hydrogen was forced in the solution with a pressure of 20 to 100 kg/cm^2 while the temperature was gradually increased to 80° to 150°C. In some cases, the pH of the solution drops and the reaction slows down. If such is the case a decomposition of maltose will take place and the yield of maltitol is decreased to about 90% with the production of about 8% sorbitol.

In some experiments, therefore, the pH of the solution was adjusted midway in an effort to improve the yield. However, the yield was only slightly increased to 93%. However, when calcium carbonate was added as a moderate neutralizing agent to the starting maltose in an amount of 0.06 to 0.3%, sorbitol production was 1.6%, thus indicating that the decomposition of maltose with an increase of maltitol yielded more than 96%. The result was satisfactory in both purity and yield.

Example: Hydrogenation of Maltose — The composition of the starting material was 96.6% maltose, 0.4% dextrose, and 3.0% maltotriose. This decolored and purified maltose was used in the form of a 40 to 60% aqueous solution. The reduction catalyst used was Raney nickel as developed in alkali in an amount equivalent to 8 to 10% of the amount of the starting material. The temperature was between 90° and 100°C, at which two-thirds equivalent of hydrogen was absorbed. Finally the temperature was increased to 125°C to reach the equilibrium point of absorption. The pressure used ranged from 20 to 100 kg/cm^2. Calcium carbonate was added in an amount of 0.3 to 0.05% of the amount of the starting material.

Under these conditions, the reaction was carried out with stirring. The compositions of the reaction products obtained with or without the addition of calcium carbonate were as tabled below. On completion of the reaction, the catalyst Raney nickel was removed and each product was concentrated, decolored and purified through ion exchange in the usual manner.

	Without CaCO$_3$		With CaCO$_3$ (% per maltose)		
	pH adjusted	pH not adjusted	0.06%	0.13%	0.25%
pH at the start of the reaction	4.0 – 6.2	7.0 – 7.5	7.3	7.4	7.0
pH at the end of the reaction	4.2 – 3.7	4.3 – 3.8	4.8	5.7	6.1
Unreacted direct sugar/material(%)	0.4 – 1.8	6.2 – 1.2	0.63	0.28	0.25
Reaction time (hr.)	8 – 13	8 – 10	8	8	8
Composition: (%)					
Sorbitol	7.3	4.6	2.1	1.8	1.6
Maltitol	90.8	93.5	95.9	96.2	96.4
Others	1.9	1.9	2.0	2.0	2.0

Sugar Alcohols from Syrup Blends

Starch syrups having a DE below 30 are used extensively as additives or basic lubricants for nutrient infant foods, carriers for synthetic sweeteners, granulating assistant agents for flavor powder and powdered coffee, dispersing agents and bodying agents for soups, puddings, other confectioneries and artificial cream. Although such starch syrups have hardly any sweetness, the syrups are noncrystallizable and stable to heat and nitrogeneous compounds. Their extremely high viscosity makes handling occasionally difficult, and moreover they deteriorate the properties of the products and the palatabilities of the foods in which they are used.

M. Hirao and M. Mitsuhashi; U.S. Patent 3,692,580; September 19, 1972; assigned to Hayashibara Company, Japan has prepared a starch syrup having a low DE, a low viscosity and good chemical stability by adding to a starch syrup prepared by an acid or an enzyme hydrolysis of starch, a second starch syrup prepared from a starch hydrolysis involving the action of α-1,6-glucosidase. The second syrup may be hydrogenated or the syrup mixture may be hydrogenated.

Studies on the properties of the products, obtained by blending 30 to 50% of linear chained low viscosity starch syrups of DE 30 to the low acid conversion starch syrups of DE of about 25, showed that the viscosities of the starch syrup mixtures were significantly decreased and the procedures of heating, concentration, mixing etc. during the production of foods were greatly facilitated. Thus coloration and discoloration were completely prevented. The properties of non-crystallization, homogeneous dispersion etc. were greatly improved. By varying the blending rate, the viscosities of the products were controllable. The tendency of improvement of other characteristics of the products was found effective in expanding the applications for common low conversion starch syrups; simultaneously, the limit of blending was extended. These are unneglectable advantages effective for quality improvement of the final products and reduction of production costs. Such effects of blending can be displayed considerably not only in extremely low conversion starch syrup but also in moderate DE starch syrups with DE of 30 to 40, thus effective in varying the delicate properties of foods.

The above effects can be emphasized by using nonreducing starch syrups obtained by hydrogenating the linear chained and low viscosity starch syrups. In the cases of these nonreducing starch syrups, their viscosities are reduced, and owing to their nonreducing properties, they are chemically very stable and of course heat stable. Admixing of these nonreducing starch syrups lowers the viscosities of common low conversion starch syrups. This mixture is also effective in improving their properties of noncrystallization, flavor retention and moisture retention, as well as stabilizing them.

A much more effective method than the above mixture is to hydrogenate starch syrup mixtures comprising common low conversion starch syrups and linear chained low conversion starch syrups. Thus the constituents of the starch syrup mixture can be converted to nonreducing alcohols and to chemically stable starch syrups. The resultant products are ideal low conversion starch syrups.

In the following examples, the syrups used are:

　1a　oxalic acid converted cornstarch
　1b　α-amylase converted sweet potato starch
　1c　α-amylase converted waxy cornstarch
　1d　liquefied starch treated with α-1,6-glucosidase
　1e　hydrogenated product of 1d

Example 1: Various Blended Starch Syrups and Their Characteristics — To the common starch syrups (1a), (1b), (1c) were mixed linear chained starch syrups and nonreducing starch syrups (1d), (1e), at a rate of 30 to 50% as dry substances respectively and the properties of the mixtures were compared. The results are given in the table below. Viscosities were determined with a B-type rotating viscosimeter. Candy tests to determine their heat resistance, were performed by placing 200 grams starch syrup in a regulation copper pan, heating and concentrating the syrup. The results given in the following table are the temperatures that caused coloration.

As shown in the following table, (1a) and (1b) were starch syrups with very high viscosities and when added to other foods, their low hygroscopicity helped make these syrup mixtures effective stabilizers and their viscosities were excessively high. The additions of (1d) and (1e) to these starch syrups lowered their viscosities and made these starch syrups easy to handle as in the case of moderate conversion starch syrups. Thus used as additives for powdered foods, or as ingredients of soups they imparted desirable viscosities and in a powdered state facilitated drying and shortened the drying time. In addition these products imparted desirable viscosities to soups. With regard to sweetness of this mixture, there was imparted practically no increase of sweetness and even the hygroscopicity of the obtained products was improved. When used as ingredients for infant foods the syrups accelerated their dissolving rates and displayed their efficiencies as stable flavor retention agents. In addition they accelerated the drying speed of jellies and their effects of preventing deformation were great.

| No. | Variety of— | | Added syrup's rate (percent) | Moisture content (percent) | D.E. | Viscosity (cp.) | Candy test (° C.) |
	Common starch syrup	Added starch syrup					
	1-a$_1$	----------	0	25	25.0	2,300	130
1	1-a$_1$	1-d	30	25	26.8	1,700	135
2	1-a$_1$	1-d	50	25	28.0	1,500	140
3	1-a$_1$	1-e	50	25	13.0	1,350	145
	1-a$_2$	----------	0	25	35.0	1,200	135
4	1-a$_2$	1-d	30	25	33.8	1,150	140
5	1-a$_2$	1-d	50	25	34.0	1,130	143
6	1-a$_2$	1-e	50	25	17.5	1,050	145
	1-b	----------	0	25	31.0	1,900	130
7	1-b	1-d	30	25	31.0	1,500	140
8	1-b	1-d	50	25	31.0	1,400	142
9	1-b	1-e	50	25	16.0	1,200	145
	1-c	----------	0	25	25.0	2,500	125
10	1-c	1-d	30	25	26.8	1,800	135
11	1-c	1-d	50	25	28.8	1,500	140
12	1-c	1-e	50	25	13.0	1,300	145

Example 2: The Production of Nonreducing Mixed Sugars — The pH of starch syrup No. 5 described in Example 1 was adjusted to a concentration of 50% by addition of a small amount of calcium carbonate. The syrup was completely hydrogenated using Raney nickel catalyst at less than 150°C and hydrogen pressure within 130 kg/cm². Upon removal of the catalyst, decoloration and deionization, a colorless and transparent starch syrup was obtained. Since the syrup was absolutely nonreductive and contained large amounts of linear chained dextrins, the viscosities of the product were low and the syrup exhibited most excellent properties of low hygroscopicity, noncrystallization and moisture retention.

Use of Sugar Alcohols in Foods

Conventional types of sugar alcohol sweetners, for example, sorbitol, are digestible in human bodies, therefore they are quite unsuitable for the production of sweetened foods and noncaloried sweetened foods for diabetics or for those who are conscious of and anxious about weight or obesity. The present process disclosed by *M. Mitsuhashi, M. Hirao and K. Sugimoto; U.S. Patent 3,705,039; December 5, 1972; assigned to Hayashibara Company, Japan* covers the production of foods and drinks, in which a sweetener consisting mainly of noncaloried maltitol and maltotriitol is used. This sweetener has a better taste than cane sugar, a desirable viscosity, high solubility and water retention capacity, and is noncrystallizable, and also imparts to the products flavor and color stabilities.

This sweetener is a suitable additive for carbonated beverages such as cola drinks, ciders, etc., lactated beverages, such as Calpis, concentrated fruit juice, etc. Use of this sweetener as a substitute for glucose, cane sugar and other syrups makes even higher concentrated solutions possible, and all the carbohydrate contents may be converted into noncaloried substances. Due to its ability to impart viscosity and flavor stability to the products, first rate drinks can be produced utilizing this sweetener, which has good flavor with no undesirable aftertaste. Similarly, when it is used in sponge cakes and others, the calorie content can be maintained at a minimum and its moisture retentive property prolongs the shelf life greatly, and improves the product's texture, which is a vital problem for these products.

This sweetener can be added to bakery products such as biscuits, cookies, etc., to produce nonsugar or low-caloried foods. Heat resistance of the sweetener prevents over-baking caused by heating, cracking and deformation which are usually observed upon cooling the baked products, increases yield, prevents dispersion of flavor and increases the durability of foods. When it is used in jellies, etc., the sweetener not only imparts sweetness to the products but makes production of the products with noncaloried components possible. It prevents discoloration and crystallization of the products, imparts moisture and flavor retentive properties to the products and fully exhibits its efficiency of preserving the initial qualities of the products after prolonged periods.

Example 1: Process for Canning Fruits — The sugar solution to be charged into canned oranges was prepared by dissolving 50 kg of the present sweetener which contained 85% of maltitol (dry base), 25 grams of saccharin in water to a total volume of 100 kg. After the solution was charged to the peeled oranges, the products were canned and sterilized according to usual methods. The syrup produced according to this process, had a suitable viscosity, a mild sweetness that harmonized well with the acid taste of the oranges, retained its flavor satisfactorily and the syrup itself was noncaloried.

Example 2: Process for Production of Sponge Cakes — A popular formula for production of sponge cakes is shown below.

	Grams
Maltitol-maltotriitol sweetner (maltitol content 80%, dry base)	1,000
Egg	1,100
Flour	500
Honey	50

Dough was prepared according to the above formula and the usual method. The dough was poured into an iron pan, covered with paper, and baked in an oven heated at 180° to 190°C. The products had an appealing baked color and a soft, spongy and improved texture. Delicious products with suitable moisture were available even after a prolonged storage period, due to the fact that their degradation or drying are delayed by the addition of the sweetener. Also a reduction of carbohydrates to half makes production of low caloried foods possible.

FRUCTOSE FROM STARCH HYDROLYSATES

The monosaccharide known as fructose, levulose or fruit sugar is obtained commercially by the hydrolysis of sucrose (to glucose plus fructose) or by the isomerization of glucose (dextrose) or corn syrups using alkaline catalysts or enzymes.

Fructose is generally known to be sweeter than dextrose. A dextrose solution, after isomerization followed with purification treatments becomes a solution of both dextrose and fructose. The sweetness is thus increased. The measurement of sweetness is still a controversial subject in the scientific field, so that the increase in sweetness of an isomerized solution is subjective.

In the syrup making business, it is generally known that there is an upper limit on the dextrose content of the final syrup product if the product is to retain the characteristics of a liquid. If the final product contains, say, about 80% dissolved solids (DS), the dextrose content of the syrup should not be more than 40 to 44% (based upon the total dissolved solids).

Syrups of about 80% DS and 40 to 44% or more dextrose always have the tendency of becoming solidified due to crystallization. This tendency of crystallization can be avoided if a part of the dextrose is transformed into fructose. Thus isomerization of at least some of the dextrose to fructose provides a syrup of high saccharose content with reduced tendency to crystallize.

FRUCTOSE BY ALKALINE ISOMERIZATION

The first method developed to produce fructose from glucose was by means of alkaline catalysts. Exemplary of processes using alkaline catalysts are those disclosed, for instance, in U.S. Patents 2,487,121; 2,354,664 and 2,746,889. However, there are a number of distinct disadvantages associated with alkaline isomerization in previous processes. For instance, due to the nonselectivity of the alkaline catalysts various objectionable by-products are formed, such as large amounts of colored bodies and acidic materials. Refining alkaline-isomerized liquors to remove the objectionable by-products to produce an acceptable product requires rather complicated and costly procedures. The following processes disclose improved methods for the alkaline isomerization process.

Preparation and Purification of High DE Fructose Syrups

B.L. Scallet and I. Ehrenthal; U.S. Patents 3,305,395; February 21, 1967; and 3,285,776; November 15, 1966; both assigned to Anheuser-Busch, Incorporated disclose the production of sugar syrups by hydrolysis of starch with subsequent alkali treatment and partial separation of the resulting mixture of sugars.

Briefly, the process of U.S. Patent 3,285,776 comprises the isomerization of a high DE, high glucose starch conversion syrup with a limited amount of alkali under continuously controlled and maintained pH conditions of 8.5 to 9.5 to produce a high DE syrup having less than 45% glucose and at least 10% fructose. The syrup is then passed through an ion exclusion and/or an ion exchange column to produce a noncrystallizing very sweet tasting syrup having a DE of 70 to 85, containing a substantial portion of rapidly fermentable sugars, and being free from objectionable color and impurities.

Example 1: A 44 DE acid converted liquor containing 84.6 lb of dry substance at 29.3°Bé is treated with 0.1% (DSB) of Diazyme powder at pH 5.0 and at 57°C for 22 hr. The resultant liquor has a DE of 83.1 and a glucose content of 70.5. The temperature of the liquor is raised to 65°C to inactivate the enzyme. 0.3% carbon (DSB) is added to the foregoing liquor and is filtered. The temperature of the high glucose liquor is raised to 65°C and the pH adjusted to 9.0 and maintained at that temperature and pH for 4½ hr.

A total of 0.0079 g of sodium hydroxide per gram of dry solids is added to the liquor during the isomerization. The finished product contains 30.2% ketose (DSB) of which 21.0% is fructose. The finished product also contains 44% glucose. A portion of this liquor is then passed through a monobed column containing XE-168 and Amberlite 200 at the rate of 0.2 to 0.5 gal/ft^3 of resin. The liquor from the column is substantially free of ash and organic acids and is light in color. The liquor is concentrated to 43°Bé.

In U.S. Patent 3,305,395, the process comprises the isomerization of a high DE, high glucose starch conversion syrup with a limited amount of alkali under continuously controlled and maintained pH conditions of 8.5 to 9.5 to produce a high DE syrup having less than 80% glucose and at least 10% fructose, which syrup is then passed through an ion exclusion and/or ion exchange column to produce a syrup having a DE of 80 to 100 (compared to 70 to 85 of U.S. Patent 3,285,776) and free from objectionable color and impurities, and finally passing this syrup through a molecular exclusion column or columns to remove a portion or all of the glucose and leave a syrup high in fructose.

Example 2: A starch slurry of 23°Bé is hydrolyzed with hydrochloric acid to 22 DE. The neutralized liquor is filtered with a filter aid and adjusted to 40% solids. The pH is adjusted to 4.8 and the temperature to 59°C. The 22 DE syrup is then treated with 0.5% (DSB) of the amyloglucosidase enzyme Diazyme L-30 for 72 hr at the foregoing temperature and pH. The resultant syrup has a DE of 93.3 and a glucose content of 89.5% (DSB). A treatment with 0.5% (DSB) of activated carbon is then made at a temperature of 75°C for 30 min, followed by filtration and cooling to 65°C.

At the latter temperature the syrup is treated with a concentrated solution of sodium hydroxide to raise the pH to 9.0, and both temperature and pH are maintained at these values for 4 hr to produce a syrup having glucose, 70.6% DSB; ketose, 21.4%; disaccharides and higher sugars 9%.

The syrup is passed immediately through an ion exchange column operating in the sodium cycle to remove heavy metal ions. This operation is sometimes called softening, and involves use of any of the customary cation exchange resins such as Dowex 50X8 of 20 to 50 mesh in the sodium form. The effluent from the column is heated to 82°C and introduced into an ion exclusion column where substantially all of the salts and a major portion of the color and organic degradation products of the earlier alkaline isomerization reaction are removed in one pass. The column is jacketed for circulation of hot water, and the resin bed is 2" in diameter and 3' high.

The resin used is, for example, Dowex 50 WX 4, a strong acid cation exchange resin in the sodium form. The resin is polystyrene crosslinked with 4% divinyl-benzene and is 50 to 100 mesh. The liquor is passed into the column at the rate of 0.5 gal/min/ft^2 of cross-sectional area of the column. Total volume of feed is 450 ml at 40% solids. Soft water is then introduced at the rate of 0.5 gallon per minute per square foot of cross-sectional area. Effluent is collected in successive fractions of 100 ml each. The data are given in Figure 11.1. The salt and other impurities appear early in the effluent, while the sugar solids come off in the later fractions.

If further purification is desired, it is accomplished by passing the effluents from the ion exclusion column through ion exchange equipment, either of the mono-bed type, where resins such as Rohm and Haas XE 168 and Amberlite 200 are used, or of the separate cation and anion type, where the above or other standard types of ion exchange resins are used in series, to remove final traces of salts and coloring matter.

A composite of several fractions of the purified syrup is then cooled to 25°C and passed into a molecular exclusion column which contains an ion exchange resin such as Dowex 50 WX4 at 50 to 100 mesh in the strontium form. The column is 2" in diameter and 3'6" high. Flow rate is 0.1 gal/min/ft^2 of cross-sectional area. Volume of feed is 350 ml.

Deionized water is then passed into the column at the same rate. Fractions of 100 ml each are removed, the early ones consisting primarily of glucose, disac-charides, and higher sugars. Compositions of the effluent fractions are given in the table. While the data shown do not indicate disaccharides and higher sugars, these appear predominantly in the early fractions (2 through 6). However, there are traces of these in all portions of the effluent. Fractions 7 through 14 are now mixed and concentrated to a solids concentration of 82%, at which point the Baumé is 43° and the syrup is transparent, water-white, viscous, highly sweet, not subject to microbiological spoilage, and is stable against crystallization for an indefinite period at room temperature.

Composition of the syrup on the dry solids basis is 43.9% glucose, 45.7% fructose, 2.3% other ketoses, and 8.1% disaccharides and higher sugars. The effluent frac-tions from the molecular exclusion column containing predominantly glucose and disaccharides are mixed, concentrated to 40% solids and added to the subsequent batch of high glucose liquor prior to isomerization.

FIGURE 11.1: PREPARATION AND PURIFICATION OF HIGH DE FRUCTOSE SYRUPS

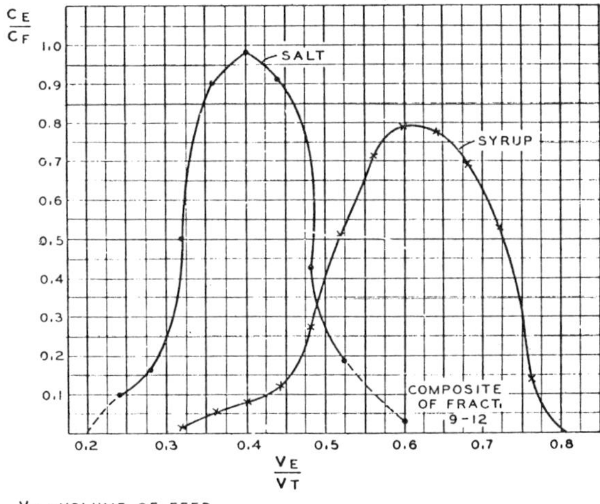

Source: B.L. Scallet and I. Ehrenthal; U.S. Patent 3,305,395; Feb. 21, 1967

Fraction (100 cc per fraction)	DS Fraction (grams)	DS Cumulative (grams)	Ketose/Fraction (grams)	Ketose/Fraction (percent)	Fructose Cumulative (grams)
1	–	–	–	–	–
2	0.29	0.29	–	–	–
3	3.75	4.04	–	–	–
4	10.60	14.64	–	–	–
5	17.40	32.04	–	–	–
6	19.85	51.89	0.457	2.3	0.46
7	16.05	67.94	1.11	6.9	1.57
8	9.20	77.14	2.69	29.2	4.26
9	5.70	82.84	4.40	77.2	8.66
10	5.15	87.99	4.94	96.0	13.60

(continued)

Fraction (100 cc per fraction)	DS Fraction (grams)	DS Cumulative (grams)	Ketose/Fraction (grams)	Ketose/Fraction (percent)	Fructose Cumulative (grams)
11	4.18	92.17	4.10	98.0	17.70
12	2.40	94.57	2.40	100	20.10
13	0.71	95.28	0.71	–	20.81
14	0.07	95.35	0.07	–	20.88

Feed = 21.4% ketose

A multistep purification process is further disclosed by *B.L. Scallet and I. Ehrenthal; U.S. Patent 3,383,245; May 14, 1968; assigned to Anheuser-Busch, Incorporated* for an alkali isomerized corn conversion syrup of above 70 DE. The ash is removed by electrodialysis, color is removed by an ion exchange resin in the chloride form and the syrup is treated with carbon. The purification steps follow.

(1) Resin Decolorization — Isomerized liquor is produced as per U.S. Patent 3,285,776. The isomerized liquor is first passed through a strong base quaternary ammonium decolorizing anion exchange resin in the chloride form. Preferred flow rate is 1 gpm/ft^3 bed volume and a preferred temperature range of 90° to 120°F. Higher temperatures within allowable resin limits permit better resin decolorizing efficiency but also risk further degradation of isomerized liquor. Therefore, an upper limit in the 120°F area is preferred. This decolorizing step is necessary to remove some of the organic degradation products and color bodies so as to improve electrodialysis stack efficiency and maximize stack membrane life in the next step of the subject process.

(2) Electrodialysis — The liquor from step (1) is passed through an electrodialysis stack system to remove most of the ash present in the isomerized liquor. The desired final ash in the product is less than 0.5% and preferably less than 0.25% of a syrup which is about 80% solids. Assuming an initial ash of 0.75% on 40% solids, 83% of the ash must be removed to yield 0.25% ash at 80% solids and 67% of the ash must be removed to yield 0.5% ash in the final product. The level of ash can be controlled primarily by the number of passes made through the electrodialysis system at a given set of conditions.

The preferred current density is ie_0 = 4 to 6 and the preferred temperature is 110° to 140°F. Production rate for 90% ash removal of 0.91% ash from 40% solids liquor is 0.33 gal/hr/ft^2 at 110°F and ie_0 = 4 and 0.53 gal/hr/ft^2 at ie_0 = 4 and T = 140°F. Yields are better than 97% and energy consumption for 90% ash removal of 0.91% ash from 40% solids feed liquor at ie_0 = 4 is 0.035 kwh per gallon liquor. The electrodialysis step has deashed the liquor, removed traces of color and yields an effluent of essentially the same sugar composition as the feed.

(3) Resin Decolorizing — The electrodialysis effluent liquor can now be passed through the same decolorizing process as in step (1) to remove further color bodies and reduce carbon requirements. This step is optional and if omitted will only involve a higher carbon consumption requirement in the next step of the process.

(4) Carbon Treatment — The liquor either from the second decolorizing step or direct from the electrodialysis process can be treated with activated carbon in one of the following two ways:

Activated powdered carbon batch treatment: The liquor can be treated with powdered activated carbon and filtered to remove remaining color bodies.

Granular carbon column: A granular carbon column containing Pitt Chem CPG 44 x 40 mesh or equivalent is used to decolorize liquor from steps (2) or (3). Liquor flow rate is preferably 0.5 to 2.0 bed volumes per hour at a temperature of 100° to 140°F. The effluent is then filtered.

(5) Evaporation — The carbon treated liquor is concentrated to a very sweet high fermentable noncrystallizing syrup.

Example: The example uses an isomerized feed liquor with the following analysis:

DE	75.3
Ketose	24.6
Dextrose	44.3
Fructose	20.0
Psicose	5.0
Solids	41.8
Ash	0.91
Color, percent transmission:	
450 mμ	None
550 mμ	6.0
620 mμ	31.0

All colors measured in 6" x ¾" optically matched colorimeter tubes with a Bausch and Lomb Spectronic 20 Colorimeter.

(A) A volume of 4,000 cc isomerized liquor having the foregoing specifications is passed through a 100 cc bed volume of IRA 401-S (Rohm & Haas) in the chloride form at a rate of 1 gpm/ft^3 at ambient temperature.

(B) The liquor effluent from the resin is passed through an electrodialysis stack pack assembly at ie_0 = 4 at 110°F. The stack pack consists of 10 cell pairs of 61 CZL and 111 EZL 9" x 10" membranes with 0.040" spacers (Ionics Inc.). The electrodes are Pt-Ta and Hastelloy C. The electrode stream is 0.1 N Na_2SO_4 and the starting concentration (waste) stream is 0.05 N NaCl. Starting liquor resistivity is 220 ohm-cm at 110°F and final resistivity is 3,050 ohm-cm at 110°F.

(C) The electrodialyzed liquor is passed through a 100 cc bed of IRA 401-S as per step (A).

(D) The liquor is then carbon treated as a final decolorizing step. An analysis of product properties at the end of each processing step is shown in the table on the following page.

Processing Description	Solids, Percent	Ash, Percent	D.E., Percent	Ketose, Percent	Fructose, Percent
Feed (isomerized liquor)	41.8	0.91	75.3	24.6	20.0
A. Resin Decolorizing	41.0	0.87	75.2	24.8	
B. Electrodialysis	41.3	0.095	77.4	26.6	19.4
C. Resin Decolorizing	38.9	0.098	77.6	27.4	
D. Carbon Treatment (2—2% carbon treatments)	40.1		78.2	27.7	20.6

Processing Description	Psicose, Percent	Dextrose, Percent	Color, Percent Transmission		
			450 mμ	550 mμ	620 mμ
Feed (isomerized liquor)	5.0	44.3	None	6.0	31.0
A. Resin Decolorizing		44.4	None	57.0	83.0
B. Electrodialysis	3.9	45.3	3.0	30.0	46.0
C. Resin Decolorizing		45.6	33.5	87.0	93.0
D. Carbon Treatment (2—2% carbon treatments)	3.7	46.5	79.5	93.0	95.0

Isomerization with Cyclohexylamine

An improved method of transforming dextrose to fructose has been developed by *G.T. Tsao, T.H. Reid, F.L. Hiller and L.H. Hubbard; U.S. Patent 3,432,345; March 11, 1969; assigned to Union Starch & Refining Co., Inc.* using cyclohexylamine. Besides the isomerization, dextrose and cyclohexylamine can undergo another reaction to form a compound of which the molecular structure is not determined. This compound can be crystallized and dried. It has a very sharp melting point of 97.5° to 98.5°C. This compound is soluble in water and in a number of organic solvents such as acetone, ethanol, etc., and also mixtures of solvents.

The dextrose solution used in the isomerization reaction is first adjusted to a proper concentration. The concentration can be as low as 5 g of dextrose per 100 ml solution, or as high as 50 g dextrose per 100 ml of solution. The amount of cyclohexylamine can be as low as 0.5 ml of cyclohexylamine per 100 ml of dextrose solution.

At 15°C, the isomerization reaction is very slow. A temperature of 30° to 35°C is considered the most desirable temperature. A temperature of 60°C or higher may cause excessive coloration of the dextrose solution. When the reaction is carried out at a low temperature and a low level of cyclohexylamine, one to two weeks may be required to reach an appreciable level of fructose. For process convenience, the length of time should be controlled between 30 min to 72 hr. A 24 hr reaction time seems to be a very convenient one for plant operation.

The treated product solution is usually colored due to side reactions. The solution can be purified to remove cyclohexylamine and by-products. The purifications can be carried out with the ordinary techniques such as distillation, ion exchange resin treatment, active carbon treatment, and/or evaporation. The solution thus obtained can contain up to 30 to 33% fructose based on the total amount of dextrose and fructose present. Known isomerization reactions with sodium hydroxide or calcium hydroxide usually produce a maximum of 28 to 32% of fructose. In the cases where still higher percentage of fructose is desired, the sugar solution after the isomerization reaction will be further processed in a different manner. To a great extent, this latter part of the treatment depends on the concentration of the sugar solution and the amount of cyclohexylamine added initially. If these amounts are fairly high to start with, by simple cooling, crystals of the dextrose-cyclohexylamine complex will be precipitated out. The crystals can easily be separated by filtration or centrifugation. The mother liquor thus obtained will be then purified to remove cyclohexylamine and other impurities. The crystals have a melting point of 97.5° to 98.5°C and are soluble in a number of solvents. The purified sugar solution containing up to 80% fructose based on the total amount of dextrose and fructose present in the solution are readily prepared in this manner.

Example 1: A dextrose solution made by dissolving 22 g of dextrose monohydrate in 100 ml of water was mixed with 10 ml of cyclohexylamine. The mixture was maintained at 30°C for 23 hr. The treated solution was analyzed and found to contain 33% fructose based upon the total amount of dextrose and fructose present.

Example 2: A syrup of 92 DE made by the hydrolysis of both acid and hydrolytic enzyme was found to contain 80% dextrose on a dry solids basis. A sample of this syrup was diluted with water to 10 g of dextrose per 100 ml solution. To a 100 ml portion of this solution, 25 ml of cyclohexylamine was added. The mixture was then maintained at 30°C for 24 hr. The treated solution was analyzed and found to contain 16% fructose.

Example 3: A solution containing 33% fructose made according to the procedure of Example 1 was adjusted by evaporation under vacuum to 50 g of reducing sugar as dextrose per 100 ml of the concentrated solution. To 1,000 ml of the concentrated solution, 1,000 ml of cyclohexylamine was added. The mixture was stirred for 2 hr in an ice bath. Crystals were precipitated at the end of the 2 hr period. The crystals are removed by filtration. The mother liquor was analyzed and found to contain 77.2% fructose based on the total amount of dextrose and fructose present.

The crystals obtained were washed and purified. The dry crystals have a melting point of 97.5° to 98.5°C. They are the dextrose-cyclohexylamine complex described above. A portion of the crystals was dissolved in water. The solution was allowed to stand overnight. The solution was then evaporated under vacuum to remove a part of cyclohexylamine which can be recycled. The evaporated solution was incubated at 30°C for 24 hr. The solution was then analyzed and found to contain 33% fructose based on the total dextrose and fructose. This solution containing 33% fructose can be concentrated by another crystallization treatment carried out in the same way.

Use of Deionized Corn Syrup

The process disclosed by *R.G.P. Walon; U.S. Patent 3,475,216; October 28, 1969; assigned to Corn Products Company* is based upon the discovery that deanionization of the dextrose-bearing starting material prior to interconversion is highly desirable. This deanionization removes the mineral anions, e.g., Cl^-, $SO_4^=$, normally present in the dextrose starting material and replaces them with OH^- ions. The deanionization also adjusts the pH of the material to the range required for effective interconversion. The deanionization is achieved by the use of a strongly

basic anion exchanger or, preferably, by the use of an electrodialysis unit.
Broadly speaking, dextrose-bearing materials may be interconverted to fructose
containing sweet syrups according to this process which involves three steps. In
the first step, the starting material is deanionized yielding an effluent with a pH
in the desired range. In the second step this effluent is heated to effect isomerization and in the third step the interconversion reaction products are refined.
The starting material may be any dextrose-bearing solution, preferably a corn
syrup. The starting material should have a DE of 30 to 100, preferably 70 to 95.

In general, the starting material will contain anions such as Cl^- and $SO_4^=$ in about
1,600 to 2,900 ppm. In the first step of the process, these anions are substantially removed. In general, the deanionization is continued until the starting material has a pH of 8.5 to 10, preferably 9 to 9.5, and the anions present are
about 200 to 900 ppm, preferably less than 600 ppm. The deanionization is
achieved by passing the starting material through a strongly basic anion exchanger
such as a Dow 2X8, a Dow 21R, an Amberlite 1R or 401, or a Permutit
MP600. Preferably, the deanionization is accomplished by the use of an electrodialysis unit as described in Figure 11.2.

FIGURE 11.2: USE OF DEIONIZED CORN SYRUP

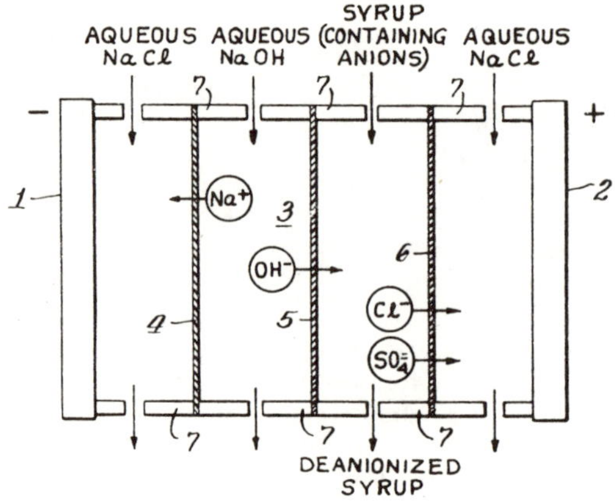

Source: R.G.P. Walon; U.S. Patent 3,475,216; October 28, 1969

Figure 11.2 is a cross-sectional view showing the arrangement of the elements
used in the electrodialysis unit. The unit will comprise a plurality, for example,
10 to 100 or more, of identical cells. The number of cells used in any particular
process will of course depend upon the capacity of the cells and the size of the
process. However, only one cell has been shown in Figure 11.2. Each cell of
the unit comprises three membranes which are separated by spacers to obtain
circulation through the cell and uniform distribution at the surface of the membrane. In Figure 11.2 there is shown an anode **1** and a cathode **2**. Positioned
between the anode and the cathode is a cell **3** composed of membranes **4, 5** and
6. Membrane **4** is a cationic permeable membrane.

Membranes **5** and **6** are both anionic permeable membranes. All membranes and electrodes are separated by spacers **7**. The starting material, e.g., corn syrup, to be deanionized is circulated through the cell between anionic membranes **5** and **6**. Between cationic membrane **4** and anionic membrane **5** an aqueous solution of sodium hydroxide is circulated. Between cationic membrane **4** and its associated electrode (anode **1**) or the adjacent anionic membrane of the next cell (not shown) and between anionic membrane **6** and its associated electrode (cathode **2**) or the adjacent cationic membrane of the next cell (not shown) an aqueous solution of salt (NaCl) is circulated.

When electric voltage is applied across the electrodes of the cell, by a conventional power source (not shown) OH^- ions move into the product (i.e., deanionized syrup) stream while Cl^- and $SO_4^=$ and other mineral anions move out of the product stream into the salt stream. The membranes used are commercially available cationic and anionic membranes. Examples of such cationic membranes are Ionics Nepton 61 A2L/61A2G and examples of such anionic membranes are Ionics Nepton 711 B2L/717E2L, or any other suitable types.

Example: A corn syrup was deanionized by passing it through an electrodialysis unit of the type described above. The electrodialysis unit consisted of five cells arranged as shown in Figure 11.2. The corn syrup starting material had the following analysis: 92.5 DE, 88.8% dextrose, 0.37% ash, 1,450 ppm chloride, and color 4.1.

The syrup was supplied to the unit at a pressure of 15 psi and circulated through the unit at a rate of 200 cc/min. The aqueous caustic stream to the unit comprised 0.5 N NaOH and the salt stream supplied to the cell comprised 0.5% NaCl. The cell was operated at an amperage of 1.5 and a voltage of 13.5. During the deanionization the pH of the syrup changed from 7.5 to 9.7 and the chloride content was decreased from about 1,450 to about 300 ppm.

The deanionized effluent from the electrodialysis unit was then sent to an isomerization unit. This unit consisted of a jacketed, continuous heat exchanger maintained at a temperature of about 92°C. The deanionized syrup was maintained at this temperature for about 6 to 7 min. Following isomerization the interconverted syrup was passed through a cooler to bring the syrup to about room temperature.

The cooled syrup was then again circulated through the electrodialysis unit where small amounts of organic anions formed during the interconversion were removed. The effluent from the electrodialysis unit was passed through a Dusarite cationic exchanger. The resulting product was a corn syrup having the following characteristics.

DE	92.5
Dextrose content	70 %
Fructose content	18.8 %
Color	2.5
Chloride	180 ppm
Ash	0.09 %

Other sweet syrups were made according to this process by changing slightly the starting material and the operating conditions. Analyses are shown in Tables 1 and 2.

TABLE 1

DE	71.4	73.9	70.3	68.1
Dextrose (percent)	43.4	43.1	41.4	41.3
Fructose (percent)	10.6	13.7	14.2	12.1
Ash (percent)	0.08	0.004	0.09	0.041

TABLE 2

DE	94.8	95.1	91.7	96.8	92.2
Fructose (percent)	22.9	23.8	21	24.6	18.9
Dextrose (percent)	67.5	67.8	67.2	71.1	69.2
Ash (percent)	0.12	0.08	0.11	0.06	0.21

Isomerization with Anion Exchange Resins

A line of fructose-containing, water-white, ash-free, high DE syrups is made by *E. Katz, I. Ehrenthal and B.L. Scallet; U.S. Patent 3,690,948; September 12, 1972; assigned to Anheuser-Busch, Incorporated* using a macroreticular resin in the OH^- form. The high dextrose (95%) feed is recirculated through a strong base anion macroreticular resin column and then passed through a strong acid cation column.

At the end of the isomerization a typical analysis range would be 94 to 97 DE and 20 to 35% ketose. Exact specifications are determined by specific isomerization conditions. Lower DE feeds produce final syrups of lesser ketose percent. Figure 11.3 is a flow diagram of this process for producing a syrup.

Initially starch is acid hydrolyzed to 10 to 20 DE by conventional means and then is enzyme converted to produce a 95+ DE liquor, preferably 98+ DE with 95%+ dextrose. The specific process illustrated uses a 98 DE liquor. The liquor is then purified by conventional methods and ion exchanged to yield a water-white, low ash 98 DE liquor. This liquor is the raw material for the isomerization step.

The specific purification involves a precoat filter for removal of muds, followed by a carbon treatment for color removal. The syrup then is pressure filtered for removal of carbon and all solids followed by an ion exchange treatment to remove ash. The ion exchange consists of a strong acid cation resin (C-25, IR200, or Dowex 50) and weak base anion resin (A-6). Thus the feed to the column isomerization is water-white of low ash (less than 0.05% and high DE and dextrose. The specific preferred feed is 98+ DE and 95%+ dextrose. The feed goes to a recirculation tank **10** where it is pumped through the resin isomerization column **12** in a continuous recirculating path.

The resin column contains a strong base anion macroreticular resin (styrene-divinylbenzene copolymer containing quaternary ammonium groups) in the OH^- form, such as IRA 900. A macroreticular resin is essential because of its porosity and superior cycle life properties. The feed liquor has 20 to 60% DSB (Dry Solids Basis) and preferably 40% DSB. The feed volume is 2 to 12 bed volumes 40% DSB liquor per volume of strong base anion resin in the column **12**.

FIGURE 11.3: ISOMERIZATION WITH ANION EXCHANGE RESINS

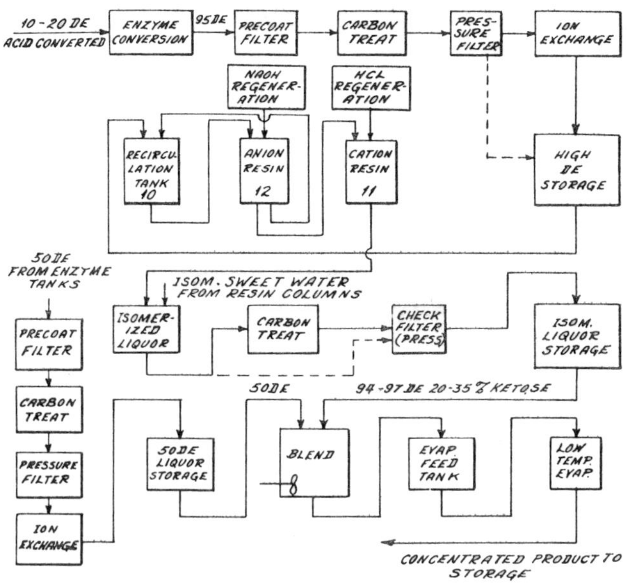

Source: E. Katz, I. Ehrenthal and B.L. Scallet; U.S. Patent 3,690,948; Sept. 12, 1972

A range of 4 to 8 bed volumes is preferred. The surface area of the macroreticular resins can vary between 20 and 70 or even higher square meters per gram of dry resin. The porosity is 0.4 to 0.6 or greater ml/ml of resin. The average pore diameter is 600 to 1400 or more angstroms. Each batch of feed is recirculated through the column **12** until the desired degree of isomerization is obtained. This may involve repeated passes through the tank **10** and the column and may take 4 to 24 hr in the preferred form.

After isomerization, the syrup is passed through a cation resin column **11** which contains a strong acid cation resin, such as, Duolite C-3, H⁺ form (methylene sulfonic functional group in phenolic matrix). Other suitable resins include Duolite C-25, etc. The column **11** contains 1 bed volume resin per 1 to 50 bed volumes 40% DSB isomerized feed liquor and preferably contains 1 bed volume resin for each 10 to 20 bed volumes of feed.

After completion of each batch, the columns **11** and **12** are regenerated before another batch of syrup is isomerized. The reaction temperature of the system should be between ambient and 150°F, preferably at 110° to 140°F. At the end of the isomerization a typical analysis would be 94 to 97 DE and 20 to 35% ketoses. These ketoses consist predominantly of fructose, but include small amounts of psicose, as was shown in U.S. Patent 3,383,245 (page 360). The rate of circulation of the feed liquor through the system is 0.01 to 30 bed

volumes anion resin basis per minute. The preferred rate is 0.05 to 1.0 bed
volumes/min. The amount of fructose obtained depends on a number of factors,
such as resin contact time, temperature, solids level, type of resin, method of
resin regeneration, etc. The anion resin is regenerated by using a strong base reagent such as NaOH solution. The acid resin is regenerated by using a strong
acid such as sulfuric or hydrochloric. The regeneration is done after every batch
has completed its passes through the column.

As each batch is removed from the isomerization column, the columns **11** and
12 are regenerated. The column **12** may be given a prewash with approximately
1 to 2 N acid to strip color bodies and degradation products from hydroxyl sites
in the resin. The spent acid from the regenerated cation resin may be used as a
part of the prewash acid used to regenerate the anion column **12**. After the initial acid treatment the anion column is rinsed and given its base treatment to
regenerate the isomerizing hydroxyl sites in the resin to prepare the resin for the
next batch of syrup.

The product from the isomerization has about 94 to 97 DE and 20 to 35%
ketose. The small loss in DE during isomerization is more than compensated
for by the increased sweetness of the fructose produced. This product may be
carbon treated and pressure filtered before blending. In a preferred process, the
isomerized product made from the 95+ DE feed liquor is blended with a 50 DE
high maltose liquor (preferably ion exchanged) to yield a blend containing 10
to 17% fructose, 68 to 80 DE and 37 to 45% dextrose. The reason that the isomerized product is blended is to reduce the dextrose content to below about
45%. Higher amounts of dextrose tend to crystallize upon standing, particularly
in cool climates.

Example: A 1½" i.d. jacketed glass column was charged with 100 cc IRA-900
OH⁻ form macroreticular strong base anion resin. A glass aspirator bottle immersed in a 140°F water bath is connected by hose to the suction of a submerged
pump, the discharge of which is connected to the top of the jacketed resin column. The column bottom discharge runs into the top of the aspirator bottle to
provide a recirculating system.

400 cc of ion exchanged 95 DE 43.0% DSB feed liquor is introduced into the
aspirator bottle. A pump is used to recirculate 140°F water through the jacket
of the resin column. The feed liquor pump is started up when the feed liquor
is at 140°F. Liquor is recirculated through the system at 1,000 cc/min at 140°F.
No nitrogen or inert gas is required for purging the system. Conversion is as
follows.

Time (hours)	Ketose (percent)
0	0
3	22.6
6	27.3
9	27.4
12	28.5
24	29.2

A similar method has also been developed by *E. Katz, I. Ehrenthal and B.L.*

*Scallet; U.S. Patent 3,684,574; August 15, 1972; assigned to Anheuser-Busch,
Incorporated* involving a loop arrangement in the overall process. In this modification, the sections **10, 11** and **12** of Figure 11.3 of U.S. Patent 3,690,948 on
page   are replaced by those of Figure 11.4.

FIGURE 11.4: MODIFIED METHOD FOR ISOMERIZATION WITH ANION EXCHANGE RESINS

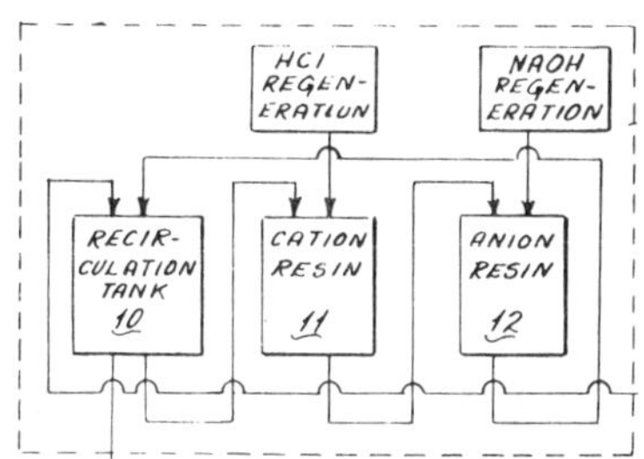

Source: E. Katz, I. Ehrenthal and B.L. Scallet; U.S. Patent 3,684,574; Aug. 15,
1972

Using the same material as in U.S. Patent 3,690,948, the feed goes to a recirculation tank **10** where it is pumped through the isomerization loop. It first passes
through the cation resin column **11**, then through the second resin column **12**
which contains the strong base anion macroreticular resin. Each batch of feed
is recirculated through the loop until the desired degree of isomerization is obtained. This may involve repeated passes through the tank **10** and the columns
11 and **12** and may take 4 to 24 hr. After isomerization of each batch, the columns are regenerated before another batch of syrup is isomerized.

Example: Conditions: temperature of feed is 125°F. Resin columns: 100 cc
IRA 900 (OH⁻ form) + 20 cc Duolite C-3 (H⁺ form) in series; flow rate is 10 bed
volume per minute anion resin basis; and feed is 400 cc ion exchanged 99 DE
liquor, 97% dextrose and 40% DSB.

Time (hours)	Ketose (percent)
1	12.7
2	20.1
4	27.4
6	30.0
8	32.0
12	33.9
24	34.4

PRODUCTION OF GLUCOSE ISOMERASE

Although several enzyme preparations are known which will convert D-glucose
to D-fructose and which involve one or more chemical intermediates (e.g., D-

glucose-6-phosphate), the most promising enzymes appear to be those which promote the direct conversion of D-glucose to D-fructose (i.e., the isomerases). Such isomerase preparations have been obtained from a number of microorganisms including the genera Lactobacillus, Pseudomonas, Pasteurella, Leuconostoc, Streptomyces and Aerobacter. A recent, brief literature review relating to this subject matter is given by Yamanaka in *Biochim. Biophys. Acta* 151, 670-680 (1968).

Use of Xylan in Culture Medium

Xylan can replace the more expensive xylose in the process disclosed by *Y. Takasaki and O. Tanabe; U.S. Patent 3,616,221; October 26, 1971; assigned to The Agency of Industrial Science and Technology, Japan* for the production of a glucose isomerizing enzyme. Microorganisms of the Streptomyces genus are used including *Streptomyces flavovirens* ATCC No. 3320, *Streptomyces achromogenus* ATCC No. 12767, *Streptomyces echinatus* ATCC No. 21133, *Streptomyces wedmorensis* ATCC No. 21230, and *Streptomyces albus* ATCC No. 21132.

These Streptomyces microorganisms, such as *Streptomyces wedmorensis,* may be grown in a medium containing plant materials having available xylan, such as straw, wheat, bran, corn cobs, corn husks, rice bran, pulp waste, etc. and produce glucose isomerase. It is more economical to use xylan to cultivate Streptomyces than to use a medium which contains as the principal carbon source the more expensive D-xylose.

The glucose isomerizing enzymes are prepared by cultivating the Streptomyces organisms in an aqueous medium, such as disclosed in the example. At the completion of cultivation, the cellular material which may be used as the enzyme source is separated from the supernatant liquid. The glucose is transformed to fructose by incubating a glucose solution with the enzyme. The enzyme has superior characteristics for use in commercial processes since it is active over a wide pH range, temperature range and range of glucose concentration.

If desired the process may be operated to convert up to about 50% of the initial glucose into fructose. This produces a syrup having substantially the sweetness characteristics of invert syrup. The isomerization may be carried out at a pH of between 5.5 and 8, and preferably between 6.8 and 7.2, with 7 being the preferred. The reaction temperature may be between 45° and 80°C, with a preferred range of between 60° and 70°C. The preferred Streptomyces species are *Streptomyces wedmorensis* ATCC No. 21230 and *Streptomyces albus* ATCC No. 21132 whose characteristics and morphology are included in detail in the complete patent.

Example: A seed medium containing 1% of xylose, 1% of polypeptone, 0.3% of K_2HPO_4 and 0.1% of $MgSO_4 \cdot 7H_2O$, was introduced into a Kolbe flask and sterilized. Then *Streptomyces wedmorensis* ATCC No. 21230 was inoculated into the medium and cultivated at 30°C for 48 hr in order to provide an inoculant. In order to increase cell production, an inoculant from the seed stage was used to inoculate 50 l of medium and cultivated; this was used to inoculate 500 liters of medium and cultivated, and finally this was used to inoculate 5,000 l of medium and cultivated. The composition of the media used in this operation (excluding the seed medium) was 3% wheat bran, 4% corn steep liquor, 0.1%

MgSO$_4$·7H$_2$O and 0.025% CoCl$_2$·6H$_2$O and water. After this medium had been sterilized, the initial pH was adjusted to 6.5 with NaOH, the cultivation was carried out at 30°C, aeration, one-third volume of broth per minute, and 200 rpm agitation. After 24 hr, enzyme activity reached a maximum. Enzyme solution prepared from 50 ml of broth had an activity of 750 units. Wheat bran in the broth was removed by the use of a screen sieve. Cells were harvested by the use of filter press to obtain 1,000 kg of cells. The harvested cells were then suspended in water. The slushy cells were disintegrated by the use of a high pressure homogenizer. The disintegrated cells were used as the enzyme source.

As a substrate for isomerization, a 92.5 DE converted starch syrup in a concentration of 50% (consisting of 87% of glucose and 13% of oligosaccharide containing disaccharide) was used. 15,000 kg (calculated as solid) of the above converted starch syrup was fed into an isomerizing tank having 30 m^3 capacity. 0.0025 M Mg(OH)$_2$ was added and after the syrup had been heated to 65°C the enzyme source was added. The pH was maintained in the neighborhood 7.0 with NaOH.

After 90 hr, the conversion ratio reached about 45%. The reaction solution was then purified by the use of activated carbon and ion exchange resin, as in the conventional purification process. The solution was concentrated to a 25% water content. The yield of the manufactured product was 19,600 kg. An analysis of this product showed 29.6% fructose, 36% glucose, 9.4% oligosaccharide containing disaccharide, and 25% water.

Heat Fixing Streptomyces Produced Enzymes

The glucose isomerizing enzyme, an intracellular enzyme, can be disadvantageously extracted from the cells by autolysis at 15° to 50°C. The autolysis is a phenomenon caused by the action of other enzymes present in the system.

Y. Takasaki and A. Kamibayashi; U.S. Patent 3,753,858; August 21, 1973; assigned to Agency of Industrial Science & Technology, Japan have improved the enzymes produced in U.S. Patent 3,616,221 by treating the microorganism containing the enzyme at a temperature of 60° to 85°C to fix the microorganism in the cell. The glucose isomerizing enzyme produced in the cells of a strain of Streptomyces exhibits extremely high heat stability at an optimum temperature of 80°C, whereas enzymes which participate in the autolysis are inactivated upon heat treatment at 60°C for 10 min. Thus, it is possible to fix the glucose isomerizing enzyme intracellularly in an insoluble state by treating the glucose isomerizing enzyme-containing cells at a temperature above 60°C but lower than 85°C thereby inactivating only the enzymes which participate in the autolysis.

Fixing the glucose isomerizing enzyme in the cells may preferably be effected by heat treatment of an aqueous suspension of the cells, but, if the broth of the glucose isomerizing enzyme-producing microorganism is heat treated immediately after completion of the culturing, the treatment is easier, and, moreover, it becomes possible to recover the enzyme in high yield, since the dissolving-out of the enzyme caused by autolysis during the filtration step of the broth can be avoided. Furthermore, when the heat treatment is conducted after the addition of a cobalt salt such as CoCl$_2$ in an amount of 0.001 to 0.01 M to the broth to be heat treated, the heat stability of the glucose isomerizing enzyme can be increased thereby minimizing the loss of the enzyme caused by heat inactivation.

Example: 100 ml of the medium comprising 3% wheat bran, 2% corn steep liquor and 0.024% $CoCl_2$ and having pH 7.0 were charged into a 500 ml Sakaguchi flask and, after being sterilized by a standard technique inoculated with *Streptomyces albus* ATCC 21132. The microorganism was then cultured at 30°C for 24 hr with shaking. After completion of the culturing, the broth was centrifuged to collect the cells, which were then washed with water and suspended in water (25.4 mg/ml on the dry basis).

Using the above cell suspension (20 ml) the amount of the glucose isomerizing enzyme extracted from the cells was determined with respect to the suspension heat treated at 60°C for 10 min and the untreated suspension by adding ethyl acetate in an amount of 1% to each of the suspensions and incubating the suspensions under the optimum conditions for autolysis, i.e., at a temperature of 40°C and at pH 6.5 to 7.0, for 6, 12 and 24 hr, respectively. The results obtained were as shown in Table 1. The total amount of the glucose isomerizing enzyme is referred to as the amount of the enzyme extracted after disrupting the cell suspension by a sonic oscillator for 10 min with a wavelength of 10 kilocycles.

TABLE 1

Extraction time (hour)	Control			Heat-treated cell		
	Extracted G.I. (A) (unit/ml.)	Total G.I. (B) (unit/ml.)	Extraction ratio (A)/(B) ×100 (percent)	Extracted G.I. (A) (unit/ml.)	Total G.I. (B) (unit/ml.)	Extraction ratio (A)/(B) ×100 (percent)
6	1.0			1.3		
12	2.9			1.0		
24	9.5	11.3	84	2.1	13.0	16

As is clear from Table 1, where the cell suspension was subjected to the heat treatment, only 16% of the glucose isomerizing enzyme was extracted after 24 hours incubation as compared with the control (no heat treatment) wherein the glucose isomerizing enzyme was extracted increasingly as the time elapsed. Table 2 indicates the amount of glucose isomerizing enzyme extracted from the cells when 0.1% of cetyl pyridinium chloride which possesses the action to promote the extraction of the glucose-isomerizing enzyme by autolysis was added to the above heat treated cell suspension and the untreated cell suspension, respectively, and the resulting cell suspensions were incubated for 6, 12 and 24 hr, respectively, at 40°C in the same manner.

TABLE 2

Extraction time (hour)	Control			Heat-treated cell		
	Extracted G.I. (A) (unit/ml.)	Total G.I. (B) (unit/ml.)	Extraction ratio (A)/(B) ×100 (percent)	Extracted G.I. (A) (unit/ml.)	Total G.I. (B) (unit/ml.)	Extraction ratio (A)/(B) ×100 (percent)
6	2.1			1.0		
12	6.8			2.5		
24	15.6	15.2	103	2.4	14.0	17

NOTE.—G.I.=Glucose isomerizing enzyme.

The results in Table 2 indicate that, in the case where the cell suspension was subjected to the heat treatment, only about 17% of the enzyme was extracted as compared with the control (no heat treatment) in which the enzyme was completely extracted after 24 hr incubation.

Arthrobacter Strains for Enzyme Production

It has been discovered by *C.K. Lee, L.E. Hayes and M.E. Long; U.S. Patent 3,645,848; February 29, 1972; assigned to R.J. Reynolds Tobacco Company* that microorganisms belonging to the diphtheroidic genus Arthrobacter are capable of producing an isomerase which effects the direct conversion of either D-glucose or D-xylose to the corresponding ketose. Production of an isomerase by Arthobacter has not been previously reported.

Furthermore, it has been found that aldose isomerization by an Arthrobacter produced enzyme will proceed in the absence of arsenate or arsenite and that the concentration of the aldose in the isomerization medium is not critical. The enzyme shows unusual properties in that temperatures as high as 80° to 90°C can be used in the isomerization without a significant loss in enzyme activity. The presence of divalent cations is beneficial as shown by the improved activity of the enzyme in the presence of magnesium ions.

A particularly surprising aspect of this process is the finding that certain of the Arthrobacter strains isolated are capable of producing comparable amounts of the isomerase enzyme using glucose instead of xylose as the sole carbohydrate source in the growth medium. Whereas all of the strains disclosed here are able to produce the isomerase in the presence of xylose or xylan, three of the strains *(Arthrobacter nov.* sp NRRL B-3726, NRRL B-3727, NRRL B-3728) are capable of producing the isomerase in the complete absence of xylose and xylan.

Stock cultures are maintained on Nutrient Agar (Difco) and are generally cultivated in a medium containing sources of carbon, nitrogen and inorganic salts. The carbon source chosen is dependent upon the particular strain selected. Additionally, a yeast extract, a protein source (e.g., tryptone) and sources of phosphate, ammonium and magnesium ions may be included in the medium for enrichment. A typical inoculum medium contains diammonium hydrogen phosphate, potassium dihydrogen phosphate, magnesium sulfate, a brewers' yeast fraction, hydrolyzed animal protein and the carbon source. The inoculum may be prepared by growing the organism in the medium on a rotary shaker at 20° to 35°C and preferably 25° to 30°C in the presence of air from 1 to 3 days during which time good growth occurs.

The glucose-isomerizing enzyme is obtained by transferring the inoculum to a sterile medium containing sources of carbon, nitrogen and inorganic salts. Glucose is preferred as the carbon source when using strains such as the three mentioned previously. For other strains such as NRRL B-3724 and NRRL B-3725, xylose or xylan is preferred.

It is to be understood that materials containing the desired carbohydrate (such as starch or cellulose hydrolysates) may also be used. The particular carbon source chosen will be determined largely by economic factors. In any case, the inoculated medium may be incubated with agitation in a suitable fermentor at 25° to 30°C while maintaining an airflow of approximately 0.015 ft^3/min/gal.

Maximum isomerase activity levels are normally attained under these conditions after 50 to 60 hr. Activity levels of approximately 120 microunits per milliliter of culture broth have been reached within 15 hr. A microunit of activity is defined as that quantity of enzyme which will produce 1 microgram of fructose from glucose in 1 min at 60°C using 1 M glucose in 0.1 M phosphate buffer at pH 7.0 and containing 0.01 M magnesium chloride. The whole cells are then harvested and used as such or, alternatively, the enzyme may be extracted from the cells using techniques known to the art.

The isomerase activity levels obtained vary, depending on the particular strain used and the growth media. Activity levels of at least 800 microunits per milliliter of culture broth have been observed from NRRL B-3726 and NRRL B-3728 grown in a glucose medium. Suitable activity levels are also obtained with *Arthrobacter nov.* sp NRRL B-3724 and NRRL B-3725 grown in a xylose or xylan medium.

The effects of temperature and pH on the conversion of glucose are shown in Tables 1 and 2. The data in Table 1 were obtained using an isomerization medium containing 1.0 M glucose, 0.1 M phosphate buffer and 0.01 M magnesium chloride at pH 7.0. Data in Table 2 were obtained using an isomerization medium containing 1.0 M glucose, 0.01 M magnesium chloride and a 0.1 M phosphate or glycine-sodium hydroxide buffer at 60°C. The enzyme used for these studies was derived from the NRRL B-3724 organism.

TABLE 1

Temperature (°C)	Fructose Formed mg/min/10 mg of Enzyme
50	8
60	94
65	140
70	210
75	270
80	290

TABLE 2

pH	Fructose Formed mg/min/15 mg of Enzyme
6.0	50
6.5	100
7.0	150
7.5	210
8.0	280
8.5	240
9.0	240
10.0	240

The quantities of fructose shown in both tables have been corrected for any amounts resulting from nonenzymatic conversion of glucose. Although enzyme activity is still significant at 80°C (Table 1), conducting the isomerization at temperatures higher than about 80°C results in a significant increase in side reactions such as thermal degradation of the fructose. In the following example the meat protein, O.M. HAP, is an enzyme-hydrolyzed animal protein. The yeast extract BYF-100 is a water-soluble brewers' yeast fraction.

Example: Arthrobacter nov. sp NRRL B-3728, was rinsed from a Nutrient Agar petri plate in which the Nutrient Agar was supplemented with 0.01% $MgSO_4 \cdot 7H_2O$ and 0.5% dextrose. The growth was rinsed from the plate (under sterile conditions) into 500 ml Erlenmeyer flasks containing 100 ml of the following aqueous inoculation medium (the medium was previously sterilized in an autoclave for 20 min at 121°C).

	Grams/Liter
Dextrose	20.0
Meat protein (O.M. HAP)	3.0
Yeast extract (BYF-100)	1.5
$(NH_4)_2HPO_4$	6.0
KH_2PO_4	2.0
$MgSO_4 \cdot 7H_2O$	0.1

The flasks were incubated on a rotary shaker at 30°C for 24 to 48 hr during which time considerable growth occurred. This inoculum was used in the subsequent enzyme production step. The glucose-isomerizing enzyme production medium was prepared and sterilized in a 166 liter fermentation vessel by heating for 1 hr at 121°C. This aqueous medium contained the following components.

	Grams/Liter
Meat protein (O.M. HAP)	6.0
Yeast extract (BYF-100)	1.5
$(NH_4)_2HPO_4$	6.0
KH_2PO_4	2.0

To this sterile medium was added a previously sterilized dextrose-$MgSO_4 \cdot 7H_2O$ solution (121°C for 20 minutes) to give final concentrations of 2% dextrose and 0.01% $MgSO_4 \cdot 7H_2O$. The pH of the resulting medium was 6.9. The sterile enzyme production medium (total volume, 113 liters) was cooled to 30°C and to it was added 1 liter of the previously prepared inoculum. The fermentation was allowed to proceed at 30°C while air was introduced at the rate of 0.015 ft^3/gal of broth per minute and while the medium was agitated at 300 rpm.

After 55 hr it was found by standard assaying methods that the cells contained isomerase activity of approximately 800 microunits per milliliter. The pH of the medium was 5.6. The cells were then harvested by centrifugation and were frozen at –5°C. The glucose isomerization was carried out by adjusting at 95 DE corn syrup (32 to 35% solids) to pH 7.5. To the syrup were added 20 g (net weight) of the harvested cells and 2 g of $MgSO_4 \cdot 7H_2O$ for each liter of syrup. The syrup was incubated for 26 hr at 60°C after which time it was found to contain 41% fructose based on the total weight of the solids.

Enzyme from *Aerobacter levanicum*

A glucose isomerizing enzyme is produced by *K.K. Shieh, B.J. Donnelly and H.A. Lee; U.S. Patent 3,813,320; May 28, 1974; assigned to Anheuser-Busch, Incorporated* when *Aerobacter levanicum* is grown on a substrate containing xylose, preferably unpurified birchwood sulfite liquor. The *Aerobacter levanicum* was obtained from the Northern Regional Research Laboratory of the United States Department of Agriculture at Peoria, Illinois, and identified as deposit NRRL B-1678.

The *Aerobacter levanicum* can be grown on a medium containing carbon and nitrogen. In order to produce converting enzyme, however, xylose must be present in the substrate. For isomerization to take place, Co^{++} must be present in the isomerizing medium or the cells or enzyme must have been grown on a substrate containing Co^{++}. For best results K^+ or Mg^{++} also are present, either during cell growth or in the isomerizing medium.

Hardwood sulfite liquor, specifically birchwood sulfite liquid is a suitable source of xylose. The birchwood sulfite liquor (BSL) is not purified or sterilized as is required in the growth of other microorganisms which produce glucose isomerase. This is an important commercial consideration as this syrup (the isomerized glucose-fructose syrup) is intended to replace conventional invert sugar in certain syrups, candy and baking applications, and thus must be competitive in price.

In producing the enzyme, the *Aerobacter levanicum* cells can be propagated on a suitable growth medium and then added to a xylose medium (BSL) to induce enzyme or the growth medium can contain xylose so that enzyme is induced as the cells are grown. The enzyme can be added to the glucose substrate in the form of air dried cells or the enzyme can be separated from the cells by sonification and only the enzyme added to the glucose substrate. The following examples illustrate the one and two step production of the enzyme and use of the enzyme in converting glucose to fructose.

Example 1: The process is carried out in a 250 ml Erlenmeyer flask containing 100 ml of a suitable growth medium. The flask is shaken by a rotary shaker (200 rpm with a 1" orbit) at a temperature of 23°C for 24 hr. The flask is inoculated from a YM broth (pH 7.0) with *Aerobacter levanicum* NRRL B-1678. 2% v/v of *Aerobacter levanicum* is added to the flask. The composition of the growth medium is the following:

	Percent
Yeast extract	2.0
Glucose	0.2
Xylose	0.8
$CoSO_4 \cdot 7H_2O$	0.009
KCl	0.18
Tap water	96.8

The pH is adjusted to 7.5 with a 50% KOH solution and the flask containing the medium is sterilized for 10 min at 121°C in an autoclave before being inoculated with *Aerobacter levancium.* After 24 hr the fermentation is stopped and the cells are harvested. 7.3 mg cells/ml growth medium are recovered and the enzyme recovered has 0.3 units/ml activity based on the quantity of growth medium.

Example 2: The process is carried out in 30 1-liter Erlenmeyer flasks in a two stage process. The first stage is a growth process in which 300 ml of medium is incubated at 22° to 25°C for 16 hr on a rotary shaker with a 2" orbit at 175 rpm. The flasks are each inoculated with 3 ml of cultured *Aerobacter levanicum.* The composition of the growth medium is the following:

	Percent
Yeast extract	2.0
Glucose	0.5
Tap water	97.5

The pH is adjusted to 7.5 with a 50% KOH solution and the flasks containing the medium are sterilized for 15 min at 121°C in an autoclave before adding the *Aerobacter levanicum.* The second stage of the process is an induction phase in which the desired glucose isomerase enzyme is produced in response to the introduction of xylose into the fermentation. A cheap source of xylose is the hardwood sulfite liquors from pulp and paper mills.

The particular source used in this example is a birchwood sulfite liquor concentrated to 50 to 57% solids. The medium prepared with the birchwood sulfite liquor is introduced into the fermentation process by adding 300 ml of nutrient medium to each of the flasks and incubating an additional 24 hr at 22° to 25°C with the shaking speed increased to 225 rpm. The composition of the added medium is given below. After addition the pH is adjusted to 7.5 with 50% KOH.

	Percent
Birchwood sulfite liquor (as xylose)	4
$CoSO_4 \cdot 7H_2O$	0.0036
KCl	0.024
Tap water	95.9724

The cells are harvested and washed by centrifugation and the resulting cell paste, 240 g, from 18 liters of total medium (both stages), may be sonified to solubilize the enzyme, or the cells may be spread out and allowed to air dry at room temperature. In either procedure, the yield of enzyme is 0.97 units/ml of culture medium.

Example 3: One gram of air dried cells having an activity of 480 units/g are added to 100 ml of 36% weight/volume glucose solution containing 6×10^{-4} M $CoCl_2$ and 6×10^{-3} M $MgSO_4$. Conversion is allowed to continue for 72 hr at 55°C with pH controlled automatically at 6.8 by addition of 1 N NaOH.

The conversion is stopped by centrifugation. The syrup is purified by activated carbon treatment, ion exchanged and concentrated. The resultant product has the following composition: 40% weight/volume solids with 50.3% fructose and 49.7% glucose.

Isomerase of High Activity

Although glucose isomerizing enzymes are commercially used in converting glucose to fructose, there are a number of problems associated with their use. The principal problem is that the yield of glucose isomerase produced by microorganisms is ordinarily relatively small. Also, since the glucose isomerase is produced primarily intracellularly by microorganisms and because it is a common practice to use the whole cells for the isomerization reaction, large quantities of the cellular material must be used.

These large quantities of cellular material result in an undesirable amount of handling and transportation, promote undesirable side reactions during the isomerization reaction and physically take up space in the reactor in which the isomerization is performed that could otherwise be used for the glucose solution.

B.L. Bengtson and W.R. Lamm; U.S. Patent 3,654,080; April 4, 1972; assigned to Standard Brands Incorporated have disclosed a process for enzymatically converting glucose to fructose in which microbial cells containing large quantities of glucose isomerase are used as the enzyme source.

This is done by treating viable microorganisms which contain intracellular glucose isomerase with an amount of a toxic agent which destroys at least 95% of the viable microorganisms, culturing the treated viable microorganisms under conditions to promote their growth, incorporating the cellular material into a glucose-containing solution and maintaining the cellular material under conditions where a quantity of the glucose is converted to fructose.

A number of toxic agents may be used in this process, such as nitrogen mustard, β-propionolactone, ethyl carbamate, ethyleneimine, hydrogen peroxide, N-methyl-N'-nitro-N-nitrosoguanidine, acriflavine hydrochloride, 8-ethoxy caffeine, nitrosomethylurea and radiation from radioactive isotopes and ultraviolet light. Preferably these toxic agents should be used in amounts and under conditions so that at least 99% of the viable microorganisms are destroyed. The microorganisms which have not been destroyed by such a treatment generally produce at least 30% and possibly 50% more glucose isomerase than microorganisms which have not been so treated.

In commercial processes for propagation of microorganisms, it is necessary to proceed by stages. Propagation is started by inoculating spores from a slant of a culture into a presterilized nutrient medium in a shake flask. Growth of the microorganisms is encouraged by various means, e.g., shaking for aeration and maintenance of suitable temperature. This step or stage may be repeated one or more times in flasks or vessels containing the same or larger volumes of nutrient medium. These stages may be conveniently referred to as culture development stages.

The microorganisms with or without accompanying culture medium, are introduced or inoculated into a large scale fermentor to produce commercial quantities of the microorganisms or their by-products. In the process it is preferred to treat the microorganisms with the toxic agent in the initial culture development stage, for instance, in a shake flask containing a suitable nutrient medium. After treatment, cells of the microorganisms may be placed in Petri dishes containing, for example, an agar medium and the cells allowed to sporulate.

Spores may then be removed and cultivated to produce microorganisms containing large amounts of glucose isomerase. Other methods of treatment may be used, for instance, spores may be treated with a toxic agent, the toxic agent removed by centrifugation, filtration or the like, the spores washed and transferred into a suitable medium where germination is promoted.

After the microorganisms are cultivated, they are separated from the nutrient medium by filtration, centrifugation or the like. Generally a small amount of filter aid is used to facilitate the separation. Treating the microorganisms with a toxic agent according to this process results in microorganisms which are easily separated from the nutrient medium. Without such a treatment separation is difficult, filtration being prolonged and large quantities of filter aid required.

The separated material may then be incorporated directly into a glucose-containing solution and glucose isomerizing conditions maintained. After the desired amount of fructose is formed, the cellular material is separated from the solution by filtration, centrifugation, etc., and the isomerized solution subjected to a refining technique which removes objectionable color and odiferous materials. In this process it is preferred to use microorganisms of the Streptomyces genus and most preferred to use Streptomyces ATCC 21175 or 21176.

Example: This example illustrates the use of ethyleneimine as the toxic agent for producing a microorganism which forms large quantities of glucose isomerase. A culture of Streptomyces ATCC 21175 was sporulated on a sterile agar medium containing 0.4% xylose, 0.4% yeast extract, 1.0% malt extract and 2% agar. The pH of the medium was 7.3. Spores from this slant were inoculated into a shaker flask containing 100 ml of a sterile medium composed of 1% peptone, 1% yeast extract, 0.1% $MgSO_4 \cdot 7H_2O$, 0.2% agar and 1% xylose. The pH of the medium was 7.0.

The flask was maintained at 30°C and shaken at 200 rpm for 16 hr, and then 3.5 ml of a 0.5% aqueous solution of ethyleneimine was added and the flask shaken for 30 min. Cells were removed from the medium by centrifugation and the cells were plated onto Petri dishes containing the sterile agar medium described above. The cells were allowed to grow and sporulate.

A loop of these spores was transferred into a 300 ml Erlenmeyer flask containing 100 ml of sterile medium, at a pH of 7, composed of 1% peptone, 1% yeast extract, 1% xylose, 0.1% $MgSO_4 \cdot 7H_2O$, 0.3% K_2HPO_4, and 0.2% agar. The inoculated flask was incubated for 48 hr at 30°C on a rotary shaker at 200 rpm. 20 ml of the fermented broth was removed from the flask and transferred into a 2 liter baffled Erlenmeyer flask containing 800 ml of the medium described immediately above.

Another 20 ml of fermented broth containing Streptomyces ATCC 21175 which was not treated with ethyleneimine was transferred into another 2 liter baffled Erlenmeyer flask containing 800 ml of the same medium. These flasks, at 30°C, were shaken for 48 hr at 180 rpm and the entire contents of each flask were separately inoculated into two 40 liter fermentors containing 25 liters of a sterilized medium, at a pH of 7, composed of 1,000 g corn steep liquor (29° Bé), 6 g $CoCl_2 \cdot 6H_2O$, 250 g sorbitol, 190 g glucose, 1,344 ml cottonseed hydrolysate, 25 ml of an antifoaming agent (Pluronic L-61), and the remainder water. The cottonseed hydrolysate was prepared by treating a suspension containing 16%

by weight cottonseed hulls and 2.5% by weight H_2SO_4 for 10 min at 300°F. The hydrolysate was cooled to about 120° to 140°F, neutralized to a pH of about 4, filtered and concentrated to 30° Bé. The fermentors were maintained at a temperature of 30°C, agitated at 200 rpm and aerated with 1 volume of air per volume of medium per minute. After 65½ hr the cellular material was removed by filtration.

The cellular material containing the microorganism treated with the ethyleneimine was observed to filter faster than did the cellular material containing the microorganism not treated with ethyleneimine. The glucose isomerase activity of the filter cake was determined and the results are shown below.

Microorganism	Glucose Isomerase Activity (IGIU/g dry substance)
Streptomyces ATCC 21175	318
Streptomyces ATCC 21175 treated with ethyleneimine	518

To 800 ml of corn syrup containing 528 g dextrose, 0.001 M $CoCl_2$, 0.005 M $MgSO_4$ at pH 6.5 was added a sufficient amount of cellular material of the microorganism treated with ethyleneimine to achieve a level of 560 IGIU. The mixture was maintained at 70°C and at pH 6.5. After 69 hr the corn syrup contained 44.13% fructose dry basis.

Isomerase from Nocardia, Micromonospora and Others

R.O. Horwath and G.W. Cole; U.S. Patent 3,829,362; August 13, 1974; assigned to Standard Brands Incorporated have found that microorganisms of the genera Nocardia, Micromonospora, Microbispora and Microellobospora produce glucose isomerase. These microorganisms are characterized as having the ability to assimilate xylose and produce glucose isomerase. It should be understood that these microorganisms may assimilate other carbon or carbohydrate sources and produce glucose isomerase.

The preferred microorganisms used in the process are *Nocardia asteroides* ATCC 21943 (IMRU #3148), *Nocardia dassonvillei* ATCC 21944 (IMRU #509), *Micromonospora coerula* ATCC 21945 (IMRU # 12328), *Microbispora rosea* ATCC 21946 (IMRU #37485) and *Microellobospora flavea* ATCC 21947 (IMRU #3857R). A typical medium for propagation of these microorganisms may contain xylose, dextrose, sorbitol, corn steep liquor and a source of metal ions such as cobaltous ions and the like. The metal ions apparently activate or stabilize the glucose isomerase.

The glucose isomerase derived from the microorganisms may be contacted with or incorporated into a glucose-containing solution and conditions maintained to isomerize a portion of the glucose to fructose. The glucose isomerase used may be in the form of whole broth, i.e., the propagation medium, cellular material containing the glucose isomerase or a purified extract of the broth or the cells.

Example: This example illustrates the propagation of microorganisms of the genera Nocardia, Micromonospora, Microbispora and Microellobospora and the use of glucose isomerase derived therefrom to convert glucose to fructose. The following microorganisms were propagated under aerobic conditions: *Nocardia asteroides* ATCC 21943 (IMRU #3148), *Nocardia dassonvillei* ATCC 21944 (IMRU #509), *Micromonospora coerula* ATCC 21945 (IMRU #12328), *Microbispora rosea* ATCC 21946 (IMRU #37485) and *Microellobospora flavea* ATCC 21947 (IMRU #3857R).

The propagation media contained 5.0 g/l xylose, 0.5 g/l dextrose, 2.0 g/l sorbitol, 40 g/l corn steep liquor dry basis and 1.0 ppm $CoCl_2$. The propagations were performed in Erlenmeyer flasks at a temperature of 30°C on a rotary shaker to provide aerobic conditions. After the propagations were completed, cellular material was removed from the flasks by filtration, washed, suspended in deionized water and sonicated for about 40 sec using a Branson Sonifier (Model S-125). The sonicated material was diluted to a known volume and centrifuged for about 5 min at 16,000 rpm. The potencies of the supernatants were determined and found to be as follows.

Microorganism	Potency (GIU/ml)*
Nocardia asteroides ATCC 21943 (IMRU #3148)	4.28
Nocardia dassonvillei ATCC 21944 (IMRU #509)	4.37
Micromonospora coerula ATCC 21945 (IMRU #12328)	3.46
Microbispora rosea ATCC 21946 (IMRU #37485)	1.71
Microellobospora flavea ATCC 21947 (IMRU #3857R)	1.73

*GIU is the abbreviation for Glucose Isomerase Unit and is that amount of enzyme which will convert 1 mg of glucose to fructose per hour in a solution initially containing 4% of dextrose per·liter, 20 ml of a 1 molar solution of $MgSO_4$ per liter at a pH of 6.7 (maleic acid) and a temperature of 70°C.

Matrix Immobilized Enzymes

The conversion of enzymes into immobilized or insolubilized products possessing specific catalytic activity has been of interest for some time. In general, these enzyme products have been prepared by various methods, including polymerization onto organic polymer lattices and attachment to polymers of amino acids, coupling to cellulose, dextran and other carbohydrate or polystyrene derivatives, immobilization in starch, and acrylamide gels, and coupling to inorganic carriers such as bentonite, kaolinite and porous glass.

These immobilized or insolubilized enzyme products are usually formed through the aid of three general types of reactions: covalent bonding, ionic bonding and physical adsorption. But it has not been reported previously that a practical or useful immobilized or insolubilized glucose isomerase can be prepared by any of these methods.

T. Sipos; U.S. Patent 3,708,397; January 2, 1973; assigned to Baxter Laboratories, Inc. has produced an immobilized glucose isomerase enzyme having improved stability to heat and capable of increased conversion of glucose to fructose. This is prepared by mixing glucose isomerase obtained from *Streptomyces phaechromogenes* or *Lactobacillus brevis,* with basic anionic exchange cellulose in aqueous buffer solution at pH 7 to 10 and recovering the enzyme complex from the reaction mixture.

It has been found that the above glucose isomerase enzyme complex exhibits excellent heat stability and can be used repeatedly (e.g., 5 to 10 times) before 50% of its enzyme activity is lost at 70°C . In conventional practice by comparison, the unbound, soluble enzyme loses substantially all of its activity in only one conversion. Moreover, it has been found that in the matrix supported form, the pH optimum of the glucose isomerase changes to one unit lower and the percent conversion of glucose to fructose substantially increases.

In general, from 12 to 15% more of the available glucose in syrups can be converted to fructose per conversion by use of the enzyme complex instead of the unbound, soluble enzyme. This improved stability to heat and increased yield of glucose conversion represent outstanding advantages of this process.

In the preparation of this immobilized glucose isomerase enzyme it is important to use only certain types of cellulose derivatives, namely, the basic anion exchange celluloses. By the term "basic anion exchange cellulose" is meant any cellulose containing basic anion exchange groups bonded to the cellulose molecule and capable of entering into an exchange reaction with anionic groups of other compounds. Preferably, the cellulose is an alpha cellulose such as cotton, wood pulp, paper, or cotton cloth. In particular, these anion exchange cellulose compounds include the di- and triethylaminoethylated celluloses, e.g., DEAE-cellulose and TEAE-cellulose, and the cellulose derivatives of epichlorohydrin and triethanolamine, e.g., ECTEOLA-cellulose.

It was found that other carbohydrate or cellulose derivatives, e.g., CM-cellulose, phosphocellulose, dextran, Sephadex and DEAE-Sephadex, and various other ion exchangers such as Dowex 1-X8, and Dowex 2-X8, which are also strongly basic anion exchangers (quaternary ammonium type), do not yield active immobilized or insolubilized glucose isomerase enzyme preparations.

Although the glucose isomerase enzyme or enzyme preparation can be obtained from numerous microorganisms, for this process it is preferable to obtain the enzyme or enzyme preparation from Streptomyces and Lactobacillus species, and most preferably, from *Streptomyces phaechromogenes* and *Lactobacillus brevis.*

In general, the preparation of the immobilized glucose isomerase enzyme complex is achieved by mixing the enzyme or enzyme preparation with the basic anion exchange cellulose in aqueous buffer at a pH of 7 to 10, and preferably at pH 8. After thoroughly mixing these components, at normal temperature (about 25°C) or other convenient temperature conditions, the resulting enzyme complex is recovered such as by filtration, centrifugation and the like separation procedures and then preferably washed or sparged with additional buffer. The filtered and washed enzyme complex can be used in its wet form as is, or the cake can be dried by conventional protein or enzyme drying techniques, such as by shelf, rotary drum or spray drying, but preferably, freeze drying. A preferred method of purifying the enzyme complex comprises selective gradient elution. Since undesirable constituents of the fermented liquor can also be adsorbed to the cellulose exchanger, a selective elution with buffer containing 0.2 M NaCl to remove these impurities without eluting the enzyme can be employed. Selective elution of the enzyme from the cellulose exchanger can be achieved by resuspending the cake in buffer containing about 0.6 M NaCl and then remixing eluted enzyme, after suitable dilution, with the new exchange cellulose, as before, to reform the enzyme complex.

Suitable buffers for use in the preparation of the immobilized enzyme complex are tris(hydroxymethyl)aminomethane, boric acid-borate, glycinamide and cholamine chloride. A 0.02 M tris buffer, pH 8 to 8.5, is preferred. Phosphate buffers which might lead to phosphorylation of starch fragments, glucose, or the fructose are preferably avoided in the practice of this process. Other metal salts, preferably monovalent metal chlorides, e.g., LiCl or KCl, can be used in place of the NaCl for the selective elution described above.

Example: A 1,000 lb culture growing of *Streptomyces phaechromogenes* is provided by inoculating a medium containing 1% D-xylose, 0.3% K_2HPO_4, 0.1% $MgSO_4 \cdot 7H_2O$, 0.02% $CoCl_2 \cdot 6H_2O$, 1% peptone, 3% wheat bran and 2% corn steep liquor, with *Streptomyces phaechromogenes* from an agar slant, and fermenting 48 hr at 30°C. The crude growth product contains 125,000 GI units by passing the ferment through a 20 mesh screen to remove residual bran, adding 4% toluene and then 4% Speedex filter aid, and filtering.

The clear supernatant is buffered with 0.02 M tris buffer, pH 8.5, and then 500 grams of preequilibrated, wet DEAE-cellulose is added. The pH of the medium is reduced to about pH 8.0 by the addition of the cellulose exchanger. After mixing thoroughly by stirring for 1 hr, the mixture is filtered, the filter cake is sparged with 0.02 M tris buffer, pH 8.0, and the washed cake is retained as the active glucose isomerase enzyme complex. This complex has excellent heat stability and can be used for the conversion of D-glucose to D-fructose with substantially increased conversion compared to the conversion with the unbound, soluble glucose isomerase.

ENZYME PRODUCTION OF FRUCTOSE

Although glucose-isomerizing enzymes are more selective in converting glucose to fructose than is an alkaline catalyst, there are still a number of problems associated with the commercial use of these enzymes. For example, appreciable quantities of colored bodies are produced during enzymatic isomerization, which make the resulting products difficult to refine. Also there is a tendency for the isomerizing enzyme to be inactivated in a shorter period than is desired.

The formation of colored bodies and the inactivation of the isomerizing enzyme are largely dependent upon the conditions under which the isomerization reaction is carried out. If the reaction is performed for relatively long periods of time and/or at high temperatures, in order to obtain high yields of fructose, there will be greater amounts of colored bodies formed and the enzyme will be inactivated to a greater degree.

The purity of the glucose isomerase preparation also affects the formation of colored bodies in the isomerized liquor. If relatively large amounts of extraneous materials are present in the glucose isomerase preparation, there is a greater tendency for larger amounts of colored bodies to be formed.

Use of Borates in Isomerizing Glucose

Y. Takasaki; U.S. Patent 3,689,362; September 5, 1972; assigned to Agency of Industrial Science & Technology, Japan has developed a process for the enzymatic production of fructose from glucose or glucose containing solution by effecting the reaction in the presence of a borate compound. The borate compound gives an exceptionally high yield of fructose and, with recovery of the borate compound used, the process has economic advantages.

The borate compounds which can be used for this process include such water-soluble borates as sodium borate and potassium borate of various types (such as orthoborate, metaborate and tetraborate), such borates insoluble or sparingly soluble in water as magnesium borate, lithium borate, barium borate, strontium borate, calcium borate and manganese borate, and esters formed from organic compounds such as alcohol or phenol and boric acid. Anion exchange resins of the borate form can be used as well.

Magnesium and barium borate are insoluble or sparingly soluble in water and do not inhibit the reaction even when large quantities of these salts are added to the reaction solution. Therefore, the isomerization reaction can be carried out at high glucose concentration. Thus, use of these salts proves more advantageous in applying this method to industrial operation. As to the time that the borate compound is added to the reaction system, it is more desirable to add it from the beginning. It is otherwise possible to add the compound at the time that the formation of fructose has progressed to some extent or at the time that the reaction has reached its equilibrium.

The quantity of the borate compound to be added must be changed with the kind of borate compound to be used but has a close relationship with the concentration of the glucose. For example, at 0.3 M of glucose concentration, 54% of the glucose was converted to fructose when the isomerization was carried out in the absence of sodium tetraborate, while at the same glucose concentration, 87% of the glucose was converted to fructose when the reaction was carried out in the presence of 0.08 to 0.2 M of sodium tetraborate.

The quantity of fructose produced decreased when the borate was added in the same concentration as glucose (0.3 M). When the isomerization reaction was carried out at 1 M of glucose concentration without the addition of sodium tetraborate, about 55% of the glucose was converted to fructose. About 72% of the glucose was converted to fructose when the reaction was carried out in the presence of 0.1 M of sodium tetraborate.

When 0.2 M of sodium tetraborate was added at the same glucose concentration, the isomerization ratio rose to about 87%. However, the formation of fructose was no longer increased beyond this ratio when sodium tetraborate was added further. Namely, it was found that there is an optimum range of concentration of borate to be added for obtaining the maximum isomerization ratio.

Table 1 shows the minimum quantity of sodium tetraborate added to obtain the maximum isomerization ratio at different concentrations of glucose.

TABLE 1

Glucose concentration A, M	Minimum concentration of sodium tetraborate (M) required for obtaining maximum production of fructose at indicated concentration of sugar B	$(B)/(A)\times100$ (percent)
0.1	ca. 0.04	ca. 40
0.2	ca. 0.07	ca. 35
0.3	ca. 0.08	ca. 27
0.5	ca. 0.11	ca. 22
0.6	ca. 0.13	ca. 22
1.0	ca. 0.22	ca. 22

Table 2 shows the quantity of different borate compounds which must be added for isomerizing glucose to fructose at the concentration of 1 M at an isomerization ratio of 80 to 90%. The data clearly indicate that the required quantity of a given borate can be determined on the basis of the number of the negative charges (borate ions) or the normal concentration of the borate.

TABLE 2

Kind of borate	Negative charge of borate	Concentration, M, of borate for obtaining 80–90% isomerization ratio	Normal concentration of borate
Sodium tetraborate	$B_4O_7^=$	0.1–0.4	0.2–0.8
Potassium tetraborate	$B_4O_7^=$	0.1–0.4	0.2–0.8
Sodium metaborate	BO_2^-	0.2–0.8	0.2–0.8
Potassium metaborate	BO_2^-	0.2–0.8	0.2–0.8
Magnesium metaborate	$(BO_2)_2^-$	0.1–0.2	0.2–0.8
Barium tetraborate	$B_4O_7^=$	0.1–0.4	0.2–0.8

From the fact that since free boric acid has a low solubility in water and the complex formed between the boric acid and the sugar is unstable on the acid side, it was found that more than 80% and up to 90 to 98% of the boric acid can be separated as a precipitate from the solution by treating the sugar solution containing borate with a cation exchange resin or by acidifying the sugar solution containing borate to a pH below 3 by adding mineral acid to cause the borate to be decomposed into boric acid and the corresponding inorganic salt. Since the inorganic salt produced is soluble, it helps to decrease the solubility of boric acid and increase its recovery.

For example, a mixture containing 8.3% of fructose, 0.92% of glucose and 4.0% of sodium borate ($Na_2B_4O_7$) was divided into 100 ml portions which were adjusted to pH 4, 3, 2, 1 and 0.5 with the addition of sulfuric acid, respectively. Each portion was concentrated under a reduced pressure to 25 ml and maintained at 0° to –5°C to form the precipitates of boric acid, which was removed by means of suction filtration.

The content of glucose, fructose and boric acid in the resultant sugar solution were examined. The results are given in Table 3.

TABLE 3

pH adjustment of sugar solution	Boric acid content (mg./ml.)	Fructose content (mg./ml.)	Glucose content (mg./ml.)
Original pH value	46.4	32.6	9.6
4	7.3	72.4	8.5
3	5.3	68.4	8.2
2	3.0	69.6	8.4
1	3.0	69.4	8.3
0.5	1.9	70.8	7.9

It is clear from Table 3 that more than about 90% of the boric acid could be removed by the treatment adjusting the pH value of the sugar solution to below 3. At this stage, about 10% of the sugar content is removed together with boric acid. This boric acid containing sugar is neutralized with alkali and put to re-use in isomerization reaction.

Example: Five reaction mixtures consisting of 5 g of anhydrous glucose, 0.024% of cobalt chloride and 2,300 units of glucose isomerase (measurement of activity being based on the method described in *Agricultural Biological Chemistry,* Vol. 3, page 1247, 1966) were prepared. To four of the five mixtures, magnesium metaborate was added to a concentration of 0.42, 0.83, 1.25 and 1.69% respectively. The mixtures were made up to a total volume of 100 ml with water and incubated at 70°C.

At definite time intervals, a prescribed amount of the test specimen was taken from each reaction solution and the contents assayed for fructose and glucose. Table 4 shows the fructose content, glucose content and isomerization ratio in the state of equilibrium of each reaction mixture. It is clear from this table that the isomerization ratio was about 56% in the absence of magnesium metaborate, while the isomerization ratio was raised to 70 to 88% when magnesium metaborate was added to the reaction solution.

TABLE 4

Amount of magnesium borate, $Mg(BO_2)_2$, added (percent)	Fructose content (g./100 ml.)	Glucose content (g./100 ml.)	Fructose plus glucose content (g./100 ml.)	Isomeriza-tion ratio (percent)
0	2.80	2.20	5.00	56.0
0.42	3.48	1.46	4.94	70.4
0.83	4.21	0.98	5.19	81.1
1.25	4.21	0.67	4.88	86.3
1.67	4.25	0.58	4.83	88.0

pH Regulation of Isomerization

A method is provided by *Y. Takasaki and A. Kamibayashi; U.S. Patent 3,715,276; February 6, 1973; assigned to Agency of Industrial Science & Technology Governmental, Japan* for enzymatically converting glucose into fructose characterized by using calcium carbonate, magnesium carbonate, an anion exchange resin or an amphoteric ion exchange resin as a pH regulator. In the isomerization reaction, the organic acid produced in the course of reaction is neutralized by calcium or magnesium or removed by the ion exchange resin and, consequently, the pH value of the reaction solution is maintained at a pH value desirable for the isomerization reaction and the isomerization reaction is carried out efficiently.

Particularly when calcium carbonate is used as the pH regulator, the calcium carbonate not only serves as a pH regulator but also proves advantageous in that the resultant calcium salt of the organic acid is sparingly soluble, with the result that only a very small amount of ions are dissolved in the reaction solution and the burden of the ion exchange resin used in the refining treatment of the resulting sugar solution will be lessened to a great extent.

According to the conventional method, the total ion concentration in one liter of 50% glucose solution is from 4,000 to 5,000 ppm as calcium carbonate. By contrast, the total ion concentration in the sugar solution which has undergone the isomerization reaction by this method is 2,524 ppm, a value which is nearly equal to the ion concentration 2,644 ppm of the glucose solution prior to the reaction. This indicates that virtually no increase occurs in the ion concentration in the course of isomerization reaction.

The ion exchange resin to be used as pH regulator is either an anion exchange resin or an amphoteric ion exchange resin. When the reaction is carried out in the presence of such ion exchange resin, the sugar solution obtained consequently will have a very low ion concentration and a very low coloration degree as compared with the conventional method.

Example 1: Each of several lots was prepared by adding to 36 g of glucose, 123 mg of $MgSO_4 \cdot 7H_2O$, 24 mg of $CoCl_2 \cdot 6H_2O$, 180 units of glucose isomerase and a different amount of $CaCO_3$ in the range of 0.25 to 1.5 g as a pH regulator. Each lot was made up to a total volume of 100 ml with water and incubated at 70°C for 46 hr. The fructose content and pH value were determined. The results are shown in Table 1. For comparison, the results obtained with a lot using phosphate buffer solution (0.05 M) are also given in the table.

TABLE 1

Calcium carbonate (%)	pH value prior to start of reaction	pH value after reaction	Fructose formed (g/100 ml)	Isomerization ratio (%)
0	7.5	4.9	4.79	13.3
0.25	7.0	6.0	7.81	21.7
0.50	6.9	6.1	7.16	19.9
0.75	7.0	6.2	8.21	22.8
1.00	7.0	6.2	8.57	23.8
1.50	7.1	6.2	8.53	23.7
0.05 M phosphate buffer solution	7.0	6.4	11.5	32.0

Isomerization ratio is obtained by the following equation.
Isomerization ration = (Fructose/Total Sugar) × 100

Example 2: Each of several lots was prepared by adding to 40 g of glucose, 123 mg of $MgSO_4 \cdot 7H_2O$, 24 mg of $CoCl_2 \cdot 6H_2O$, cells containing glucose isomerase (280 units) and a different amount of $MgCO_3$ as a pH regulator in the range of 0.25 to 1.5 g. Each lot was made up to a total volume of 100 ml with water and incubated at 70°C. The fructose content and pH value of the fractionated solution were determined. The results are shown in Table 2.

TABLE 2

Magnesium carbonate (%)	pH value prior to start of reaction	pH value after reaction	Fructose formed (g/100 ml)	Isomerization ratio (%)
0	7.5	5.7	11.9	29.8
0.25	8.0	6.68	14.4	36.1
0.50	8.15	6.80	15.4	38.5
1.00	8.20	6.90	16.0	40.0
1.50	8.30	7.08	15.4	38.5

Example 3: Each of several lots was prepared by adding to 42 g of glucose, 123 mg of $MgSO_4 \cdot 7H_2O$, 24 mg of $CoCl_2 \cdot 6H_2O$, 1,200 units of glucose isomerase (extracted enzyme) and 10 g of a different ion exchange resin as enumerated in Table 3. Each lot was made up to a total volume of 100 ml with water and incubated at 60°C. The fructose content and pH value of the fractionated solution were determined.

The results are shown in Table 3 which show that, in the absence of ion exchange resin, the pH fell below 6 and the reaction failed to continue efficiently. In the reaction carried out in the presence of an ion exchange resin, particularly, an amphoteric ion exchange resin, the pH value was maintained above 6 throughout the entire reaction and the isomerization was carried out efficiently.

TABLE 3

Ion-exchange resin used	Form of resin	Presence or absence of enzyme (+, −)	pH value prior to reaction	pH value after reaction	Isomerization ratio (percent)
No ion-exchange resin		−	8.1	4.6	2
		+	7.5	5.8	26.4
Retardation 11A-8	H, OH	−	8.6	7.4	20
		+	8.3	6.9	42.2
Amberlite I RA–400	OH	−	9.4	6.2	14.6
		+	9.6	6.1	41.6
Dowex 1–X8	OH	−	9.6	5.9	18
		+	9.8	6.0	43.0
Amberlite I RA–400	OH	−	8.4	5.8	22.2
Amberlite I R–120	Na	+	8.3	6.3	50.4

Use of Sulfite to Prevent Color Formation

W.P. Cotter, N.E. Lloyd and C.W. Hinman; U.S. Patent 3,623,953; November 30, 1971; assigned to Standard Brands Incorporated have found that the presence of relatively small amounts of water-soluble salts of sulfurous acid during the enzymatic isomerization of glucose reduces color formation and increases the stability of the glucose-isomerizing enzyme. In this process, the salts of sulfurous acid may be provided in the glucose-containing liquors by any convenient method. For instance, sulfite or bisulfite salts, or other substances which will generate sulfite or bisulfite ions, e.g., SO_2 or H_2SO_3 solution, may be incorporated directly into the glucose-containing liquor before the isomerization is carried out or may be incorporated during isomerization.

Also sulfite or bisulfite ions may be provided in the glucose-containing liquors by passing these liquors through ion exchange resins in the sulfite form. Preferably, however, the bisulfite and sulfite salts are provided in the glucose-containing liquors before the isomerization is initiated since the full benefit of the presence of these salts will be obtained. The preferred microorganisms used to produce glucose isomerase for this isomerization are those belonging to the Streptomyces genus. The most preferred microorganism is Streptomyces sp ATCC 21175.

The amount of bisulfite or sulfite salts used in the glucose-containing liquor may vary, but under the preferred conditions sufficient amounts are added to provide an SO_2 content of 0.02 to 0.3% by weight based on the dry substance content of the liquor, most preferably from 0.03 to 0.07%. Although at greater concentrations of SO_2 there will be a relatively long period during the isomerization reaction when less color is produced, than in the case of an isomerization reaction without the sulfites or bisulfites, after this initial period, the rate of color formation will increase very rapidly until the color formed will exceed that formed when the isomerization is carried out without the sulfites or bisulfites.

Therefore, when these salts are used, the isomerization reaction should be terminated before the color formed reaches a point where the subsequent removal is difficult. Although the glucose-isomerizing enzyme is relatively stable at high temperatures it is subject to thermal denaturation normal to all proteins. The presence of the sulfite salts during the isomerization reaction, especially at high isomerization temperatures, surprisingly exerts a protective effect towards the glucose-isomerizing enzyme. This provides the benefit that lesser quantities of the enzyme are needed to achieve the same yield of fructose when sulfites or bisulfites are present, or conversely for the same quantity of enzyme, higher yields of fructose can be obtained.

Example: This example illustrates the enzymatic isomerization of glucose in glucose-containing liquors in the presence and absence of sulfite salts. Streptomyces sp ATCC 21175 was grown under aerobic submerged fermentation conditions at a pH of 7 in a presterilized aqueous medium containing 1% sorbitol, 0.75% dextrose, sufficient corncob hydrolysate to provide 1% xylose, 4% steep water at 29° Bé and 0.024% cobaltous ion.

The fermentation was carried out at 30°C, an airflow of 1 volume of air per volume of medium per minute and a back pressure of 10 psi. The fermenting broth was mechanically stirred at 200 rpm and after 65 hr 4% filter aid was admixed into the broth and the cellular material harvested from the broth by filtration with suction. The filter cake was washed with demineralized water, broken into small pieces and dried for 5 hr in a forced air oven at an air temperature of 140°F. The activity of the air dried filter cake was 660 GIU/g. A series of four glucose-containing liquors prepared from hydrolysates of cornstarch were prepared having the compositions shown in the following table.

| | -----Sample----- | | | |
	1	2	3	4
Glucose content (percent dry basis)	55.5	55.5	55.5	67.5
$CoCl_2 \cdot 6H_2O$ (molarity)	0.001	0.001	0.001	0.001
Na_2SO_3 (percent dry basis)	0.25			
$MgCl_2 \cdot 6H_2O$ (molarity)	0.005		0.005	
$MgSO_3 \cdot H_2O$ (percent dry basis)		0.25		0.25
Total SO_2 (percent dry basis)	0.13	0.12	none	0.12
Glucose isomerase (GIU/g dry basis)	2.3	2.3	2.3	4.6

These samples were isomerized at a temperature of 70°C for 92 hr with the pH being maintained at 6.5 by the addition of a 0.5% solution of sodium hydroxide. An atmosphere of nitrogen was maintained over the three samples which contained the sulfites. The color and the fructose content of the liquors were determined throughout the isomerization. The results of these determinations are shown in the following table.

| | Sample Number | | | | | | | |
| | 1 | | 2 | | 3 | | 4 | |
Isomerization time (hours)	Percent fructose	Color	Percent fructose	Color	Percent fructose	Color	Percent fructose	Color
0	0	6	0	6	0	6	0	7
8	2.9	9	5.4	9	3.4	11	9	9
20	9.4	10	12.1	11	8.3	36	18	10
26	9.8	7	12.7	8	7.4	53	20.2	6
44	16.7	10	19.1	14	13.2	96	26.8	17
68	21.0	17	22.1	71	14.6	251	30.5	211
92	25.0	98	25.3	409	17.5	434	32.5	707

As seen from the table, as the amount of fructose increased the color of the isomerized liquor also increased. In each of the isomerization reactions carried out in the presence of sulfites less color was formed than in the liquor which contained no sulfites on an equal fructose formed basis. Also it is seen that more fructose formed in the samples containing the sulfite salts indicating that the sulfites reduced the degree of inactivation of the enzyme during the isomerization reaction. The color of the glucose-containing liquor of the example was determined spectrophotometrically by measuring the absorbance at 450 and 600 mμ of an appropriately diluted liquor in a 1 cm cell versus water as a reference. The color was calculated by the following:

$$\text{Color} = \frac{(100)(109)(A_{450} - A_{600})}{C}$$

In the above C = concentration in grams of dry substance per 100 ml of liquor; A_{450} = absorbance at 450 mμ; and A_{600} = absorbance at 600 mμ.

Five Step Purification of Fructose Syrups

A method of purifying enzymatically isomerized glucose syrups containing fructose in five separate steps has been developed by *W.J. Nelson and C.W. Hinman; U.S. Patent 3,784,409; January 8, 1974; assigned to Standard Brands Incorporated.* These syrups contain salts which contribute to the ash content, colored bodies, odoriferous materials, and other impurities.

To remove these impurities, the isomerized, glucose syrups are treated (1) with a strong acid cation exchange resin in the hydrogen form which removes a portion of the colored bodies and appreciable quantities of the metallic constituents of the salts. Following this treatment, the syrups are subjected to (2) a carbon treatment which removes the major portion of the colored bodies and odoriferous materials. The syrups are then treated with (3) a weak base anion exchange resin in the free base form, (4) a strong acid cation exchange resin in the hydrogen form, and with (5) a strong base anion exchange resin in the hydroxyl form.

Enzymatically isomerized glucose syrups containing fructose develop color upon standing, due to the color-forming bodies present. When the pH of the isomerized syrup is adjusted, for instance, prior to filtration, to values from 4 to 5, and preferably 4.5, color development is reduced. This pH adjustment may be accomplished by the addition of an acid, or, preferably by adding to the syrup a portion of an isomerized syrup which has been acidified by hydrogen ion exchange, in a subsequent step of this process of purification.

This method is preferred since no additional impurities are added to the isomerized syrup which must subsequently be removed. The filtered or unfiltered isomerized syrup is next treated with a strong acid cation exchange resin in the hydrogen form. Examples of such resins are Duolite C-3 and Duolite C-25, Dowex 50-W and Amberlite 200. This treatment removes most of the metallic constituents which contribute to the ash, some of the colored bodies, and substantially all the remaining proteinaceous impurities, such as amino acids.

The isomerized syrup is then acidified to a pH below 3. If this pH adjustment is not accomplished by the strong acid cation exchange treatment in the preceding purification step, sufficient acid should be added to lower the pH to below 3. Generally, however, the preceding cation exchange treatment is performed under conditions such that the syrup is acidified solely by hydrogen ion exchange. If the pH of the syrup is partially adjusted with an acid such as HCl, the weak base ion exchange resin used in a subsequent step will be exhausted more rapidly than if the adjustment is made solely by hydrogen ion exchange.

The pH adjusted, isomerized syrup is next passed through a bed of activated granular carbon or slurried with activated powdered carbon and the carbon removed by filtration. This carbon treatment will remove a major portion of the colored bodies, odoriferous and flavoring materials in the syrup.

Typically, when using activated granular carbon a burn ratio of from 1 to 9, and most preferably 7, is used. Burn ratio is defined as the pounds of granular carbon that must be reactivated per 100 lb of syrup solids processed to obtain the desired degree of purification. Examples of suitable activated carbons are Pittsburg types CPG and SGL, and NuChar types CEE and WV-L.

The carbon refined, isomerized syrup is next treated with a weak base anion exchange resin in the free base form. This step adjusts the pH from 3 to 7, and typically to 6. As a consequence of this treatment, various acids generated by the prior pH adjustment step are removed. Examples of weak base anion exchange resins which may be used are Duolite A-6 and Amberlite IRA-93.

Next, the syrup is treated with a strong acid cation exchange resin in the hydrogen form. This treatment removes a portion of the remaining colored bodies and substantially all the remaining metallic constituents. Suitable strong acid cation exchange resins are those mentioned for use in the first step. Finally, the syrup is treated with a strong base anion exchange resin in the hydroxyl form. This removes all the remaining undesirable, odoriferous and flavoring materials from the syrup and improves the color stability. Suitable strong base anion exchange resins are Amberlite IRA-900 and Dowex 11.

Depending upon the character of the strong base anion exchange resin used in the preceding step, it may be desirable to treat the syrup with a strong acid cation exchange resin in the hydrogen form in a sixth step. In some cases, as evidenced by the odor and flavor of the syrup, a small amount of amines may be present due to amine leakage of the strong base cation exchange resin. The presence of these amines is, of course, objectionable, and may be removed by treatment with a strong acid cation exchange resin in the hydrogen form.

Preferably, the syrup is purified at a temperature of from 37° to 60°C. At higher temperatures excessive color development may be observed in the syrup during purification, and at lower temperatures the syrup will be rather viscous, and will be difficult to process through beds of ion exchange resins and granular carbon. It is preferred that the enzyme used to isomerize a glucose syrup be derived from microorganisms of the genus Streptomyces, particularly Streptomyces sp ATCC 21175 and Streptomyces sp ATCC 21176.

The microorganisms of the Streptomyces genus specifically identified above primarily produce glucose isomerase intracellularly. The glucose isomerase may be separated from the cells by a sonic treatment in an aqueous medium and the cells removed by filtration. The filtrate containing the glucose isomerase may be used to isomerize glucose in a glucose syrup. In commercial practice, however, it is economically undesirable to use such a costly procedure. Consequently, it is preferred that the cells be separated from a fermentation broth and be incorporated directly into a glucose syrup.

In an example of this process, sufficient quantities of filtered cells of Streptomyces sp ATCC 21175, to provide the necessary glucose isomerase activity, are incorporated directly into a glucose syrup, for instance a corn syrup. The glucose is isomerized by maintaining the temperature of the syrup at 70°C and at a pH of 6.8 to 7.2.

During the isomerization, suitable quantities of glucose isomerase activators and stabilizers are present. When the desired yield of fructose is obtained, the isomerized syrup is purified in accordance with the purification process detailed above. The table below shows the effectiveness of this purifying process in terms of removing the color bodies from five samples of syrups isomerized by the general procedure described above.

Treatment Step	Color of the fructose-containing syrups after each treatment step as measured by Corn Refiners Association Incorporated Method F-14 at pH of 4.8				
			Samples		
	1	2	3	4	5
Isomerized crude fructose-containing syrup (filtered)	5.370	2.540	1.830	6.940	3.050
Powdered carbon treatment (Nuchar CEE) (filtered)	2.190	1.490	0.880	—	3.150
Strong acid cation exchange treatment (Duolite C-3)	1.870	1.190	0.570	2.670	1.820
Granular carbon treatment (Pittsburgh Type CPG)	0.102	0.097	0.030	0.142	0.131
Weak Base anion exchange treatment (Duolite A-6)	0.123	0.099	0.057	0.132	0.114
Strong acid cation exchange treatment (Amberlite 200)	—	—	—	—	—
Strong base anion exchange treatment (Amberlite IRA-900)	—	—	—	—	—
Strong acid cation exchange treatment (Amberlite 200)	0.000	0.006	0.003	0.005	0.001

Treatment Step	pH of the fructose-containing syrups after each treatment step				
			Samples		
	1	2	3	4	5
Isomerized crude fructose-containing syrup (filtered)	4.7	4.3	4.2	4.9	4.7
Powdered carbon treatment (Nuchar CEE) (filtered)	4.6	4.4	4.2	5.0	4.7
Strong acid cation exchange treatment (Duolite C-3)	1.2	1.2	1.2	1.2	1.2
Granular carbon treatment (Pittsburgh Type CPG)	1.5	1.4	1.4	1.3	1.3
Weak Base anion exchange treatment (Duolite A-6)	5.8	6.0	6.8	6.0	6.5
Strong acid cation exchange treatment (Amberlite 200)	—	—	—	—	—
Strong base anion exchange treatment (Amberlite IRA-900)	—	—	—	—	—
Strong acid cation exchange treatment (Amberlite 200)	4.3	4.5	5.2	5.3	4.2

Use of Cell-Bound Glucose Isomerase

Two processes of enzymatically converting glucose to fructose are disclosed by *N.E. Lloyd, L.T. Lewis, R.M. Logan and D.N. Patel; U.S. Patents 3,694,314; September 26, 1972; and 3,817,832; June 18, 1974; both assigned to Standard Brands Incorporated.*

In these processes a glucose-containing solution is passed, under specific conditions, through a bed of cells of microorganisms containing cell-bound glucose isomerase, the bed has a depth to width ratio of less than 2. Because of the economics involved in producing glucose isomerase, it is of the utmost importance to use the isomerase under conditions so that maximum yields of fructose are produced using minimum quantities of glucose isomerase. Moreover, the conditions for isomerization should be such that minimal quantities of objectionable by-products are produced.

This may be attained by forming a glucose-containing solution having a viscosity of from 0.5 to 100 cp, a pH in the range of 6 to 9 and containing 5 to 80% dry substance glucose, heating the solution to a temperature of 20° to 80°C and passing the solution through a bed containing cells of microorganisms which have naturally fixed isomerase or have been treated to inhibit the extraction of the isomerase from the cells, having a glucose isomerase activity of at least 3 IGIU per cubic centimeter of bed and a stability value of at least 50 hr, thereby converting up to 54% of the glucose to fructose.

The ratio of the depth to the width (or diameter) of the bed should be less than 2. The flow rate of the glucose-containing solution passing through the bed should be such that the color of the solution exiting the bed is increased by less than two color units and there is no substantial production of psicose.

The stability value is determined by placing a sufficient amount of the fixed glucose isomerase in a column to obtain from 1,000 to 4,000 IGIU. A solution that is 3 molar in glucose at pH 7, 0.001 molar in $CoCl_2$ and 0.005 molar in $MgSO_3$ is passed through the column at a rate of 20 to 80 ml/hr. The column is maintained at 60°C. The fraction of glucose converted to fructose in the effluent is determined after 20 hr to insure that the bed of fixed glucose isomerase is under equilibrium conditions. The activity index of the fixed glucose isomerase is calculated using the following formula:

$$\text{Activity Index} = (R/E)\log(0.504/0.504 - I)$$

where I is the fraction of glucose converted to fructose, R is the flow rate (ml per hour) and E is the number of IGIU initially in the column. The Activity Index is determined periodically and the time it takes for the Activity Index to reach one-half the initial value (value after 20 hr) is the stability value in hours.

IGIU is the abbreviation of an International Glucose Isomerase Unit and is that amount of enzyme which will convert 1 micromol of glucose to fructose per minute in a solution containing 2 mols of glucose per liter, 0.02 mol of $MgSO_4$ per liter, and 0.001 mol of $CoCl_2$ per liter at a pH of 6.84 to 6.85 (0.2 M sodium maleate) and a temperature of 60°C.

Cellular material containing fixed or stabilized isomerase is held in an aqueous suspension containing 0.001 mol of cobalt chloride per liter at a temperature of 58°C and a pH of 6.5 (0.05 M sodium maleate buffer). The cellular material is held under these conditions for 20 hr. A portion of the suspension is sonicated at 20 kilocycles by the use of a Branson S75 Sonifier. The sonicated material is centrifuged and the centrifugate analyzed for isomerase activity.

Another portion of the suspension (not sonicated) is centrifuged and the isomerase activity of this centrifugate is determined. The activity of extracted isomerase in the latter centrifugate divided by the activity of the isomerase in the centrifugate from the sonication treatment multiplied by 100 is the extractability coefficient of the treated cellular material.

Example 1: This example of U.S. Patent 3,694,314 illustrates the use of fixed or stabilized glucose isomerase to continuously convert glucose to fructose. Streptomyces ATCC 21175 was grown under submerged aerobic conditions.

The pH of the broth containing the cellular material was adjusted to 7.5, and the broth heated for one-half hour until it reached 75°C and was maintained at this temperature for 5 min to fix the isomerase to the cells. 3% by weight of filter aid was incorporated into the broth and then the broth was filtered on a rotary drum vacuum filter precoated with diatomaceous earth. The cellular material was washed on the filter. The resulting filter cake was dried in a forced air drier at a temperature of 49°C for 3.5 hr to yield a dried, fixed isomerase preparation comprising the cellular material containing the fixed isomerase admixed with filter aid. The extractability coefficient of the dried, fixed isomerase preparation was 15% and the stability value was 645 hr.

256 lb of the dried, fixed isomerase preparation was blended for 2 min and was then slurried with a glucose-containing solution consisting of a refined cornstarch hydrolysate at a pH of 7, containing 60 g of glucose/100 ml, 4.5 g of maltose, isomaltose and oligosaccharides/100 ml, 0.001 mol of $CoCl_2$/l, 0.01 mol of $MgSO_4$/l and 0.006 mol of $NaHSO_3$/l. The temperature of the glucose solution was 65°C. The slurry was pumped through a pressure leaf filter having 6 filter leaves and each leaf was coated with a 1 to 1.5 inch layer of the fixed isomerase preparation.

The total activity of the isomerase on the leaves was 1.26×10^7 IGIU. The filter leaves had a total surface area of 74 ft^2. The glucose solution was pumped continuously through the filter at a rate such that 45% of the glucose was converted to fructose. The flow rate was gradually reduced as the activity of the fixed isomerase on the leaves diminished, so that the percent conversion of glucose to fructose was maintained constant at 45%.

At the end of about 40 hr, the flow rate through the leaves was 1.0 gal/min. At the end of 200 hr the flow rate was 0.5 gal/min and the pressure drop across the leaves was 1 psi. The color of the glucose-containing solution entering the bed was 0.01 color unit and the color of the solution exiting the bed was 0.2 color unit.

Example 2: This example of U.S. Patent 3,817,832 illustrates the use of naturally fixed or stabilized glucose isomerase to continuously convert glucose to fructose. Microorganisms of the Arthrobacter sp ATCC 21748 were grown under submerged aerobic conditions and the cells harvested. The cells had a glucose isomerase activity of 49 IGIU/g and an extractability coefficient of 2.5%.

0.644 g of the cells were mixed with 3.0 g of filter aid (Dicalite Superaid), and 40 to 45 ml of a glucose-containing solution containing 61 g of glucose/100 ml at a pH of 7.5 containing 5×10^{-4} mols of $CoCl_2$/l, 5×10^{-3} mols of Na_2SO_3/l and 5×10^{-3} mols $MgSO_4$/l. The glucose-containing solution having the above identified ingredients was used throughout this example.

This slurry was stirred for 1 hr under reduced pressure (about 23" Hg) to deaerate the same and then transferred to a water jacketed column (diameter of 2.5 cm) which was maintained at 65°C. The bottom of the column was equipped with an Adjusta-Chrom column plunger which was precoated with Dicalite. A glucose-containing solution was pumped through the column to form a bed of the cells. The bed formed had a height of 2.2 cm. Then a fine mesh stainless steel screen was placed on the top of the bed and the screen was covered with a layer of glass beads (2 to 3 cm diameter).

Glass wool soaked in the glucose-containing solution was packed loosely above the glass beads to a depth of 4 to 5 cm. A flow adapter connected to a proportioning pump was placed in the column immediately above the glass wool and glucose-containing solution was pumped downwardly through the bed at a rate of 0.090 ml/min. After 22 hr the conversion of glucose to fructose was 16.1% and after 238 hr the conversion was 10.3%.

Microorganisms of the Arthrobacter sp NRRL B-3726 and Arthrobacter sp NRRL B-3728 were grown under submerged aerobic conditions and the cells harvested. The cells of the former microorganism had a glucose isomerase activity of 124 IGIU per gram and an extractability coefficient of 4.1%. The cells of the latter microorganism had an activity of 93 IGIU/g and an extractability coefficient of 7.2%. Beds of cells of these microorganisms were prepared in the manner described above using 0.611 g of cells of Arthrobacter sp NRRL B-3728 and 0.607 gram of cells of Arthrobacter sp NRRL B-3726.

In the case of the former microorganism, the bed depth was 2 cm and in the case of the latter the bed depth was 2.1 cm. A glucose-containing solution was continuously passed through the bed of cells of Arthrobacter sp NRRL B-3726 at a rate of 0.080 ml/min. After 21 hr the conversion of glucose to fructose was 26.6% and after 361 hr the conversion of glucose to fructose was 18.2%. Glucose-containing solution was pumped through the bed of cells of Arthrobacter sp NRRL B-3728 at a rate of 0.090 ml/min. After 18 hr the conversion of glucose to fructose was 22.4% and after 378 hr the conversion was 11.6%.

Isomerase Fixed on DEAE Cellulose

A fixed isomerase is also used in the process disclosed by *K.N. Thompson, R.A. Johnson and N.E. Lloyd; U.S. Patent 3,788,945; January 29, 1974; assigned to Standard Brands Incorporated* for converting glucose to fructose. The isomerase used is fixed on an inert carrier such as anion exchange cellulose or synthetic anionic exchange resin.

In the overall process, a glucose solution having a viscosity of 0.5 to 100 cp, a pH of 6 to 9 and containing 5 to 80% glucose by weight is heated to a temperature of 20° to 80°C and passed through a bed containing bound glucose isomerase having a glucose isomerase activity of at least 3 IGIU/cc of bed and a stability value of at least 50 hr, thereby converting up to 54% of the glucose to fructose. The flow rate of the glucose solution passing through the bed should be such that the color of the solution exiting the bed is increased by less than 2 color units and there is no substantial production of psicose.

Forming the bound glucose isomerase may be accomplished in any convenient manner so long as the bound glucose isomerase has the characteristics set forth above. To effect binding, the glucose isomerase must be removed from the cells and there must be no interfering substances present during binding.

The binding may be accomplished in an aqueous medium or in a sugar solution, e.g., corn syrup. Also the glucose isomerase may be bound within inert carriers either along with cellular material or in the relatively pure state. Various polymeric materials may be suitable for this purpose but of course the porosity of such materials must be such that allows the glucose to contact the glucose isomerase.

When the isomerase is bound to a finely divided carrier such as DEAE cellulose, it is preferred that the process be performed by passing the glucose-containing solution through a relatively shallow bed of the same, from a fraction of an inch to 5". The cross sectional area of the bed, however, will be large. It is preferred that the ratio of the depth to the width of the bed be from 0.01 to 0.1 and most preferably from 0.02 to 0.05. This provides that the pressure drop across the bed is small and that compaction of the bed will be minimal. However, since the bed is relatively shallow, there is a greater tendency of channeling to occur.

If at least two beds, preferably six beds, of bound glucose isomerase are positioned in series and there is provision made for mixing the effluent from a previous bed before it is passed through a subsequent bed, channeling would not have a serious effect on the efficiency of the process. An apparatus which can be used for this purpose is a pressure leaf filter, an assembly of flat filtering elements (leaves) supported vertically or horizontally in a cylindrical tank.

The glucose isomerase bound to an inert carrier may be slurried in a glucose-containing solution and this slurry pumped through the pressure leaf filter in such a manner as to cover each leaf evenly with the bound enzyme. The pressure applied to the solution will hold the bound glucose isomerase to the leaves. A glucose-containing solution may then be pumped through the pressure leaf filter and while it passes through each bed of bound glucose isomerase, isomerization will occur.

Example 1: This example illustrates the use of glucose isomerase bound on DEAE cellulose to continuously convert glucose to fructose. Streptomyces sp ATCC 21175 was grown under submerged aerobic conditions and then separated from the fermentor broth by filtration. 1 kg of filter cake was slurried in 5 l of deionized water to which was added 50 ml of 0.1 M CoCl, and 8 g of a cationic detergent (Arquad 18-50). The temperature of the slurry was maintained at 60°C, and the pH at 6.7 to 6.8. After 3.75 hr of stirring, the slurry was filtered under vacuum through Whatman #1 filter paper. The filtrate was concentrated by evaporation under vacuum to 30 IGIU/ml.

500 ml of the concentrated filtrate was diluted to 1,500 ml with deionized water. 5 g of DEAE cellulose (Cellex-D) was added to the filtrate, stirred one-half hour and filtered through Whatman #1 filter paper on a vacuum filter. The filter cake was washed with water while still on the filter and the washings were collected with the filtrate. The filter cake contained mostly nonisomerase material. The purified filtrate contained 9.2 IGIU/ml.

30 g of DEAE cellulose was suspended in 1,500 ml of water, stirred vigorously and the suspension allowed to set for 30 to 45 min. The supernatant suspension was decanted to remove the DEAE cellulose fines. This was repeated four times and then the suspension was filtered under vacuum.

20 g (dry basis) of the DEAE cellulose filter cake were added to the purified filtrate containing 9.2 IGIU/ml and stirred for one-half hour at room temperature. The suspension was filtered and a moist filter cake comprising the DEAE cellulose with the isomerase absorbed thereon was recovered. The moist filter cake contained 215 IGIU/g. 6.25 g of the moist filter cake was slurried in 50 ml of an aqueous solution containing 0.001 mol of $CoCl_2/l$, 0.005 mol of $MgSO_3/l$ and 0.1 mol of $NaCl/l$.

This slurry was poured into a column having a diameter of 1 cm and allowed to settle. The bottom portion of the column contained a shallow layer of glass wool to retain the DEAE cellulose. The bottom outlet of the column was opened and when the bed height of the DEAE cellulose was 25 cm, 350 ml of an aqueous solution containing 0.001 mol of $CoCl_2$/l, 0.005 mol of $MgSO_3$/l and 0.1 mol of NaCl/l passed through the bed to remove colored materials and other impurities from the DEAE cellulose.

A glucose solution at a pH of 6.5 containing 3 mols of glucose per liter, 0.001 mol of $CoCl_2$/l and 0.005 mol of $MgSO_3$/l was passed through the column at a rate of 0.2 ml/min. The temperature of the column was maintained at 60°C. The conversion of glucose to fructose after 6 hr was 49.6% and after 186 hr was 45.0%. The stability value was 198 hr.

Example 2: This example illustrates the use of glucose isomerase bound to a porous synthetic anion exchange resin to continuously convert glucose to fructose. 100 g of Amberlite IRA-938 were placed in a column having a diameter of 2.6 cm. The bed height of the Amberlite IRA-938 was 37 cm. The column was maintained at 60°C. The column was treated by passing through, in sequence, 500 ml of 1.5 N NaOH, 1,000 ml of deionized water, 1,000 ml of 2 M HCl, and 1,000 ml of deionized water.

To 1,480 ml of a partially purified solution of glucose isomerase purified according to Example 1 at a temperature of 50°C were added 100 g of Amberlite IRA-938. The solution was stirred for 2 hr and then filtered. The filtrate was percolated through the bed of Amberlite IRA-938 at a rate of 3 ml/min. The bed was then washed with 100 ml of deionized water. The bed cortained 3,430 IGIU.

A glucose-containing solution at a pH of 6.5 containing 3 mols of glucose/liter, 0.001 mol of $CoCl_2$/l and 0.005 mol of $MgSO_3$/l was passed through the bed at a rate of 3.6 ml/min. The temperature of the column was maintained at 60°C. The degree of conversion of glucose to fructose was 21.9% after the column had been operating for 41 hr and was 19.5% at 210 hr. The stability value was estimated at about 600 hr.

Use of Isomerase Sorbed on $MgCO_3$

R.E. Heady and W.A. Jacaway, Jr.; U.S. Patent 3,847,740; November 12, 1974; assigned to CPC International, Inc. have also disclosed an immobilized xylose isomerase preparation having improved retained isomerase activity and flow rate characteristics, as compared to the prior enzyme preparations. It has been found that cell-free, soluble xylose isomerase enzyme preparations, when contacted with basic magnesium carbonate, will become bound or sorbed, to provide an immobilized enzyme preparation which has a high effective isomerase activity.

Other related materials, such as calcium carbonate, do not provide a highly active and stable xylose isomerase. Also, intracellular xylose isomerase does not become sorbed onto basic magnesium carbonate. The immobilized xylose isomerase preparation of this process is characterized as having a very high enzyme efficiency, whereby the sugar contact time necessary to produce a fructose-bearing syrup of at least 45% by weight fructose from a dextrose-containing solution is less than 2 hr. Due to the lower contact time occasioned by the improved retained enzyme activity of the composition, the resulting fructose-bearing syrups have lower color, lower organic acid, and lower psicose content, i.e., generally less than 1% by weight, dry basis, and quite often less than 0.3% by weight, dry basis, psicose. These advantages are quite significant from the standpoint of capital investment and inventory in commercializing a process for enzymatically preparing fructose-bearing syrups from dextrose-containing solutions.

The stabilized xylose isomerase enzyme preparation is preferably prepared by contacting cell-free xylose isomerase with particulate basic magnesium carbonate. The particulate basic magnesium carbonate may be either in the form of a powder or a granular structure. However, the granular structure is preferred from the standpoint of flow properties when used in a deep bed converter.

The stabilized enzyme preparation can be conveniently recovered by conventional means, such as by filtration and the like. Alternatively, the particulate basic magnesium carbonate may be first placed in a converter which is later to be used in the enzymatic isomerization reaction and, secondly, a solution of cell-free xylose isomerase may then be pumped through the column until no more enzyme is sorbed from the solution by the particulate basic magnesium carbonate. Following the preparation of the stabilized enzyme preparation, the converter is ready for use simply by supplying to the converter a dextrose-containing solution.

The xylose isomerase enzyme preparation used is in its soluble form; in other words, the xylose isomerase is freed from the microbial cell where it is formed. The enzyme may be released from its cellular material by any conventional means, and the enzyme in the crude material may be then sorbed on the basic magnesium carbonate. The xylose isomerase becomes bound to the basic magnesium carbonate.

The immobilized enzyme preparation can then be recovered by filtration, centrifugation and the like, and then washed with a buffer. The filtered and washed immobilized enzyme composition can be used in its wet form, as is, or the cake can be dried by conventional drying techniques useful with enzyme preparations, such as by cell, rotary drum or spray drying or freeze drying techniques. Superior results are obtained, however, when the enzyme is purified before it is sorbed.

The immobilized xylose isomerase preparation of this process is characterized as having at least about 100 units of effective isomerase activity per gram of immobilized enzyme preparation, dry basis (i.e., 100 u/g effective isomerase activity), preferably at least about 175 u/g. Another striking characteristic of the immobilized and stabilized enzyme preparation is its long half-life. This enzyme preparation is characterized as being capable of producing 45% fructose, DB, at flow rates of greater than 0.5 bed volume per hour with an enzyme half-life in excess of 15 days.

The xylose isomerase enzyme is preferably derived from a microorganism belonging to the genus Streptomyces, and more preferably, a microorganism strain that is selected from the group of mutant strains consisting of *S. olivochromogenes,* ATCC No. 21713, *S. olivochromogenes,* ATCC No. 21714, and *S. olivochromogenes,* ATCC No. 21715.

Example: Continuous Enzymatic Conversion of Dextrose to Fructose Using Immobilized Xylose Isomerase — A granular basic magnesium carbonate material having a particle size of –12 to +20 (55.8 g) was placed in a column having a volume of 73.3 ml. A substantially pure, soluble, cell-free xylose isomerase enzyme derived from *S. olivochromogenes* ATCC No. 21713 in a 0.01 M $MgSO_4$ solution, having a 50 u/ml activity, was contacted with granular basic magnesium carbonate to load the column with 5 million u/ft^3 of enzyme. The immobilized xylose isomerase enzyme had more than 100 u/g of effective activity per gram of immobilized enzyme. An aqueous dextrose feed liquor was then supplied to the column at 27° Bé (50% DS). The feed liquor contained $MgCl_2$ (0.005 M with respect to $MgCl_2$). The pH of the feed was adjusted to the range from 8.4 to 8.8 with 4 N NaOH at 58°C.

The column containing the immobilized enzyme preparation was maintained at a temperature of 58°C. The residence time in the column was less than 1 hr. The initial bed volume per hour for the production of 45% fructose, dry basis (BVH_{45} fructose) was about one. The effluent syrup contained about 45% by weight fructose, dry basis, and less than about 0.3% by weight, dry basis, psicose. The following is an analysis of the pH of the feed syrup and effluent syrup after a prolonged continuous conversion campaign.

Hours of Operation	Feed Syrup pH at 25°C	Effluent pH at 25°C
261	8.60	8.60

The above experiment was repeated, except that the xylose isomerase was bound on an alumina carrier. The same dextrose feed liquor, preadjusted to a pH of 8.3 to 8.4 with 4 N NaOH, was pumped through the column, which was maintained at a temperature of 58°C at a rate to provide 45% by weight fructose, dry basis, in the effluent. The residence time was less than 1 hr. The following is an analysis of the pH of the feed syrup and effluent syrup after a prolonged continuous conversion campaign.

Hours of Operation	Feed Syrup pH at 25°C	Effluent pH at 25°C
16	8.3	8.4
376	8.15	8.25
428	8.4	8.3

In the operation of a column where the residence time may exceed about 4 hr, it is particularly preferred to feed an alkaline material such as sodium hydroxide or $Mg(OH)_2$ solutions or a combination of the two into the column at different elevations. A suitable means for accomplishing this is to provide alkaline inlets at one or more places along the column, so that the pH of the reaction solution is maintained at a pH of at least about 7.5.

In another example of the process, the pH of the enzymatic isomerase solution can be programmed by the use of one or more alkaline inlets along the column. In such a technique, the initial pH of the feed liquor can be 7.5 or greater, preferably from 7.5 to 8.2. By introducing alkaline materials such as sodium hydroxide or $Mg(OH)_2$ or a combination of the two into the reaction solution via inlets along the column, the pH of the reaction solution can be conveniently increased to 9.5. Preferably, the pH will be increased by at least 0.1 from supply liquor to effluent, although more beneficial results in terms of stability of the basic magnesium carrier and quality of the isomerase are realized when the pH is increased by at least 0.2.

Two Temperature Process Using Sorbed Enzymes

R.E. Heady and W.A. Jacaway, Jr.; U.S. Patent 3,847,741; November 12, 1974; assigned to CPC International Inc. have also disclosed an improved conversion of dextrose to fructose using two temperature stages, preferably using a fixed enzyme. Generally, the process involves conducting the isomerization at an initial, fairly constant temperature of at least 50°C, followed by increasing the operating temperature, either in several increments or in a single step, to a value that is 5°C or more higher than the initial operating temperature, to an operating temperature not exceeding 80°C.

In one preferred mode, the process involves subjecting a stream of a solution containing dextrose to the action of an immobilized enzyme preparation at a pH of at least 7.0 and preferably in the range from pH 7.5 to 8.5, and at a temperature level of at least 50°C but that is at least 10°C below the temperature at which rapid inactivation of the enzyme occurs, during an initial isomerization phase.

For enzyme preparations derived from microorganisms of the Streptomyces genus, ordinarily the temperature above which rapid inactivation of the enzyme preparation can be observed is about 70°C, so that the preferred operating temperature during the initial phase of isomerization is 60°C or lower, that is, preferably 50° to 60°C. At temperatures in the range from 50° to 60°C, some enzyme inactivation occurs, but the rate of inactivation is low relative to the rate of inactivation at 70°C.

After some material loss of enzyme activity is observed after operating in the range from 50° to 60°C or so, the temperature is raised at least about 5°C, either gradually, in increments of 2° to 3°C or less, as needed to maintain the desired ketose level in the product, or alternatively, the temperature may be raised at least 5°C in a single step.

Preferably, however, the increase in temperature for the second phase of operation is at least about 10°C, preferably through gradual or small incremental increases in operating temperature. Among the advantages of this process are better enzyme stability, more desirable carbohydrate content of the isomerized product, improved efficiency as compared to batch conversions, and operating economy.

Example: Continuous Isomerization Using a Fixed Bed of Microbial Cells — Single Step Temperature Adjustment: For this example a jacketed column was used. The jacket was connected to a source of hot water, for controlling column temperature during isomerization. A strain of a microorganism of the Streptomyces genus was grown under submerged, aerobic conditions on a medium containing xylose, to produce intracellular isomerase.

After fermentation, magnesium hydroxide was added to the fermenter broth in the ratio of 2 parts by weight of magnesium hydroxide for each 1 part by weight of the cell mass in the broth. The slurry thus obtained was filtered, and the filter cake was then dried in an open pan at room temperature. The activity of the dry enzyme preparation obtained was 330 units per gram. The dry enzyme preparation was dispersed in a 50% w/v solution of dextrose.

The slurry was then placed in the jacketed column, and as the slurry was added to the column, small glass beads, approximately 3 mm in diameter, were added simultaneously. The glass beads served as a support and also prevented the enzyme preparation from packing and thus plugging up the column. In this manner, approximately 750 units of the enzyme were charged to the column. A dextrose syrup at 50% w/v concentration was adjusted to a pH of 7.0 to 7.5 by adding magnesium hydroxide. The syrup was sparged with nitrogen, and was then fed to the top of the column under a nitrogen atmosphere.

The isomerized product was collected in aliquots of 15 ml in test tubes. The test tubes each contained 5 ml of 0.5 N perchloric acid to inactivate any soluble isomerase that might be present in the product. The temperature of the column was maintained at 60°C during an initial phase of operation. The flow rate of dextrose solution through the column was maintained at a substantially uniform rate, and the ketose content of the effluent was 40 to 50% on a dry solids basis.

The isomerization was conducted in this manner for 9 days before a substantial decrease in enzyme activity became apparent as evidenced by a dropping off in the ketose value observed in the effluent. During that initial phase of operation, the average ketose content of the effluent was 38.7%. At the end of this initial phase of operation, the temperature of the column was increased (from the initial level of 60°C) to 70°C in a single step. Isomerization was then continued, at the increased temperature, for an additional period of 24 hr. The ketose content of the product averaged out at about 49%, for the second phase of operation. The results are summarized in the following Table 1.

TABLE 1: CONTINUOUS ISOMERIZATION: SINGLE STEP TEMPERATURE ADJUSTMENT

Time Days	Temperature °C	Throughput Bed Volumes per Hour	Output of 42% DB Fructose Product (lb DS/ft³ of bed)	
			Expected at 60°C	Obtained
1	60	1.27	951	951
2	60	1.27	951	951
3	60	1.27	951	951
4	60	1.27	951	951
5	60	1.25	936	936
6	60	1.17	876	876
7	60	1.10	823	823
8	60	1.03	771	771
9	70	1.65	734	1,235
10	70	1.65	681	1,235

When the isomerization was repeated with an enzyme preparation prepared from a mixture of diatomaceous earth with the microbial cells, closely comparable results were obtained.

When a similar reaction was done using small step temperature adjustment, the results obtained are shown in Table 2 on the following page.

TABLE 2: FIXED BED ISOMERIZATION WITH SMALL STEP TEMPERATURE INCREASES

Days	Temperature °C	Output of 42% d.b. Fructose Product (lbs. d.s. per cu. ft. of bed)	
		Expected at 60°C	Obtained
1	60	787	787
2	60	787	787
3	60	787	787
4	60	787	787
5	60	622	622
6	60	516	516
7	62	418	590
8	62	343	516
9	62	286	516
10	64	233	509
11	64	187	509
12	66	158	473
13	68	127	674
14	70	106	622
	TOTAL	6144	8695

CHROMATOGRAPHIC SEPARATION OF FRUCTOSE

Various prior methods claimed to be useful for the individual separation of glucose and fructose from mixtures containing such sugars. Examples of these methods are: (1) method for separating fructose from glucose by converting fructose into calcium-fructose complex by treatment with calcium hydroxide; (2) methods for effecting the desired separation by using a calcium form cation exchange resin bed (U.S. Patent 3,044,904); (3) method involving the use of a strontium form cation exchange bed (U.S. Patent 3,044,905); (4) method involving the use of an Ag form cation exchange resin bed (U.S. Patent 3,044,906); (5) method involving the use of a borate form anion exchange resin bed (U.S. Patent 2,818,851; (6) method involving the use of a bisulfite form anion exchange resin bed (O. Samuelson, *Ion Exchangers in Analytical Chemistry,* pages 198 to 199, 1953).

Continuous Ion Exchange Separation

Fructose and glucose are separated using a crosslinked sulfonated polystyrene cation exchange resin in the process developed by *C.B. Mountfort, B. Cortis-Jones and R.T. Wickham; U.S. Patent 3,416,691; December 17, 1968; assigned to The Colonial Sugar Refining Company Limited, Australia.*

A syrup containing fructose and glucose is admitted to a column charged initially with a water-immersed bed of an alkaline earth metal salt of a suitably crosslinked nuclearly sulfonated polystyrene cation exchange resin, whereby the fructose is preferentially absorbed by the resin and is subsequently eluted after displacing the surrounding glucose enriched solution. The eluate from the column is divided into at least six fractions, at least two of which are then recycled to the column in a specified sequence with respect to additional syrup feed and water feed. This process comprises the following steps:

(1) sequentially admitting predetermined volumes of the syrup and water to a column charged with a water-immersed bed of the alkaline earth metal salt of a crosslinked nuclearly sulfonated polystyrene cation exchange resin;

(2)   separating the effluent from the column sequentially into at least
      the fractions:
      sweet water (I) consisting of dilute glucose-rich solution,
      concentrated glucose-rich solution,
      recycle (I) consisting of concentrated glucose-rich solution
          highly contaminated with fructose,
      recycle (II) consisting of concentrated fructose-rich solution
          highly contaminated with glucose,
      concentrated fructose-rich solution,
      dilute fructose-rich solution;

(3)   admitting sequentially to the column at least the fractions:
      recycle (I),
      predetermined volume of the syrup,
      recycle (II),
      predetermined volume of water;

(4)   repeating steps (2) and (3) in a cyclic manner.

The process is carried out in an apparatus which can be illustrated in Figure 11.5.
In the figure, **1** is a vertical column adapted to be charged with a cation exchange
resin of the type described, and fitted with a top distributor for admitting solutions uniformly to the upper surface of the resin and with a bottom distributor
for collecting effluent solutions uniformly from the bottom of the resin bed.

FIGURE 11.5: SEPARATION OF FRUCTOSE AND GLUCOSE

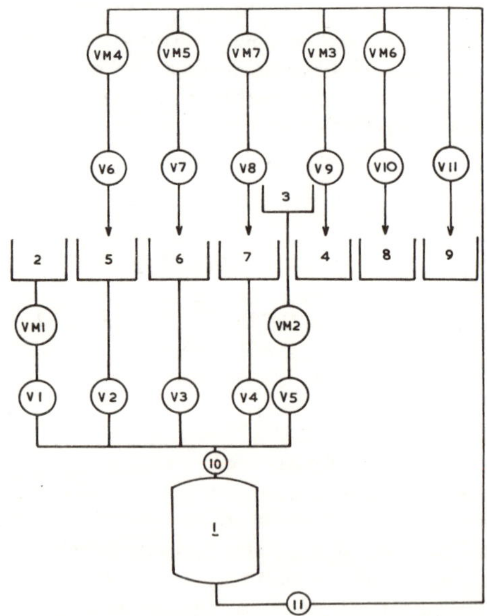

Source: C.B. Mountfort, B. Cortis-Jones and R.T. Wickham; U.S. Patent
          3,416,961; December 17, 1968

2 is a storage tank for syrup feed solution (e.g., invert syrup), and 3 is a storage tank for feeding water while 4 is a storage tank for receiving concentrated glucose-rich solution. A storage tank for recycle (I) solution is shown at 5, for recycle (II) at 6, and for recycle (III) at 7. A storage tank for concentrated fructose-rich solution is shown at 8. 9 is a storage tank for sweet waters (I) consisting of dilute glucose-rich solution and (II) consisting of very dilute fructose-rich solution.

A feed pump for transferring syrup, water, recycle (I), recycle (II) and recycle (III) solutions from their respective tanks to the top distributor in the column is shown at 10. An effluent pump 11 transfers the effluent from the bottom distributor in the column to the concentrated glucose-rich solution tank, the recycle (I) tank, the recycle (II) tank, the concentrated fructose-rich solution tank, the recycle (III) tank and the sweet waters (I) and (II) tank.

Valves V1 through V11 and volume control means VM1 through VM7 are for directing the flow of solutions to and from the column and the various tanks in accordance with the desired method of operation. In addition to the above, there are also provided (but not shown on Figure 11.5): a heating means for keeping the contents of the various feed tanks at approximately 60°C and means for insulating the resin column to minimize heat losses; a means to allow the resin bed to be backwashed with water periodically to remove accumulated dirt, etc.; a feed control system; and an effluent control system.

Suitably, the resin bed depth in the column is 6', but depths greater and smaller than this can be used with satisfactory results. The resin undergoes a volume change during the operating cycle (shrinking with increasing sugar concentration within the resin) and the resin surface rises and falls. At the start of the operation the column is half full of a water-immersed resin bed. The water level is lowered to the upper surface of the resin and a predetermined volume of syrup (50% by weight soluble solids) is fed to the top of the column.

The effluent consists first of water only, but is gradually enriched in glucose. This effluent is directed to the sweet water tank 9 until a density measuring instrument or other suitable device detects the presence of a predetermined concentration of glucose, whereupon the effluent (containing 60 to 90% by weight glucose based on dry solids) is directed to the concentrated glucose-rich solution tank 4.

After a predetermined volume of this concentrated glucose-rich fraction is collected in tank 4, the effluent is switched to recycle (I) storage tank 5 where there is collected a predetermined volume of a solution containing more than 50% by weight glucose based on dry solids. Then the effluent is directed to recycle (II) storage tank 6 where there is collected a predetermined volume of a solution containing more than 50% by weight fructose based on dry solids.

The effluent is then switched to the concentrated fructose-rich solution tank 8 where there is collected a predetermined volume of solution (containing 60 to 90% by weight fructose based on dry solids). Volumes of effluent collected in recycle (I) and recycle (II) storage tanks, identified above as predetermined, will be defined subsequently. In a preferred cycle of operations, the next step is to direct the effluent flow to storage tank 7 to collect recycle (III) solution. This solution is actually richer in percentage fructose (based on dry solids) than the

preceding fructose rich-fraction, but it is much more dilute. For reasons of
economy in evaporation, it is preferred therefore to return it to the column as
a recycle fraction rather than to allow it to mix with the previous fructose-rich
fraction. Finally, the effluent is directed to the sweet water tank **9**, and the effluent cycle is repeated. The process may, of course, be operated successfully
without separating sweet water from recycle (III) solution, however, by collecting
recycle (III) as a separate fraction, the concentration of total solids in the fructose-rich effluent fraction is increased and the economy of the process is improved.

It is preferred not to submit sweet water fractions per se to recycling, however,
this very dilute solution may be disposed of advantageously by using it to dilute
invert syrup feed solution to the required concentration. At the start of the
feeding cycle, when the required amount of invert syrup has been applied to the
top of the resin column, this is followed by a predetermined volume of water.
After the water feed, the contents of recycle (I) storage tank are fed to the column followed by a further predetermined volume of invert syrup.

After this stage of the feeding cycle (and after all subsequent equivalent stages),
invert syrup feed is not followed by water but by the contents of the recycle (II)
storage tank then the contents of the recycle (III) storage tank (if the effluent
has been divided into such a fraction). Only then is a water feed introduced,
followed by the contents of the recycle (I) tank. After a number of successive
cycles an equilibrium is reached at which the following material balances apply.

Mass Balance (solids basis):

Invert syrup = concentrated glucose-rich fraction + concentrated fructose-rich fraction + sweet water

Volume Balance:

Invert syrup + water = concentrated glucose-rich fraction + concentrated fructose-rich fraction + sweet water

In Figure 11.6, the graph relates effluent volumes at equilibrium to the concentrations of fructose and glucose. The relative slope and disposition of the fruc-
tose/glucose curves are constant under these conditions for a given resin and a
given flow rate of solution. An effluent flow rate of 40 imperial gallons per hour
per square foot of resin (cross-sectional area) has been found satisfactory, though
rates greater or less than this may be used. By suitable adjustment of the feed
water volume, successive effluent cycles may be made to follow each other closely
with a small overlap between them as shown in the figure.

Example: Apparatus as shown in Figure 11.5 was used to obtain a fructose enriched syrup from a syrup having a 50% by weight soluble solids total solids concentration (comprising 50% fructose and 50% glucose). The resin used in the
column was Dowex 50W having a crosslinkage content of 4% divinylbenzene,
and a particle size in the mesh range 35 to 70 (U.S. Standard Sieve).

At the start of the process the resin bed height was 6 ft. Heating and insulation
means were used to maintain a temperature of 60°C in the feed solutions and
the column. The effluent flow rate was 40 imperial gal/hr/ft^2 of resin (cross
sectional area). At equilibrium, fractional feed and effluent volumes (expressed
as fractions of the resin bed volume) were as shown in the following tables.

FIGURE 11.6: EFFLUENT VOLUME vs CONCENTRATION OF GLUCOSE AND FRUCTOSE

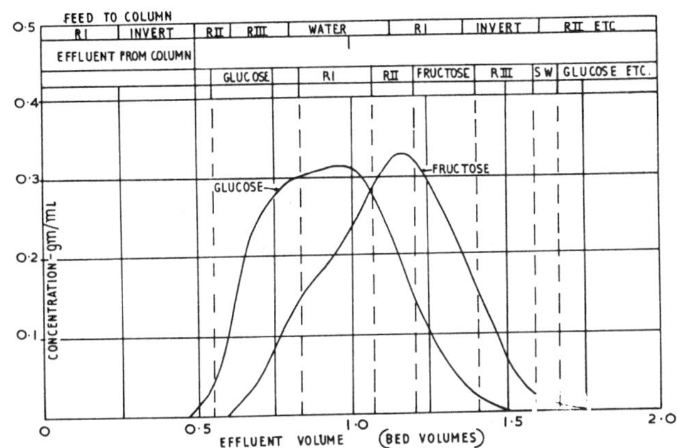

Source: C.B. Mountfort, B. Cortis-Jones and R.T. Wickham; U.S. Patent 3,416,961; December 17, 1968

Feed Volume to Column

Recycle (I)	0.241
Invert syrup	0.241
Recycle (II)	0.130
Recycle (III)	0.185
Water	0.332
	1.129

Effluent Volume from Column

Concentrated glucose-rich	0.298
Recycle (I)	0.241
Recycle (II)	0.130
Concentrated fructose-rich	0.201
Recycle (III)	0.185
Sweet water	0.074
	1.129

The effluent analysis was as follows: the concentrated glucose-rich fraction had a total solids concentration of 24% by weight, of which 78% was glucose and 22% was fructose; the concentrated fructose-rich fraction had a total solids concentration of 29% by weight, of which 82% was fructose and 18% was glucose; and the sweet water had a total solids concentration of 1% by weight, substantially all of which was fructose. This effluent analysis corresponds to that given in Figure 11.6.

Use of Hydrazine Treated Cation Exchange Resins

The process disclosed by *H. Quietensky and E. Nitsch; U.S. Patent 3,471,329; October 7, 1969; assigned to Laevosan-Gesellschaft Chem. Pharm. Industrie Frank & Dr. Freudl, Austria* provides an efficient process for the separation of the individual sugars in a mixture by reacting a cation exchange resin with hydrazine. The hydrazine compound becomes bonded to the active acidic group of the cation exchange resin. An aqueous or an aqueous-alcoholic solution of the mixture of sugars is then contacted with the hydrazine loaded exchange resin. The sugars are removed from the solution and become bonded to the hydrazine. The individual sugars are then obtained by washing the sugar-containing cation exchange resin to fractionate the sugars and obtain the individual sugars in the separate wash fractions.

The cation exchange resins useful in this process are those which are substantially insoluble in the water or water-alcohol solutions used. They are polymeric materials and contain active acidic groups. To obtain the maximum surface area and the maximum number of active groups to efficiently utilize the process, it is preferred that the exchange resin should be porous and be in the form of relatively small particles. Suitable resins that meet these necessary criteria are the widely available and economic polystyrene resins, and phenolic (generally phenol-formaldehyde) resins.

The preferred cation exchange resins are those which contain highly acid active groups. These include sulfate groups, carboxy groups, and phosphite groups. Hydrazine will react with the active group on the resin, and is still able to react with sugars to form compounds corresponding to the sugar hydrazones. Hydrazine and equivalent functioning hydrazine derivatives are useful in the process.

It has been found that the effectiveness of the separation is improved and the volume of wash water reduced, if the resin is only partially loaded with hydrazine ions, and also contains H^+ ions. Preferably, the hydrogen ions are in approximately equal molar amounts with the hydrazine. The preferred procedure for preparing this resin loaded with hydrazine and H^+ ions is to treat the cation exchange resin with a strong acid such as hydrochloric acid, to convert all of the active group to the acidic form. The resin is then thoroughly washed and all the hydrochloric acid removed. Hydrazine is then reacted with the acidic resin; the amount of hydrazine used being only about half the molar amount necessary to react with all the active acidic groups.

This process is useful for separating individual sugars commonly found in sugar mixtures. It is most usefully applied to the very common mixture of glucose and fructose known as invert sugar. The examples furnish additional illustrative sugar mixtures which are separated into their individual component sugars.

The process is carried out in a column packed with the resin. The aqueous solution of the sugars, or the aqueous-alcoholic solution of the sugars, is contacted with sufficient of the loaded cation exchange resin to remove all the sugar from the solution. The column is then washed. The first washing usually produces a forerun which contains substantially no sugar. Individual relatively small fractions are then taken from the subsequent washings, as described in more detail below. The different sugars in the original mixture are heavily concentrated in different fractions.

The fractions may be concentrated by distilling off the water. The sugar may be crystallized from the washing liquid or from a suitable solvent. The mother liquors may be recycled to the feed, resulting in a substantially 100% yield. The process may be carried out in a cyclic manner.

The temperature used during the process is of importance since the reaction rates and the position of the hydrolytic equilibrium are dependent upon it. For the separation of the components of invert sugar using hydrazine, the process is preferably carried out between 40° and 80°C, with 65°C being the optimum. As is apparent, the specific temperature utilized during the separation will be dependent upon the specific mixture of sugars and the components.

Example 1: Separation of Glucose and Fructose from Invert Sugar — Into a glass tube having a 33 mm bore and 2 m length, are placed 1.6 liters of cation exchange resin Duolite C 27. A solution of hydrochloric acid is passed through the column. Then a bimolar solution of hydrazine is passed through until an alkaline effluent appears. The column is then washed neutral with water. The temperature of the column is maintained at 65°C by a heating jacket with pumped circulating water controlled by a thermostat.

The inlet to the column is heated in the same manner. By means of a metering pump there is fed into the head of the column, 460 ml invert sugar solution 50% w/v, corresponding to 115 g of glucose, and of fructose, at a rate of 15.5 milliliters per minute. It is then washed with distilled water at the same rate. A sugar-free first run of 720 ml appears and is rejected. The following fractions each of 160 ml (equal to $\frac{1}{10}$ of the column volume) are collected separately and the sugars collected were determined by optical rotation and refraction.

Fraction No.	Total amount of sugar, percent	Fructose, percent	Glucose percent
1	7.9	7.0	0.9
2	24.1	23.5	0.6
3	29.0	25.0	4.0
4	17.7	12.25	5.5
5	11.2	3.2	8.0
6	9.3	0.7	8.5
7	8.1	0.1	8.0
8	7.0	0.05	7.0
9	5.9		5.9
10	5.8		5.8
11	3.9		3.9
12	3.2		3.2
13	2.8		2.8
14	1.8		1.8
15	1.1		1.1

The fractions, 1 to 4, which contain the greatest quantity of fructose are combined, concentrated and crystallized from methanol in a conventional manner. Another crystallization is obtained from the mother liquor of the first crystallization. Both crystallisates, after washing with methanol, are pure white and consist of pure fructose. Yield is 65 g. The remaining mother liquor is only weakly colored and can be returned to the process without any further treatment. Likewise, from fractions 5 to 15, 53 g of pure glucose is obtained. The glucose mother liquor also can be returned to the process.

Cyclic Column Chromatographic Separation

An apparatus has been devised by *K. Lauer, H.-G. Budka and G. Stoeck; U.S.*

Patents 3,785,864; January 15, 1974; and 3,686,117; August 22, 1972; both assigned to Boehringer Mannheim GmbH, Germany which permits pure fractions to be obtained from multicomponent mixtures on a large scale by means of cyclic column chromatography. In prior use of this separation method, great difficulties arose when, in addition to the separation of the main components, by-products are also to be separated which result in the cycles being so widely spaced apart that the economy of the separation process is rendered questionable.

In this process the time needed for one cycle can be considerably shortened and the abovementioned disadvantages in prior processes avoided when the separation column is divided into two sections, the contaminated fraction is removed from the column at the end of the first section, simultaneously an equivalent amount of elution agent is passed into the second section of the column and, after removal of the contaminated fraction, the two column sections are connected together until, at the end of the first column section, the contaminated fraction of the next cycle emerges.

In this manner, the complete length of the column can be utilized for the cyclic separation of multicomponent mixtures. Furthermore, the contaminants have no further influence upon the length of the cycle, i.e., the separation of the multicomponent mixture takes place just as economically and quickly as in the case of the separation of a mixture which contains only the components of interest. The process has been investigated with epimerized and nonepimerized starch syrups and has provided to be of extraordinary technical advantage. For the separation of such syrups into their components, it is preferred to use an ion exchanger loaded with calcium ions.

This process can be explained in more detail with reference to Figure 11.7 which is a schematic representation of an apparatus for carrying out the process. From a supply tank S, a solution of a multicomponent mixture is pumped by pump **1** through a pipe **2** into a first column section **3**. After a predetermined period of time, the pump is switched off and, simultaneously, from an elution agent reservoir **W**, elution agent is pumped by pump **4** through pipe **5** into the first column section, the supply capacity of pumps **1** and **4** being the same.

After termination of the elution phase, pump **4** is switched off and pump **1** is again switched on. Pumps **1** and **4** are operated by a switch clock **6** after empirically determined and rigidly maintained time intervals. The liquid leaving the first column section passes through an analyzer **7** which, depending upon the concentration of the individual fractions, operates valve **8** via a control device **12**. When the fraction which only contains impurities reaches the analyzer **7**, valve **8** is switched in such a manner that liquid leaving the column section **3** is passed into container **D**.

Simultaneously, pump **9** is switched on, which pumps elution agent from elution agent container **W** into column section **10**, the supply capacity of pump **9** corresponding exactly to those of pumps **1** and **4**. Above a definite concentration in the liquid passing through the analyzer **7**, i.e., when the impurities are removed and the concentration of the first main fraction has reached a definite value, valve **8** is switched in such a manner that the analyzer **7** is in direct connection with pipe **11**. In this position, pump **9** is switched off so that there is no connection between tank **D** and tank **W**. By means of a logic element incorporated into the analyzer, this valve position remains unchanged until the concentration in the liquid flowing through the analyzer **7** increases from 0 to a low limiting value. This is the case when, after an intermediate flow of pure elution agent, impurities of the next cycle again appear in the analyzer. By means of this process only the main fractions, which are to be further separated, pass into the second and longer column section **10**.

The further separation is carried out in the following manner. An analyzer **13** operates valve **15**, via a control device **14**, in such a manner that the first fraction is collected in storage tank **G** and the second fraction in storage tank **F**. Intermediate fractions, which contain both components, are passed into tank **G/F**. The separation plant contains three control units **I**, **II** and **III** which are encircled by broken lines. Control unit **I**, containing the switch clock **6** and pumps **1** and **4** operates completely independently and, in the case of the operation of the separation plant, is empirically adjusted in such a manner that the fractions enter control unit **III** without interruption and with the smallest possible overlapping, one after the other.

Control unit **II**, containing the analyzer **7** and control device **12**, as well as valve **8** and pump **9**, serves for the removal of the fractions containing the impurities. The analyzer consists of a simple concentration measuring device, for example, a flow–through refractometer, the measurements results of which are passed to control device **12** as a proportional voltage. This control device possesses a logic element which consists, for example, of a sequence relay and enables valve **8** and pump **9** to operate at a definite concentration only when, at the point of time of operation, the concentration in the liquid flowing through the analyzer **7** is on the increase.

FIGURE 11.7: CYCLIC COLUMN CHROMATOGRAPHIC SEPARATION

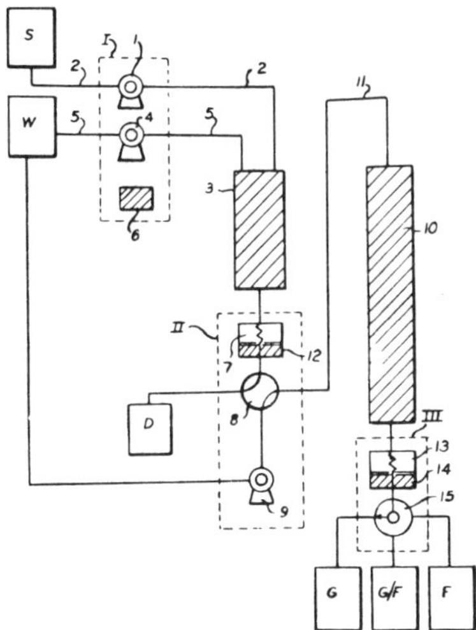

Source: K. Lauer, H.-G. Budka and G. Stoeck; U.S. Patent 3,785,864; Jan. 15, 1974

The control device also starts upon the operation of the water pump **9** during the take-off of the fractions containing the impurities. The known control device III for the separation of fructose and glucose (see British Patent 1,095,210) consists of an analyzer **13** with control device **14** and valve **15**. It serves to conduct the main fractions into tanks **G**, **G/F** and **F**. The analyzer **13** consists of measuring devices for the angle of rotation and for the refractive index. The measuring devices are equipped with flow-through cuvettes and provide the control device **14** with measured values in the form of proportional voltages.

This control device contains an analog calculator which calculates the partial concentrations of glucose and fructose, and, upon exceeding the permissible limiting concentrations, operates the valve **15** via a switch relay. For control of the separation process, a multicolor recorder can be attached to the analog calculator of the control device **14**, this recorder recording the concentrations of glucose and fructose as a continuous elution diagram.

Example: Large Scale Production of Pure Fructose from Epimerized Starch Syrup — The column measurements are: column section **3**, 10 cm diameter and 4.5 m length; and column section **10**, 10 cm diameter and 9.0 m length. At a rate of flow of 6 l/hr, 2 l of an approximately 15% by weight epimerized starch syrup solution were applied to column section **3**. After 20 min, pump **1** was switched off and, via pump **4**, water was pumped for 2.6 hr at a rate of flow of 6 l/hr, from tank **W** into column section **3**.

Pumps **1** and **4** now alternately supply the column section **3** with a constant rhythm. After about 90 min, the analyzer **12** indicated the first changes in the refractive index and valve **8** was switched in such a manner that the eluate flowed off into tank **D**. The column section **10** was now supplied with water by pump **9**. After a further 90 min, the impurities were removed from the column section **3**.

At a concentration of about 190 g glucose/liter, valve **8**, was switched over the column section **3** connected with column section **10**. The eluate was now further separated in the column section **10** in the usual manner and fed into tanks **G**, **G/F** and **F**. The fructose solution obtained had a specific rotation of about –92°.

Use of Bisulfite Anion Exchange Resins

Y. Takasaki; U.S. Patent 3,806,363; April 23, 1974; assigned to Agency of Industrial Science and Technology, Japan has found that fructose of high purity can be separated effectively and economically from a sugar solution containing glucose and fructose or from an inexpensive raw material containing contaminating substances in addition to glucose and fructose by using a bisulfite form anion exchange resin bed maintained in a fixed range of temperatures.

The separation of sugar by the use of a bisulfite form anion exchange resin is based on the theory that the carbonyl group of sugar forms an addition compound upon union with the bisulfite group present in the bisulfite form anion exchange resin. Fructose and glucose, however, are sugars which do not readily form such addition compounds in conjunction with the bisulfite group. The quantity of these sugars capable of being treated for separation by a unit volume of bisulfite type anion exchange resin has been extremely small i.e., less than

100 mg of sugar (glucose plus fructose) per 100 ml of the ion exchange resin.
Quantity production of fructose by this method, therefore, has required use of
extremely large facilities for the size of yield. According to this process, the
quantity of sugar which can be treated per unit volume of anion exchange
resin can be increased greatly. To be specific, the quantity of sugar (glucose
plus fructose) which can be treated per 100 ml of the ion exchange resin is on
the order of 5 to 20 g. Since fructose and glucose are eluted both in a high
concentration, the expense incurred for the concentration can be amply lowered.

An anion exchange resin having a sulfite group or bisulfite group to be used
for this process may be prepared from any of various kinds of anion exchange
resins, for example, Dowex 1-X4 and Dowex 1-X8, and preferably Amberlite
IRA-400. The bisulfite form anion exchange resin is prepared by treating a
selected anion exchange resin with sodium hydroxide thereby converting the
resin into the OH form, allowing sodium bisulfite or potassium bisulfite to
to react upon the resultant OH form anion exchange resin for necessary substitution of radicals, and finally washing the substitution product with water.

The bisulfite form anion exchange resin thus produced is put to use. The
sulfite form anion exchange resin is prepared by following the procedure mentioned above, except that sodium sulfite or potassium sulfite is used in place
of the bisulfite. When sodium or potassium bisulfite of an impure grade is
used for the substitution of the anion exchange resin, there is simultaneously
produced a small proportion of sulfite form anion exchange resin besides the
bisulfite form anion exchange resin.

The separation of fructose and glucose by means of the bisulfite (and/or sulfite) form anion exchange resin can be accomplished more advantageously by
using two or more anion exchange resin beds arranged in series connection.
The efficiency of separation can be increased when the two or more anion exchange beds are arranged in series connection in the decreasing order of bed
width relative to the direction of the flow of treatment.

From experimental data it was determined that the temperature of the separation step is a critical factor. At temperatures below 40°C, little separation
occurred. At temperatures of 40° to 60°C, separation of fructose and glucose
proceeded quite smoothly and glucose flowed out of the column at a high concentration. This means that the effluence of not only fructose but also glucose
from the column interior can be accomplished in a short period of time and
by using a smaller volume of water.

Both fructose and glucose can be recovered completely from the column without leaving any portion thereof behind. Once the column has been adapted
for the treatment, it can be used repeatedly without being regenerated and
desired continuous separation of fructose can be carried out on a commercial
scale.

Example: This example describes a case in which fructose was separated from
the fructose-containing glucose isomerized syrup derived from starch as the raw
material. Glucose isomerase was added to react upon the hydrolysate of starch
of DE 97.5 which had been obtained by dextrinizing starch with α-amylase
and saccharifying the liquefied starch with glucoamylase. The isomerization
reaction was carried out at 70°C, pH 6.5 to 7.0.

The obtained fructose-containing glucose isomerized syrup (containing 35.4%
of fructose, 36.9% of glucose, and about 5% of oligosaccharides) consequently
obtained was decolored and refined with active carbon and anion exchange resin
and then used as the raw material for separation of fructose. To a column packed
(1.5 x 60 cm) with sulfite form Dowex 1-X8 and kept at about 40°C, there was
fed 5 ml of the isomerized syrup. The effluence from the column was effected
by using water. The flow rate was 10 ml/hr and the fraction volume was 3 ml.

The fractions were assayed for fructose, glucose, and oligosaccharides. The results are shown in Figure 11.8. The oligosaccharides content was expressed by
the amount of the reducing sugar (expressed in terms of glucose) which was obtained by decomposing the oligosaccharides with 2.5% hydrochloric acid for one
hour at 100°C. As is clearly shown in the graph, the effluence from the column
occurred first on the oligosaccharides (fraction numbers 12 through 20) and then
on fructose and glucose.

By collecting the fractions, numbers 21 through 39, there could be obtained
sugar solution containing glucose (2.39 g) and fructose (2.30 g) and no oligosaccharides. Then, this mixed sugar solution containing glucose and fructose was
concentrated. The concentrated solution was fed to a column packed (2 x 30 cm)
with bisulfite form Dowex 1-X8 maintained at about 40°C and then water was
supplied. Consequently, there were contained 2.23 g of pure fructose and a
mixed sugar solution containing 2.29 g of glucose and 0.06 g of fructose.

FIGURE 11.8: USE OF BISULFITE ANION EXCHANGE RESINS

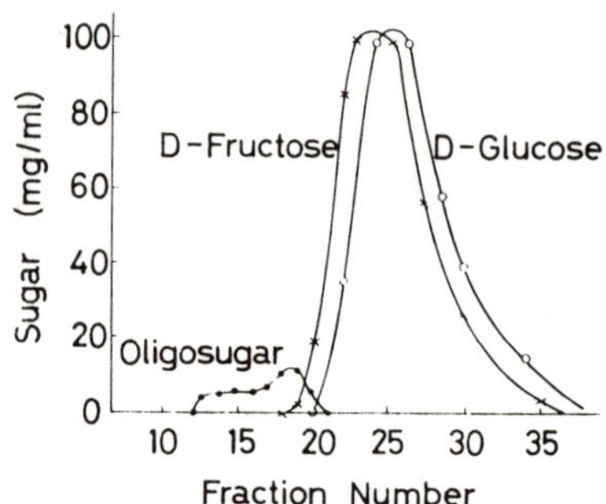

Source: Y. Takasaki; U.S. Patent 3,806,363; April 23, 1974

DRYING FRUCTOSE SYRUPS

Fructose-Starch Films

A process has been disclosed by *E.F. Glabe; U.S. Patent 3,718,484; February 27,
1973; assigned to Food Technology, Inc.* for preparing a dry, flowable powder,
free of gummy and lumpy characteristics from a high fructose syrup. This is
accomplished by intimately mixing the syrup with an ungelatinized starch having

a gelatinization temperature of at least 150°F in sufficient amount to form a slurry, heating the slurry at a temperature 15° to 30°F below the gelatinization temperature of the starch until the starch is conditioned so that it will only partially gelatinize when subsequently heated above the gelatinization temperature, and subsequently heating the slurry above the gelatinization temperature to dehydrate the slurry and to partially gelatinize the starch.

The quantity of the starch and the degree of gelatinization should be sufficient to give a product which when ground forms a dry appearing flowable powder, free of gummy, sticky and lumpy characteristics. If desired, a portion of the high fructose corn syrup can be replaced by honey. Deaerating agents (emulsifiers) are preferably added and antihumectants can be added to the product. The product consists essentially of the following ingredients.

	Parts by Weight
High fructose corn syrup solid	40 - 70
Partially gelatinized starch of a type having a gelatinization temperature of at least 150°F	60 - 30
Honey solids	0 - 60
Water	0.5 - 4.0
Emulsifier	0 - 1
Antihumectant	0 - 1

The ungelatinized starch used in the process is preferably wheat starch which has a gelatinization temperature around 160°F. Other suitable starches are corn and rice starch. Wheat and rice starches are superior to cornstarch because they are bland in flavor, whereas cornstarch carries a definite flavor characteristic which it imparts to the finished dried product.

The method of drying is very important. This process preferably uses thin-film drying. This can be accomplished in a number of ways on commercial drying equipment. These are a double drum hot roll dryer, either operated at atmospheric pressures or in a vacuum chamber, and tray dryers or conveyor dryers, again operated either at atmospheric pressures or in a vacuum chamber.

The essential feature of the drying method is to subject a thin film of the high fructose corn syrup slurry (with or without honey) to a heated surface. The temperature of this surface is usually controlled in a range of 325° to 375°F. Exposure of the film to the surface is brief, consisting of approximately 5 to 30 seconds. The principle involved in this method of drying is to combine the high fructose corn syrup with starch. During the exposure of the syrup and starch slurry to the highly heated drying surface, the starch becomes partially gelatinized in situ.

Under these conditions, the water in the syrup is rapidly transferred from the sugar molecules to the starch molecules, permitting the latter to become partially gelatinized and then forcing the starch granules to release the water by evaporation. The resulting dried composition, if properly made, shows the presence of partially gelatinized starch. There is some evidence to indicate that the starch and the sugar have combined chemically, since the sugar cannot again be extracted from the dried product and returned to its original form.

Similarly, the starch cannot be extracted or separated from the sugar. Microscopic examinations show that the starch is partially gelatinized and not totally gelatinized or disorganized. In carrying out this process, the use of a deaerating substance, e.g., hydroxylated lecithin, glycerin monostearate and combinations of glycerin mono- and distearates, is important from a commercial manufacturing standpoint where hot roll drying is used by providing much greater ease of handling of the partially gelatinized dried product as it is being peeled off the hot rolls by the knife blade.

When the high fructose corn syrup, with or without honey, and the starch are being mixed together, the result is a rather viscous slurry in which air is easily entrapped by the agitator of the mixer. If this air is allowed to remain in the slurry, the sheet of dried material which results from the film of slurry picked up by the hot rolls tends to be porous and discontinuous at the knife blade. The hydroxylated lecithin (or other emulsifier) acts, therefore, as a deaerating agent in preventing the retention of air in the slurry and also as an aid in the release of the dried material at the knife blade.

Example: The following ingredients were combined.

Ingredients	Pounds
High fructose corn syrup (HFCS)	65.0
Wheat starch	34.5
Hydroxylated lecithin	0.5
Total	100.0

The mixer used in this example has an agitator turning on a horizontal axis. The mixer tank is U-shaped on the cross section. The mixer is shaped as a partial spiral. Other mixers may be used provided that the agitator blades are sturdy and powered by a motor of sufficient strength to blend viscous materials. The mixer tank must be jacketed. The agitator may be of such construction that steam can be transmitted through it, in which case it serves as a source of heat rather than the jacket on the mixer tank.

The HFCS is placed in the mixer. The wheat starch is added in increments while the mixer is turning. After all the starch has been added, mixing is continued until a smooth slurry is obtained in which all of the starch is uniformly distributed throughout the corn syrup. At this point, steam is permitted to enter the mixer jacket or the agitator and the temperature is brought to 130°F. At this point, the hydroxylated lecithin is added and mixing is continued to deaerate the slurry. The time required for the first step in mixing depends upon the size of the mixer, but for the 100 lb batch shown in this example, approximately 7 to 10 min are required. An additional 5 min is required to deaerate.

When mixing and deaeration is complete, the steam is shut off and the mixer stopped. It is then allowed to stand for a minimum of 8 hr and as long as 16 hr. During this standing, the temperature will have dropped below 130°F. The mixer is then started and heat is applied to return the temperature to 130°F. At this point, the slurry is ready for pumping to a hot roll dryer. Upon being pumped to the dryer, the slurry is permitted to fall into the nip between the two counter-turning rollers.

A level of approximately 6 to 10 inches of slurry depth is maintained in the nip of the rollers on a dryer having rollers which are 48 inches in diameter and 10 feet long. The steam pressure in the rollers is maintained at an ideal of 85 psi. The roller speed is ideally 2.5 rpm. It should be understood that steam pressure may be slightly increased if the roller speed is slightly increased. Conversely, steam pressure may be lowered by a few pounds from the ideal if the roller speed is reduced correspondingly.

Adjustment of the aperture between the rollers is an important feature and must be observed along with the other features which control the degree of gelatinization. Obviously, the aperture may vary relative to the amount of starch versus the amount of HFCS in the slurry. When all of the above features have been carefully controlled, the film, when it reaches the knife blade, will be very easily shaved away from the surface of the hot rolls.

The appearance of the sheet at the knife blade is that of a piece of thin white paper. It is limber because the temperature is still quite high. As cooling takes place, however, the sheet becomes rapidly fragile and shatterable. This is a matter of seconds. The sheet is easily then passed through a hammermill or other milling device to reduce the particle size to that desired. The ideal particle size is 100% through 30 mesh.

Fructose-Soy Protein Films

A dried product is also obtained by *E.F. Glabe, P.W. Anderson and S. Laftsidis; U.S. Patent 3,833,413; September 3, 1974; assigned to Food Technology, Inc.* from high fructose syrups using a soy protein as additive.

In order to be effective, the soy protein must have a high protein content of at least 45% by weight and the water-soluble protein content should be at least 20% by weight of the total protein content, preferably within the range of 20 to 90% by weight. A soy protein should be used which has been at least partially defatted and the weight ratio of water-soluble protein to fat content should be at least 1.5:1 and preferably within the range of 1.5:1 to 90:1.

In accordance with this process high fructose corn syrup is dehydrated by intimately mixing the syrup with the soy protein, with or without an ungelatinized starch having a gelatinization temperature of at least 150°F, in sufficient amount to form a slurry, and subsequently subjecting the slurry in a thin film to a heated surface for a period of time sufficient to dehydrate the slurry. The product obtained consists of the following ingredients.

Ingredients	Parts by Weight
High fructose corn syrup solid	40 - 70
Partially gelatinized starch of a type having a gelatinization temperature of at least 150°F	60 - 0
Honey solids	0 - 60
Soy protein of the type described	0 - 30
Water	0.5 - 4.0
Emulsifier	0 - 1
Antihumectant	0 - 1

The product contains at least 0.25 part of the soy protein when it also contains added starch, and at least 30 parts of the soy protein when it contains no added starch.

Example: The following ingredients were combined.

Ingredients	Parts by Weight
High fructose corn syrup (HFCS)	70.0
Wheat starch	0.5 - 29.5
Soy protein	29.5 - 0.5

The HFCS was warmed from approximately room temperature to approximately 110°F. The starch and soy protein were then added and the mixture stirred until a smooth slurry was obtained. This slurry was then further warmed and held at an elevated temperature of 130°F. At this point 0.5 part by weight of hydroxylated lecithin was added and mixing was continued to deaerate the slurry. The time required for the first step was approximately 7 to 10 min and an additional 5 min was required to deaerate the slurry.

When mixing and deaeration was complete the steam was shut off and the mixer stopped. It was then allowed to stand for a minimum of 8 hr and as long as 16 hours. During this standing the temperature will have dropped below 130°F. The mixer is then started and heat is applied to return the temperature to 130°F. At this point the slurry is ready for pumping to a hot roll dryer. The drying process is then the same as given in the example of U.S. Patent 3,718,484.

Using this general procedure on a 10 inch double drum hot roll dryer with equipment for cooling the dehydrated product and with different types of soy proteins and different proportions of soy protein and starch, and by rating the characteristics of the dehydrated films in categories of good, fair and poor, it was established that only soy proteins having a high protein content and a high water-soluble protein content of at least 40% by weight received a rating of good. Examples of such proteins are the following.

Soy Protein	Percent Protein	Percent Water-Soluble Protein *	Percent Fat
Soya Fluff 200 W	53.0	70.0	1.0
Soya Fluff 200 C	53.0	40.0	1.0
Soyalose 105	52.0	65.0	6.0
Soyarich 115	45.0	65.0	16.0
Pro-Fam 70 H/S	70.0	90.0	1.0
Pro-Fam 90 H/S	90.0	88.0	1.0

*Based on percent protein.

COMPANY INDEX

The company names listed below are given exactly as they appear
in the patents, despite name changes, mergers and acquisitions
which have, at times, resulted in the revision of a company name.

INVENTOR INDEX

 Edible Starches and Starch-Derived Syrups

3,623,953 - 388
3,625,701 - 311
3,628,966 - 131
3,628,969 - 150
3,630,773 - 37
3,630,774 - 54
3,630,844 - 254
3,630,845 - 256
3,632,475 - 182
3,635,741 - 49
3,639,389 - 230
3,644,126 - 323
3,645,848 - 373
3,650,770 - 197
3,650,829 - 300
3,652,294 - 88
3,652,295 - 44
3,653,922 - 241
3,654,080 - 378
3,654,081 - 211
3,654,082 - 338
3,655,442 - 307
3,655,443 - 47
3,655,644 - 146
3,657,010 - 293
3,658,588 - 21
3,663,369 - 225
3,666,492 - 107
3,666,511 - 129
3,666,557 - 71
3,668,007 - 272
3,669,674 - 26
3,669,687 - 152
3,669,739 - 14
3,674,555 - 291
3,674,556 - 286
3,677,896 - 330
3,684,574 - 369
3,685,999 - 166
3,686,177 - 410
3,689,362 - 384
3,690,948 - 366
3,691,013 - 346
3,692,542 - 343
3,692,580 - 352

3,692,581 - 204
3,694,314 - 393
3,695,933 - 232
3,697,378 - 210
3,699,095 - 113
3,701,714 - 270
3,702,254 - 97
3,702,847 - 145
3,703,440 - 269
3,705,039 - 354
3,705,891 - 140
3,706,598 - 294
3,706,731 - 138
3,708,396 - 351
3,708,397 - 382
3,709,731 - 298
3,713,978 - 331
3,715,276 - 386
3,716,455 - 172
3,717,475 - 35
3,718,484 - 414
3,719,661 - 158
3,719,662 - 121
3,720,662 - 102
3,720,663 - 103
3,721,571 - 30
3,721,605 - 184
3,725,386 - 141
3,728,332 - 105
3,729,380 - 179
3,730,840 - 177
3,743,539 - 304
3,748,151 - 49
3,748,175 - 302
3,751,268 - 191
3,751,410 - 159
3,753,858 - 371
3,754,935 - 31
3,756,853 - 227
3,756,854 - 18
3,756,919 - 228
3,766,011 - 185
3,767,826 - 48
3,769,038 - 89

3,777,039 - 192
3,783,100 - 260
3,784,409 - 391
3,785,864 - 410
3,786,159 - 233
3,788,910 - 336
3,788,945 - 396
3,791,865 - 340
3,792,183 - 244
3,795,584 - 326
3,799,805 - 206
3,800,050 - 193
3,803,311 - 94
3,804,715 - 327
3,804,716 - 215
3,804,717 - 317
3,804,828 - 156
3,806,363 - 412
3,806,415 - 251
3,809,773 - 279
3,812,011 - 219
3,813,297 - 20
3,813,298 - 9
3,813,320 - 376
3,817,832 - 393
3,819,484 - 235
3,829,362 - 380
3,830,697 - 181
3,830,949 - 187
3,832,285 - 319
3,832,342 - 161
3,833,413 - 417
3,835,226 - 33
3,836,680 - 79
3,838,006 - 349
3,838,149 - 122
3,839,320 - 99
3,842,071 - 124
3,843,805 - 87
3,847,740 - 398
3,847,741 - 401
3,849,194 - 214
3,853,706 - 212
3,862,342 - 281

NOTICE

SUGAR SUBSTITUTES AND ENHANCERS 1973

by Roger Daniels

Food Technology Review No. 5

Because of recent restrictions on the use of synthetic sweeteners, there has developed considerable interest and research activity concerning the use of naturally occurring materials to induce or expand the sweetness of natural sugars, so that lower sugar levels can be used for lower caloric content of foods and beverages.

All of these approaches are covered in this volume, as are new synthetic sweeteners with very low toxicity figures, such as chemical modifications of natural amino acids.

The appendix contains important directives by the U.S. Food and Drug Administration on the status, use, and labeling requirements of effective combinations of such sugar substitutes and enhancers.

Except for the appendix, the book is based on 105 recent U.S. patents. Numbers in ( ) indicate the number of processes per topic. Chapter headings are given here, followed by examples of important subtitles.

1. MIRACULIN, GLYCYRRHIZIN, AND JERUSALEM ARTICHOKE SWEETENERS (8)
Stable, Solid Product
Solubilizing Method
Chewing Gum Coating
Combination with Sucrose
Citrus Juice Sweetener
Basic Extraction Methods
Artichoke Flour for Bread
Nigerian Berries

2. DIPEPTIDES, CHALCONES, MALTOLS (13)
Aspartic Acid Lower Alkyl Esters
L-Aspartyl-L-phenylalanine Esters
L-Aspartyl-L-(β-cyclohexyl)alanine
Dihydrochalcone Derivatives
Preparation from Flavanone Glycosides
Hesperetin Dihydrochalcone
Saccharin and Dihydrochalcone
Maltol Derivatives
Isomaltol Derivatives

3. STRUCTURAL VARIATIONS AND SYNTHETICS (11)
Diacetone Glucose
5-(3-Hydroxyphenoxy)-1H-tetrazole
Kynurenine Derivatives
8,9-Epoxyperillartine Sweeteners
6-(Trifluoromethyl)tryptophan

4. COMBINATIONS WITH SACCHARIN (15)
Saccharin and Dipeptides
Saccharin + Glucono-delta-lactone
Sodium Gluconate Buffer
Saccharin + Lactose
Saccharin and Pectin
Saccharin + Maltol

5. INCREASING BULK OF MIXES (16)
Pillsbury Low Calorie Drink Mix
Mix with Fumaric Acid
Peebles Process
McKesson Process
Malto-Dextrin Agglomeration

6. REDUCED CALORIE PRODUCTS (11)
Maltol Sweetness Potentiator
Sugars and Saccharin
Arabinogalactan
Arabinogalactan and Corn Starch
Lactitol Sweeteners
Maltitol Sweeteners

7. PRODUCTS: DRINKS, JELLIES, FRUITS (16)
Dry Cola Beverage Mix
Basic Effervescent Process
Glycerol Sweetener
Carragheenan-Pectin Base
Honey-Malt Flavoring
Freezedried Fruit Sweeteners

8. FROZEN DESSERTS, DAIRY PRODUCTS, BAKED GOODS, CONFECTIONS (15)
Use of Polyoses
Dietary Dry Cake Mix
Insoluble Protein Approach
Gum Acacia + Corn Syrup
Sugarless Hard Candy

9. APPENDIX
Exemption of Certain Food Additives from the Requirement of Tolerances.
Combinations of Nutritive and Non-nutritive Sweeteners in Canned Fruits.
Food Additives Permitted in Food for Human Consumption, or in Contact with Food, for Limited Periods of Time.

ISBN 0-8155-0492-6

275 pages

SUGAR ESTERS 1974

Preparation and Applications

by J. C. Johnson

Chemical Technology Review No. 32

Sugar esters have many potential end uses. As hydroxy compounds which are also natural food products, sugars form nontoxic esters which can be used in pharmaceuticals, cosmetics, and food products.

In addition to these applications, the chemical process industry has found uses for sugar esters in plasticizers, polymerizable monomers, flame retardants, adhesives, tanning agents, and detergents. This is very important from a commercial point of view, since sugars are products of agriculture. These raw material supplies are renewed annually, and no danger of eventual depletion of natural resources exists.

This process review covers 147 U.S. patents and two British patents, mostly issued since 1950. The term "sugar" here includes some of the higher polysaccharides such as dextran and the phosphomannans. Inorganic acid esters as well as carboxylic acid esters are covered. Also esters of sugar ethers, particularly the hydroxyalkyl ethers used in polyester or polyurethane condensations are included in this survey.

A partial and condensed table of contents follows here. Numbers in ( ) indicate the number of processes per topic. Chapter headings are given, followed by examples of important subtitles.

ISBN 0-8155-0537-X

310 pages

EDIBLE GUMS
AND RELATED SUBSTANCES 1973

by A. A. Lawrence

Food Technology Review No. 9

This book concerns itself with edible gums from synthetic and biosynthetic sources, as well as with the hydrophilic polysaccharides obtained from land and sea plants.

A wide range of processes is disclosed, including many modifications of techniques for the extraction and preparation of the gums, methods for enhancing the desired properties of particular gums, as well as food technology processes for individual gums or combination of gums.

Edible gums find wide application in the food industry, where they are used as water binders, suspending agents, thickeners, and emulsion stabilizers. They are essential in improving "mouth feel" in many products. Gums are important in stabilizing ice creams, sherbets and other frozen desserts. Their waterbinding action prevents an undesirable grainy texture and the growth of ice crystals.

In the baking industry gums are useful in obtaining doughs of constant properties, regardless of gluten variations in flour. They are accepted additives for beer, sauces, cheeses, jams, jellies and other confections, soups, as well as pharmaceuticals. A conservative estimate of their consumption in the USA alone approaches the 1,000 million pounds per year figure.

This patent-based book describes 202 processes. A partial and condensed table of contents follows. Numbers in parentheses indicate the number of processes per topic. Chapter headings are given, followed by examples of important subtitles.

1. GALACTOMANNANS (21)
Controlled Hydration of Guar Gum
Making Locust Bean Gum Soluble in
   Cold Water
Hydroxy Lower Alkyl Ethers of
   Galactomannans
Ca + Na Carboxymethylgalactomannans
Egg White Icing with Guar Gum
Whipped Egg White with Guar and
   Okra Gums
Stabilized "Italian Ices"

2. ARABINOGALACTAN (7)
Process Modifications for Larch
   Gum
Use as a Low Calorie Sugar
   Substitute
Use as a Flavor Fixative

3. GUMS FROM OTHER LAND PLANTS (12)
Gum Tragacanth
Tamarind Polysaccharide
Gum Arabic
Gum from Crotalaria Intermedia
Constituent of Dehulled Barley

4. PECTINS (18)
From Citrus Peel
High Methyl Ester Pectin
Use in Jams and Jellies
With Food Acids
Medicinal Uses
Protopectins in Baked Goods

5. CARRAGEENANS—MARINE PLANT EXTRACTS (17)
Extraction from Gigartinaceae
   and Solieriaceae
Eucheuma Gels
Locust Bean Gum + Ca Carragenate
Carraglucan Polysaccharides

6. ALGINATES (22)
Preextraction of Marine Brown
   Algae
Polysaccharide Extraction by
   Irradiation
Improving the Properties of Alginates
Alginate Dessert Gels
Extending Shelf Life of Foods

7. AGAR AND FURCELLARAN (7)
Separation of Agarose and
   Agaropectin
Use in Cultured Milk Products
Use in Candy and Other Confections

8. CARBOXYMETHYLCELLULOSE [CMC] (18)
Preparation and Etherification
CMC Properties and Their
   Improvement
Food and Drug Use of CMC

9. OTHER CELLULOSE ETHERS (17)
Hydroxypropyl Cellulose
Hydroxyethyl Cellulose
Mixed Cellulose Ethers

10. XANTHOMONAS HYDROPHILIC COLLOID (31)
Recovery from Fermentation Media
Deacetylated Polysaccharide
   Manufacture
Sulfate Esters and Ester Salts
Use in Foods and Beverages

11. OTHER MICROBIAL POLYSACCHARIDES (6)

12. CYCLODEXTRINS (14)
Production from Modified Starch
Derivatives and Uses

13. MISCELLANEOUS EDIBLE COLLOIDS (12)
Starch Gels
Various Plastic Gels
Lactyl Lactylate Gels

ISBN 0-8155-0511-6

342 pages

CONFECTIONS AND CANDY TECHNOLOGY 1974

by M. E. Schwartz

Food Technology Review No. 12

The scope of the technology covered in this volume is generally taken to include products whose major ingredient is sugar. Thus the range of products includes not only candy per se, but also chewing gum, desserts, sweet snacks, icings and whipped toppings, jellies and syrups. Also included are many processes designed to produce some of the basic ingredients of the industry such as sugars, syrups, fats, milk, and whey products. Technology of additives is similarly included, as far as such additives come within the scope of this survey.

Current concerns of the industry with regard to dietetic demands of the consumer have received special attention. Thus candy and confectionery processes geared to the production of low calorie or nonsugar sweets are covered fully.

The technology of the industry is large as evidenced by the appearance of well over 200 U.S. patents in the last ten years. This vast storehouse of information has been organized so that it may be of maximum use to practitioners in the art of candymaking.

A partial and condensed table of contents follows here. Numbers in parentheses indicate a plurality of processes per topic. Chapter headings are given, followed by examples of important subtitles.

1. CANDIES (55)
Core Formulations
Hard Candies
Marshmallows
Fudge Mixes
Gelled Confections
Use of Low Fat Starches
Aerated Confections
Dietetic Sorbitol Formulations
Molding Methods
Chocolate Chips
Novelty Products

2. PRODUCTION OF CONFECTIONERY INGREDIENTS (31)
Free-Flowing Sugars
Starch Conversion Sugars and Syrups
Low DE Starch Hydrolyzates
Dehydration of High Fructose Corn Syrup

Fats
Lauric-Type Hard Butters
Partially Hydrogenated Glycerides

3. CONFECTIONERY ADDITIVES (27)
Low Density Lactose
Electrodialyzed Whey Solids
Dry Instant Sugar-Fat
Butter Flavors
Specialty Sugars (Moisture Resistant)
Adjuvants

4. FROZEN DESSERTS (38)
Cream-Type
Water-Based Products
Dry Mixes
Dietary Formulations

5. MILK PUDDINGS (16)

6. GELLED DESSERTS (24)
Gelatin Gels
Alginate Gels
Pectin Gels
Adjuvants

7. ICINGS (23)
Meringues
Bakery Items

8. WHIPPED TOPPINGS (17)
Dry Mixes
Aerosol Products

9. JELLIES (8)

10. NOVELTY PRODUCTS (20)
Sweetened Peanut Butter Compositions
Hydrophilic Additives and Emulsifiers
Butter-Syrup Emulsions
Honey Butter
Sweet Snack Foods
Candied Produce

11. CHEWING GUMS (18)
Suitable Polymer Mixtures
Plaque-Removing Chewing Gum
Sugarless Gums
Candy-Coated Gum Balls
Miraculin-Containing Formulation

ISBN 0-8155-0524-8

338 pages

FOOD ADDITIVES
TO EXTEND SHELF LIFE 1974

by Nicholas D. Pintauro

Food Technology Review No. 17

Aside from freezing, canning and sophisticated methods of packaging, food is preserved by dehydration, salting, sugaring, smoking, curing, and certain types of fermentation.

A newer effective approach toward prevention of spoilage is by the use of chemical additives other than sugar, salt, vinegar, and spices.

Food additives, as defined by the National Academy of Sciences are those relatively nontoxic chemicals that may be incorporated into foodstuffs during the growing, processing, or storing periods. Every chemical added must serve one or more of these general purposes: Improve or maintain nutritional value, enhance quality, increase consumer acceptability, and facilitate preparation. In modern applications food additives are combined with established, classical methods of food preservation to maximize stability for extended shelf life. There is a great demand for additives to prevent or retard food deterioration. These additives include antioxidants, antibacterial agents, mold inhibitors, color stabilizers, anticaking agents, antibrowning agents, cloud stabilizers, metal scavengers, enzyme inhibitors.

This book describes over 140 processes involving the newest technology available in the U.S. patent literature using food additives. A partial and condensed table of contents follows here. Chapter headings are given, followed by examples of important subtitles.

1. ANTIOXIDANTS FOR FATS & OILS (20)

BHA-BHT Synergistic Combinations
Plicatic and Thiodipropionic Acids
Cystine as Stabilizer
Metal Deactivators
Citric Acid Esters
EDTA Esters
Stable Antisplattering Oil

2. MEAT PRODUCTS (22)

Salt-Antioxidant Combinations
Color Stabilization
p-Aminobenzoic and Ascorbic Acids
Nicotinic Acid for Frozen Meats
Glycerin Infusion
Low pH
Use of Antibiotics

3. FISH PRODUCTS (10)

Use of Sulfur Dioxide
Ammonia Preservative
Use of Peroxides
Use of Chelating Agents

4. EGGS AND EGG PRODUCTS (7)

Antifungal Antibiotic Treatment
Phosphates for Fresh Eggs
Antioxidants for Dried Eggs
Epoxide Vapors for Eggs in Shell
Enzyme System for Mayonnaise

5. DAIRY PRODUCTS (6)

Bioflavanoids
Sorbates for Cheese
Sorbic Acid Powder Coatings

6. BREAD SOFTENERS AND MOLD INHIBITORS (13)

Antistaling Additives
Stearyl-2-Lactylic Acid Salts
Nonionic Surfactants
Amylopectins for 9-Month Shelf Life
Acetates + Coated Organic Acids

7. BAKED GOODS AND CEREAL PRODUCTS (12)

Color Stabilization with Nisin & Subtilin
Polyvinylpyrrolidone Additives
Shelf Stable Pancakes & Waffles

8. FRUITS AND VEGETABLES (20)

Sodium Bisulfite + Phosphoric Acid
Stannous Chloride
Amino Acids as Preservatives
Enzyme Inactivation

9. VITAMINS AND NATURAL COLORS (10)

Vitamin A Fat Antioxidant
Stabilized Carotenoids
Dihydroquinoline Antioxidants

10. SALT, BEVERAGES & FLAVORS (18)

Potassium-Enriched Conditioning
Agent for Salt
Antioxidant Salt with BHA
Coffee Bean Extract
Glucose Oxidase System

11. ALL-PURPOSE PRESERVATIVES AND ANTIOXIDANTS (18)

Bisphenols
Isoascorbic Acid Salts
Benzohydroxamic Acid
Pyrocarbonic Acid Esters

APPENDIX

FDA Policy Statements, etc.

INDEX of additives arranged alphabetically by their chemical names.

ISBN 0-8155-0548-0

402 pages

MICROBIAL ENZYME
PRODUCTION 1974

by Sidney J. Gutcho

Chemical Technology Review No. 28

Enzymes are catalysts of biochemical origin and, unlike metal catalysts, are capable of extraordinary specificity and reactivity in biological and chemical systems.

Enzymes produced by microbes are being used in a large number of industries. The recognition of their commercial applications and advantages has given their production an added impetus. Not the least is the fact that they do not impart any toxicity to their substrates under controlled conditions. Also, they can be inactivated readily by heat or a change in pH, when their services are no longer wanted, and usually need not be removed from the material in which they have produced a chemical reaction. The crystallization of enzymes has led to their identification as proteins.

The production of enzymes has become very important and the search for new biological sources is going on every day. Since microbial cultures can be controlled closely and are independent of seasons, and their nutrients can be varied at will, the production of microbial enzymes has a bright future. Such enzymes are used as pharmaceuticals, in the manufacture of pharmaceuticals, in detergents, as additives to foods, in the manufacture of a variety of food products, as diagnostic reagents, as reagents for the production of chemicals, and in the treatment of industrial wastes.

The production of microbial enzymes, as described in this book, is proceeding at an ever-increasing pace. This book is written for microbiologists, enzymologists, biochemists, food technologists, and others who are interested in the economical production and efficient application of microbial enzymes.

A partial and condensed table of contents follows. Numbers in parentheses indicate a plurality of processes per topic. Chapter headings and some of the more important subtitles are given.

1. BACTERIAL ENZYMES (50)
α-Amylase by B. stearothermophilus
Asparaginase
Catalase
Collagenase
Dextranase
Glucose Isomerase by Streptomyces spp.
α-1-6-Glucosidase by Klebsiella, Aerobacter, & Pseudomonas
Hyaluronidase
Lipase
Lytic Enzymes by B. cereus
Zymolyase
Polynucleotide Phosphorylase by Micrococcus lysodeikticus
Proteolytic Enzymes
  by Arthrobacter
  by B. licheniformis
  by Thermopolyspora
  by Serratia marcescens
  by Streptomyces griseus
Streptokinase & Streptodornase
Sucrase

2. FUNGAL ENZYMES (53)
Amyloglucosidase
  by Aspergillus foetidus
  by Aspergillus niger
  by Aspergillus phoenicis
  by Aspergillus awamori
Aminopeptidase
Cellulase
Dextranase
Galactose Oxidase
Glucose Oxidase
Milk Clotting Enzymes
Proteolytic Enzymes
Ribonuclease

3. YEAST ENZYMES (6)
Invertase
Lipase
Lactase
Uricase

4. SPECIAL TECHNIQUES (18)
Production of Microorganisms
Effect of Phosphatidyl Inositol on Exoenzyme Production
Chromatographic Techniques
Hydroxyapatite for Isolation of Neutral Protease
Precipitation Processes
Tannic Acid & Lignin
Precipitation of Enzymes by Synthetic Tanning Materials
Removal of Organic Impurities with Cations
Inorganic Salts as Precipitants
Quaternary Ammonium Compounds

ISBN 0-8155-0532-9

272 pages

FRUIT AND VEGETABLE

JUICE PROCESSING 1975

by J. K. Paul

Food Technology Review No. 21

The large market for fruit and vegetable juices, because of their relatively low cost and high nutritional value, coupled with modern diet appeal, accounts for the sustained interest in these products and the continued development of improved methods of technology.

It is a primary object of this technology to prepare fruit juices, vegetable juices and related beverages that are able to maintain their natural flavor and aroma characteristics under ordinary, unspecified storage conditions over prolonged periods of time.

Considerable progress has been made in this direction: practically every process in this book is oriented toward the ultimate goal of every juice manufacturer—to retain natural flavors and aromas and to retard indefinitely the formation of unnatural flavors, aromas or colors.

Gone are all those preservatives of the past which, while stabilizing a beverage, also introduced undesirable off-flavors. Approved modern additives not only maximize stability, they also enhance the natural flavors. How to use them in connection with advanced extraction and concentration processes, which were specially designed for juice processing, is the know-how of this book: 164 patent-based processes are described.

A partial and condensed table of contents follows here. Chapter headings are given, followed by examples of important subtitles. Numbers in parentheses indicate the numbers of processes per topic.

1. MANUFACTURING TECHNIQUES (26)
Pome Juice with Pulp
Removal of Bitterness Precursors
Separation of Juice from Pulp
Increasing Viscosity
Clarified Juice Manufacture
Dialysis Methods
When to Change pH
Ultrasonic Treatments

2. CONCENTRATION PROCESSES (42)
Brix to Acid Ratio
Osmotic Processes
Freeze Concentration
Use of Immiscible Coolants
Crystal Purification Columns
Sonic Defoaming

3. DEHYDRATION (16)
Drum Drying
Addition of Amylolytic Enzymes
Foam-Mat Processes
Spray Drying
Vacuum Drying
Other Drying Processes

4. FREEZE DRYING (28)
Fluidized Bed Processes
Continuous Processes
Desiccation of Frozen Particulates
Foam Drying Processes
Slush Freezing
Thermal Shock Process

5. STABILIZATION (19)
Adding O-Methyltransferase
Use of Stannous Ions
Benzhydroxamic Acid
Disubstituted Benzoic Esters
    and Ketones
Use of Polyphosphates
Ion Exchange Treatments

6. FLAVORS FROM JUICES (14)
Essence Recovery
    by Continuous Condensation
    by Distillation
Flavor Extraction from Fruit

7. JUICE ENHANCERS (8)
Flavor Enhancers
Glutaminase Hydrolysis of
    Glutamine to Glutamic Acid
Addition of β-Hydroxybutyrates
    to Enhance Grape Flavor
2-Ethylpyromeconic Acid
2-Alkylthiazoles
Coloring Agents
Carotenoids in Abietic Acid Melt
Benzopyrilium Compounds
Enrichment of Orange Juice
    with Chromoplasts

8. VARIOUS PROCESSES (11)
Pear Beverage
Drink from Citrus Juice and
    Fermented or Acidified Milk
Citrus Juice with Energy Supplement
Licorice-Containing Juices
Juice from Green Leaves
    of Wheat and Barley
Aeration with Volatile Additive
Aeration with Inert Gas
Dispensing of Semifrozen Comestible

ISBN 0-8155-0565-5

277 pages

BAKERY PRODUCTS 1975

—Yeast Leavened—

by D. J. De Renzo

Food Technology Review No. 20

Bakery products vary greatly in appearance and taste, yet they can be classified readily by the way they are leavened. In the processes described in this volume this is always done with yeast. Its single fungal cells ferment simple sugars and produce alcohol, thereby evolving carbon dioxide which is entrapped by the gelatinous gluten of the dough.

Commercial baking processes generally require flour, water, salt, milk, sugar, eggs, shortening, yeast and yeast foods (such as ammonium chloride and potassium bromate), additional enzymes, mold inhibitors, flavors and enrichment ingredients (thiamine, riboflavin, niacin, iron) as required by law for white bread.

How this is done economically in practical ways with the aid of additives and special procedures is the subject of this book. More than 200 patent-based processes are described.

A partial and condensed table of contents follows here. Numbers in parentheses indicate the number of processes per topic. Chapter headings are given, followed by examples of important subtitles.

ISBN 0-8155-0559-0

456 pages